Asymmetric Synthesis of Natural Products

Asymmetric Synthesis of Natural Products

Third Edition

ARI M.P. KOSKINEN

Department of Chemistry and Material Sciences, Aalto University,
School of Chemical Engineering, Espoo, Finland

Registered Offices
John Wiley & Sons, Inc., 111 River Street, Hoboken, NJ 07030, USA
John Wiley & Sons Ltd, The Atrium, Southern Gate, Chichester, West Sussex, PO19 8SQ, UK

Editorial Office
The Atrium, Southern Gate, Chichester, West Sussex, PO19 8SQ, UK

For details of our global editorial offices, customer services, and more information about Wiley products visit us at www.wiley.com.

Wiley also publishes its books in a variety of electronic formats and by print-on-demand. Some content that appears in standard print versions of this book may not be available in other formats.

Library of Congress Cataloging-in-Publication Data applied for

Paperback: 9781119707028

Cover design by Wiley
Cover image: Courtesy of Ari M.P. Koskinen

Set in 10/12pt TimesLTStd by Straive, Chennai, India

SKY10036235_092922

Contents

Preface to the First Edition

This book is based on a one-semester, 24 hour lecture course given over the past six years at the University of Helsinki, Finland, University of Surrey, England, and University of Oulu, Finland. The course is intended for senior undergraduate and beginning graduate students. It is also hoped that the book will be useful for practicing research workers who want to refresh their knowledge on the field.

The basic idea of a course combining asymmetric synthesis and natural product chemistry came from Professor Tapio A. Hase early in 1987 when discussing how best to cover both the fundamentals and latest developments in asymmetric synthesis in a stimulating way. As natural product synthesis is the logical field of application for asymmetric transformations, I decided to try out the concept. Over the years it has worked well, and the course has developed into an enjoyable one, both for the students and the teacher. For the evolution, I must thank the many students at the three universities for their helpful comments and suggestions.

The book begins with a brief introduction to the general field and its allied applications. Chapter 2 covers the basic thermodynamics and terminology as well as processes for asymmetric synthesis. Chapter 3 forms the main body of the individual asymmetric reactions which are covered both in terms of theory and applications. The rest of the book, Chapters 4 to 10, covers the individual natural product classes. I have tried to give a brief overview of the structural varieties and biosynthetic pathways leading to these compounds, as well as the practical (mainly pharmacological) importance of a number of representative compounds. To keep the reading lighter, I have also included some rather amusing anecdotes from the past. The syntheses of the individual natural product types are covered with examples, giving some general methods for the particular natural products. I have deliberately not included repetitions of long sequences of reactions which are not pertinent to the subject-these can be found in the references and in recent literature. The references are not exhaustive; quite the contrary, I have tried to keep the number as low as possible without sacrificing the context. For all omissions of important work, or references, I express my apologies. I will also warmly welcome all comments for possible future editions.

I wish to thank Professor Tapio A. Hase for the impetus for coming up with the course, and my mentors Professor Mauri Lounasmaa (Helsinki University of Technology, Finland) and Professor Henry Rapoport (University of California, Berkeley, USA) for leading me to the wonderful world of natural product chemistry and asymmetric synthesis.

Finally, my wife Päivi, and three daughters, Tiina, Joanna and Heidi, have taken a lot of grief during the writing process. Without their everlasting understanding and love, the whole project could not have been accomplished.

A.M.P.K.

Preface to the Second Edition

The second edition of this book took a long time to come about. Maybe too long - perhaps not, since the intervening time actually gave me a lot of perspective, which I try to convey to you, too. In 1992, when the first edition was just finished, the chemistry scene was very much different from today. Organometallic chemistry had only recently become a mainstream practice for synthetic chemists; asymmetric synthesis was still a field mainly practiced by specialists, and practically nobody talked about combinatorial chemistry. In the analytical field, a 500 MHz NMR was considered a luxury equipment, and 2D experiments were usually limited to COSY and NOESY spectra. This made the analysis of conformations much more tedious than today. And of course, computer modeling was still in its infancy, force field methods were slow, cumbersome, and could only handle small molecules.

The publisher asked for a second edition quite some time ago, about ten years ago, but honestly I was too busy with changes in my career. While the book project was on a back burner, asymmetric synthesis kept developing at an increasing pace, as did other specialist fields of chemistry. Organometallic chemistry turned asymmetric in a big way in the late 1990's, and metal catalyzed reactions won many followers. In the 1990's there was a huge thrive to combinatorialize practically everything, until the hype for combinatorial chemistry faded as quickly as it came about. It still remains a tool, a useful one, but only one tool among many. Methods for structural analysis, both experimental and computational, have given us unprecedented means to rationalize and predict the stereochemical outcomes of reactions. Organic molecules have been used as catalysts for reactions at least for as long as one has written records. Organocatalysis is a term coined between the two editions of this book, and it will be interesting to see how the concept will eventually evolve. One thing is for sure, our increasingly comprehensive understanding of chemistry and its adjoining fields requires a new perspective in the development of new chemical reactions, which is the core of chemistry.

This book forms the core for a continuum of two smaller lecture courses, each of a length of half a semester, that is about six weeks each (it used to be a single full term course, but was split in two a few years ago). Since the course is for advanced level students, I have done away with traditional tutorials, exams, and essays, and instead together with the students we usually try to do something differently every year. Whether you use this book for teaching a course or use it to learn yourself, I strongly recommend you to experiment with learning methods, too. Some suggestions that have been very popular, are the following: the students should keep a learning diary - and compare them with their class mates. We have divided the students to small task groups, where in pairs, sorry, in doublets or triplets, they analyzed for instance the syntheses of a particular natural product from two different eras, like 40 years apart. Or they compared the mechanistic explanations of a particular reaction based of two papers several decades apart. This can be challenging, but will teach the student to read and understand old papers, where even the structures we drawn differently! Among the favorites are also student lectures, where the students mine the literature and databases and construct their own interpretation for discussion BEFORE I preach the orthodox teachings. When the students come up with a new explanation, they can also share the moment of creating new knowledge, which culminates in the dissemination to their colleagues.

I simply must share the way this book finally came to be finished. In 2004 we had a wonderful meeting, under the auspices of the EU COST actions, in the beautiful old monastery of Certosa di Pontignano near Siena, Italy. At some point during the discussions over very palatable Tuscan wines, I said to Maurizio and Bruno Botta that I really would love to write the second edition in Tuscany. Of course, the brothers warmly welcomed me to Italy to take up the job. After the meeting, reality struck back, and everyday chores kept me plodding the well-trodden paths of academic life. The dream of writing under olive trees remained a warm joke with friends and family, and a hazy dream in my solitary moments. Things started rolling in the summer of 2010, first verbally during the ICOS meeting in Bergen, and then later in the autumn with a written contract ready for signature. By the end of 2010 I was committed, but the logistics were still a complete mess, until in May 2011 Aalto University granted me a sabbatical leave for the end part of the year, without which the whole project would never have been finished. Olive trees, Italy, and writing the book would turn a dream come true. With the help of La Sapienza professor Bruno Botta, we managed to find a wonderful house in Tuscania, a small town in northern Lazio. The tranquility and the atmosphere of the house provided the best surrounding imaginable for doing the job. Waking up in the morning to the sounds of a nearby waterfall in the background and the early morning baahs of Francesco's sheep in the foreground. Our landlord, Francis Kuipers, a well-known composer and music professional, and his "assistant" Ione Kerr Ciccioli, made sure that we were completely safe - mobile phones did not pick up transmission, but the Wi-Fi was strong, so I could do all my database searches speedily.

It was a truly rewarding experience to be able to concentrate on this project so totally without other worries, and I hope this is also reflected in the outcome. I am greatly indebted to those who made it all possible. First of all, all the students at Oulu University, Helsinki University of Technology, presently Aalto University, and Helsinki University where I have given the lecture courses regularly. My students in my research group have all contributed over the years, providing a perfect sounding board for new ideas and insights. My colleagues both at the Universities I have been affiliated with as well as my friends around the world have all contributed to my professional development, and deserve my warmest thanks.

The first and second editions are separated by two decades, which makes this edition a second generation. On personal level, that's what has also happened: my daughters, whom I thanked in the preface to the first edition, have moved out, and have been replaced with Sara, and finally a boy, Mikko. They and my muse and lifeline Päivi have given me the strength and inspiration to finish this work.

Finally to you, my readers, I wish to convey the same message I finish my guest lectures with. When I was flying out of the nest of my main mentor professor Henry Rapoport of UC Berkeley, a pioneer of asymmetric synthesis and a great protagonist for experimentation, gave me the message to be passed on: "Keep the Bunsen burning!"

Ari Koskinen

Karkkila, February 2012

Preface to the Third Edition

The second edition of this book was published a decade ago, but the developments in organic chemistry, organic synthesis, and natural product chemistry have been so overwhelming that a new addition is fully warranted. This is reflected in this edition: organocatalysis has established its place among the three pillars of synthesis together with organometallic catalysis and biocatalysis. Several examples of organocatalysis have been included in the text. At the same time, older examples have had to be left out due to space restrictions. However, I have tried to retain enough older material to give the reader a comprehensive picture of the evolution of natural product chemistry. Colored graphics hopefully illustrate both mechanistic aspects as well as structural aspects more clearly than the figures in the previous black and white editions. The structure of the book has remained almost unchanged except for the fact that the old chapter 9 Shikimates is now being completely rewritten as chapter 7 Phenylpropanoids to cover more widely this subject area.

This book continues to form the core for a continuum of two smaller lecture courses, each of a length of half a semester, that is about six weeks each (Asymmetric synthesis covering Chapters 1-3, and Natural Product Synthesis covering Chapters 4-10). Since the courses are for advanced level students, traditional tutorials, exams, and essays, are replaced with more self-study materials. Learning diaries have become very popular: the students are expected not to just repeat what has been discussed in the text or lectures, but rather reflect on their learning process. The students have formed smaller study groups, where in pairs, sorry, in doublets or triplets, they analyze the literature: for instance syntheses of a particular natural product from two different eras, like 40 years apart, or compare the mechanistic explanations of a particular reaction based of two papers several decades apart. This can be challenging, but will teach the student to read and understand old papers, where even the structures we drawn differently! At the end of the courses the students also present a seminar presentation on a topic of their choice together with a portfolio of relevant literature. For the past two years the COVID-19 pandemic has forced us to think of alternative ways of education. All the lecture sessions have been recorded and made available to the students for the duration of the lecture courses. Also shorter videos on particularly challenging topics have been developed. These have been extremely popular among the students as they can watch the shorter clips on their mobile devices without time or site restrictions.

As before, I am greatly indebted to those who made it all possible. First of all, all the students at Oulu University, Helsinki University of Technology, presently Aalto University, and Helsinki University where I have given the lecture courses regularly. My students in my research group have all contributed over the years, providing a perfect sounding board for new ideas and insights. My colleagues both at the Universities I have been affiliated with as well as my friends around the world have all contributed to my professional development, and deserve my warmest thanks.

On personal level, my home team, although the children are already out of the nest, and especially Päivi, have given me the strength and inspiration to finish this work.

Finally to you, my readers, I wish to convey the same message I finish my guest lectures with. When I was flying out of the nest of my main mentor professor Henry Rapoport of UC Berkeley, a pioneer of asymmetric synthesis and a great protagonist for experimentation, gave me the message to be passed on: "Keep the Bunsen burning!"

Ari Koskinen
Brussels, January 2022

List of Common Abbreviations

AA	asymmetric aminohydroxylation	CB	catecholborane
Ac	acetyl	CBS	Corey-Bakshi-Shibata
acac	acetylacetonate		oxazaborolidine
Ad	adamantyl	Cbz or Z	benzyloxycarbonyl
AD	asymmetic dihydroxylation		(carbobenzyloxy)
Adoc	adamantyloxycarbonyl	CCK	cholecystokinin
AIBN	azoisobutyronitrile	CD	circular dichroism
Alloc or AOC	allyloxycarbonyl	CDI	1,1′-carbonyldiimidazole
An	*p*-anisyl	*c*Hex	cyclohexyl
anh.	anhydrous	CMPI	2-chloro-1-methylpyridinium
aq.	aqueous		iodide
atm	atmosphere	Cod	cyclooctadiene
ATP	adenosine triphosphate	Cp	cyclopentadienyl
AZADO	2-azaadamantane N-oxyl	Cp*	pentamethylcyclopentadienyl
9-BBN	9-borabicyclo[3.3.1]nonane	CSA	camphorsulfonic acid
BINAP	2,2′-(bisphenylphosphino)-	Cy	cyclohexyl
	1,1′-binaphthyl	DABCO	1,4-diazabicyclo[2.2.2]octane
BIPHEP	biphenylphosphine	DAIB	3-*exo*-(dimethylamino)
bipy	2,2′-bipyridine		isoborneol
BMDA	bromomagnesium	DAST	diethylamino sulfur trifluoride
	diisopropylamide	dba	dibenzylideneacetone
BMS	borane-dimethyl sulfide	DBAD	di-*t*-butyl azodicarboxylate
Bn	benzyl	DBN	1,5-diazabicyclo[4.3.0]non-5-ene
BOC (or Boc)	*tert*-butoxycarbonyl	DBU	1,8-diazabicyclo[5.4.0]undec-
BOM	benzyloxymethyl		7-ene
BOP reagent	benzotriazol-1-yloxytris	DCC	dicyclohexylcarbodiimide
	(dimethylamino)phosphonium	DDQ	2,3-dichloro-5,6-dicyano-1,4-
	hexafluorophosphate		benzoquinone
BOP-Cl	bis(2-oxo-3-oxazolidinyl)	DEAD	diethyl azodicarboxylate
	phosphinic chloride	DEIPS	diethylisopropylsilyl
Bt	benzotriazol-1-yl	DET	diethyl tartrate
n-Bu	*n*-butyl	DHP	3,4-dihydro-2H-pyran
s-Bu	*sec*-butyl	DIAD	diisopropyl azodicarboxylate
t-Bu	*tert*-butyl	DIBAL-H	diisobutylaluminum hydride
Bz	benzoyl	DIOP	2,3-O-isopropylidene-
18C6	18-crown-6		2,3-dihydroxy-1,4-bis
CA	chloroacetyl		(diphenylphosphino)butane
CAL	*Candida antarctica* lipase	DIPAMP	1,2-bis[(2-methoxyphenyl)
CAN	ceric ammonium nitrate		(phenylphosphino)ethane]
cat	catalytic amount		

DIPEA	diisopropylethylamine (Hünig's base)
DIPT	diisopropyl tartrate
DMA	dimethylacetamide
DMAD	dimethyl azodicarboxylate
DMAP	4-*N*,*N*-dimethylaminopyridine
DME	1,2-dimethoxyethane
DMF	*N*,*N*-dimethylformamide
DMP	Dess-Martin periodinane
DMPU	*N*,*N'*-dimethylpropyleneurea
DMS	dimethyl sulfide
DMSO	dimethyl sulfoxide
DMTr or DMT	di-(*p*-methoxyphenyl) phenylmethyl or dimethoxytrityl
DNP	2,4-dinitrophenyl
DNs	2,4-dinitrobenzenesulfonyl
DPIPS	diphenylisopropylsilyl
DPM or Dpm	diphenylmethyl
Dpp	diphenylphosphinyl
DPPA	diphenylphosphoryl azide
dppb	1,4-bis(diphenylphosphino) butane
dppe	1,2-bis(diphenylphosphino) ethane
dppf	1,1'-bis(diphenylphosphanyl) ferrocene
DTBMS	di-*t*-butylmethylsilyl
DTT	dthiothreitol
EDC or EDCI	1-ethyl-3-(3-dimethylamino-propyl)carbodiimide
EDTA	ethylenediaminetetraacetic acid
EE	1-ethoxyethyl
Et	ethyl
Fmoc	9-fluorenylmethoxycarbonyl
HATU	O-(7-azabenzotriazol-1-yl)-*N*,*N*,*N'*,*N'*-tetramethyluronium hexafluorophosphate
HBTU	*O*-benzotriazol-1-yl-*N*,*N*,*N'*,*N'*-tetramethyluronium hexafluorophosphate
HMDS	hexamethyldisilazane
HMPA	hexamethylphosphoric triamide (Me$_2$N)$_3$P=O
HMPT	hexamethylphosphorous triamide (Me$_2$N)$_3$P
HOAt	7-aza-1-hydroxybenzotriazole
HOBt	1-hydroxybenzotriazole
HOMO	highest occupied molecular orbital

IBX	*o*-iodoxybenzoic acid
Im	imidazole
IPA	isopropyl alcohol
Ipc	isopinocamphenyl
(Ipc)$_2$BH	diisopinocampheylborane
KAPA	potassium 3-aminopropyl amide
KDA	potassium diisopropylamide
KHMDS	potassium hexamethyldisilazide
L-Selectride	lithium tri-*sec*-butylborohydride
LA	Lewis acid
LAH	lithium aluminum hydride
LDA	lithium diisopropylamide
LDBB	lithium 4,4'-di-*t*-butylphenylide
Lev	levulinoyl
LiTMP	lithium 2,2,6,6-tetramethylpiperidide
LUMO	lowest unoccupied molecular orbital
Lut.	2,6-lutidine
MAD	methoxyaluminiumbis (2,6.di-*t*-butyl-4-methylphenoxide)
mCPBA	*m*-chloroperoxybenzoic acid
Me	methyl
MEM	methoxyethoxymethyl
Mes	mesityl or 2,4,6-trimethylphenyl
MMTr or MMT	*p*-methoxyphenyldiphenylmethyl or methoxytrityl
MOM	methoxymethyl
MoOPH	oxodiperoxymolybdenum (pyridine)hexamethyl-phosphoramide
MPM see PMB	
Ms	mesyl (methanesulfonyl)
MTBE	*t*-butyl methyl ether
MTHP	4-methoxytetrahydropyranyl
MTM	methythiomethyl
MTPA	α-methoxy-α-trifluoromethylphenylacetic acid
Mtr	2,3,6-4-methoxybenzenesulfonyl
Mts	2,4,6-trimethylbenzenesulfonyl or mesitylenesulfonyl
NADH	nicotinamide dinucleotide hydride
NADPH	nicotinamide adenine dinucleotide phosphate
nbd	norbornadiene

NBS	*N*-bromosuccinimide
NCS	*N*-chlorosuccinimide
Ni(acac)$_2$	nickel acetylacetonate
NIS	*N*-iodosuccinimide
NMM	*N*-methylmorpholine
NMMO or NMO	*N*-methylmorpholine *N*-oxide
NMP	*N*-methylpyrrolidinone
Nosyl or Ns	(2- or) 4-nitrobenzenesulfonyl
Nu	nucleophile
OBO	2,6,7-trioxabicyclo[2.2.2]octyl
OP, OPP	pyrophosphate (in biosynthetic schemes)
OTf	trifluoromethanesulfonate
PALP	pyridoxal phosphate
PCC	pyridinium chlorochromate
PDC	pyridinium dichromate
Pd$_2$(dba)$_3$	tris(dibenzylideneacetone) dipalladium
Pf	9-phenylfluorenyl
PG	protecting group
Ph	phenyl
Pht	phthalimidyl
Pim	phthalimidomethyl
Piv (or Pv)	pivaloyl
PLE	porcine liver esterase
PMB or MPM	*p*-methoxybenzyl
PMP	*p*-methoxyphenyl
PNB	*p*-nitrobenzyl
POM	4-pentenyloxymethyl (carbohydrates)
POM	pivaloyloxymethyl
PPL	porcine pancreatic lipase
PPTS	pyridinium *p*-toluenesulfonate
Pr	propyl
Proton sponge	1,8-bis(dimethylamino) naphthalene
PTSA	*p*-toluenesulfonic acid
Pv	pivaloyl
Pyr	pyridine
Px or pixyl	9-(9-phenyl)xanthenyl
RAMP	*(R)*-1-Amino-2-methoxymethylpyrrolidine
Red-Al	sodium dihydrobis(2-methoxyethoxy) aluminate, see also Vitride
SAMP	*(S)*-1-Amino-2-methoxymethylpyrrolidine
(R)-SEGPHOS	*(R)*-(+)-5,5′-Bis(diphenyl-phosphino)-4,4′-bi-1,3-

	benzodioxole, [4(*R*)-(4,4′-bi-1,3-benzodioxole)-5,5′-diyl]bis [diphenylphosphine]
SEM	trimethylsilylethoxymethyl
SMEAH	sodium bis(2-methoxyethoxy)aluminium hydride
STABASE	1,1,4,4-tetramethyldisilylaza-cyclopentane
Su	succinimidyl
TADDOL	α,α,α′,α′-tetraaryl-2,2-dimethyl-1,3-dioxolane 4,5-dimethanol
TASF	tris(dimethylamino) sulfur (trimethylsilyl)difluoride
TBAF	tetrabutylammonium fluoride
TBAI	tetrabutylammonium iodide
TBDMS or TBS	*t*-butyldimethylsilyl
TBDPS	*t*-butyldiphenylsilyl
TBHP	*t*-butyl hydroperoxide
TDS	thexyldimethylsilyl
TEA	triethylamine
TEBAC or TEBA	triethylbenzylammonium chloride
TEMPO	2,2,6,6-tetramethyl-1-piperidinyloxy
Teoc	2-(trimethylsilyl)ethoxycarbonyl
TES	triethylsilyl
Tf	trifluoromethanesulfonyl
TFA	trifluoroacetic acid
TFAA	trifluoroacetic anhydride
TfOH	trifluoromethanesulfonic acid
Thexyl	2,3-dimethyl-2-butyl
THF	tetrahydrofuran(yl)
THP	tetrahydropyranyl
TIBS	tripsobutylsilyl
TIPS	triisopropylsilyl
TMANO	trimethylamine *N*-oxide
TMEDA	*N,N,N′,N′*-tetramethylethyl-enediamine
TMGA	tetramethylguanidinium azide
TMOF	trimethyl orthoformate
TMS	trimethylsilyl
TMSE or TSE	2-(trimethylsilyl)ethyl
Tos or Ts	*p*-toluenesulfonyl
TPAP	tetra-*n*-propylammonium perruthenate
TPP	triphenylphosphine
TPS	triphenylsilyl

TPS	triisopropylbenzenesulfonyl	Voc	vinyloxycarbonyl
Tr	trityl	xyl	xylene
Trisyl	trimethylphenylsulfonyl	%ee	enantiomeric excess
Troc	2,2,2-trichloroethoxycarbonyl	%de	diastereomeric excess
Vitride	sodium dihydrobis(2-methoxyethoxy) aluminate, see also Red-Al		

1

Introduction

Die Geschichte einer Wissenschaft ist diese Wissenschaft selbst.

J.W. von Goethe

The number of known chemical compounds increases at a stunning rate. At the time of writing this manuscript, roughly 180 million chemical compounds have been thoroughly characterised and approximately 40 000 new compounds find the daylight every single day. Many of these compounds have originally been isolated in the nature but of course, today most of the new compounds emerge through the action of chemical synthesis. This book covers natural products, that is, compounds that are formed by living systems and occur in plant or animal life.

As a scientific discipline, natural product chemistry covers the chemistry of naturally occurring organic compounds; their biosynthesis, function in their own environment, metabolism, and more conventional branches of chemistry, such as structure elucidation and synthesis. Natural products can be divided into three broad categories based on their role. The first group is the *primary metabolites* which are compounds occurring in all cells and are involved in the metabolism and reproduction of the cells. These compounds include nucleic acids, amino acids, and sugars. The second group is the *high molecular weight polymeric materials* that constitute the structural framework of all cells. These include materials such as cellulose, proteins, and lignin. The final group is the *secondary metabolites*, which are formed from the primary metabolites. They typically have biological effects on other organisms, and it is this third group of compounds that we are interested in.

The purpose of this text is to familiarise the reader with the common classes of natural products, their typical structures, and a brief introduction to their biogenesis, that is, the way the compounds are synthesised within the cells. Particular emphasis is placed on the methods currently available for their asymmetric synthesis.

1.1 Terminology

We shall begin by defining a few common concepts often encountered in connection with natural product chemistry: endogenous and exogenous substances, primary and secondary metabolism, toxicity, and detoxification.

Endogenous substances are compounds produced as a result of the normal functioning of an organism. Amino acids, many carbohydrates, peptide and steroid hormones, neurotransmitters, etc. produced by the body are typical endogenous substances. *Exogenous compounds* similarly refer to compounds coming from the outside of the organism, such as drugs and many environmental pollutants. Exogenous compounds are also known as *xenobiotics*.

Asymmetric Synthesis of Natural Products, Third Edition. Ari M.P. Koskinen.
© 2023 John Wiley & Sons Ltd. Published 2023 by John Wiley & Sons Ltd.

Primary metabolism is the system of biochemical reactions whose products are vital for the living organism. Photosynthetic plants convert carbon dioxide to primary metabolites carbohydrates, amino acids and other compounds ubiquitous to all forms of life. Primary metabolic pathways often function in cycles, such as the *Calvin cycle* (after *Melvin Calvin*, 1911–1997, Nobel Prize in Chemistry in 1961) for carbon fixation (Scheme 1.1), which converts carbon dioxide into glyceraldehyde-3-phosphate, the building block for carbohydrates, fatty acids and amino acids.

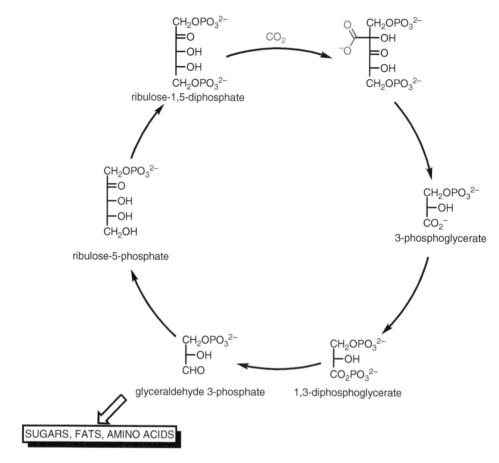

Scheme 1.1   *Calvin cycle*

Secondary metabolism refers to the functions of an organism yielding products that are not necessary for the essential biochemical events. Thus, secondary metabolites are compounds, which are often species dependent. Although earlier believed to be waste products of the (plant) metabolism, the actual role of secondary metabolites is still largely unclear. As plants cannot defend themselves against predators and pathogens such as animals through immune defense or movement, secondary metabolites are important for the survival of the plant: they often exhibit toxicity against exogenous pathogens, herbivores, or parasites. Secondary metabolites are often produced in a specific organ, cell, or tissue type, or during a specific growth stage. Secondary metabolites are highly characteristic for the plant and can thus be used as *chemotaxonomic* markers. In chemotaxonomy, one uses the similarities and differences in the chemical composition for the classification of species.

Photosynthetic plants convert carbon dioxide and water into simple carbohydrates (monosaccharides), which can be combined to make more complex polysaccharides and glycosides (Scheme 1.2). Further breakdown of the simple carbohydrates leads to pyruvic acid, which itself functions as the precursor to shikimic acid and thereby the aromatic compounds present in nature. Decarboxylation of pyruvic acid gives acetic acid, which functions as the biogenetic precursor

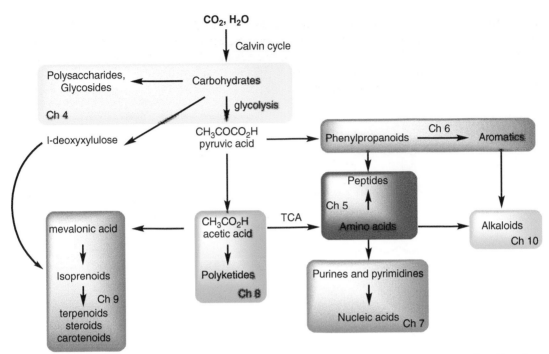

Scheme 1.2   *Secondary metabolites*

to practically all the remaining types of natural products. Condensation reactions lead to polyketides, mevalonic acid acts as a link between acetate and isoprenoids (terpenes), and amino acids and thereby peptides and alkaloids are formed from acetic acid. Tricarboxylic acid cycle (TCA, also known as citric acid cycle or Krebs cycle), is used by all aerobic organisms to generate energy through the oxidation of acetate into carbon dioxide. We will return to this in Section 4.1.2.

Prostaglandins are an example of mammalian secondary metabolites (Scheme 1.3). They are biosynthesised from an unsaturated fatty acid, arachidonic acid, via enzymatic oxygenation producing an intermediate prostaglandin endoperoxide, prostaglandin H_2. An alternative dioxygenation leads to 5-hydroperoxyeicosatetraenoic acid, 5-HPETE, the precursor of leukotrienes. Prostaglandins and leukotrienes participate in several physiological events in cells and organs.

The distinction between primary and secondary metabolites is not always straightforward: pyrrolidine-2-carboxylic acid (proline) is a primary metabolite, whereas piperidine-2-carboxylic acid (pipecolic acid) belongs to the secondary metabolites (Figure 1.1).

The reaction path leading to a particular natural product is called the biosynthetic path, and the corresponding event is known as the biogenesis. Different plant and animal species can employ dramatically different biosynthetic pathways to produce the same metabolite. This feature can be employed in the classification of plants in terms of their chemotaxonomy.

Natural products are very often associated with adverse effects, and many of them have become famous for being poisons, venoms, or toxins, as well as useful as pharmaceuticals. But what do these terms actually mean? Poison refers to a substance that is capable of causing the illness or death of a living organism when introduced or absorbed. Venom is a poison that is delivered mechanically through a bite or sting by for instance snakes and bees. The word toxin derives from the Greek word *toxicon*, which referred to a poison arrow or any sharp object that could be used to carry the poison. In ancient Greece, *pharmakon* was the word used for toxin, like the poison hemlock that Socrates was forced to ingest. The same word is the root for the modern words pharmacology and pharmaceuticals, revealing the intricate equilibrium between a compound holding equal powers for health and illness.

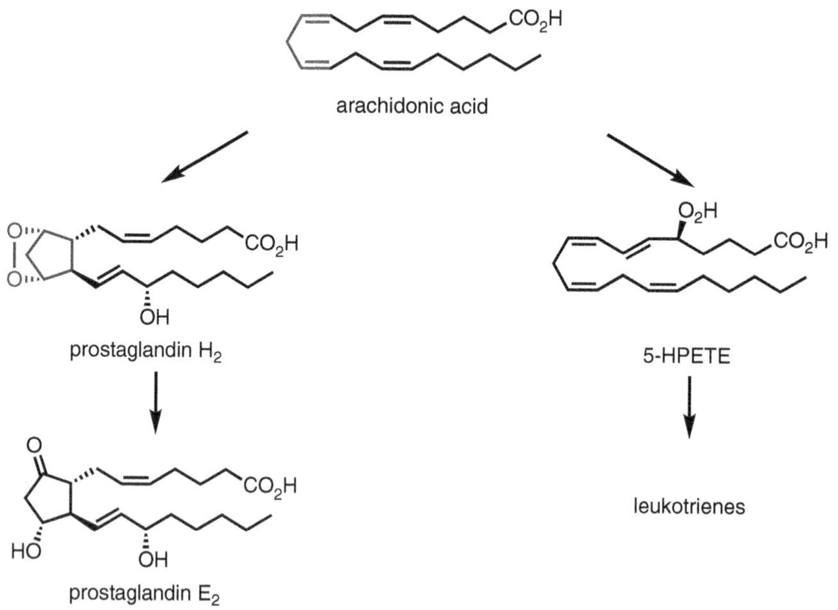

Scheme 1.3   *Formation of prostaglandins and leukotrienes*

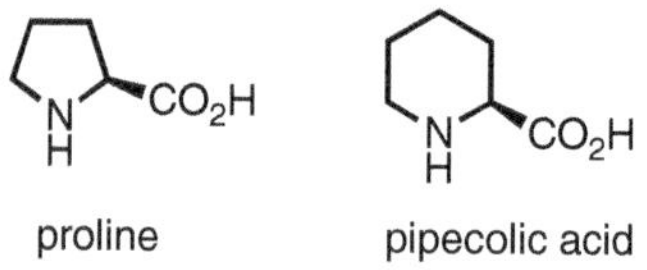

Figure 1.1   *Primary and secondary metabolites can be structurally quite similar*

Elimination of foreign, often toxic, compounds employs metabolic reactions. In such cases, one commonly speaks of *detoxification processes*. A major part of detoxification requires oxidative transformations, and in this respect, the liver plays a central role. The liver contains a multitude of oxidases capable of converting many different types of compounds to more highly oxidized products, which are broken down and eventually secreted.

Biochemical reactions and their control with all the associated intricacies do not belong to the subject matter of this text. To give the reader a sufficient general understanding, enzymatic reactions of special value to natural product chemistry will be discussed in connection with the individual natural product types.

1.2   Some Properties of Natural Products

Throughout our known history, humankind has been interested in naturally occurring compounds. Simple aqueous extractions of flowers, plants, and even insects have been used to isolate compounds whose taste, color, and fragrance properties could be used for various purposes. In fact, the Mesopotamian women, who isolated ethereal oils from plants, can be considered the first known chemists. The oils were apparently used as perfumes, and the art of distillation was probably learned as early as 3600 BC. The Greek *Theophrastus* (ca. 370–285 BC) is renowned as giving the first explicit details on the distillation of fragrants. Healing creams and liniments were also produced from plant extracts in practically all ancient cultures.

South American Indian hunters used, and still use, animal and plant extracts as arrow poisons. With all these applications, it is no surprise that with the development of the chemical methods the natural products have gained increasing importance in various aspects of human endeavor.

Originally, most natural products were isolated from plant origins, mainly due to the ease of the isolation process. The most common procedure still in use is, in outline, as follows: The plant material is divided according to the plant parts (the leaves are separated from the roots and stem, etc.), and the material is dried and ground to suitable particle size. This dry material is then extracted with a suitable solvent (e.g. methanol or chloroform), and the organic extract is then concentrated. The crude extract may contain hundreds of compounds, and earlier, their separation was based on crystallisation or distillation techniques. The development of modern chromatographic methods has facilitated the separation processes, and in practice, nearly all the components can be isolated in pure form. With the advent of more sophisticated analytical techniques, the isolation process can be guided by biological activity. Gene technology also allows incorporation of gene fragments from different species to produce 'hybrid' compounds. Of course, one can ask whether such a compound is really a natural product. In mutational biosynthesis or mutasynthesis [1], one can also administer synthetically modified intermediates, so-called mutasynthons, to genetically engineered organisms, where the formation of key biosynthetic intermediates is blocked to produce new secondary metabolites. Incorporation of these mutasynthons leads to the generation of novel metabolites, which can be isolated and evaluated for their therapeutic potential [2].

Natural products are usually given a trivial name derived from the plant origin (e.g. muscarine from the mushroom *Amanita muscaria*, fly agaric) (Figure 1.2). In some cases, the name describes the physiological action. Putrescine and cadaverine were described by *Ludvig Brieger* (1849–1919) as early as 1885 as substances produced during the putrefaction of cadavers, and hence their names [3]. They are protein degradation products of decarboxylation of amino acids lysine and arginine [4]. The name of the isoquinoline alkaloid emetine leaves no doubt as to its physiological effects.

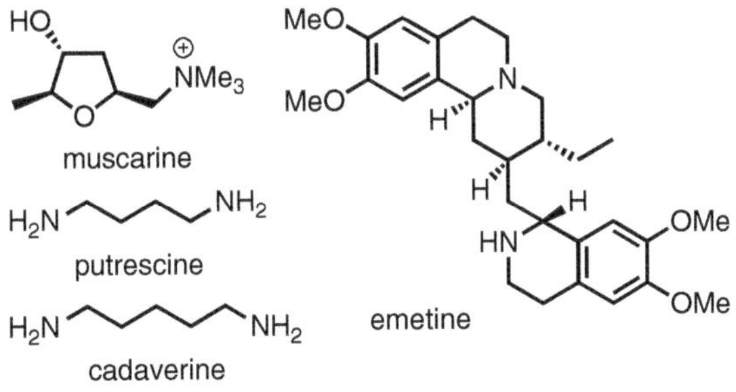

Figure 1.2 *Trivial names for natural products*

Taste, odor, and color of organic natural products are usually the properties most easily detected. The relationships between aroma compounds can often be quite surprising, and chemically closely related structures can have quite different properties (Figure 1.3). Strawberries and pineapple contain large quantities of furaneol, whose close relative mesifurane can be found in arctic bramble (*Rubus arcticus*) and (canned) mango [5].

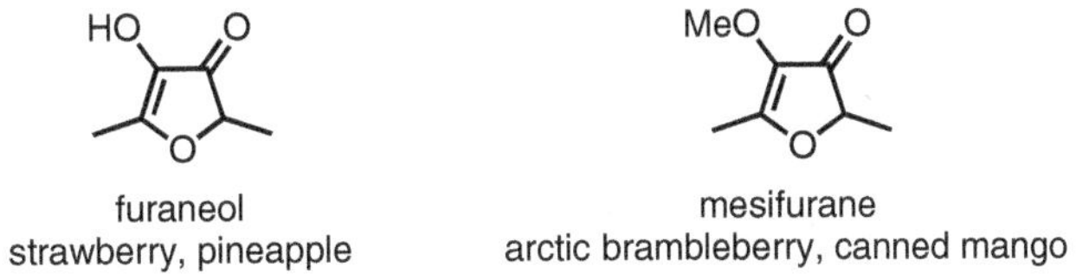

furaneol
strawberry, pineapple

mesifurane
arctic brambleberry, canned mango

Figure 1.3 *Related furans with distinct flavors*

Sulfur-containing compounds are abundant in onions (*Allium* species) (Figure 1.4): The strong lachrymatory action of onions is caused by decomposition products of S-allylcysteine [6], and the aroma compounds of fresh and boiled onions are chemically different. The aroma compounds of garlic (*Allium sativum*) are closely related to those of onion (*Allium cepa*) [7].

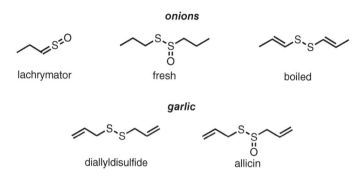

Figure 1.4   *Sulfur containing compounds in onions*

The chiral (+)-*cis*-2-methyl-4-propyl-1,3-oxathiane is the main aroma compound found in the yellow passion fruit (*Passiflora edulis f. flavicarpa*) (Figure 1.5) [8]. It has recently been synthesised utilising an organocatalytic conjugate addition over thiol onto an unsaturated aldehyde in 84% *ee* [9]. Tropical fruits are particularly rich in volatile sulfur compounds, which often elicit characteristic odor and flavor properties for the fruit [10].

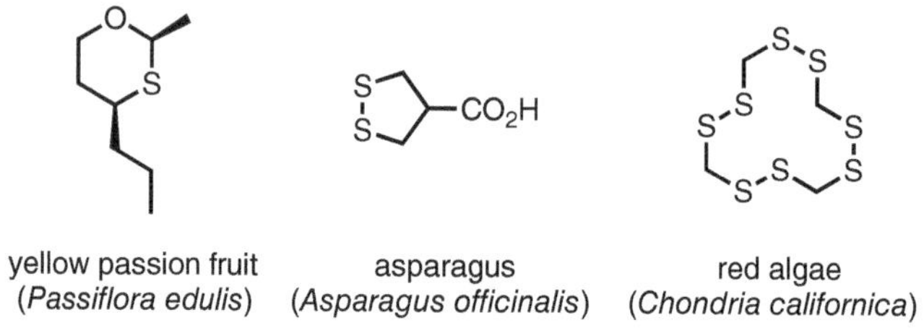

Figure 1.5   *Sulfur-containing natural products in foodstuffs*

Asparagusic acid is a sulfur-containing compound that has been only found in asparagus [11]. It has been suggested the breakdown of this compound to volatile sulfur-containing compounds like dimethyl sulfide causes the odor effects sometimes observed with people who have consumed asparagus. Perhaps the organic compound richest in sulfur occurs in red alga *Chondria californica* [12].

Examples of related structures with different aroma are the pyrazine aroma compounds isolated from potato chips [13], bell peppers [14], and Cabernet sauvignon grapes [15] (Figure 1.6).

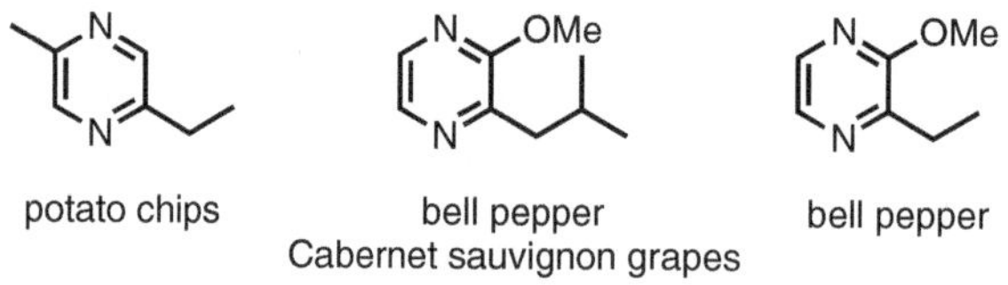

Figure 1.6   *Pyrazines in foodstuffs*

Some very unpleasant smell and taste effects may also be caused by organic natural products (Figure 1.7). When beer is left in the open sunlight, humulone is photolytically degraded, and the degradation product reacts with hydrogen sulfide to produce 3-methylbut-2-ene-1-thiol, the "aroma" compound of sun-burnt beer [16]. It is structurally related to the odor substances of North American skunk [17] and cat's urine (cat ketone, 4-mercapto-4-methylpentan-2-one) [18], which also occurs in Sauvignon grapes [19]. Sulfur-containing volatile compounds in fragrance chemistry have been reviewed [20].

The taste effects of many organic compounds can also depend on the solvent. For instance, 2-acetyl-3-methyl thio-phene dissolved in syrup induces a honey-like taste, whereas in coffee it gives a nutty flavor.

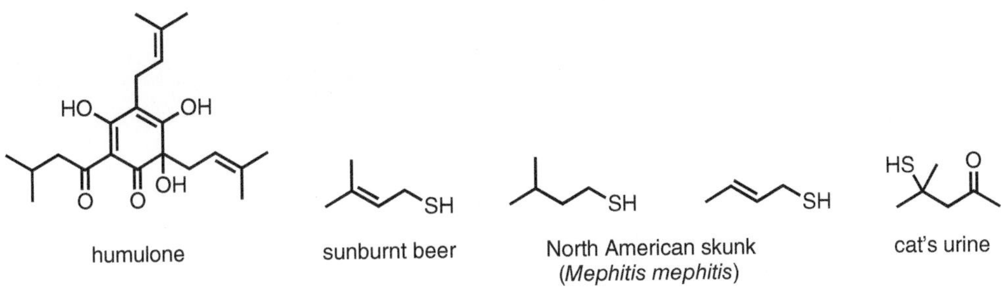

Figure 1.7   *Some aroma compounds*

1.3   Roles of Natural Products

The reason why plants produce secondary metabolites is still largely unknown and subject to speculations. In many cases, the importance of a particular substance to the plant is not known. It has often been suggested that the plant simply excretes a part of its waste products in the form of natural products. This is not an appealing suggestion since the natural products often exhibit very complicated structures. Recent developments in biology have given us some hints to understand the importance of these compounds. Many natural products have a regulatory role (e.g. growth hormones). Some function as chemical defense agents against pests; the most potent ones may be lethal. The role of certain compounds is to act as chemical messenger molecules between species of the same genus.

In the 1960s, an abscission-accelerating substance was isolated from cotton fruit and it was named abscisin II. The same substance was subsequently isolated from sycamore leaves as the result of a search for a "dormin" [an endogenous substance inducing dormancy]. The chemical structure of abscisin II was determined by synthesis in 1965, and the compound was renamed abscisic acid (ABA) (Figure 1.8) [21]. When plants wilt, ABA levels increase considerably and cause stomatal closure. ABA is thus important in mediating responses of vegetative tissues to environmental stresses such as drought, high salinity, and low temperature. Despite its name, ABA is not a major regulator of abscission, which is primarily controlled by ethylene. The role of ethylene as a plant grows regulator will be discussed in Chapter 5. Zeatin is a cytokinin derived from adenine (see Chapter 6), it stimulates seed germination and seedling growth. *cis*-Jasmonic acid is a representative of fatty acid derived plant growth factors. It is produced from linolenic acid through a lipoxygenase pathway, which we will encounter in Chapter 7. These compounds are named after the initial isolation and characterisation of methyl jasmonate from jasmine oil *Jasminum grandiflorum* in the early 1960s [22].

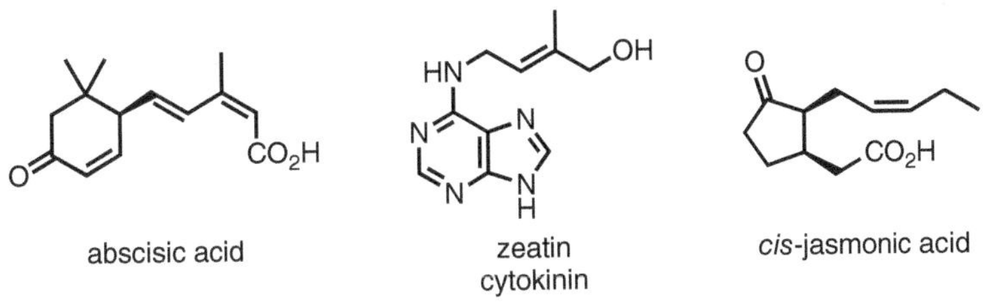

Figure 1.8   *Plant growth regulators*

Many green plants produce gibberellins, which function as growth hormones. As soon as the structures of the compounds became known, the synthetic efforts led to practical preparations of many of their congeners. Currently, several gibberellins are produced in bulk for agriculture and plant development. Certain gibberellins are also widely used in brewing to shorten the time needed for malting (Figure 1.9).

Some natural products act as chemical police. Aphidicolin, a complex terpene, reduces the appetite of aphids. This has been put to use in crop protection. The most effective compounds of this class are so potent that 10 grams of material is sufficient to treat one hectare for the entire growth period.

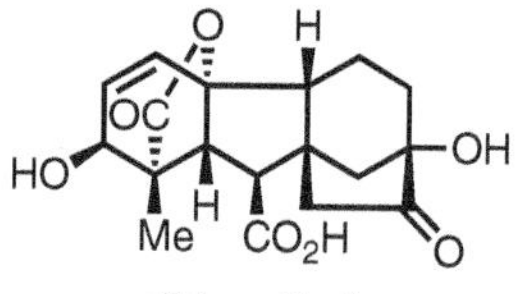

gibberellin A_3

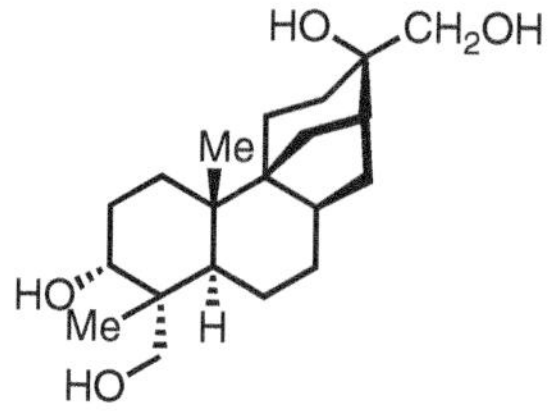

aphidicolin

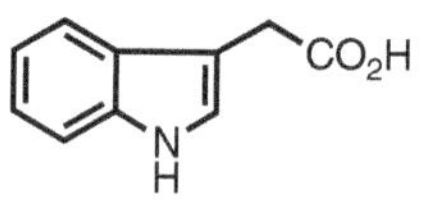

indole-3-acetic acid
(hetero)auxin

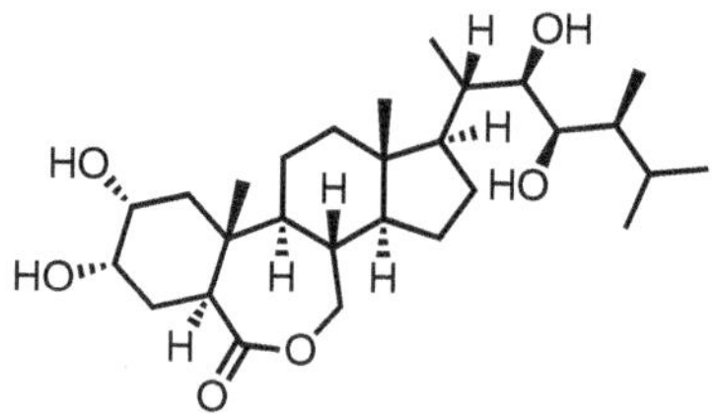

brassinolide

Figure 1.9 Plant growth regulators

Indole-3-acetic acid (IAA) or heteroauxin is the best-known plant growth hormone belonging to the class of auxins. Its biosynthesis is not clearly known but at least three biosynthetic pathways are known to emerge from tryptophane and one further that is tryptophane independent.

Brassinolide is a steroidal plant hormone discovered when it was shown that pollen from rapeseed (*Brassica napus*) could promote stem elongation and cell division [23].

All animals can communicate with members of the same species, but voice, touch, and sight are not the only possible methods for transferring messages. Perhaps the most sensitive sense is based on chemical recognition: the senses of smelling and tasting are fairly sensitive among humans, but for instance dogs have considerably better abilities to handle information transduced in these ways. Even small insects transmit messages dressed in chemical form. These often rather simple organic molecules are called pheromones. Many such compounds affecting the sexual behavior of insects are widely used in plant protection (Figure 1.10). For instance, European spruce bark beetle (*Ips typograpus*) is the economically most costly pest, destroying especially trees of the genera *Picea* (spruce), *Abies* (fir), *Pinus* (pine), and *Larix* (larch) in Scandinavia. Pest traps with *cis*-verbenol can be used to restrict the outbreak of beetle invasions. Bombykol is a sexual attractant pheromone excreted by female silkworms, the primary producers of silk. *Adolf Butenandt* (1903–1995) discovered bombykol in 1959, the first pheromone isolated and characterised chemically [24].

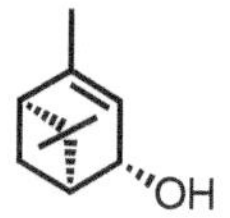

(*S*)-*cis*-verbenol
(*Ips typographus*)

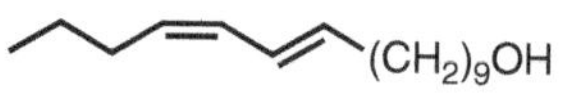

bombykol
(*Bombyx mori*)

Figure 1.10 Pheromones are used for chemical communication

Many plants and animals produce chemicals with toxic or even lethal action on other species. Snake venoms are usually rather small peptides, but small organic molecules are also known with such characteristics (Figure 1.11). The brightly colored small South American frogs of *Phyllobates* (literally "leaf walkers") and *Dendrobates* (literally "branch walkers") species have been used by Colombian hunters to envenomate their poison darts. The Columbian Golden poison frog *Phyllobates terribilis* is perhaps the most poisonous land animal on the planet. An average adult frog carries 200 to 500 micrograms of each of the three deadly toxins on its skin: batrachotoxin or homobatrachotoxin, and batrachotoxinin-A. This would be enough to kill an animal of several hundreds of kilograms, even a hippopotamus.

The frogs do not synthesise the toxins themselves but they get their arsenal from the insects they eat. Interestingly, homobatrachotoxin is the same steroidal alkaloid (Chapter 10) that is found on the skin and feathers of the New Guinean hooded Pitohui birds (*Pitohui dichrous*), the only known poisonous birds [25]. The birds and the frogs live 10 000 miles apart, and it is believed that they both consume the same soft-winged flower beetles in the family of *Melyridae* which are loaded with batrachotoxins.

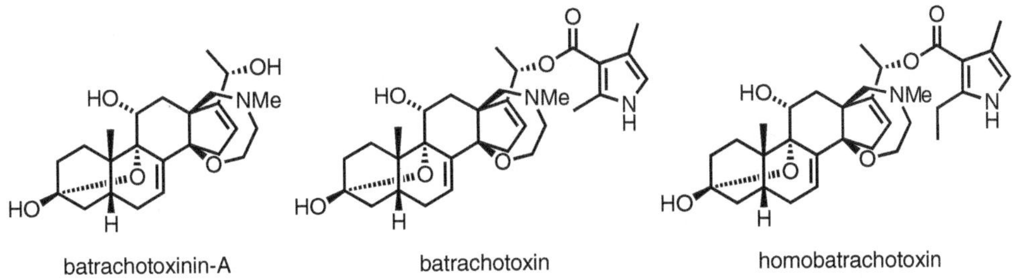

batrachotoxinin-A   batrachotoxin   homobatrachotoxin

Figure 1.11   *Frog poisons*

The poison frogs of the genus *Dendrobatidae*, particularly *Oophaga histrionica* (formerly *Dendrobates histrionicus*), secrete in their skin the much less poisonous alkaloids pumiliotoxins, histrionicotoxins, and gephyrotoxins (Figure 1.12). All these affect neural activity, but their modes of action differ slightly. Pumiliotoxins affect calcium channels, causing muscle contraction, which can lead to partial paralysis, difficulty in moving, or death. Histrionicotoxins are nicotinic acetylcholine receptor antagonists, blocking the action potentials and causing slowed neural activity. Gephyrotoxins in turn are muscarinic antagonists.

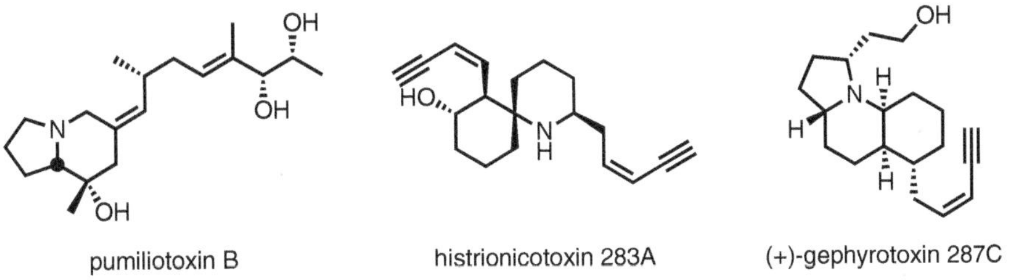

pumiliotoxin B   histrionicotoxin 283A   (+)-gephyrotoxin 287C

Figure 1.12   *Poison dart frog alkaloids*

Sea cucumber (*Bohadschia argus*) produces holothurines (Figure 1.13), which can cause blindness in humans. These toxic glycoside compounds are concentrated inside the tubular structures called Cuvierian organs, which the sea cucumbers will blast out of their bodies when threatened. Although sea cucumbers are considered delicacies in many Asian diets (*iriko*, *namako* in Japanese, Chinese *hai shen*, Malaysian *gamat*, and Filipini *balatan*), one has to be careful to remove the tubules when preparing the food.

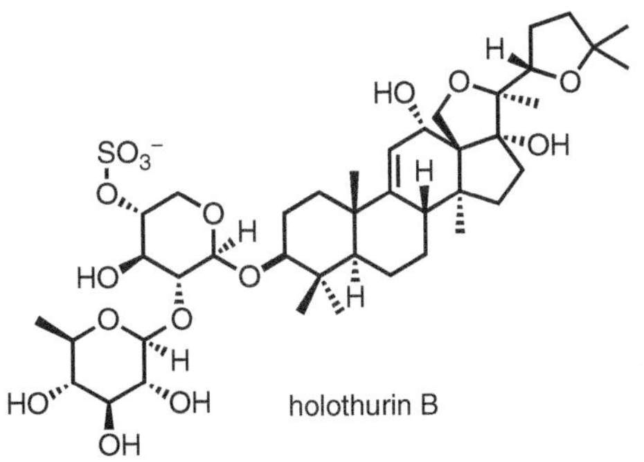

holothurin B

Figure 1.13   *Holothurines are toxins from sea cucumbers*

Tetrodotoxin (TTX, Figure 1.14) is a potent neurotoxin which is present in many fish of the order *Tetraodontiformes*. Several of these species carry the toxin, including pufferfish, porcupinefish, ocean sunfish, and triggerfish. Tetrodotoxin has also been discovered in several other animals including blue-ringed octopus (*Hapalochlaena lunulata*) and sea slug (*Pleurobranchaea maculate*), but it is actually produced by certain infecting or symbiotic bacteria like *Pseudomonas* and *Vibrio* [26].

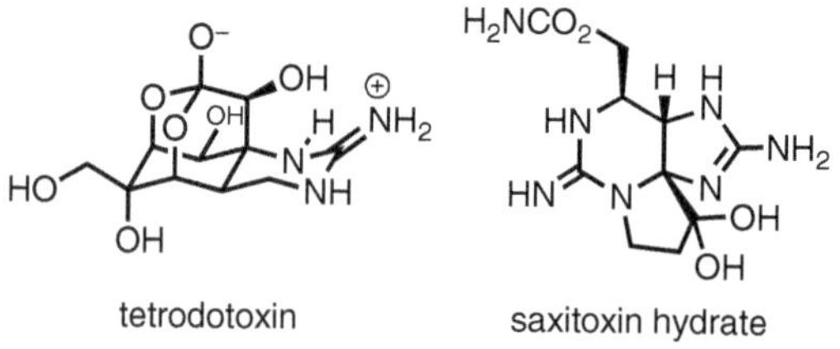

tetrodotoxin   saxitoxin hydrate

Figure 1.14 *Marine neurotoxins*

Saxitoxin (STX) is a potent neurotoxin originally isolated from the butter clam (*Saxidomus*). The consumption of shellfish (e.g. mussels, butter clams, and oysters) that have been contaminated by toxic algal bloom may cause paralytic shellfish poisoning (PSP) in humans (mytilotoxism). The term saxitoxin also refers to the more than 50 structurally related neurotoxins (known collectively as "saxitoxins") produced by algae and cyanobacteria [27]. Saxitoxin has a questionable reputation of being known as chemical weapon agent TZ.

1.4 Natural Products as Drugs

Natural products have played a key role in the development of medicinal chemistry. Even today, a large number of new chemical entities are arrived at through the help of natural products [28]. In the early days, medicines were isolated from plant material. Later, humans learned to utilise organic synthesis and fermentation to produce the medicinal agents, and within the last decades of the twentieth century, the methods of molecular biology enabled the programming of the cells to produce several variants of earlier known compounds. These different approaches are not alternatives; rather, they complement each other giving the medicinal or natural product chemist access to a wide spectrum of tools to work with.

We shall take a brief look at the history of drugs as far as natural products are concerned. The oldest information on drugs goes back to China in the Bronze Age. The first medical herbal, *Shen-nung pen ts'ao-ching* (Divine husbandman's materia medica), describes 365 drugs used in those days. It is believed to have been compiled by *Shen Nung* (ca. 2800 BC), also known as the Yan emperor. The Chinese are renowned for being the people who first familiarised themselves with the noble teachings of alchemy. The honor of being the first alchemist is often attributed to *Li Shao-chun*, and the first textbook of alchemy is *Ts'an t'ung Ch'i* from *ca.* 120 BC. Thus, it is no surprise that the first known drug comes from China: ephedra (the 'horsetail' plant), isolated from a plant known as *ma huang* and described by Shen Nung. Ephedra has been used for thousands of years as a stimulant, a remedy for respiratory diseases, to induce fever and perspiration, and to depress cough. Ephedra was also included in the Greek pharmacopoeia. In Western medicine, the active principle (ephedrine, Figure 1.15) was isolated in chemically pure form in 1887 by the Japanese *Nagayoshi Nagai* (1844–1929) [29]. Nagai was among the first Japanese students who were allowed to travel abroad for further education. He spent 12 years in Berlin with *August Wilhelm von Hofmann* (1818–1892) and returned to Japan in 1883 to start his work on ephedra. Ephedrine played a major role in the elucidation of the adrenergic nervous system and eventually led to the development of salbutamol as an asthma medication in inhalers.

Ginseng (*Panax ginseng*), rénshēn (Chinese for "human"), genus of 12 species of medicinal herbs of the family *Araliaceae*, was also known by the Yan emperor. The root of Asian ginseng (*Panax ginseng*), native to Manchuria and Korea, has long been used as a drug and is made into a stimulating tea in China, Korea, and Japan. More than 30 ginseng saponins or ginsenosides have been identified. Among other effects, the root tea was supposed to delay aging and restore sexual powers. Ginseng also ameliorates the symptoms of diabetes and stabilises blood pressure [30].

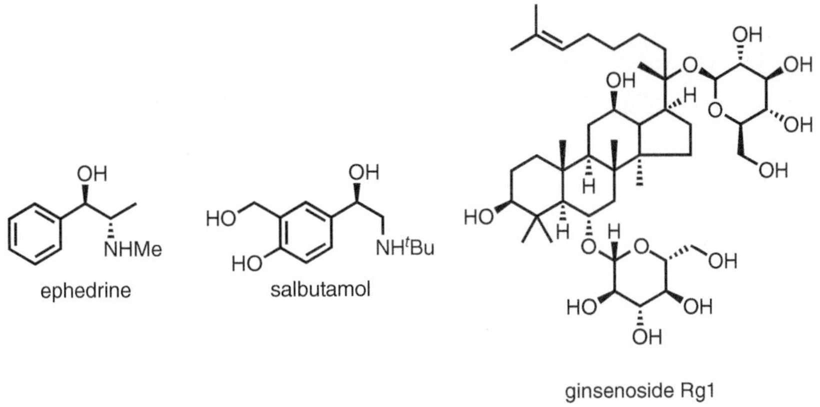

Figure 1.15   *Drugs from traditional Chinese medicine*

The next important step comes from the Egyptians, the so-called Papyrus of Ebers (from *ca.* 1500 BC, named after the German Egyptologist Georg Ebers) which was not transcribed until 1937 by the Norwegian doctor Bendix Ebbell. The Papyrus of Ebers describes several practical preparations still in use today such as opium, castor oil, and liver (vitamin A). The Egyptians also used 'rotten bread' to treat infections – this clearly has a striking connection to our current understanding and use of compounds produced by molds and fungi (such as penicillin) as antibacterial agents.

On the north shores of the Mediterranean, we meet the Greek father of medicine, *Hippocrates* (ca. 460–370 BC), and the Roman physicians Dioscorides and Galen; the fathers of the doctrine of signatures, according to which nature itself has marked the shapes of plants to indicate suitable medicines for all illnesses: liver shaped leaves sign that they can be used to treat the illnesses of liver. *Pedanius Dioscorides* (ca. 40–90 AD) was an army physician who studied the medical uses of hundreds of plants and probably wrote the first systematic pharmacopoeia; the five-volume *De materia medica* during the first century AD. Already *Diocles of Carystus* (in the fourth century BC) had collected similar information on medicinal plants, but the works of Dioscorides are attributed to be the first thorough and systematic studies in their kind. Together with *Pliny* (*Caius Plinius Secundus*, the 'Elder Pliny,' 23–79), Dioscorides also described the medicinal properties of wines. Among the effects noted in those days were quickening of heart pulse and the injurious effects of its continued use. *Galen of Pergamon* (*Claudius Galenus*, 129– *ca.* 199 or 217) is considered to be the forebear of experimental physiology. He was convinced that all illnesses could be cured with mixtures of concoctions, as long as one can find out the necessary proportions. A typical concoction from antiquity was teriak (Gr. counterpoison) which often contained more than 60 components, such as opium, dried snake meat, cinnamon, pepper, onions, fennel, and cardamom. Teriak reached other parts of the world very slowly, for instance Scandinavia got its first teriak in the eighteenth century, and as late as the twentieth century one could still find such concoctions from some small chemists' shops in Central Europe.

The colorful Middle Ages brought back the signature theory. *Philippus Aureolus Theophrastus Bombastus von Hohenheim* (1493–1541), better known as *Paracelsus*, was the town doctor of Basel from 1526, and a lecturer of medicine in the University of Basel. The much-disputed writings of Paracelsus were the first in their kind in the field of *iatrochemistry*, which aimed solely at curing illnesses. Some earlier indications towards similar goals can be found in the writings of *Raymond Lull* (*Raymundus Lullus*), (1232–1315) and the Arabian *Abu bakr al-Razi* (865–925). Paracelsus, however, so furiously resisted traditional practices of medicine that he was forced to leave Basel. After living a life of a vagabond, he died in Salzburg in 1541.

During the seventeenth century, the Jesuits brought with them from South America the bark of the quina (*Cinchona officinalis*) tree (1632, for the treatment of malaria) and some plant concoctions developed by the Quechua (Inca) people. As a reflection of these events, one can today find a large proportion of medicinal agents being derived from alkaloids. During the same century, chemistry also started to gain respect as a natural science. The first chemistry university laboratory was opened in 1609 in the University of Marburg where *Johannes Hartmann* (1568–1631) was appointed

Professor of Chymiatria. Together with his students, he started to produce pharmaceutical products and published the first publication in this field *Disputationes Chymico-Medicae* in 1611, which included his inaugural address and seven disputations by his students [31].

In the nineteenth century, the development of organic chemistry rapidly took off, and the isolation and identification of natural products started to be more systematical. In 1820, Pelletier and Caventou isolated the alkaloid quinine from the quina tree, the active compound against malaria (Chapter 10). This sparked a rapidly growing interest in isolating the chemical constituents of the medicinal plants.

The art of organic synthesis was transmitted from apothecaries to the expert chemists, and at the same time the quality of the products improved. Pure chemical entities started to replace old dried isolates and decocts (extracts). The first such compounds were naturally occurring nitrogenous compounds, alkaloids, which were easy to isolate by repetitive extractions and could be purified in their salt form by crystallisation.

Gerardus Johannes Mulder (1802–1880) at the University of Utrecht had first observed the existence of proteins in the 1830s. *Felix Hoppe-Seyler* (1825–1895) was able to obtain crystals of a protein, hemoglobin, in 1864. Five years later, in 1869, *Friedrich Miescher* (1844–1895) found the chemical carrier of heredity, nuclein, whose deoxyribonucleic acid structure remained obscure until the works of *Albrecht Kossel* (1853–1927) in the 1890s.

Toward the end of the nineteenth century, microbiology developed into a separate scientific discipline. *Robert Koch* (1843–1910) showed that living organisms may cause an infection (1876). *Louis Pasteur* (1822–1895) and *Jules-Francois Joubert* (1834–1910) showed in 1877 that bacteria may antagonise each other's growth. Pasteur's student *Paul Vuillemin* (1861–1932) defined the concept of antibiosis (*Gr. anti* = against, *bios* = life) in 1889.

The development of antibiotics in the early part of the twentieth century literally changed history. Several compounds were instrumental in this change, and the advent of synthetic chemistry played a major role. Natural products were at the helm, and a few chemicals display the importance here (Figure 1.16). The widest known story of antibiotics produced by microbes must be that of penicillin, which was first isolated by the Scotsman *Alexander Fleming* (1881–1955) in 1929. The development of penicillin into a drug was slow, and only during World War II did the efforts of the Australian born *Howard Florey* (1898–1968) and the German born *Ernst Boris Chain* (1906–1979) make the drug available in larger quantities to the Allied soldiers. However, this was not the first commercial antibacterial drug produced by microbes. *Bartolomeo Gosio* (1863–1944) had already shown in 1893 [32] that mycophenolic acid, produced by the mold *Penicillium brevicompacticum* inhibited the growth of the bacterium *Bacillus anthracis* [33]. The first hospital use of an antibiotic was a crude extract from *Pseudomonas aeruginosa*, which contained an enzyme, pyocyanase, by Emmerich and Loew in 1899 [34]. The structure of the active antibiotic pyocyanine was not elucidated until 1929.

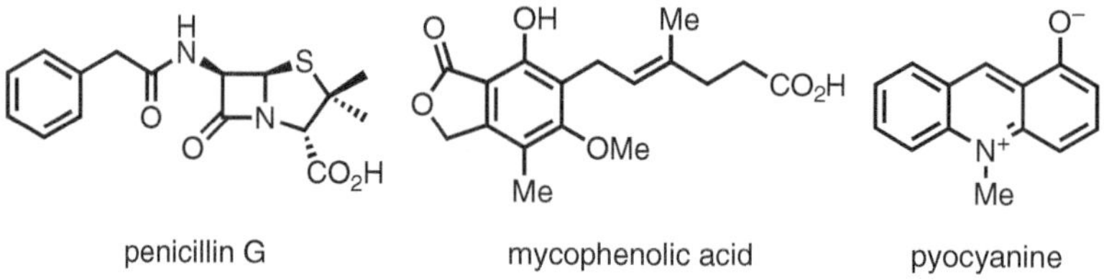

penicillin G   mycophenolic acid   pyocyanine

Figure 1.16 *Antibiotic, small molecule, natural products isolated in the early twentieth century*

The Ukrainian scientist *Selman Abraham Waksman* (at Rutgers University, USA, 1888–1973) and *H. Boyd Woodruff* (1917–2017) isolated actinomycin in 1940 (Figure 1.17) [35]. Twelve years later, this compound became the first cytostatically active compound for humans. Cytostats are agents that inhibit the growth of tumors. Waksman isolated also streptomycin in 1944, a compound that is particularly useful for the treatment of tuberculosis. As an anecdotal piece of information, you might be interested to know that antibiotics isolated from *Streptomyces* species are called –mycins, whereas antibiotics from other genera (e.g. *Micromonospora*) are called –micins.

As a source for drugs, nature provides a wide spectrum of compounds, which themselves can be used to treat diseases. These compounds can be classified as fine chemicals with typically high degree of refining, and thus a high price. The *Catharanthus* alkaloids, important for the treatment of several forms of cancer, can cost hundreds of thousands of

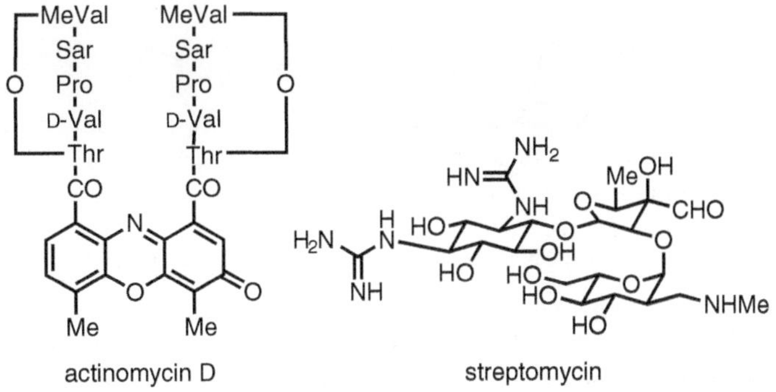

Figure 1.17 *Mycin antibiotics isolated in the 1940s (Sar = sarcosine = N-methylglycine)*

dollars per kilogram. However, one should not forget the importance of simpler compounds produced by nature, as these can be valuable starting materials or intermediates for the synthesis of other products. Many amino acids, lipids, and carbohydrates are being produced in ton quantities, often for the price of a few dollars per kilogram or less. Table 1.1 lists some commercially important drugs whose production is based on natural products isolated from plants.

Table 1.1 *Drugs of plant origin*

Compound	Origin	Medicinal use
Steroids:		
Hormones (95% of diosgenin)	*Dioscorea* (mexican yams)	Contraceptives, anabolic steroids, corticosteroids
***Digitalis*-glycosides** (digitoxin, digoxin)	*Digitalis purpurea* (foxglove)	Cardenolides
Alkaloids:		
***Opium*-alkaloids** (morphine, codeine)	*Papaver somniferum* (opium poppy)	Pain relief
***Catharanthus*-alkaloids**	*Catharanthus roseus* (Madagascan periwinkle)	Cancer
pilocarpine	*Pilocarpus* species	Glaucoma
colchicine	*Colchicium autumnale* (autumn crocus)	Gout
cocaine	*Erythroxylon coca* (coca leaves)	local anaesthesia

1.5 Structures of Natural Products

The elucidation of the biosynthetic pathways, varying properties, and medicinal uses of natural products would be in themselves good enough reasons to study the synthesis of natural products. However, added momentum is gained by the enormous variations in the structures, and especially by the occurrence of structures whose complexities have surpassed the wildest imagination of the chemist the day when these compounds were isolated. The diversity of the chemical structures is exemplified by the two structures in Figure 1.18. Ryanodol is a highly oxygenated diterpene containing eight oxygen atoms and altogether eight contiguous quaternary carbon atoms. Silphinene is a sesquiterpene, which contains only carbon and hydrogen atoms and has three quaternary carbons.

In the early days of the nineteenth century, the structures of natural products had to be determined by independent synthesis. From early twentieth century, the structural complexity has been a major driving force for the development of spectroscopic and spectrometric means of structure elucidation. Detailed information on UV chromophores was obtained, and efficient correlations (e.g. the Scott and Woodward rules) were formulated in connection with the work

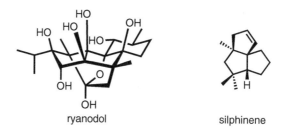

Figure 1.18 *Natural products with unorthodox structures*

on steroid structures. Infrared spectroscopy gained a similar impetus from the study of ketones and lactones of varying size. The fragmentation patterns in mass spectrometry were uncovered mainly through natural product work, and ^{1}H and ^{13}C NMR spectrometric studies on the ever more complex structures have aided the development of good correlations between structure, and chemical shifts and coupling constants. All these spectroscopic developments after the Second World War have tremendously helped structure elucidation of even complex natural products, to the extent that in many cases the structures may be assigned within days rather than months, as in the era of synthesis only. However, even X-ray crystallography is not infallible, and every year numerous spectroscopically assigned structures have to be reassigned based on synthetic work [36].

Understanding biosynthetic reaction mechanisms has played an important role in the development of the general theory of physical and mechanistic organic chemistry. For instance, the biosynthetic pathways leading to terpenes often involve rather deep-seated cationic rearrangements. A good case in point is the computationally predicted synthesis of adamantane, which was experimentally proven (Figure 1.19). The simple cyclopentadiene dimer was hydrogenated to the saturated tricycle, and simply heating this compound in the presence of a Lewis acid caused the cationic rearrangement to produce the thermodynamically most stable adamantane [37].

Figure 1.19 *Computationally predicted synthesis of adamantane*

In other cases, photochemically induced rearrangement reactions give structures with unforeseen carbon skeleta. Nucleophilic and electrophilic alkylation reactions abound in the nature. Some remarkable general patterns can be seen in the stereochemical outcome of such reactions, as evidenced by the nearly exclusive existence of a single diastereomeric series of aldol products in the biogenesis of macrolide antibiotics. Rapid developments in molecular biology have enabled us to understand the genetic basis of polyketide synthesis and even manipulate the genetic machinery, as we will see in Chapter 8.

Acylation reactions are also common, and perhaps millions of years ago nature found ways to achieve such reactions using a nucleophilic acylating reagent (umpolung), a concept only introduced to synthetic organic chemistry relatively recently [38]. The H2 of the thiazolium ring in thiamine pyrophosphate is acidic enough to undergo deprotonation [39], and reaction with pyruvate gives the hydroxyethylthiamine pyrophosphate, the biological 'active aldehyde' (Scheme 1.4). As an enamine, this can react with aldehydes to give the acetoin product and liberate thiamine pyrophosphate again.

In some cases, the mechanisms of the action of natural compounds have been worked on before the actual natural products have been isolated. One striking example of such a case is the discovery of the Masamune–Bergman cyclisation of enediynes to aromatic systems (Scheme 1.5) [40]. This reaction found its natural counterpart nearly two decades after it had been developed in the laboratory. Nature produces a number of enediyne antibiotics, which cleave the DNA with high efficiency [41]. As a result, the DNA cannot be replicated, and some of these compounds are promising anti-cancer agents. We will discuss these in Chapter 6, Section 3.1.

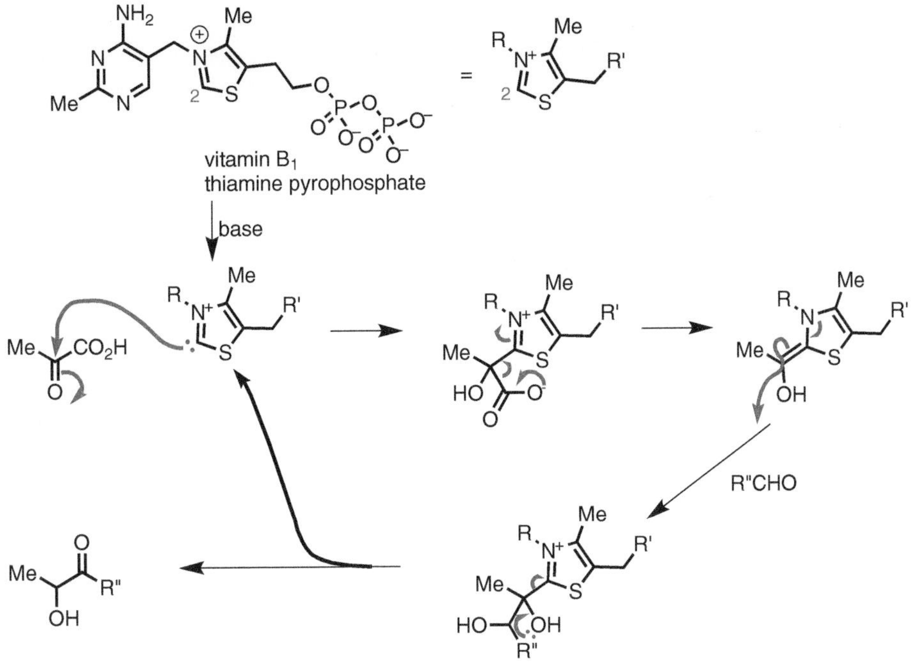

Scheme 1.4   *Natural inversion of polarity in acylation reactions*

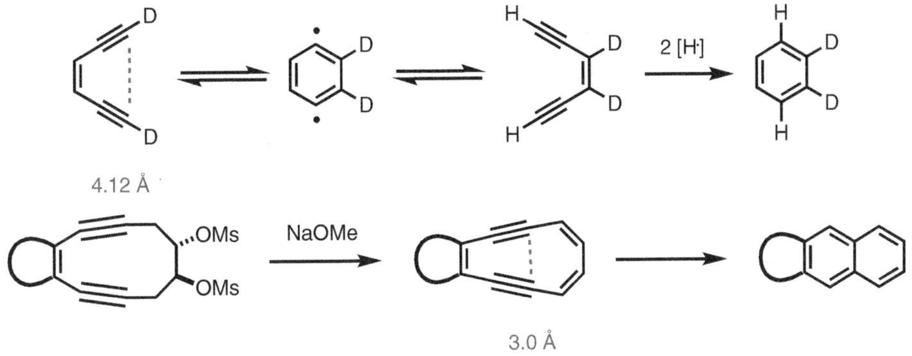

Scheme 1.5   *The Masamune–Bergman cyclisation*

From the point of view of stability, natural products often contain remarkably labile structural units. This is, of course, a complicating matter as far as isolation and structural work are concerned, but at the same time, it provides the challenges to the experimental techniques without which the development of synthetic chemistry would have missed major contributions. Only a few decades ago, few chemists would have dared even to think of synthesising a conjugated polyene with, say, a dozen double bonds. Nature does it, and these compounds are relatively stable in their respective environments, as evidenced by such compounds as polyenes and enynes. Aurantosides, which contain a dichlorinated polyene and a tricarbonyl unit, are the first tetramic acid glycosides isolated from marine organisms (Figure 1.20) [42]. These orange pigments were isolated from the marine sponge *Theonella*, and they possess cytotoxic activity.

Increasing ring strain brings with it increasing reactivity and thus also lability (Figure 1.21). Three-, four- and five-membered rings are, however, quite common in natural products. The *β*-lactam story would have suffered quite a

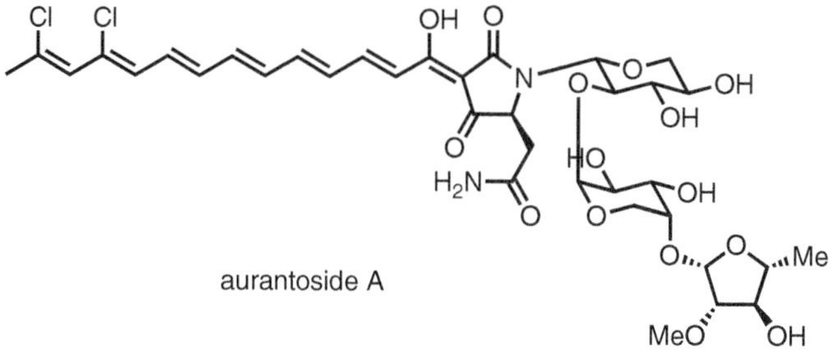

Figure 1.20   *Polyene tetramic acid*

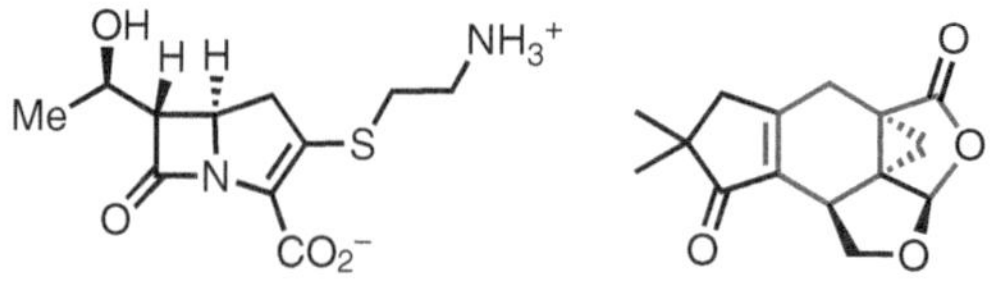

thienamycin   sterepolide

Figure 1.21   *Strained ring systems in natural products*

bit without the bold minds of the chemists who dared to attack the four-membered lactam ring structure. The sesquiter-pene sterepolide contains a highly strained [4.3.1]propellane structure (blue) and you will see its asymmetric synthesis in Chapter 9.

Peptide structures are today often considered rather mundane, being made of only twenty amino acid building blocks by rather simple condensation reactions. Not quite so! Nature often does the unexpected and provides surprises. Nature abounds peptides which contain amino acids other than the 'natural' twenty. The immuno-suppressant cyclosporin A is a case in point (Figure 1.22). The rare unsaturated hydroxyamino acid MeBmt, (4R)-4-((E)-2-butenyl)-4,N-dimethyl-L-threonine or (2S,3R,4R,6E)-3-hydroxy-4-methyl-2-(methylamino)-6-octenoic acid (blue), has been the target of several synthetic approaches, and its structural analogues have given more insight into understanding the mechanism of action of this important peptide. Nature takes this game even further; echinocandins are antifungal cyclic lipopeptides produced by *Aspergillus* species [43].

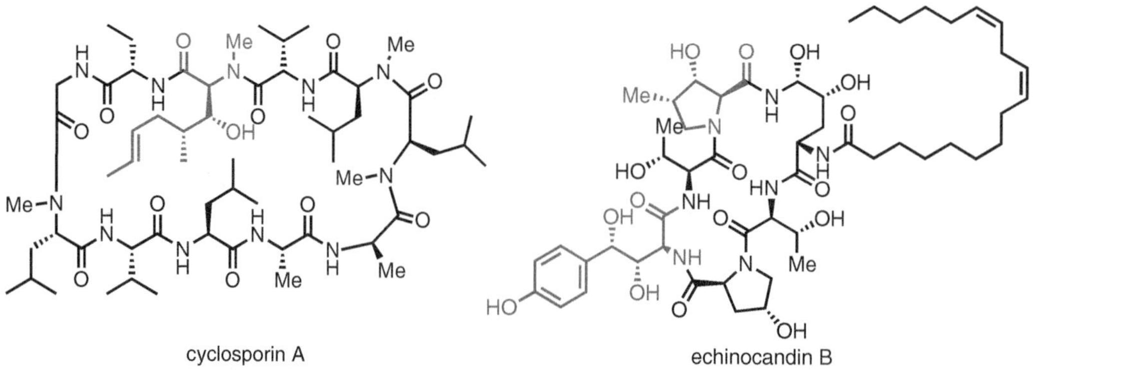

cyclosporin A   echinocandin B

Figure 1.22   *Peptides with rare amino acids*

The increasing challenge of molecular complexity is also manifested by the choice of target structures for synthesis. *Robert Burns Woodward* (1917–1979) and *Albert Eschenmoser* (ETH, Switzerland, 1925–) took on an enormous task to achieve the synthesis of vitamin B_{12} (Figure 1.23), a notoriously complex and sensitive molecule of utmost importance in the living system. They achieved the synthesis in what turned out to be perhaps the most spectacular synthetic endeavor of the twentieth century [44].

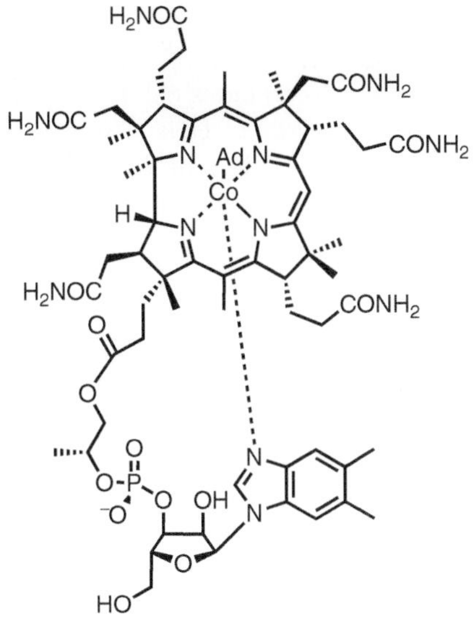

Figure 1.23 *Structure of vitamin B$_{12}$*

Structures that are much more elaborate are also found within the polyketides. Oxygenation, etherification, and several other structural modifications are often encountered. As a consequence of the biosynthetic route employed for the polyketides, oxygenation leads to 1,3- and also 1,2-dioxygenated units. This is displayed by the tremendous variation and structural complexity of the polyether, macrolide, and spiroketal antibiotics. The most remarkable structure synthesised so far is displayed by palytoxin, one of the most poisonous substances known today [45]. The toxin was isolated from a Hawaiian soft coral *Palythea toxica*, which was found in a single tidal pool of only six feet by two feet, and just 20 inches deep! Palytoxin's story began in 1961 and it took nearly three decades before the complete structure, including stereostructure, was finally established through a combination of spectroscopic and synthetic methods [46]. Palytoxin (Figure 1.24) contains no less than 64 chiral centers and seven double bonds capable of *E/Z* isomerism, giving rise to the possibility of $2^{71} = 2.4 \cdot 10^{21}$ isomers! This number is close to the Avogadro's number indeed!

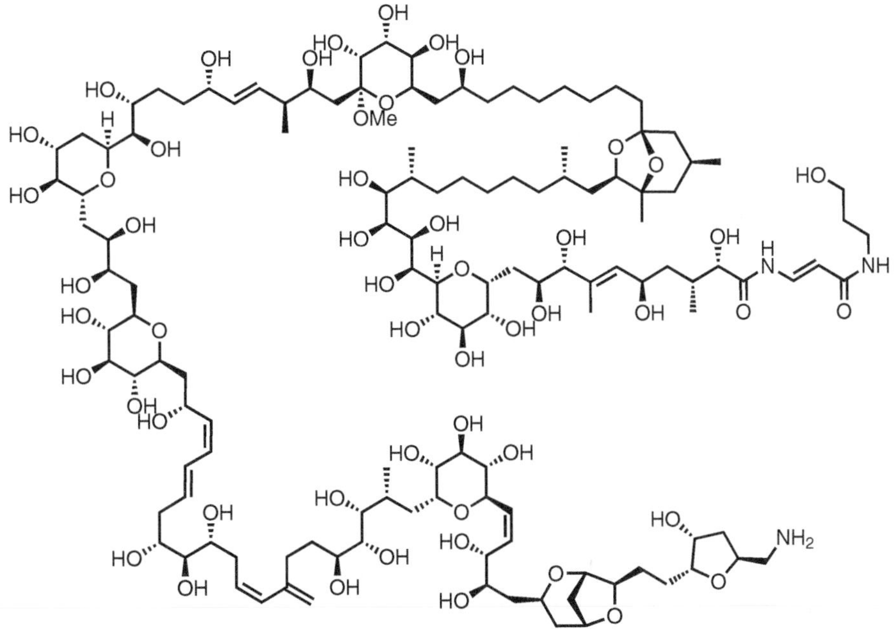

Figure 1.24 *Structure of palytoxin*

1.6   Asymmetric Synthesis of Natural Products

As a scientific concept, optical activity is barely two centuries old. In 1808, *Etienne Louis Malus* (1775–1812) accidentally studied the refraction of light: holding a crystal of Iceland spar (calcite, or calcium carbonate) in his hand, he watched the setting sun light the windows of the palace of Luxembourg. As he turned the crystal, he realised that the intensity of the light changed. Malus called this light polarised; he had invented the principle of polarimeter (Figure 1.25).

Figure 1.25   *A crystal of Icelandic spar splits the light to its components*

Soon thereafter *Jean-Baptiste Biot* (1774–1862) constructed the first working polarimeter (Figure 1.26), an instrument with which one can determine the angle of rotation of plane-polarised light caused by passing it through a sample of optically active sample (neat liquid or dissolved in non-optically active solvent).

Figure 1.26   *Biot's polarimeter*

Tartaric acid plays a key role in the early understanding of optical activity. It was discovered in 1769 by the Swedish chemist *Carl Wilhelm Scheele* (1742–1786) in the thick crusty mass called "tartar" at the bottom of wine barrels. In 1815, Biot managed to prove that tartaric acid rotates the plane-polarised light to the right. In 1820, the Alsatian wine maker Kestner at the Thann factory managed to isolate a crop of an acid with the same constitution as tartaric acid but lacking observable optical activity. Kestner kept a sample of the original, and six years later *Joseph Louis Gay-Lussac* (1778–1850) visited Kestner and decided to reinvestigate Kestner's compound. Gay-Lussac suggested that this compound be called racemic acid (from *racemus* = grape). The new acid eventually acquired several names: it was also known as vinic acid and Traubensäure. The Swedish chemist *Jöns Jacob Berzelius* (1779–1848) preferred to call it paratartaric acid, and this became the name adopted in France and England. In 1831, Berzelius showed that tartaric acid and paratartaric acid differ from each other only in their behavior towards plane-polarised light. The terms racemic and racemate have remained in general use to describe a 1:1 mixture of two enantiomers, and "paratartaric acid" is only a concept in history.

Chemist and crystallographer *Eilhardt Mitscherlich* (1794–1863), a student of Berzelius and later professor at the University of Berlin, managed to crystallise the sodium ammonium salt of tartaric acid in 1844. The question regarding the identities of tartaric acid and racemic acid was conclusively resolved in 1848 when Pasteur, while repeating the Berzelius experiments on paratartaric acid, succeeded in mechanically separating the ammonium salts of L- and D-tartrates (Figure 1.27). With the help of the polarimeter invented by Biot, he succeeded in proving the optical behavior of the two compounds. Biot wanted to repeat Pasteur's experiment himself, and having obtained the same results, he is reputed to have said: "My dear boy, I have loved Science so much during my life, that this touches my very heart," and donated his polarimeter to Pasteur.

Figure 1.27 *Pasteur's models of tartaric acid crystals*

In 1857, the Société de Pharmacie de Paris offered a prize to anyone who could produce racemic acid. The winner was (naturally) Pasteur, who obtained racemic acid through a process he called *racemization*. The gross structure of tartaric acid was established by the English chemist *William Henry Perkin* (1838–1907) in 1867. However, the ordering of the atoms remained unsolved until *William Thomas Astbury* (1898–1961) managed to obtain an X-ray structure of the compound in 1923 [47]. The absolute configuration of tartaric acid was confirmed only in 1951 by the Dutch *Johannes Bijvoet* (1892–1980) [48].

Pasteur continued his work on optically active compounds for decades. He showed that one component of a racemic mixture can be separated by adding another optically active compound which forms a crystalline complex (usually a salt). In 1848, he performed the first optical resolution, which was based on microbial transformations: in fermenting *Penicillium glaucum* in a solution of a racemate, the dextro form slowly disappeared, and the levo form became enriched in the mother liquor.

These observations were quite spectacular in their days, remembering that the Dutch *Jacobus Henricus van't Hoff* (1852–1911) and the French *Joseph Achille Le Bel* (1847–1930) proposed the tetrahedral carbon atom only in 1874. Van't Hoff also introduced the concept of *cis–trans* isomerism about the same time. Application of these ideas remained much in the hands of the German chemist *Johannes Adolf Wislicenus* (1835–1902).

Other important concepts in stereochemistry were presented by e.g. *Johann Friedrich Adolf von Baeyer* (1835–1917), whose strain theory explains ring strain on the basis of valence angles. *Victor Meyer* (1848–1897), a student of *Robert Wilhelm Bunsen* (1811–1899) and Baeyer, proposed, in 1892, the theory of steric strain, according to which a large grouping in an organic compound can prevent a reaction at the neighboring carbon atom. He made this observation while investigating the esterification of the recently discovered *o*-iodobenzoic acid with hydrochloric acid and alcohol.

At the end of the nineteenth century, the English chemist *William Jackson Pope* (1870–1939) demonstrated that optically active compounds can be prepared from nitrogen, sulfur, and selenium compounds. The Swiss chemist *Alfred Werner* (1866–1919) discovered that complexes of platinum, cobalt, and similar transition metals can exist as optical isomers.

The history of asymmetric synthesis can be considered to have begun with *Emil Fischer* (1852–1919). He was a student of *Friedrich August Kekulé von Stradonitz* (1829–1896), but soon transferred to study with Baeyer in Strasbourg, where he became interested in carbohydrates. Soon after moving to the university in Erlangen in 1882, Fischer started his studies on synthesis of sugars, which culminated in the total synthesis of glucose in 1890 [49]. The first recorded example of asymmetric synthesis comes from Fischer during his studies of cyanohydrin formation. When beet pulp (L-arabinose) was treated with HCN under acidic conditions and the cyanohydrins hydrolysed, two acids were formed: L-mannonic acid and L-gluconic acid. Fischer himself commented this:[50]

> *The simultaneous formation of the two stereoisomeric products on the addition of hydrogen cyanide to aldehydes, which was observed here for the first time, is quite remarkable in theory as well as in practice.*

In 1894, Fischer clearly presented the concept of asymmetric synthesis based on his explorations on homologation of sugars through the cyanohydrin reaction (Kiliani–Fischer synthesis) (Scheme 1.6) [51]. *Heinrich Kiliani* (1855–1945), a student of Emil Erlenmeyer, was promoted to professorship of organic chemistry at the Technical university of Münich in 1892.

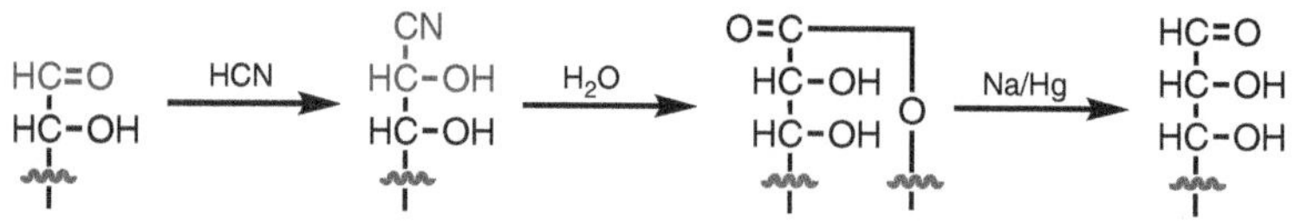

Scheme 1.6   *Kiliani–Fischer synthesis*

The next important step in the development of asymmetric synthesis is the discovery made by *Paul Walden* (1863–1957). In 1896, he reported that by treating levorotatory malic acid with phosphorus pentachloride, one obtains dextrorotatory chlorosuccinic acid. Its hydrolysis in acid solution gives dextrorotatory malic acid, whereas under basic

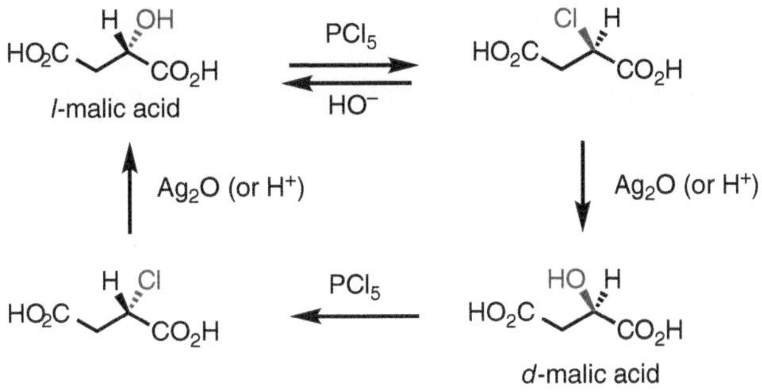

Scheme 1.7 *Walden cycle*

conditions, the levorotatory form is obtained [52]. Fischer was the first one to explain the reaction, and he called it the Walden inversion [53]. Scheme 1.7 displays the Walden cycle with the relationships between individual isomers [54].

Soon thereafter, in 1898 *Francis Robert Japp* (1848–1928) suggested that

> *Only the living organism with its asymmetric tissues, or the asymmetric products of the living organism, or the living intelligence with its conception of asymmetry, can produce this result. Only asymmetry can beget asymmetry [55].*

The concept of asymmetric induction was presented by the German physical chemist *Gustav Kortüm* (1904–1990) in 1932:

> *The action of a force in all living systems exerted by asymmetric molecules on molecules capable of changing from a symmetrical into an asymmetrical configuration.*

The first asymmetric synthesis was reported by *Wilhelm Marckwald* (1864–1942, University of Berlin) in 1904, who showed that when the brucine hydrogen salt of methylethylmalonic acid was heated at 170 °C, it lost carbon dioxide and upon neutralisation "a mixture of dl- and l-methylethylacetic acids, containing 10 per cent. of the latter," was obtained (Scheme 1.8). In his words,

> *By asymmetric synthesis is meant the synthesis of a substance containing an asymmetric carbon under such conditions that there shall be formed an excess either of the dextrorotatory or of the laevorotatory compound. The author has now made the first synthesis of this class by heating the acid brucine salt of methyl ethyl malonic acid [56].*

In a subsequent paper, he explained that

> *Asymmetric syntheses are those reactions which produce optically active substances from symmetrically constituted compounds with the intermediate use of optically active materials but with the exclusion of all analytical processes [57].*

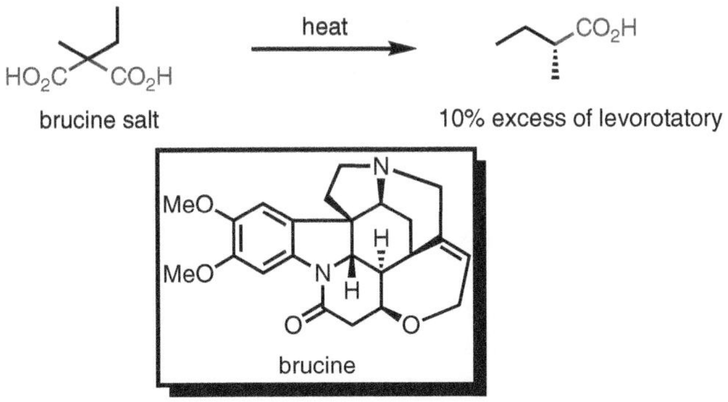

Scheme 1.8 *Marckwald's first asymmetric synthesis*

Alexander McKenzie (1869–1951), a student of Marckwald who worked at the University of Birmingham, then Birkbeck College and finally University of Dundee, showed in 1905 that Grignard addition reactions to optically active esters of benzoylformate gives an optically active product after ester hydrolysis (Scheme 1.9) [58].

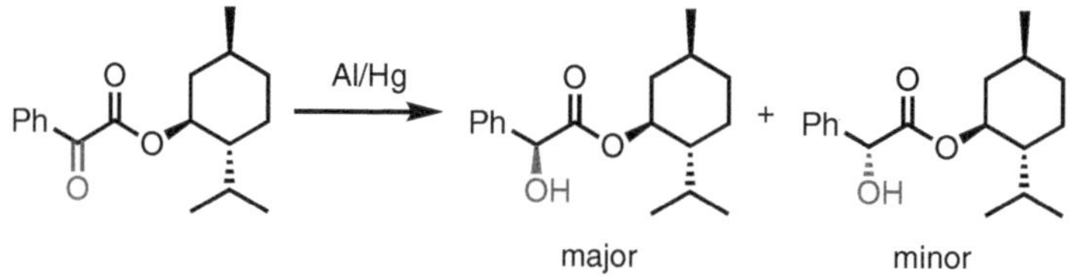

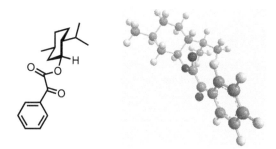

Scheme 1.9   *McKenzie's distereoselective reduction*

The development of asymmetric synthesis was slow and only in the 1950s one started to see real progress. *Gustave Vavon* (1884–1953) published the first enantioselective Grignard reduction in 1946 (Scheme 1.10) [59].

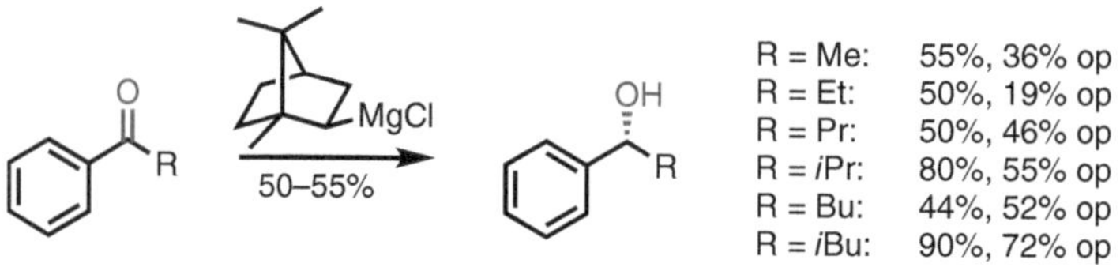

R = Me:	55%, 36% op
R = Et:	50%, 19% op
R = Pr:	50%, 46% op
R = iPr:	80%, 55% op
R = Bu:	44%, 52% op
R = iBu:	90%, 72% op

Scheme 1.10   *Vavon's first enantioselective Grignard reduction*

The result of the works by R.H. Baker and L.E. Linn [60], W. von E. Doering and R.W. Young [61], as well as L.M. Jackman, J.A. Mills, and J.S. Shannon [62] established that the Meerwein–Ponndorf–Verley reduction produces an optically active product when an aluminum alcoholate of an optically active alcohol is used.

The Croatian-born *Vladimir Prelog* (1906–1998) started about the same time to reinvestigate the studies by McKenzie on the Grignard addition reactions to optically active esters of benzoylformate. Prelog was able to correlate the stereochemistry of the product atrolactic acid with the configuration of the starting material's optically active alcohol part in what became to be known as the Prelog rule (Figure 1.28) [63].

Figure 1.28   *Prelog's rule*

Before concluding this short treatise on the history of asymmetric synthesis, we will make a few comments on definitions in asymmetric synthesis. In its broadest sense, asymmetric synthesis is a chemical reaction where an achiral unit as a substrate is transformed to a chiral unit in such a fashion that the stereoisomeric products are formed in uneven quantities. Thus, asymmetric synthesis is a process where a prochiral unit is transformed to a chiral unit producing unequal amounts of stereoisomers. Historically this is a broader definition than that presented by Marckwald in 1904. This definition requires that the chirality-inducing agent must be recoverable quantitatively after the reaction.

We can distinguish three fundamentally different processes in asymmetric induction. In the process of *internal asymmetric induction*, the chirality information is obtained from the initial starting material. Each reaction step is stereohomogeneous, dictated by its mechanism, and one typically utilises the *chiral pool* compounds as the starting materials.

In *external asymmetric induction*, the chiral information is brought into the reaction from outside the reacting molecules, typically in the form of a *chiral catalyst* (including enzymes).

In *relayed asymmetric induction*, the asymmetric information in brought in transiently, for instance in the form of *chiral auxiliaries*. We will cover more detailed examples of these different forms of asymmetric induction in Chapter 2.

1.7   Synthetic Organic Chemistry

Natural product synthesis has evolved from the early inception of organic chemistry. In the late nineteenth century, the carbochemical industry was born, with its main emphasis on aromatic compounds. By current terminology, the syntheses were based on operational transformations of functional groups, relying most heavily on associative or analogue-based planning.

Synthetic organic chemistry has been declared dead many times in its brief history. After reaching a certain stage of knowledge, it has become generally accepted that synthesis research is not warranted any more as a separate scientific discipline. Soon after Wöhler's urea synthesis, it was considered that since organic matter had been synthesised, one need not concentrate on such things anymore. The middle of the nineteenth century started the boom of flourishing synthetic chemical industry, which proved the beliefs wrong.

The early part of the twentieth century witnessed the events which may be considered as the birth of modern synthesis. Emil Fischer's outstanding achievements in carbohydrate, protein, and nucleic acid chemistry, as well as his numerous other synthetic achievements; combined with the efforts of such notable chemists as Gustav Komppa (camphor, 1903), William Perkin (α-terpineol, 1904), and Sir Robert Robinson (tropinone, 1914), sowed the seeds which in the first decade of the 19th century brought around what could be called the first total syntheses of complex natural products (Figure 1.29). Again people were writing obituaries to organic synthesis. Soon, biochemistry found its way out of the safe hems of mother organic chemistry, and again new methods were needed to make molecules of unforeseen complexity (and beauty!).

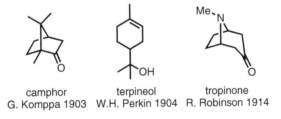

camphor
G. Komppa 1903

terpineol
W.H. Perkin 1904

tropinone
R. Robinson 1914

Figure 1.29   *Early natural product targets*

Steady development of physical organic chemistry in the era before the Second World War led to increasing understanding of detailed reaction mechanisms. During the same period, the structural theory of organic molecules evolved considerably, and this can be seen as the origin of conformational analysis [64]. It is well worth remembering that as late as the 1920s, the chair and boat conformations of cyclohexane were far from being generally accepted. Optical activity of saturated carbon compounds was already assigned to the distribution of atoms in space around a tetrahedral carbon atom in 1874 by van't Hoff and Le Bel [65]. von Baeyer's strain theory [66] taught chemists to consider cyclic carbon frameworks as planar polygons. Five years later *Hermann Sachse* (1862–1893) proposed that cyclohexane could exist in two puckered forms, in which all the valence angles are tetrahedral, and thus the cyclohexanes are free of angle strain [67]. However, conformational equilibration of the unsymmetrical (flexible) and symmetric (rigid) cyclohexane forms evaded Sachse's analysis. When the ring interconversion became accepted, Sachse's theory was put on hold, up to the level that a most popular textbook by the Finnish *Ossian Aschan* (1860–1939) promoted the view that the interconversion is so fast that one should consider only the average planar from [68]. It was only after *Ernst Mohr* (1873–1926) predicted and *Walter Hückel* (1985–1973) verified that decalin should exist in two geometric isomers, both of which are strain free [69], that the Sachse–Mohr theory of puckered cyclohexane ring forms finally became accepted. However, even this was not convincing enough for the chemists at large. The fact that there are indeed two types of bonds in cyclohexane was first experimentally demonstrated (by Raman spectra) by the Austrian *Karl Wilhelm Friedrich Kohlrausch* (1864–1953) [70], and confirmed by *Kenneth Pitzer* (1914–1997) and *Odd Hassel* (1897–1981) [71]. Hassel called the substituents parallel to the six-fold axis of symmetry ε (from Greek εστηκοζ, *estikos* = upright) and κ (from Greek κειμενοζ, *keimenos* = lying), and Pitzer distinguished them as polar and equatorial, respectively. A consensus was eventually reached, and by mutual agreement, the suggestion by *Christopher Kelk Ingold* (1893–1970) to call the 'polar' substituents 'axial' was adopted in 1953 [72]. The final chapter in the birth of conformational analysis was written by *Sir Derek H. R. Barton* (1918–1998) in the classic 1950 paper [73]. It thus took nearly fifty years before the concept of conformational analysis was established. It is fascinating to note at this stage that today planar cyclohexane compounds have actually been created (Figure 1.30) [74].

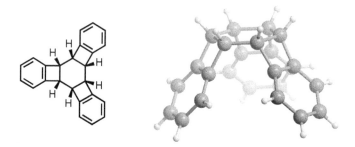

Figure 1.30   A flat cyclohexane

During the 1940s and 1950s, the development of chromatographic and spectroscopic methods, augmented with the development of new selective reagents all added to the armamentarium of the synthetic chemist, who was now able to achieve syntheses undreamed of just a few decades earlier. These included the syntheses of vitamin A (O. Isler, 1949), penicillin V (J. Sheehan, 1957), and several alkaloids, including strychnine (R.B. Woodward, 1954) and morphine (M. Gates, 1956) (Figure 1.31). This was also the era of one of the most brilliant minds in the history of synthetic organic chemistry; R.B. Woodward, whose synthetic masterpieces include such formidable target structures as quinine (1944),

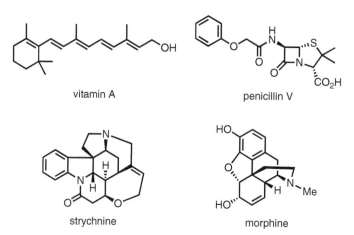

Figure 1.31   Target molecules in the 1940s and 1950s

patulin (1950), cholesterol and cortisone (1951), lanosterol (1954), reserpine (1956), chlorophyll (1960), colchicine (1963), cephalosporin C (1965), vitamin B_{12} (together with A. Eschenmoser, 1972), prostaglandin F_{2a} (1973), and erythromycin (finished post-humously, 1981) [75].

All the challenging synthetic targets seemed to be conquered, and new topics of interest were being sought. Stubbornly, synthetic chemists continued on their paths, and the last two or three decades of the twentieth century witnessed another major revolution in the art of synthesis. The practical value of chirality was realised, and this led to spectacular achievements in the development and application of reactions capable of not only distinguishing but also producing individual enantiomers and diastereomers. Mainly through and as a consequence of the efforts of Professors R B. Woodward, *Gilbert Stork* (Columbia University, 1921–2017), and *Elias J. Corey* (Harvard University, 1928–), the design of syntheses gained much from the application of logical reasoning [76]. Modern day syntheses are highly sophisticated, and the use of techniques unknown in the 1940s and 1950s are quite common today [77].

The evolution of the synthetic techniques and our understanding of the factors governing the structural and reactivity aspects of organic compounds have led to a change in the justifications of synthesis. Whereas only a few decades ago, the synthesis of a defined target compound could be a good enough justification for the execution of a multi-step synthesis, currently more and more emphasis is put on the actual design phase of a target compound. Concepts like diversity-oriented synthesis [78], and function-oriented synthesis (by *Paul Wender*, Stanford University, 1947–) [79] have been introduced to emphasise the need to incorporate the aspects of applications of the target molecule already into the design. Drug molecules, compounds valuable to biochemical or physicochemical studies, monomers for polymerisation, and other small molecules advancing the broad and rapidly developing field of materials sciences are already being designed using highly sophisticated, and readily available computer facilities. One of the present central foci in the research in synthesis is the design and execution of syntheses of defined and designed molecules, which possess defined functional properties. This also implies that the strict divisions between the different branches of chemistry, as well as the distinction of chemistry from biochemistry and molecular biology will have to be surpassed [80].

It is true that today one can, in principle, make any molecule, but the key question is: how efficient is the synthesis? This question is the culmination point of research in chemical synthesis: the state of the art of organic synthesis is defined by the complexity of target molecules that can be efficiently made. A total synthesis requiring tens of years is hardly justifiable any longer on any grounds. Therefore, synthesis research is critically dependent on the invention and improvement of new chemical transformations. The importance of the general field of synthetic organic chemistry, asymmetric synthesis, and natural product chemistry is emphasised with the high number of Nobel Prizes awarded. Table 1.2 collects the 39 Prizes given to fields related to the subject matter of this book, as we will show later in this chapter.

Figure 1.32 illustrates the state of affairs by displaying the number of publications dealing with total syntheses of natural products, and separately, the subset of papers dealing with asymmetric synthesis. It is evident that the art of total synthesis has gained momentum only during the past two decades. Even more striking is the observation that even today only a small fraction of the total syntheses of natural products are performed to produce a single enantiomer. Much more research is obviously needed in asymmetric synthesis, since one can quite justifiably question whether a racemate, let alone the enantiomer, of a natural product is by default a natural product!

Efficiency must not only be associated with financial economy. We must understand that our world, with its increasing population and diminishing food and energy production potential per capita, requires more economies. It is clear that a synthesis with fewer reaction steps is more efficient than a longer one. Convergent synthesis strategies are more powerful than simply linear synthesis, also the result of *step economy*. *Barry M. Trost* (1941–) introduced the concept of *atom economy* in 1990 (as one of my former students said in the lectures: "there are not enough atoms") [81]. Petrochemical industry utilises most of its starting materials for the production of goods for sale. Fine chemical industry and especially pharmaceutical industry are particularly notorious in producing much more waste than final product. Energy consumption is typically also not optimised in chemical industry. Although fuel and energy consumption in the European chemical industry has decreased some 24% since 1990, chemical industry is the largest single consumer of industrial energy, capturing *ca.* 20% of industrial energy consumption in Europe [82], and 37% in the US [83]. It must be clear to us that we have barely scratched the surface along our route to a sustainable future. Very often, syntheses include unnecessary reduction and oxidation steps. The ultimate source of reducing power is sunlight, and eventually all

Table 1.2 *Nobel Prizes awarded to discoveries related to asymmetric synthesis of natural products*

Year	Nobelist(s)	Awarded …
1901	Jacobus Henricus van't Hoff	in recognition of the extraordinary services he has rendered by the discovery of the laws of chemical dynamics and osmotic pressure in solutions
1902	Hermann Emil Fischer	in recognition of the extraordinary services he has rendered by his work on sugar and purine syntheses
1905	Johann Wilhelm Friedrich Adolf von Baeyer	in recognition of his services in the advancement of organic chemistry and the chemical industry, through his work on organic dyes and hydroaromatic compounds
1909	Wilhelm Ostwald	in recognition of his work on catalysis and for his investigations into the fundamental principles governing chemical equilibria and rates of reaction
1910	Otto Wallach	in recognition of his services to organic chemistry and the chemical industry by his pioneer work in the field of alicyclic compounds
1912	Victor Grignard	(VG) for the discovery of the so-called Grignard reagent, which in recent years has greatly advanced the progress of organic chemistry
	Paul Sabatier	(PS) for his method of hydrogenating organic compounds in the presence of finely disintegrated metals whereby the progress of organic chemistry has been greatly advanced in recent years
1915	Richard Martin Willstätter	for his researches on plant pigments, especially chlorophyll
1927	Heinrich Otto Wieland	for his investigations of the constitution of the bile acids and related substances
1928	Adolf Otto Reinhold Windaus	for the services rendered through his research into the constitution of the sterols and their connection with the vitamins
1930	Hans Fischer	for his researches into the constitution of haemin and chlorophyll and especially for his synthesis of haemin
1937	Walter Norman Haworth	(WH) for his investigations on carbohydrates and vitamin C
	Paul Karrer	(PK) for his investigations on carotenoids, flavins, and vitamins A and B$_2$
1938	Richard Kuhn	for his work on carotenoids and vitamins
1939	Adolf Friedrich Johann Butenandt Leopold Ruzicka	(AB) for his work on sex hormones (LR) for his work on polymethylenes and higher terpenes
1947	Sir Robert Robinson	for his investigations on plant products of biological importance, especially the alkaloids
1950	Otto Diels Kurt Alder	for their discovery and development of the diene synthesis
1952	A.J.P. Martin R.L.M. Synge	for their invention of partition chromatography
1953	Hermann Staudinger	for his discoveries in the field of macromolecular chemistry
1955	Vincent du Vigneaud	for his work on biochemically important sulphur compounds, especially for the first synthesis of a polypeptide hormone
1957	Lord (Alexander R.) Todd	for his work on nucleotides and nucleotide co-enzymes
1958	Frederick Sanger	for his work on the structure of proteins, especially that of insulin
1963	Karl Ziegler Giulio Natta	for their discoveries in the field of the chemistry and technology of high polymers
1965	Robert Burns Woodward	for his outstanding achievements in the art of organic synthesis
1968	Har Gobind Khorana (medicine)	for (their) interpretation of the genetic code and its function in protein synthesis
1969	Derek H.R. Barton Odd Hassel	for their contributions to the development of the concept of conformation and its application in chemistry
1973	Ernst Otto Fischer Geoffrey Wilkinson	for their pioneering work, performed independently, on the chemistry of the organometallic, so called sandwich compounds

(Continued)

Table 1.2   *(continued)*

Year	Nobelist(s)	Awarded …
1975	John Warcup Cornforth	(JC) for his work on the stereochemistry of enzyme-catalysed reactions
	Vladimir Prelog	(VP) for his research into the stereochemistry of organic molecules and reactions
1979	Herbert C Brown	for their development of the use of boron- and phosphorus-containing
	Georg Wittig	compounds, respectively, into important reagents in organic synthesis
1980	Walter Gilbert	for their contributions concerning the determination of base sequences in
	Frederick Sanger	nucleic acids
1981	Roald Hoffmann	for their theories, developed independently, concerning the course of chemical
	Kenichi Fukui	reactions
1984	Robert Bruce Merrifield	for his development of methodology for chemical synthesis on a solid matrix
1987	Donald J. Cram	for their development and use of molecules with structure-specific interactions of
	Jean-Marie Lehn	high selectivity
	Charles J. Pedersen	
1990	Elias J. Corey	for his development of the theory and methodology of organic synthesis
1994	George A. Olah	for his contribution to carbocation chemistry
2001	K. Barry Sharpless	for his work on chirally catalysed oxidation reactions
	Ryoji Noyori	for their work on chirally catalysed hydrogenation reactions
	William S. Knowles	
2005	Yves Chauvin	for the development of the metathesis method in organic synthesis
	Robert H. Grubbs	
	Richard R. Schrock	
2010	Richard F. Heck	for palladium-catalysed cross couplings in organic synthesis
	Ei-ichi Negishi	
	Akira Suzuki	
2016	Jean-Pierre Sauvage	for the design and synthesis of molecular machines
	Sir J. Fraser Stoddard	
	Bernard L. Feringa	
2018	Frances H. Arnold	for the directed evolution of enzymes
2021	Benjamin List	for the development of asymmetric organocatalysis
	David W.C. MacMillan	

syntheses terminate in some form of oxidised materials. Redox economy aims at reducing the number of unnecessary redox reactions in the synthesis plan [84]. Chirality is a valuable resource we are only beginning to appreciate, and the development of efficient new asymmetric transformations is an important topic in the years and decades to come.

The role of organic synthesis has evolved over the two centuries of its existence. When the target molecules of the first syntheses (ethanol from ethylene by Faraday in 1824 and urea from ammonium cyanate by Wöhler in 1828) contained no stereocenters, the most challenging modern target molecules contain tens of stereocenters. The evolution of organic synthesis can be considered in stages (Figure 1.33). The first era was that of *target-oriented synthesis* where the aim of synthetic endeavors was to produce initially natural products, and then, as the pharmaceutical industry was born in the middle of the nineteenth century, designed molecules gained more and more interest. After the Second World War, most of the reagent categories and catalysts were discovered, as was the understanding of the mechanistic principles and structural elucidation by spectroscopic means. The era of *methods-oriented synthesis* was in its heyday. Of course, during this era, the complexity of the target molecules could be increased tremendously, and by the 1990s, such target molecules as strychnine, prostaglandins, and monensin surrendered to synthesis. Target molecule complexity accelerated rapidly, and by the turn of the millennium, ingenious syntheses of such targets as ginkgolide B, calicheamicin, rapamycin, paclitaxel, brevetoxin, and even palytoxin were achieved. When Woodward had stated, in 1956, that the structure of erythromycin "*looks at present quite hopelessly complex, particularly in view of its plethora of asymmetric centers,*" [85] the chemists could now synthesise practically any molecule they could draw! Clearly,

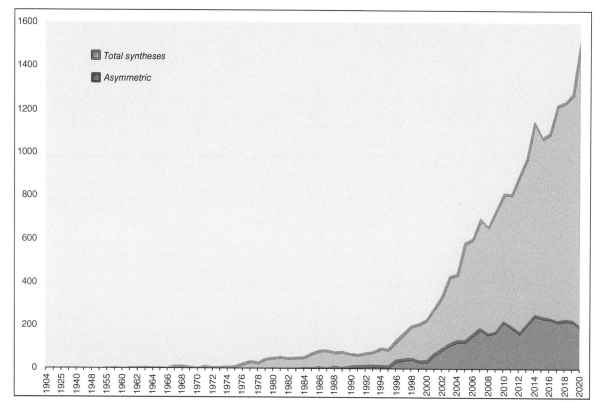

Figure 1.32   *Publications on natural product total syntheses in general and in asymmetric form*

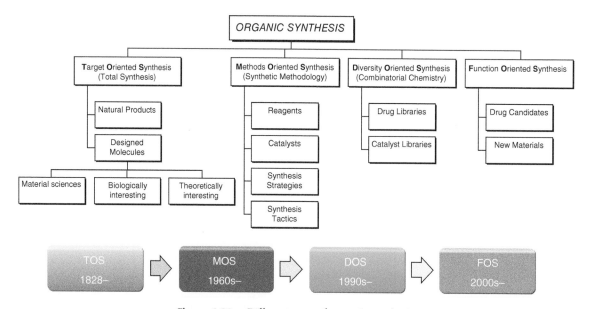

Figure 1.33   *Different eras of organic synthesis*

there was a need for a paradigm shift. Towards the end of the twentieth century, rapid developments in the efficiency of synthesis allowed chemists to synthesise large collection libraries of molecules instead of single compounds. This era of combinatorial chemistry epitomises diversity-oriented synthesis [86]. At the beginning of this century, the synthetic methodology has gained much from the daughter science of organic chemistry, i.e. biochemistry, in the sense that genes, enzymes, genetically modified enzymes, and biomimetic catalysts have been added to the synthesis toolbox. The so-called organocatalysts have become the mainstream in organic synthesis laboratories. All these developments have allowed chemists to shift the focus of their targets from individual molecules to function, where one aims to synthesise a collection of molecules in a system that will fulfill a desired or designed function (*function-oriented synthesis*) [79]. The recent development of highly successful targeted cancer treatments are among the spearheads of this era.

Finally, Table 1.2 shows a collection of chemistry Nobel Prizes related to synthesis and/or natural products. One can immediately see that the subject has evolved, and this fact is reflected in the shift of emphasis from target-oriented to function-oriented synthesis. Directed evolution of enzymes perhaps gives us a glimpse to the future [87]. The Noble Prize in 2021 amply emphasises the importance of natural product chemistry not only in synthesis and biology but also in the development of novel asymmetric catalytic synthesis methods.

This book is about asymmetric synthesis of natural products. The text is intended to give the reader a good comprehension of the general principles of asymmetric synthesis. Since natural products have functioned as demanding playgrounds for the inventive minds, it was quite natural to use this context to also introduce the developments of asymmetric synthesis in its proper historical context. The first three chapters give the foundations of asymmetric synthesis, and Chapters 4–10 then illustrate and expand on these principles. The choice of examples is very personal, it is not a beauty contest; I have attempted to find examples relevant to illustrate a certain point. A comprehensive collection of existing methods is beyond the scope of this book.

References

1. Rinehart, K.L. *Pure Appl. Chem.* **1977**, *49*, 1361–1384.
2. a) Weissman, K.J. *Trends in Biotech.* **2007**, *25*, 139–142; b) Hermane, J., Eichner, S., Mancuso, L., Schröder, B., Sasse, F., Zeilinger; C., et al. *Org. Biomol. Chem.* **2019**, *17*, 5269–5278.
3. Law, S.R. *Physiol. Plant* **2019**, *167*, 469–470.
4. Baldovini, N., Chaintreau, A. *Nat. Prod. Rep.* **2020**, *37*, 1589–1626.
5. Kallio, H.P. *J. Agric. Food Chem.* **2018**, *66*, 2553–2560.
6. Borlinghaus, J., Albrecht, F., Gruhlke, M.C.H., Nwachukwu, I.D., Slusarenko, A.J. *Molecules* **2014**, *19*, 12591–12618.
7. Ali, M., Thomson, M., Afzal, M. *Prostaglandins, Leukotrienes and Essential Fatty Acids* **2000**, *62*, 55–73.
8. Winter, M., Furrer, A., Willhalm, B., Thommen, W. *Helv. Chim. Acta* **1976**, *59*, 1613–1620.
9. Scafato, P., Colangelo, A., Rosini, C. *Chirality* **2009**, *21*, 176–182.
10. Cannon, R.J., Ho, C.-T. *J. Food Drug Anal.* **2018**, *26*, 445–468.
11. Jansen, E.F. *J. Biol. Chem.* **1948**, *176*, 657–664.
12. Wratten, S.J., Faulkner, D.J. *J. Org. Chem.* **1976**, *41*, 14, 2465–2467.
13. Buttery, R.G., Seifert, R.M., Guadagni, D.G., Ling, L.C. *J. Agric. Food Chem.* **1971**, *19*, 969–971.
14. Seifert, R.M., Buttery, R.G., Guadagni, D.G., Black, D.R., Harris, J.G. *J. Agric. Food Chem.* **1970**, *18*, 246–249.
15. Allen, M.S., Lacey, M.J., Boyd, S.J. *ACS Symposium Series* **1996**, *637*, 220–227.
16. Vermeulen, J., Bailly, S., Collin, S. *Developments in Food Sci.* **2006**, *43*, 245–248.
17. Wood, W.F. *J. Chem. Ecol.* **1990**, *16*, 2057–2065.
18. Patterson, R.L.S. *Chem. Ind. (London)* **1968**, *17*, 548–549.
19. Chenot, C., Briffoz, L., Lomartire, A., Collin, S. *J. Agric. Food Chem.* **2020**, *68*, 10310–10317.
20. Goeke, A. *Sulfur Reports* **2002**, *23*, 243–278.
21. Addicott, F.T., Lyon, J.L., Ohkuma, K., Thiessen, W.E., Carns, H.R., Smith, O.E., et al. *Science* **1968**, *159*, 1493.
22. Demole, E., Lederer, E., Mercier, D. *Helv. Chim. Acta*, **1962**, *45*, 675–685.
23. Mitchell J.W., Mandava, N., Worley, J.F.;,Plimmer, J.R., Smith, M.V. *Nature.* **1970**, *225*, 1065–1066.
24. Butenandt, A., Beckmann, R., Hecker, E. *Hoppe-Seylers Zeitschrift für Physiologische Chemie* **1961**, *324*, 71–83.
25. Dumbacher, J.P., Beehler, B.M., Spande, T.F., Garraffo, H.M., Daly, J.W. *Science* **1992**, *258*, 799–801.

26. Hanifin, C.T. *Mar. Drugs* **2010**, *8*, 577–593.

27. Wiese, M., D'Agostino, P.M., Mihali, T.K., Moffitt, M.C., Neilan, B.A. *Mar Drugs.* **2010**, *8*, 2185–2211.

28. a) Atanasov, A.G., Zotchev, S.B., Dirsch, V.M. The International Natural Product Sciences Taskforce, Supuran, C.T. *Nature Rev. Drug Disc.* **2021**, *20*, 200–216; b) Newman, D.J., Cragg, G.M. *J. Nat. Prod.* **2020**, *83*, 770–803; c) Nicolaou, K.C., Rigol, S. *Nat. Prod. Rep.* **2020**, *37*, 1404–1435.

29. Lee, M.R. *J. R. Coll. Physicians Edinb.* **2011**, *41*, 78–84.

30. Mishra, J.N., Verma, N.K. *Int. J. Pharm. Chem. Res.* **2017**, *3*, 516–522.

31. a) Partington, J.R. *A History of Chemistry*, vol 2. London, MacMillan, **1961**, p. 177; b) Debus, A. *Estudos Avançados* **1990**, *4*, 173–196.

32. Gosio, B.G.R. *Accad. Med. Torino* **1893**, *61*, 484–487.

33. Bentley, R. *Chem. Rev.* **2000**, *100*, 3801–3825.

34. Aminov, R.I. *Front. Microbiol.* **2010**, *1*, 134: 1–7.

35. Waksman, S.A., Woodruff, H. B. *Proc. Soc. Exp. Biol. Med.* **1940**, *45*, 609–614.

36. Nicolaou, K.C., Snyder, S.A. *Angew. Chem. Int. Ed.* **2005**, *44*, 1012–1044.

37. Schleyer, P. von R. *J. Am. Chem. Soc.* **1957**, *79*, 3292–3292.

38. a) Seebach, D., Enders, D. *Angew. Chem.* **1975**, *87*, 1–18; b) Lever, O.W. *Tetrahedron* **1976**, *32*, 1943–1971; c) Hase, T.A. (ed.) *Umpoled Synthons: A Survey of Sources and Uses in Synthesis* John Wiley & Sons: New York, **1987**.

39. Breslow, R. *J. Am. Chem. Soc.* **1958**, *80*, 3719–3726.

40. a) Darby, N., Kim, C.U., Salaün, J.A., Shelton, K.W., Takada, S., Masamune, S. *J. Chem. Soc. D* **1971**, 1516–1517; b) Jones, R.R.; Bergman, R.G. *J. Am. Chem. Soc.* **1972**, *94*, 660–661; c) Bergman, R.G. Accts. *Chem. Res.* **1973**, *6*, 25–31.

41. Nicolaou, K.C., Dai, W.-M. *Angew. Chem., Int. Ed. Engl.* **1991**, *30*, 1387–1416.

42. Matsunaga, S., Fusetani, N., Kato, Y. *J. Am. Chem. Soc.* **1991**, *113*, 9690–9692.

43. a) Benz, F., Knusel, F., Nuesch, J., Treichler, H., Voser, W., Nyfeler, R., et al. *Helv. Chim. Acta* **1974**, *57*, 2459–2477; b) Keller-Juslen, C., Kuhn, M., Loosli, H.R., Petcher, T.J., Weber, H.P., von Wartung, A. *Tetrahedron Lett.* **1976**, 4147–4150.

44. a) Woodward, R.B. *Pure Appl. Chem.* **1968**, *17*, 519–547; b) Woodward, R.B. *Pure Appl. Chem.* **1973**, *33*, 145–177.

45. Moore, R.E. *Progr. Chem. Org. Nat. Prod.*; Springer-Verlag: New York, **1985**; Vol. *48*, 81.

46. a) Moore, R.E., Bartolini, G., Barchi, J., Bothner-By, A.-A., Dadok, J., Ford, J. *J. Am. Chem. Soc.* **1982**, *104*, 13, 3776–3779; b) Cha, J.K., Christ, W.J., Finan, J.M., Fujioka, H., Kishi, Y., Klein, L.L., et al. *J. Am. Chem. Soc.* **1982**, *104*, 25, 7369–7371; c) Armstrong, R.W., Beau, J.-M., Cheon, S H., Christ, W.J., Fujioka, H., Ham, W.-H., et al. *J. Am. Chem. Soc.* **1989**, *111*, 7525–7530; d) Suh, E.M., Kishi, Y. *J. Am. Chem. Soc.* **1994**, *114*, 11205–11206; Kishi, Y. *Tetrahedron* **2002**, *58*, 6239–6258; e) Kishi, Y. *Pure Appl. Chem.* **1989**, *61*, 313–324.

47. Astbury, W.T. *Proc. R. Soc. Lond. A* **1923**, *102*, 506–528.

48. Bijvoet, J.M., Peerdeman, A.F., van Bommel, A.J. *Nature* **1951**, *168*, 271–272.

49. Fischer, E. *Ber. Dtsch. Chem. Ges.* **1890**, *23*, 799–805.

50. Fischer, E. *Ber. Dtsch. Chem. Ges.* **1890**, *23*, 2114–2141.

51. a) Kiliani, H. *Ber. Dtsch. Chem. Ges.* **1885**, *18*, 3066–3072; b) Fischer, E. *Ber. Dtsch. Chem. Ges.* **1889**, *22*, 2204–2205; c) Fischer, E. *Ber. Dtsch. Chem. Ges.* **1894**, *27*, 3189–3232.

52. Walden, P. *Ber.dtsch. Chem. Ges.* **1896**, *29*, 133–138.

53. Fischer, E. *Ber.* **1906**, *39*, 2894.

54. Bancroft, W.D., Davis, H.L. *J. Phys. Chem.* **1931**, *35*, 6, 1624–1647.

55. Japp. F.R. *Nature* **1898**, *58*, 452–460.

56. Marckwald, W. *Ber. Dtsch. Chem. Ges.* **1904**, *37*, 349–354.

57. Marckwald, W. *Ber. Dtsch. Chem. Ges.* **1904**, *37*, 1368–1370.

58. a) Mc Kenzie, A. *J. Chem. Soc.* **1904**, *85*, 1249–1262; b) McKenzie, A. *J. Chem. Soc., Trans.* **1905**, *87*, 1373–1383.

59. a) Vavon, G., Riviere, C., Angelo, B. *Compt. Rend.* **1946**, *222*, 959–961; b) Vavon, G., Angelo, B. *Compt. Rend.* **1947**, *224*, 1435–1437.

60. Baker, R.H., Linn, L.E. *J. Am. Chem. Soc.* **1949**, *71*, 1399–1401.

61. Doering, W. von E., Young, R.E. *J. Am. Chem. Soc.* **1950**, *72*, 631.

62. Jackman, L.M., Mills, J.A., Shannon, J.S. *J. Am. Chem. Soc.* **1950**, *72*, 4814–4815.

63. a) Prelog, V. Helv. *Chim. Acta* **1953**, *36*, 308–319; b) Prelog, V., Meier, H.L. *Helv. Chim. Acta* **1953**, *36*, 320–325.

64. Eliel, E.L., Allinger, N.L., Angyal, S.J., Morrison, G.A. *Conformational Analysis*, American Chemical Society: New York, **1965**.

65. a) van't Hoff, J.H. *Arch. néerlandaises des sciences exactes et naturelles* **1874**, *9*, 445–454; b) van't Hoff, J.H. *Bull. Soc. Chim. France* **1875**, *23*, 295–301; Le Bel, J.A. *Bull. Soc. Chim. Fr*. 1874, **22**, 337–347.

66. a) Baeyer, A. *Ber. dtsch. Chem. Ges.* **1885**, *18*, 2269–2281; b) Perrin, C.L., Fabian, M.A., Rivero, I.A. *Tetrahedron* **1999**, *55*, 5773–5780.

67. Sachse, H. *Ber. dtsch. Chem. Ges.* **1890**, *23*, 1363–1370.

68. Aschan, O. *Chemie der Alicyclischen Verindungen*, Vieweg Verlag, Braunschweig, **1905**.

69. a) Mohr, E. *Ber. dtsch. Chem. Ges.***1922**, *55*, 230–231; b) Hückel, W. *Ann*. **1925**, *441*, 1–48.

70. Kohlrausch, K.W.F., Reitz, A.W., Stockmair, W.Z. *Physik. Chem.* **1936**, *B32*, 229–236.

71. a) Beckett, C.W., Pitzer, K.S., Pitzer, R. *J. Am. Chem. Soc.* **1947**, *69*, 2488–2495; b) Hassel, O. *Tidsskr. Kjemi Bergvesen Met.* **1943**, *3*, 32–34; c) Hassel, O., Ottar, B. *Acta Chem. Scand.* **1947**, *1*, 929–943.

72. a) Barton, D.H.R., Hassel, O., Pitzer, K.S., Prelog, V. *Nature* **1953**, *172*, 1096–1097; b) Barton, D.H.R., Hassel, O., Pitzer, K.S., Prelog, V. *Science* **1954**, *119*, 49.

73. Barton, D.H.R. *Experientia* **1950**, *6*, 316–320.

74. Mohler, D.L., Vollhardt, K.P.C., Wolff, S. *Angew. Chem., Int. Ed. Engl.* **1990**, *29*, 1151–1154.

75. *Robert Burns Woodward – Architect and Artist in the World of Molecules* (Benfey, O.T., Morris, P.J.T., eds.), Chemical Heritage Foundation: Philadelphia, **2001**.

76. Corey, E.J., Cheng, X.-M. *The Logic of Chemical Synthesis*, John Wiley & Sons: New York, **1989**.

77. Nicolaou, K.C., Montagnon, T. *Molecules that Changed the World*, Wiley-VCh: Weinheim, **2008**.

78. Nielsen, T.E., Schreiber, S.L. *Angew. Chem. Int. Ed.* **2008**, *47*, 48–56.

79. a) Wender, P.A., Verma, V.A., Paxton, T.J., Pillow, T.H. *Acc. Chem. Res.* **2008**, *41*, 40–49; b) Wender, P.A., Quiroz, R.V., Stevens, M.C. *Acc. Chem. Res.* **2015**, *48*, 752–760.

80. a) Hanessian, S., Franco, J., Larouche, B. *Pure Appl. Chem.* **1990**, *62*, 1887–1910; b) Corey, E.J. *Angew. Chem., Int. Ed. Engl.* **1991**, *30*, 455–465; c) Seebach, D. *Angew. Chem., Int. Ed. Engl.* **1990**, *29*, 1320–1367.

81. Trost, B.M. *Science* **1991**, *254*, 1471–1477.

82. CEFIC, the European Chemical Industry Council. Data available at https://cefic.org/a-pillar-of-the-european-economy/facts-and-figures-of-the-european-chemical-industry/energy-consumption/ (accessed December 7, **2021**).

83. Energy Information Administration (**2001**) Manufacturing Energy Consumption Survey. Available at https://www.eia.gov/consumption/manufacturing/data/2018/#r1 (accessed December 7, 2021).

84. Newhouse, T., Baran, P.S., Hoffmann, R.W. *Chem. Soc. Rev.* **2009**, *38*, 3010–3021

85. Woodward, R.B. in *Perspectives in Organic Chemistry* (Todd, A., ed), Interscience Publishers, N.Y., **1956**, 155–184.

86. a) Schreiber, S.L. *Science* **2000**, *287*, 1964–1969; b) Burke, M.D., Schreiber, S.L. Angew. *Chem. Int. Ed.* **2004**, *43*, 46–58.

87. Qu, G., Li, A., Acevedo-Rocha, C.G., Sun, Z., Reetz, M.T. *Angew. Chem. Int. Ed.* **2020**, *59*, 13204–13231.

2

Chirality, Topology, and Asymmetric Synthesis

It is proposed that asymmetric syntheses be divided into enantioselective and diastereoselective syntheses.

Y. Izumi, 1971

2.1 Chirality

Carbon atoms carrying four different substituents possess a unique property: the substituents may be arranged in two alternate ways to bring about two forms of the molecule with the same constitution. In Figure 2.1, two molecules of the same constitution (CHXYZ) are depicted so that in each case the smallest substituent (hydrogen) lies behind the plane of the paper. The substituent X is drawn in each case in the plane, pointing up; and the other two substituents occupy positions either on the left or right of the central carbon atom. Looking along the bond from the central carbon atom towards the hydrogen at the back, one finds that the two molecules differ in the way the remaining three substituents are arranged in space: in ***A*** the substituents X, Y, and Z follow a clockwise rotation (*R*, *rectus*, right); whereas in ***B*** the rotation is counter clockwise (*S*, *sinister*, left).

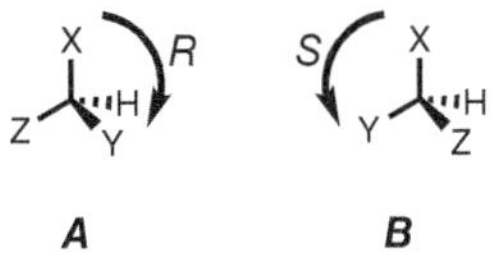

Figure 2.1 *Definition of* R- *and* S-*enantiomers*

The geometric property of a rigid object (or spatial arrangement of points or atoms) of being non-superposable on its mirror image is called *chirality* [1]. Such an object has no symmetry elements of the second kind (a mirror plane, $\sigma = S_1$; a centre of inversion, $i = S_2$; a rotation-reflection axis, S_{2n}). If the object is superposable on its mirror image the object is described as being *achiral*.

The two forms of the molecule are related as hands to each other, being non-superimposable mirror images of one another. They are called *chiral* (from Greek $\chi\varepsilon\iota\rho$, *cheir* = hand). The atom holding a set of ligands in a spatial arrangement, which is not superposable on its mirror image is called a *chirality center* (or shortened, *chiral center*). A center (usually located at an atom) in a molecule such that exchange of two ligands at the center leads to a stereoisomer of the original molecule, is called a *stereogenic center* (or shorter *stereocenter*). With tetrahedral atoms, such as carbon,

Asymmetric Synthesis of Natural Products, Third Edition. Ari M.P. Koskinen.

a chiral center is necessarily stereogenic, since the interchange of two ligands at a chiral tetrahedral atom interconverts enantiomers. However, the converse is not true: not every stereogenic center is a chiral center, since one cannot define chiral centers in achiral molecules. Thus (Figure 2.2), C1 and C3 in 1,3-dichlorocyclobutane are stereogenic, since the interchange of Cl and H at C1 or C3 leads to (diastereo)isomers, converting the *cis* isomer into the *trans* or vice versa. However, these atoms are not chiral, since the 1,3-dichlorocyclobutanes, *cis* or *trans*, are achiral.

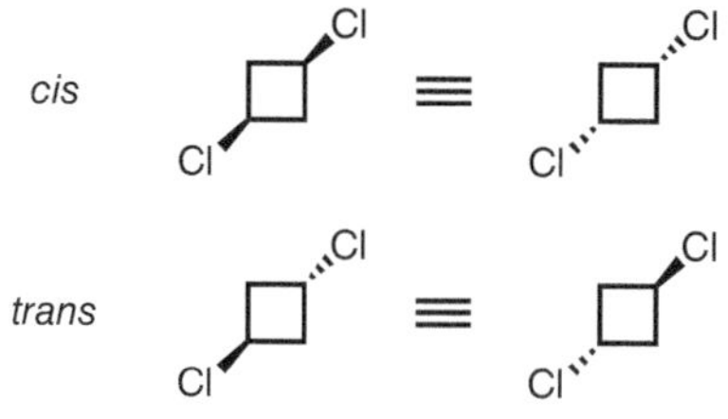

Figure 2.2   *Stereogenic centers are not always chiral centers*

Note also that the carbon atom is not asymmetric, but the spatial arrangement of the substituents on the stereocenter make the whole ensemble chiral. In this case, the whole molecule does not possess any element of symmetry (other that the trivial one of a one-fold axis of symmetry), and is therefore *asymmetric* [1]. However, asymmetry is not a necessary requirement for chirality. Dissymmetric molecules which lack one or more elements of symmetry, can also be chiral, and the requirement for chirality can be defined as follows: molecules which do not possess rotation-reflection axes (S axes, or in the most simple cases mirror planes, S_1) are chiral or dissymmetric. Based on point groups, those molecules which belong to the C_n or D_n point groups are chiral. For instance, *trans*-2,5-dimethylpyrrolidine contains a two-fold rotation axis, belongs to the point group C_2, and is chiral (Figure 2.3).

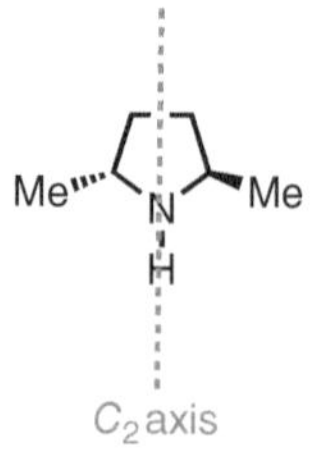

Figure 2.3   *Symmetric but chiral molecule*

If the molecule contains more than one chiral center, there emerges the possibility of another form of stereoisomerism. Stereoisomeric molecules which cannot be superimposed by any symmetry operations are called *diastereomers*. Thus, for the 2-chloro-3-hydroxybutane, one can draw four different structures; two pairs of enantiomeric compounds and two pairs of diastereomeric compounds (Figure 2.4).

Based on this definition, the *cis* and *trans* forms of 1,2-, 1,3- and 1,4-disubstituted cyclohexanes are also diastereomers, although in the last case the compounds are not chiral (and thus optical isomerism cannot exist) (Figure 2.5). It should also be pointed out that diastereomeric relationships can be found by inspecting a single molecule, whereas the property of enantiomerism always requires comparison with another molecule. Furthermore, in both cases the implication is that at least one other form (for diastereomers often several) does exist.

In reactions involving one or more chiral centers, we are interested in bringing about transformations which produce one stereoisomeric form in excess over the other possible ones. One speaks of *stereoselective reactions* if the outcome of the reaction is uneven, and *stereospecific reactions* if the product is produced in one enantiomeric form only. All stereospecific reactions are necessarily stereoselective, but selectivity alone is not a sufficient criterion for specificity.

If a sample of molecules contains molecules of only (within the limits of detection) the same chirality sense, the sample is called *enantiomerically pure* or simply *enantiopure*. One should avoid the use of the sometimes seen term

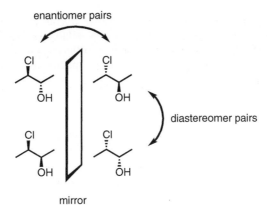

Figure 2.4   *Enantiomers and diastereomers*

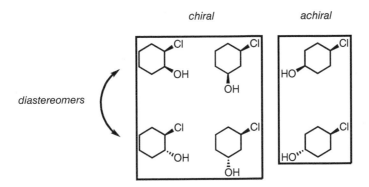

Figure 2.5   *Diastereomers and chirality*

"optically pure," as this implies optical activity, which is not a necessary property for chiral compounds. If the two stereoisomers are formed in equal amounts, the mixture is called *racemic*. A mixture containing unequal amounts of the enantiomers is called *enantiomerically enriched* or *enantioenriched*.

2.2   Enantiomeric Purity

Two measures of optical or enantiomeric purity are in common use: *optical purity*, which is based on the optical rotation of the compound; and *enantiomeric excess*, which is independent of the optical behavior of the compound. As we shall soon see, these two measures often give similar or identical results, but this is not always necessarily the case.

Optical purity is defined in terms of the optical rotation of the compound. In 1838 Biot introduced the specific optical rotatory power $[\alpha]_\lambda$ [2]. The modern definition is given by the formula

$$[\alpha]_\lambda = \frac{\alpha}{1 \bullet c}$$

where

   α is the observed rotation in degrees;
   λ is the wavelength of the light applied (usually the sodium D-line, 598 nm);
   1   is the length of the optical path in decimeters; and
   c   is the density of the liquid (given in g/100 mL).

Biot introduced the convention of giving the length of the optical path in decimetres "in order that the significant figures may not be uselessly preceded by two zeros." Sometimes the temperature of the measurement is indicated by a superscript. Note that the specific rotation is, by agreement, a dimensionless quantity!

Optical purity (% *o.p.*) is defined as the ratio between the observed specific rotation $[\alpha]_{obs}$ and the maximum specific rotation $[\alpha]_{max}$:

$$\% \ o.p. = \frac{[\alpha]_{obs}}{[\alpha]_{max}} \times 100$$

This method has its drawbacks, however. In many cases the maximum specific rotation is not known, especially in cases when one synthesises a previously unknown compound. Experimental conditions also cause ambiguities. One should always use a cell with a long path and a large diameter to avoid local concentration gradients, which can cause distorted behavior of the rotation. Particles in the sample easily modify the observed rotation, as do air bubbles, even very small ones, which refract light. In cases when the sample is colored, the absorption of the light may cause problems.

Enantiomeric excess (% *ee*) is defined as the excess of one enantiomer over the other, and this definition makes this measurement unambiguous.

$$\% \ ee = \frac{|\% \ R - \% \ S|}{|\% \ R + \% \ S|} \times 100 = 100 - 2 \times (\% \ S) (for \ R)$$

Comparison between enantiomeric excess and optical purity reveals that the latter is further hampered by the following experimental problems. The solvent, concentration, and, to a lesser extent, temperature affect $[\alpha]$. Reproduction of literature concentrations may be difficult, especially in cases where the rotation has been reported in chloroform or ethanol. One should also always specify the quality of the solvent: for instance, does the chloroform contain ethanol as a stabiliser, if so how much – the American and European standards vary!; in determinations performed in ethanol solutions, what is the percentage of ethanol? The optical rotation is not always necessarily linear with concentration due to association effects, in other words, a bimolecular complex of R–R forms of a compound does not necessarily have the same rotation as the (unimolecular) compound R. Even small amounts of impurities may adversely affect the outcome of the determination of optical rotation. For instance, if the impurity B has a relatively large rotation (e.g. 100), the measurement of the rotation of compound A (e.g. −1) is severely affected; even 1% of impurity will completely abolish the measurement. It is noteworthy that the impurities need not be optically active, due to association effects.

In 1969, *Alain Horeau* (1909–1992) had already observed that the optical purity and enantiomeric excess are not necessarily linearly related to each other, as he showed for 2-methyl-2-ethyl succinic acid (Figure 2.6) [3]. For instance, a sample which was 50% enantipure (50% pure enantiomer and 50% racemate), displayed only 35.5% optical purity. The specific rotation was noted to also depend on the concentration: at 15% concentration, the rotation was −4.4; and at 5%, the rotation was −3.4.

A case in point is the effect of acetophenone in a solution of phenylethanol (Figure 2.7) [4]. When acetophenone was subjected to lithium aluminium hydride/Darvon alcohol reduction, the presence of unreacted acetophenone significantly increased the rotation of the product (from +43.1 with no PhCOMe to +58.3 with four fold excess of PhCOMe).

These two examples pointed to the non-linear behaviour of optically active compounds, a feature that would later become important in the development of catalysts and chirality amplification, as we will see in Chapter 3.

The enantiomeric excess relates to the ratio of pure enantiomer to racemate. This is a useful quantity when one considers e.g. purification of the chiral compound from a mixture of enantiomers by physical, non-chiral means (e.g. crystallisation). Then one in fact separates the pure enantiomer from the racemate, and % *ee* represents the maximum yield obtainable in such a separation.

The enantiomeric excess is not directly related to e.g. kinetics of the reaction that produces the chiral compound. Horeau had suggested that enantiomeric purity P_E should be expressed in terms of the fraction of the more abundant enantiomer to the sum of both enantiomers [3]:

$$P_E = \frac{R}{R + S}$$

This definition, or the equivalent *enantiomeric ratio* (*er* = *R*:*S*), are more directly related to the kinetics, and therefore more useful in certain cases [5]. It is worth remembering that both expressions are useful, and one can easily interconvert

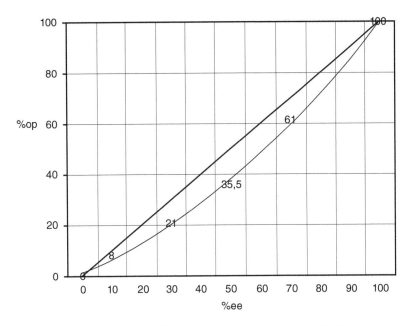

Figure 2.6   *Optical purity and enantiopurity are not necessarily linearly related*

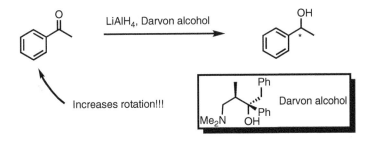

Figure 2.7   *Achiral impurity can affect optical rotation*

ee and *er*. The definition of enantiomeric ratio is equivalent to the ratio of reaction rates:

$$er = \frac{R}{S} = \frac{k_R}{k_S} = e^{-\frac{\Delta\Delta G^{\neq}}{RT}}$$

2.2.1   Determination of Enantiomeric Purity

Determination of enantiomeric composition is, in practice, performed through the determination of the relative amounts of the two enantiomers. This is rather straightforward in the case of analysing individual molecules. One should bear in mind, however, that during the course of method development, one is interested in the efficiency of the reaction itself. Thus, every operation (including chemical transformations and derivatisations, as well as separation and purification processes) subsequent to the chirality-forming reaction can also effect kinetic resolution, and thus any further treatment should be avoided before detemining the enantiomeric efficiency. The determination should be performed with a sample taken directly from the crude reaction mixture, if at all possible. The most direct method is based on the utilisation of chiral chromatographic media, where several types of chiral columns are currently available. Quantitation is simple, and no further operations are needed. In cases when this is not possible, one can revert to a number of other methods, bearing in mind the precautions concerned with kinetic resolution. Derivatisation with a chiral agent gives rise to a mixture of diastereomers, which can usually be easily distinguished either by HPLC or NMR methods. Any derivatising agent capable of achieving separation is acceptable, but the ones most commonly used include α-methoxy-α-trifluoromethyl phenylacetic acid (MTPA, Mosher's acid) [6], mandelic acid, phenylethyl amines, amino acids, and amino alcohols and their derivatives. The derivatisation need not be based on covalent bond formation. In the NMR methods, one can also use chiral solvents or chiral shift reagents.

Whatever the strategy one decides to use to establish the enantiomeric purity of the compound, one should always secure the detection limits in each particular case. A narrow peak and a broad peak will give distinctly different threshold behaviors with all detection methods.

Enantiomers do not differ in their physical properties, except when subjected to a chiral environment. For example, the *R*- and *S*- (D- and L-, respectively) forms of the alanine-derived enones (Figure 2.8) give identical UV, IR, NMR, and mass spectra, and their chromatographic mobility on normal phase and (achiral) reversed phase chromatography are also indistinguishable [7].

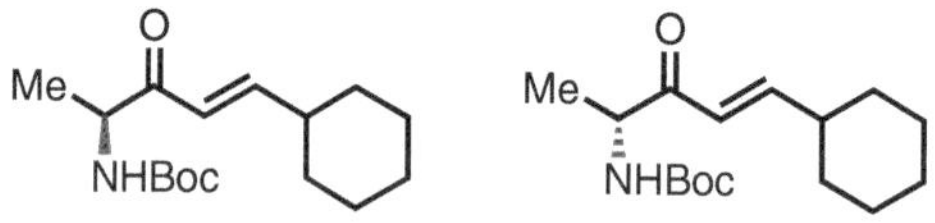

Figure 2.8   Enantiomeric separation

However, when the two enantiomers are subjected to chromatography on chiral media, the two compounds experience the chiral environment differently. In Figure 2.9, trace **A** shows the chromatogram for a racemic mixture. Trace **B** represents the chromatogram obtained from an enantiomerically enriched mixture. Similar chromatographic behaviour would be exhibited by diastereomeric derivatives (e.g. amides of chiral acids made with *R*- or *S*-phenylethylamine) even on achiral media.

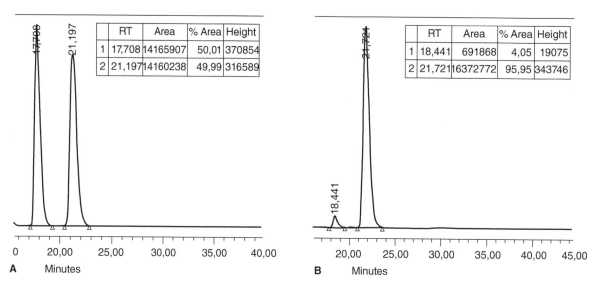

	RT	Area	% Area	Height
1	17,708	14165907	50,01	370854
2	21,197	14160238	49,99	316589

	RT	Area	% Area	Height
1	18,441	691868	4,05	19075
2	21,721	16372772	95,95	343746

Figure 2.9   Chromatograms obtained from compound of Figure 2.18: (A) racemic mixture on a chiral medium; (B) enantioenriched sample of the same (er = 96:4, or 92% ee)

The *x*-axis represents elution time, and thereby also the amount of eluent (solvent volume) to elute the compound out of the column. In calculating the time, one takes into account the void column volume (the time needed to bring the solvent front through the column, shown usually as a small notch at the left of the chromatograms). The difference in elution volumes of the two compounds is based on the affinity of the medium, being higher for the compound eluted last out of the column. The efficiency of the separation can be described by the ratio of the retention time indices t_1 and t_2 (e.g. the retention times or retention volumes),

$$\alpha = \frac{t_1}{t_2}$$

This ratio gives an estimate of the 'ease' of separation, and can be used as a guide for selecting e.g. preparative separation methods. In terms of energy, the free energy difference of retention of the two species interacting with the solid support can be estimated from this ratio of retention times using the equation for Gibbs free energy:

$$\Delta G = -R \ln \alpha$$

As the ratios α are usually quite small (typically on the order of 1.05–1.5), it is obvious that even small energy differences in interaction can be very efficiently distinguished by chromatography.

2.3 Topology

Dissolving metal reduction of 4-*tert*-butylcyclohexanone is a classic example of a stereoselective reaction (Scheme 2.1). The more stable equatorial alcohol isomer predominates in the reaction products (98% of the alcohol product) [8]. Since the *tert*-butyl group effectively locks the cyclohexane into one chair form with the bulky substituent equatorial, the reduction occurs through delivery of the hydride from the upper face of the cyclohexanone. We shall return to more thorough rationalisation of the reduction of cyclohexanones in Chapter 3. This reduction exemplifies a case of substrate-controlled process (the stereochemical bias resides in the starting material).

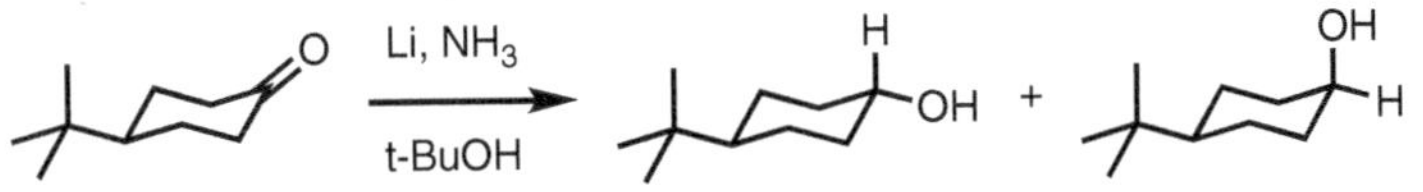

Scheme 2.1 *Stereoselective reaction*

The stereochemistry of hydroboration reactions is determined by the geometry of the starting alkene (Scheme 2.2). The *E*- and *Z*-isomers of the alkene give different products. Both reactions are stereospecific because the mechanism of hydroboration necessarily places the hydride and boron on the same face of the double bond (the attack on the upper face only is depicted, in both reactions both enantiomers are actually formed). Specificity is thus a property imposed by the mechanism of the reaction on the starting material.

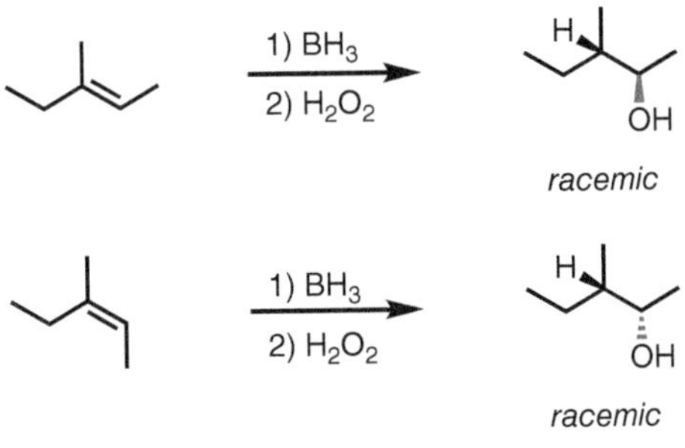

Scheme 2.2 *Pair of stereospecific reactions*

In the previous example the two faces of the double bond reacted with equal facility, and thus racemic compounds were produced. This is obvious, since we were using an achiral reagent, and thus there is no source of chiral information in the reaction. But let us consider whether the reaction could, at least in principle, be persuaded to give enantioselectivity.

We will need to consider the topology of molecules in order to be able to deduce what kinds of stereocontrolled processes a certain molecule can undergo [9]. In general, two groups are said to be topologically equivalent if they can be interchanged by rotation about any n-fold axis of rotation C_n to give a structure indistinguishable from the original one. Thus, the two methyl groups of *trans*-2,5-dimethylpyrrolidine are equivalent (Figure 2.3).

Similar definition applies for equivalent molecular faces: two faces of a molecule are equivalent if the plane defining the faces contains a coplanar axis of symmetry. Such faces are not restricted to achiral molecules, but the attack by a reagent, whether chiral or not, at equivalent faces leads to equivalent transition states, and thereby also identical results. Thus, addition of a nucleophile Nu⁻ to the equivalent faces of the chiral C_2 symmetric *trans*-2,6-dimethylcyclohexanone creates a new center which is not chiral (Scheme 2.3). The two faces of the carbonyl group are equivalent because of the two-fold axis of rotation (shown as the dashed line in Scheme 2.3).

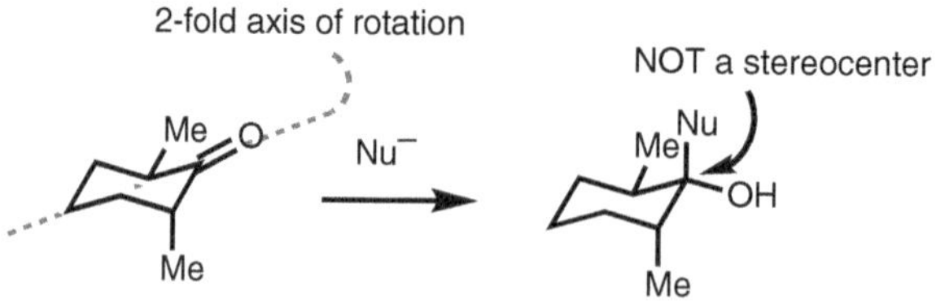

Scheme 2.3   *Not a stereocenter*

If the two groups can be interchanged by rotation–reflection, the environments of the groups are enantiomeric, and the groups are said to be *enantiotopic*. Similarly, *diastereotopic groups* are defined as groups whose environments are diastereomeric. Atoms carrying enantiotopic groups are called *prochiral*, as the replacement of one of the enantiotopic groups will produce a chiral compound [10]. In a similar fashion, the replacement of a diastereotopic atom will give rise to diastereomers. As a corollary to the definition, enantiotopic groups can only occur in achiral molecules.

Let us inspect a molecule C–XYHH, whose geminally bound hydrogen atoms are enantiotopic (Figure 2.10). The prochiral substituents are distinguished in nomenclature through an application of the sequence rules. Substitution of one of the hydrogen atoms with a dummy atom in such a fashion that it gains higher sequence rule order (H' > H) than the other one, followed by application of the normal procedure for assigning the *R/S*-descriptors, gives the labels pro-*R* and pro-*S* to the geminal hydrogen atoms.

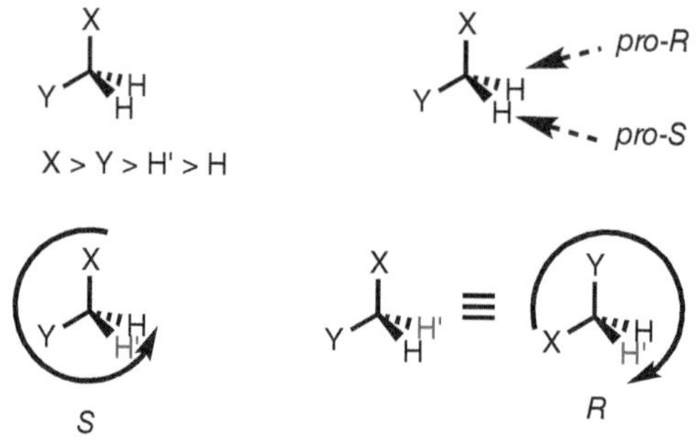

Figure 2.10   *Definition of pro-S- and pro-R- nomenclature*

Figure 2.11 shows some enantiotopic arrangements. The protons in bromochloromethane are clearly enantiotopic. In butyrolactone, exchange of protons on each of the methylene groups leads to the formation of different enantiomers, so the protons are enantiotopic in each CH_2 group. The cyclobutanone is an interesting example, showing both enantiotopic and homotopic (equivalent) groups. If we consider the planar conformation of cyclobutanone, the hydrogen atoms at C3 are equivalent. Hydrogen atoms H1 and H4 can be interconverted by rotation of the molecule by a two-fold rotation

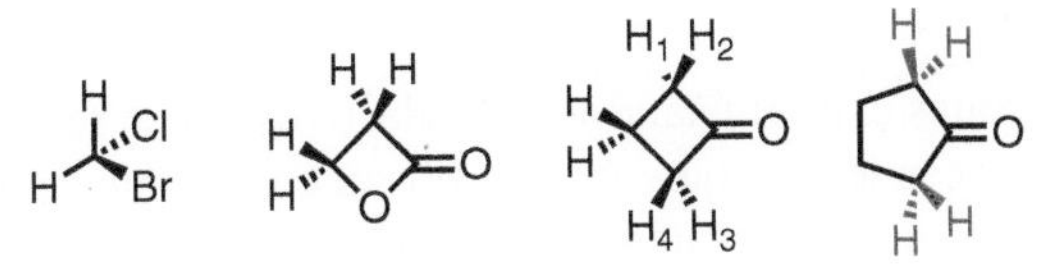

Figure 2.11 *Enantiotopic atoms*

axis, and are thus equivalent. Similarly, the hydrogens H2 and H3 are equivalent. The relationships between these pairs of equivalent atoms are enantiotopic. The pairs H1/H4 and H2/H3 form two enantiotopic sets of equivalent atoms, or *vice versa*, the pairs H1/H2 and H3/H4 form two equivalent sets of enantiotopic atoms. Cylclopentanone also has a two-fold rotation axis, and thus e.g. the blue protons and the red protons are pair-wise equivalent, and the blue are enantiotopic with the red ones. The same applies to the protons on the following C3 atoms.

It is important to emphasise the fact that the prochirality descriptor does not have any correlation with the absolute configuration of a product formed from substitution of a prochiral atom (Figure 2.12). The replacement of the pro-*R* methoxy group (red) of acetophenone dimethyl acetal gives either the *R* or *S* absolute stereochemical descriptor for the product, depending on the relative sequence rule orders of the replacing atom or group and the existing ones.

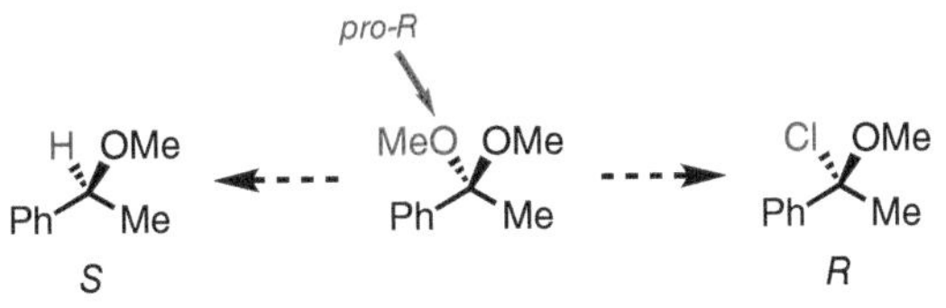

Figure 2.12 *Prochirality vs. chirality*

Groups or atoms are said to be *diastereotopic* if they reside in diastereomeric environments and cannot be interchanged by any symmetry operation. In other words, replacement of one of two diastereotopic atoms leads to the formation of diastereomers. Some typical examples are shown in Figure 2.13. Molecular dissymmetry is not a criterion for the presence of diastereotopic groups. In the chair conformation of cyclohexane ($\mathbf{D}_{3d}$), the six pairs of diastereotopic hydrogens (axial and equatorial) are interchangeable by either C_2 or C_3 and the pairs are therefore equivalent. Two examples of *N*-benzyl 2,6-dimethylpiperidine show quite drastic differences in their NMR spectra [11]. The benzylic protons in the *trans*-dimethyl compound show a clear AB system, whereas the *cis*-dimethyl compound shows a singlet for the two protons. It is a common feature of diastereotopic atoms that they can be distinguished by the appearance of their NMR signals quite easily.

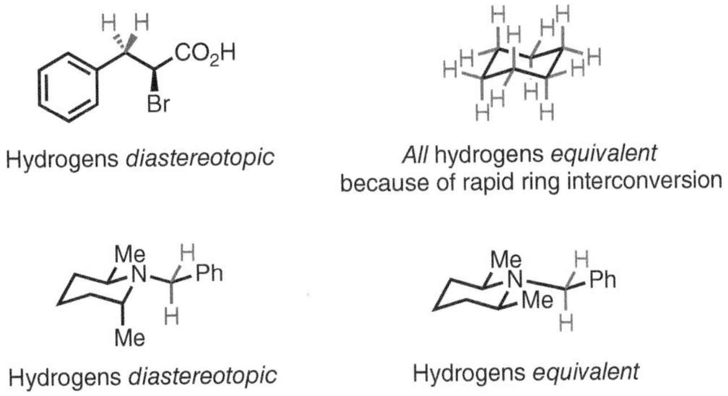

Hydrogens *diastereotopic*

All hydrogens *equivalent*
because of rapid ring interconversion

Hydrogens *diastereotopic*

Hydrogens *equivalent*

Figure 2.13 *Diastereotopicity*

2.3.1 Enantiotopic Differentiation

Reactions involving enantiotopic groups exhibit enantiotopic selectivity. Such reactions are brought about by the use of a chiral reagent, and in some cases can lead to quite high levels of enantioselectivity. The advantage of enantiotopic selectivity is that the symmetry of the starting material is converted into asymmetry of the product in the chirality generating step. The symmetry in the starting material can be taken into account in planning the synthesis of a chiral product (Figure 2.14).

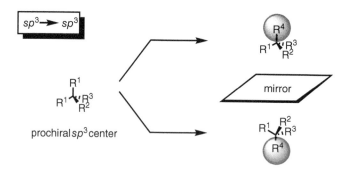

Figure 2.14 *Distinction between enantiotopic atoms*

The first example of asymmetric catalysis by enantiotopic selection in chemical reactions was shown by Marckwald in 1904 (Scheme 1.8) [12]. Albeit this gave only a modest result in current standards, it was the first recorded example of asymmetric catalysis, enantiotopic selection and organocatalysis.

The use of chiral bases provides an interesting means of achieving enantiotopic selection. An early example was provided by the conversion of cyclohexene oxide to the corresponding allylic alcohol by rearrangement brought about by the C_2 symmetric chiral base derived from phenylethylamine (Scheme 2.4) [13]. The reaction proceeds through abstraction of the proton on the same face of the cyclohexane ring as the epoxide oxygen resides. As the two protons on the methylene groups on either side of the epoxide ring are enantiotopic, a chiral base (or presumably an aggregated form of it) will be able to distinguish the two protons. Although the level of asymmetric induction (31% *ee*) was not particularly high, the reaction paved the way for subsequent developments in methodology.

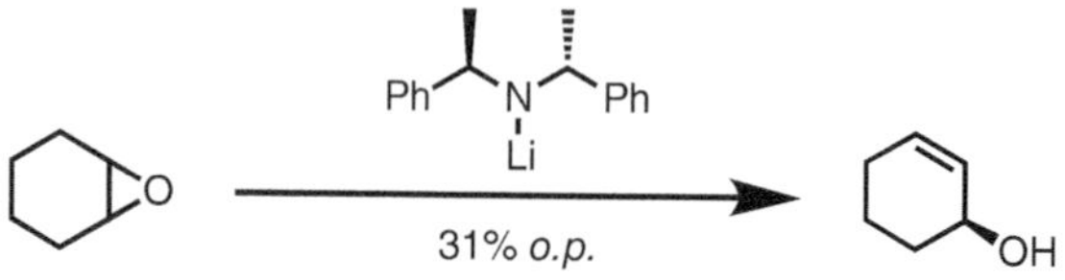

Scheme 2.4 *Enantiotopic deprotonation*

Kenji Koga (1938–2004) developed applications of enantiotopically selective enolization reactions (Scheme 2.5). The chiral amide base capable of tight chelation with lithium provides exceptionally high bias in the enolization of cyclic ketones (up to 97% *ee*) [14]. The bicyclo[3.3.0]octanone derivative was used in the synthesis of carbacyclin, a prostacyclin analogue (Scheme 2.6) [15].

Asymmetric Wittig-type olefinations have been studied for half a century [16], but a seminal practical case of enantiotopic differentiation was provided by *Stephen Hanessian* (1935–) in 1984 (Scheme 2.7) [17]. Here, the enantiopure phosphoramidate reagent preferentially attacks from the less shielded enantiotopic face of the phosphonamide

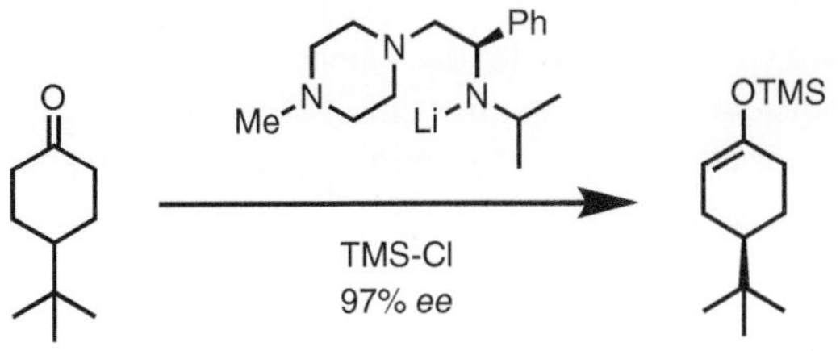

Scheme 2.5   *Enantiotopic enolization*

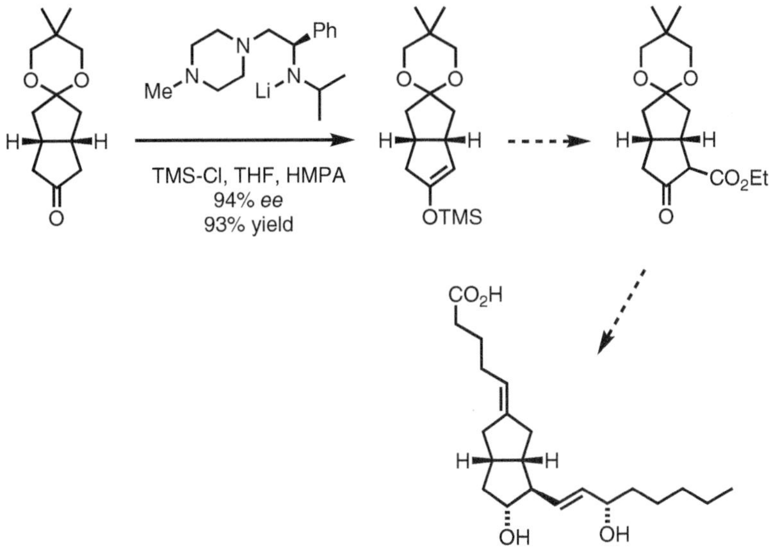

Scheme 2.6   *Enantiotopic enolization in the synthesis of carbacyclin*

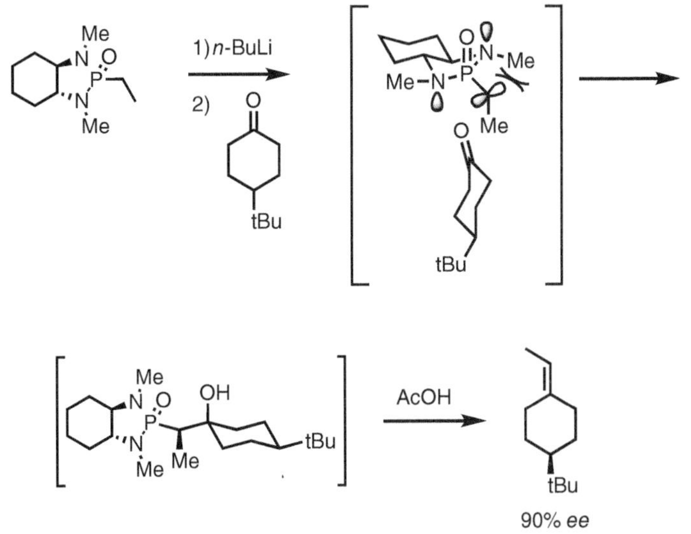

Scheme 2.7   *Asymmetric Wittig olefination*

carbanion. The cyclohexanone is attacked from the equatorial face. Since the olefin geometry is defined by the geometry of the intermediate hydroxyphosphonamide, the kinetic diastereotopic preference for the formation of the initial adduct is transferred to enantioselectivity in the formation of the olefin.

Two further examples illustrate the usefulness of enantiotopic selection. In another inaugural example, Whitesell has shown that the symmetric bicyclo[3.3.0]octadiene undergoes a clean ene-type reaction with the 8-phenylmenthyl ester of glyoxylic acid to give the product as a single diastereomer in good yield (Scheme 2.8) [18]. The dimethylphenyl group efficiently shields one of the enantiotopic faces of the glyoxalate, and the diastereomerically pure ene adduct is formed in 74% yield.

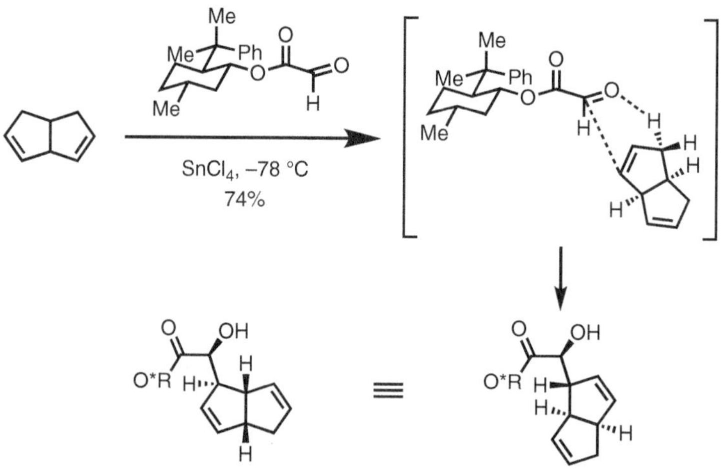

Scheme 2.8   Enantioselective ene-reaction

The last example bears some relevance to our question concerning the possibility of distinguishing the two faces of the alkene upon hydroboration (Scheme 2.9). In connection with a synthesis of prostaglandin $F_{2\alpha}$ intermediate, the cyclopentadieneacetic acid derivative was hydroborated with a chiral borane reagent, (−)-diisopinocampheylborane [(−)-Ipc$_2$BH], derived from (−)-α-pinene, to give the alcohol in high selectivity (> 92% *ee* by NMR analysis) [19]. The double bonds are enantiotopic, and the pinanylborane efficiently discriminates the faces of the double bonds. Similarly high (95–96% *ee*) enantioselectivities were obtained on the hydroboration of 5-methylcyclopentadiene followed by oxidation. This product was used in the asymmetric synthesis of loganin, a biogenetic precursor of several natural products [20].

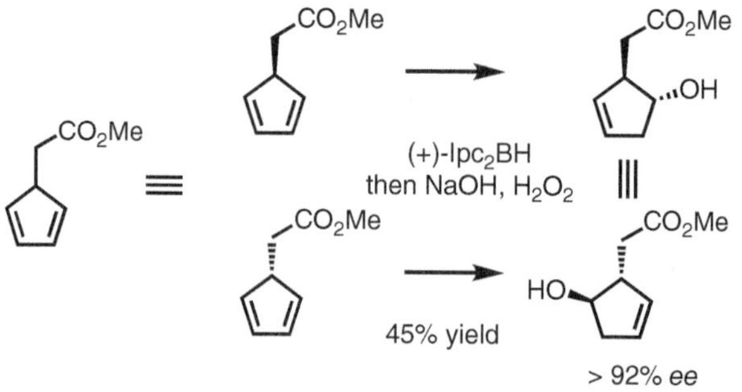

Scheme 2.9   Enantioselective hydroboration

2.3.2 Enantiofacial Differentiation

In the previous example, the attack of the reagent on different faces of the double bonds produces enantiomers, and thus the two faces are enantiotopic. Enantiotopic faces are characterised by the existence of an S_n axis perpendicular to the plane and the absence of a C_2 axis in the plane. The two faces are termed the *Re* and *Si* faces in the following way (Figure 2.15). The double bond defines a plane with three substituents at the reaction center. The substituents are given priorities according to the sequence rule. When the double bond is viewed from one face, and if the substituents rotate in a clockwise fashion the face is called *Re*, if counter clockwise the face is *Si*. The *Re/Si* nomenclature is exemplified below for a carbonyl group, and the substituent R_L is the one that gains higher preference than the substituent R_S.

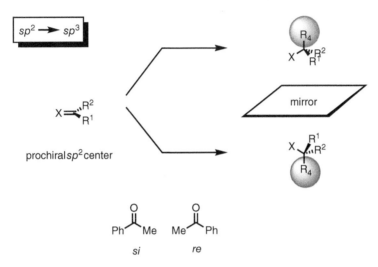

Figure 2.15 *Distinction between enantiotopic faces*

Preferential attack of a reagent on either face of the double bond is termed *enantiofacial selectivity*. In line with our previous observations on enantiotopic selectivity, enantiofacial selectivity is dependent on the reagent. An achiral reagent gives rise to transition states which are enantiomeric, and thus the free energies of the transition states will be similar. There is no difference in the activation energies, and one will obtain an equimolar mixture of the products. Approach of a chiral reagent will give rise to diastereomeric transition states whose energy contents, and thereby the activation energies leading to these transition states, will differ. This will be observed as enantiofacial selectivity. We can now answer our previous question regarding the hydroboration reaction. The two faces of the double bond can, at least in principle, be distinguished by the action of a chiral hydroborating reagent. This is clearly manifested in the several examples (including the previous ones) where extremely high enantioselectivities can be obtained [21].

Reduction of carbonyl groups by the action of a chirally modified lithium aluminium hydride reagent is a well-established method for achieving enantiofacial differentiation. Darvon (dextropropoxyphene) was originally developed as a mild analgesic of the opioid type, but was withdrawn because of severe side effects. Its optical antipode (Novrad!) is devoid of analgesic activity, and has found limited use as an antitussive. The derived alcohol (Darvon alcohol) was developed as a readily available inexpensive chiral modifier for lithium aluminium hydride reductions [22]. The following is an example of such a process where the reduction of the acetylenic ketone was achieved in high yield and high enantioselectivity (Scheme 2.10) [23]. The product was an intermediate in the synthesis of asteriscanolide, a structurally novel sesquiterpene lactone [24].

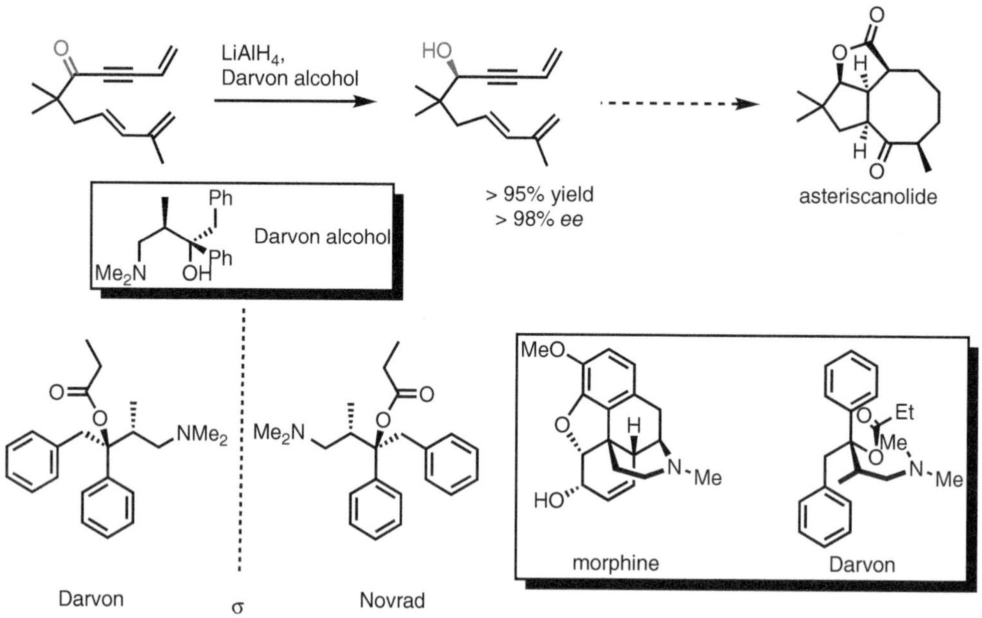

Scheme 2.10 *Enantioselective reduction: synthesis of asteriscanolide*

Both enantiotopically and enantiofacially selective reactions are *reagent-controlled*, and much effort is devoted to the development of asymmetric transformations relying on this concept. Practical realisations of the concepts outlined above will be discussed in more depth in Chapter 3.

2.4 Naming Stereocenters

A few words need to be said about the specification of the relative stereochemistry in molecules containing two or more chiral centers. As the number of chiral centers increases in a molecule, the description of the relative stereochemistry at these centers can become problematic. Naming stereoisomers has been a prevailing problem since the early days of stereochemistry: dextrorotatory tartaric acid can be correlated chemically with either D- or L-glyceraldehyde, and it was therefore once specified D by the European chemists and L by the Americans. The sequence rules (the CIP, or Cahn–Ingold–Prelog system) was introduced to avoid such ambiguities in compounds containing one chiral center [25].

In compounds containing more than one chiral center, the *threo* and *erythro* descriptors are not unambiguous either (Figure 2.16). According to the original definition, the relative configuration of two groups on the same side in the Fischer projection are called *erythro*, and those on opposite sides are *threo*. Drawing the main chain in the more convenient zigzag form, the substituents of the *erythro* isomer end up on opposite sides of the plane! The situation got even more confused when inversion of the nomenclature was suggested in 1980 [26]. The new system gained widespread acceptance, and one should bear this in mind in reading the literature.

Several alternative systems were suggested to avoid the confusion (Figure 2.17), the most comprehensive one of them being the one devised by Seebach and Prelog [27]. This system is based on the sequence rules and application of *relative topicities* of the reactions. The processes are described simply as being either like (descriptor *lk*) or unlike (descriptor *ul*). The like course of a reaction involves the combination of either *Re,Re* or *Si,Si* faces of two trigonal stereogenic atoms to give rise to *lk* relative topicity. Similarly, addition from the *Si* face of the *S*-enantiomer gives rise to *lk* relative topicity. The relative configurations of the products are similarly described as being like (*l*) for (*R,R/S,S*) or (*R*,R**) and unlike (*u*) for (*R,S/S,R*) or (*R*,S**).

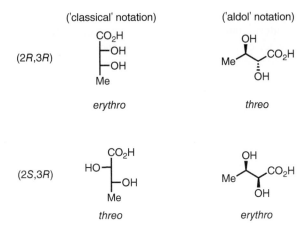

Figure 2.16   *Fischer and zigzag projections*

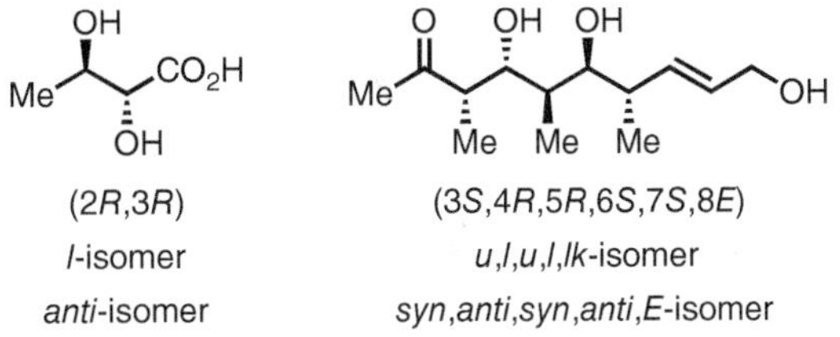

OH
Me ⟍╱⟍ CO₂H
       ÖH
(2R,3R)
l-isomer
anti-isomer

O   OH   OH
Me⟍╱⟍╱⟍╱⟍╱⟍ OH
    Me   Me   Me
(3S,4R,5R,6S,7S,8E)
u,l,u,l,lk-isomer
syn,anti,syn,anti,E-isomer

Figure 2.17   *Naming stereoisomers*

The use of *syn* and *anti* descriptors was proposed by Masamune [28] to describe two non-hydrogen substituents being, respectively, either on the same or the opposite sides of the plane defined by the zigzag main chain. This has the advantage of allowing instant recognition of the relative stereochemistry without necessitating the assignment of absolute stereochemical descriptors *R* and *S*. With several chiral centers, the system can be easily modified as shown in Figure 2.17. The *syn/anti*-descriptors are the ones we shall use throughout this text.

2.5   The Need for Enantiopure Compounds

Chirality often determines the properties and behavior of molecules in rather unexpected ways. Perhaps the oldest indication that enantiomers have different physiological properties was reported in 1886, when Piutti observed that the newly isolated (+)-asparagine ("the dextrogyrate variety") has a much sweeter taste than ordinary asparagine [29]. The odor properties of simple monoterpenes illustrates this (Figure 2.18): whereas (−)-(*S*)-limonene displays a distinguishable turpentine odor; the (+)-(*R*) isomer has a citrus fruit odor. The (−)-(*R*)- and (+)-(*S*)-carvones also differ in their flavours: the former tasting of spearmint and the latter of caraway.

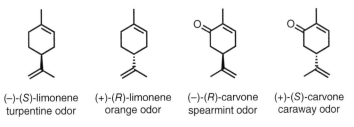

(−)-(*S*)-limonene   (+)-(*R*)-limonene   (−)-(*R*)-carvone   (+)-(*S*)-carvone
turpentine odor   orange odor   spearmint odor   caraway odor

Figure 2.18   *Differing odors of monoterpenes*

The differing behavior of enantiomers is particularly troublesome in medicines (Figure 2.19). Ethambutol has been used as medication against tuberculosis, but while the *S,S*-isomer is antituberculotic, the *R,R*-enantiomer has been associated with optic neuropathy which can lead to blindness. The enantiomers of penicillamine have opposing effects: one enantiomer is good for extracting heavy metals from the blood, whereas administration of the other enantiomer can lead to blindness. In the case of methorphan, the L-isomer, levomethorphan, is a strong opioid painkiller, whereas the D-isomer dextromethorphan is an over the counter cough medication. Thalidomide provided a most dramatic example of the difference of the enantiomers in action. The *R*-enantiomer was shown to be a good sedative agent, and was widely used in the 1950s and 1960s especially during pregnancy. However, it soon became apparent, that in some cases the foetuses developed severe malformations, and after careful investigations it became evident that the thalidomide molecule easily racemises in the body and the *S*-enantiomer is strongly teratogenic. Thalidomide was removed from the market, but has recently been reintroduced for certain applications, where the teratogenicity is not a risk.

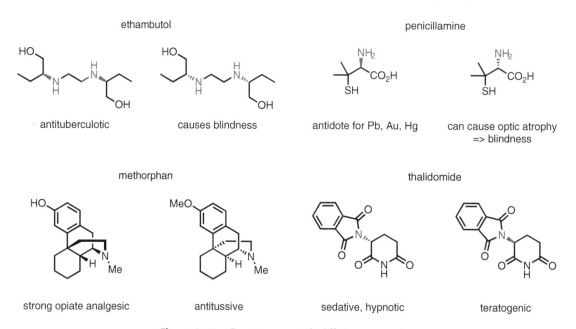

Figure 2.19 *Enantiomers with differing properties*

A prevailing misconception that all chiral natural products are enantiopure probably emerged from Louis Pasteur's view that all life processes are dissymmetric. It was only in the 1970s when the development of asymmetric synthesis methods allowed chemists to synthesise pure enantiomers and compare the properties of enantiomers and their mixtures to those of natural isolates.

Several animals communicate through chemical means, by releasing chemicals that are called pheromones. The first observation of pheromones was made by *Jean-Henri Fabre* (1823–1915), a French entomologist from Aveyron, who observed that a female Great Peacock moth (*Saturnia pyri*) attracted males. Fabre believed that this attraction was caused by scent [30].

Quite surprisingly, some bioactive compounds are not pure enantiomers, and there are even cases where the desired physiological activity requires a non-racemic mixture of enantiomers [31]. Sulcatol is a case in point (Figure 2.20). It is the aggregation pheromone of Ambrosia beetles (*Gnathotrichus sulcatus*). When the individual enantiomers were tested for their bioactivity, it was observed that the pure enantiomers were devoid of any pheromone activity. A 50:50 mixture was *more* active than the natural pheromone, and eventually it became clear that the natural pheromone is a 35:65 (*R:S*) mixture of the two enantiomers! The male Asian bull elephants (*Elephas maximus*) in *musth* secrete a fluid from their temporal glands. The secretion is rich in testosterone, but the attractive agent to females is frontalin.

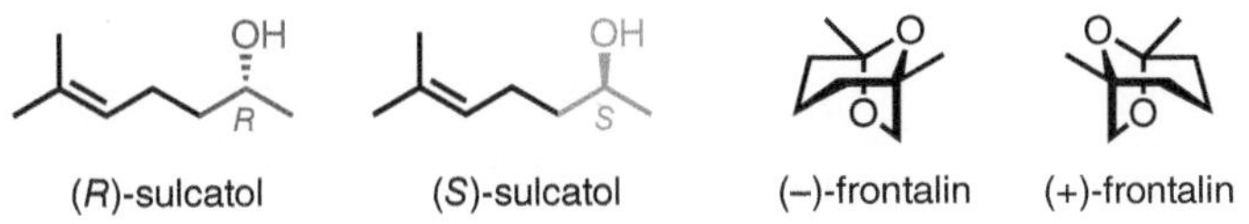

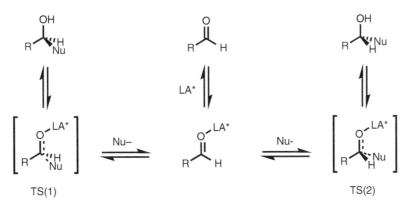

| (*R*)-sulcatol | (*S*)-sulcatol | (−)-frontalin | (+)-frontalin |

Figure 2.20 *Ambrosia beetle and Asian elephant pheromones*

The levels of (+)- and (−)-frontalin vary during the elephant's maturation from adolsecence to adulthood. Young bulls secrete more (+)-frontalin, but only the racemate is attractive to fertile females [32].

As natural products and their derivatives and analogues find wide use in our everyday life, from medicines to food additives, it is understandable that for the production of these compounds by synthetic means we need to secure them in enantiopure form. We simply cannot face risks similar to the infamous thalidomide case. In drug industry, enantiopure products are playing an increasingly more important role, and it is logical to expect that other areas of synthetic chemical activity will follow suit with rapid pace [33].

As the need for enantiopure compounds is heavily increasing, the objective of obtaining the final compound not only in chemically pure form but also in enantiopure state is usually already secured during the planning stages of the synthesis. Before looking at the various ways of achieving this goal, we shall take a brief excursion to the aspects involved in recognising chirality.

2.6   Chirality and Thermodynamic Principles of Asymmetric Induction

Let us turn our attention to the energetics of asymmetric reactions (Scheme 2.11). Not unlike the chromatographic example, a reaction producing two enantiomers from a single achiral starting material must be able to distinguish between the two emerging products. The chiral information must come from somewhere, and we shall return to the various ways of asymmetric introduction in the next section. For our present discussion on the energetics, we shall assume that the chiral information is mediated by a chiral reagent (e.g. a chiral Lewis acid LA*), although the same analysis applies to all the cases of simple enantioselection. As the starting material is a single compound, no distinction can be made at this stage – even a chiral reagent would form two enantiomeric complexes, which would be equienergetic. The products are enantiomeric with each other, and by the same principle, equienergetic. The distinction is only possible at the stage of the transition states, where the two transition states TS(1) and TS(2) are diastereomeric. The energy profiles for the two alternative reactions are shown in Figure 2.21.

Scheme 2.11   *Enantiodifferentiation occurs at the level of transition states*

Since the two transition states are diastereomeric, the substituents of the starting material occupy different positions in space. For some reactions involving cyclic six-membered transition states (such as aldol and Claisen reactions), the distinction between the two alternative transition states can be simply deduced from general principles of

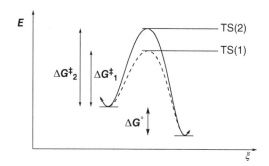

Figure 2.21   *Energy diagram for the reaction of an achiral starting material with one enantiomer of a chiral reagent to produce the enantiomeric products*

conformational analysis, such as the favored orientation of a substituent in an equatorial position rather than axial. We shall return to such cases in more depth in connection with the individual reactions in Chapter 3.

In the absence of thermodynamically driven equilibration processes, the formation of a single chiral center is typically under kinetic control, where the energy differences between the transition states determine the relative rates of the two competing reactions, and thereby also the ratio of the products formed. The formation of a second and further chiral centers is more complicated, as this can often be also thermodynamically controlled, i.e. the products are of different energies, and equilibration can alter the initial kinetic ratio. For a kinetically controlled process, the product ratio can be calculated from the free energy differences between the transition states, or vice versa, the free energy difference can be calculated from the observed product ratio. Table 2.1 shows the variation of the difference of activation energies ($\Delta\Delta G$) leading to the major and minor products in an enantioselective reaction as a function of temperature and the observed % *ee* and enantiomeric ratio *er*. The temperature effects are quite remarkable (the energy difference nearly doubles within the common temperature range −78 °C to rt). Another point to note is the fact that useful enantioselectivities (> 95% *ee*) require a substantial energy difference between the two (diastereomeric) transition states leading to the products. An energy difference of 8 kJmol^{-1} roughly matches a hydrogen bond in peptides, or the energy difference between an axial and equatorial methyl group in a cyclohexane (approximately 7.5 kJmol^{-1} repulsion for the axial conformer).

Table 2.1   *Energy differences (in kJmol^{-1}) between the two isomers at different temperatures as function of enantiomeric excess*

% ee	Temperature (°C)						er
	−100	−78	−40	−23	0	23	
99.9	10.9	12.4	14.7	15.8	17.3	17.7	99.95:0.05
99.5	8.63	9.72	11.6	12.5	13.6	14.7	99.75:0.25
99	7.63	8.59	9.39	11.0	12.0	13.0	99.5:0.5
98	6.62	7.46	8.92	9.55	10.4	11.3	99:1
95	5.28	5.95	7.12	7.63	8.34	9.05	97.5:2.5
90	4.23	4.78	5.70	6.12	6.70	7.25	95:5
80	3.18	3.56	4.27	4.57	4.99	5.41	90:10
70	2.51	2.81	3.35	3.60	3.94	4.27	85:15
60	2.01	2.26	2.68	2.93	3.14	3.39	80:20
50	1.59	1.80	2.14	2.30	2.51	2.72	75:25
40	1.22	1.38	1.63	1.76	1.93	2.10	70:30
30	0.88	1.01	1.22	1.30	1.42	1.51	65:35
20	0.59	0.67	0.80	0.84	0.92	1.01	60:40
10	0.29	0.34	0.38	0.42	0.46	0.50	55:45

2.7   Methods for Obtaining Chiral Compounds

In asymmetric synthesis, the ultimate goal is to produce the product in 100% enantiopurity. This section will discuss the different ways this goal can be reached. Let us first spend some time considering key practical requirements for the synthetic protocol developed or adopted. The requirements are naturally dependent on the exact situation we are dealing with: whether the goal is to make a specific compound or to develop general methodology to access a wide variety of target structures. The first question is related to the use of enantiopurity or enantiomeric ratio. Although one can argue that even a partially enriched compound can be enhanced to purity by e.g. crystallisation, one should remember that each operation, whether a chemical reaction or a physical action (such as crystallisation, distillation, chromatography, even extraction), represent unit operations. Each unit operation has an associated yield, which in real life is very seldom 100%. As you can see from Figure 2.22, the overall yield falls off quite rapidly as a function of yields of individual unit operations AND the number of unit operations (steps in the figure). Thus, the higher the enantioselection in the chirality generating step, the better for the overall process. Currently, levels of at least 95% *ee* (*er* 97.5:2.5, that is roughly within NMR detection limits) should be achievable in optimised processes.

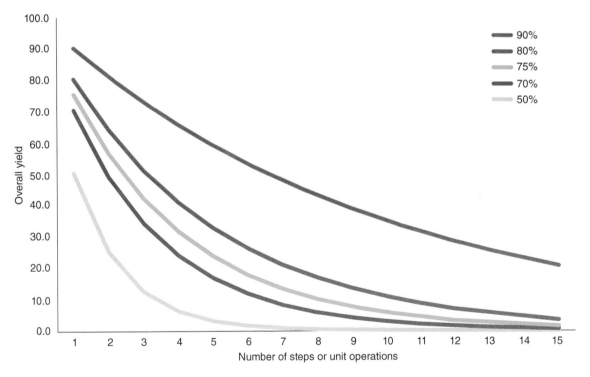

Figure 2.22   *Overall yields decrease rapidly with increasing length of synthesis*

For the development of synthetic methodologies, it is imperative that the source of chiral information (be it starting material, catalyst, or a chiral auxiliary or modifier) is readily available in both enantiomeric forms, so that one can gain access to both configurations. The transformations should be general, so that a broader selection of chemical space can be accessed.

Since the attainment of chiral compounds in enantiopure form is desired, we need to consider the various ways available for achieving this goal. If the desired compound is commercially available, then simply purchasing it may prove a trivial solution to the problem. This is true for a growing number of compounds, as many large and small companies now provide a wide selection of optically active and enantiopure compounds. Contract suppliers do this on industrial scale, and sometimes it is actually easier to buy hundreds of kilograms of a compound rather than just a few grams!

Another method, which is often the more tedious one, is to isolate the compound from natural sources. This can be a competitive route in those cases where the synthesis has not been achieved and the compound is desperately needed. One encounters this situation often during the early stages of drug development, when a drug candidate is being searched and no practical synthetic access has been developed yet. A variation of this method is the development of fermentation processes capable of producing large quantities of either the desired final compound or an intermediate. Several beta-lactam antibiotics are produced in such a fashion, by *semisynthesis*, and widespread efforts are directed at developing new fermentation avenues for compounds which are currently produced by total synthesis via a long and expensive sequence.

An economically somewhat less satisfactory method relies on resolution. One synthesises the target compound in racemic form, and then breaks the racemate to obtain the desired enantiomer. The attractiveness of this method is diminished by the fact that a maximum of 50% yield can be obtained in what often is the final step in the synthesis. A further disadvantage is that the other enantiomer often ends up in waste. In some cases the unwanted isomer can be epimerised and thus used nearly quantitatively in the synthesis. Resolution can sometimes be achieved by direct crystallisation, although this is mainly used for improving the optical purity of many crystalline compounds [34].

Menthol is an important industrial chemical, with an annual production of about 14 000 metric tonnes per year [35]. It is used mainly for oral care products (36%), in pharmaceutical industry (22%), and as flavors (17%). It is a naturally occurring compound, and approximately 75% of the production is based on extraction from the cornmint plant (*Mentha canadensis*). The extraction process was imported from China to Japan by Enzan, a Japanese priest, and for quite some time Japan dominated the world menthol production. After the major earthquake in Tokyo in 1923, some 300 000 Japanese emigrated to Brazil and started menthol production there. In the heydays of the 1970s up to 3000 metric tonnes of menthol originated from Brazil and Paraguay.

The extraction process is currently particularly popular in India and China (approximately 65% of all menthol production), and the anticipated shortages in natural supplies in the 1960s and 1970s led to the development of two industrial processes for the chemical synthesis of menthol. We first take a look at the sole European process by Haarmann and Reimer (H&R, part of Symrise since 2003), which is based on separation of enantiomers by crystallisation on industrial scale. We will return to the second process at the end of this chapter (Scheme 2.22), as well as the process implemented in 2012.

The H&R process uses the thymol production of Bayer AG as the starting point (Scheme 2.12). Thus, *m*-cresol is reacted with propene to produce thymol, which is reduced to a mixture of all eight isomers of menthol. Fractional distillation allows for the separation of (+)/(−)-menthol from the other thermodynamically less stable isomers (in menthol, all the substituents are equatorial). The remaining isomers are isomerized and recycled back to distillation. The racemic menthol fraction is converted to benzoate esters, and (−)-menthol benzoate is obtained by selective seeded crystallisation. The benzoates are hydrolysed and the (+)-isomer is aromatised and sent back to the process. The H&R process went on line in 1973 in Holzminden, Germany, and in South Carolina in 1978, and their combined capacity is 4 000 metric tonnes per year. In a tribute to German engineering, the H&R process gives an overall yield > 90%, thanks to extensive recycling [35].

Crystallisation via diastereomeric derivatives (e.g. phenylethylamine or mandelate salts) is common practice even on industrial scale processes. In chromatography, one can use either direct or indirect methods. In the direct methods, one can choose from a large number of chiral columns [36] to perform resolution. This is particularly attractive for analytical separations, and is widely used in connection with methods development. In the indirect methods, the compound to be resolved is derivatised with a suitable chiral agent, and the derivatising agent is again cleaved after chromatographic separation. The synthesis of gibberellic acid A$_3$ by E.J. Corey is an early successful example of such a process (Scheme 2.13) [37].

In a process called *kinetic resolution*, the two enantiomers react with different rates with a given chiral reagent (including enzymes!). This leads to a surplus of the less reactive enantiomer in the reaction mixture, and this phenomenon can be used to kinetically separate the enantiomers from each other. Kinetic resolution was first observed by Marckwald and McKenzie in 1899, when they esterified racemic mandelic acid with optically active (−)-menthol (Scheme 2.14) [38]. When the reaction was stopped before all mandelic acid had reacted, the reaction mixture became enriched in the less reactive (S)-mandelic acid.

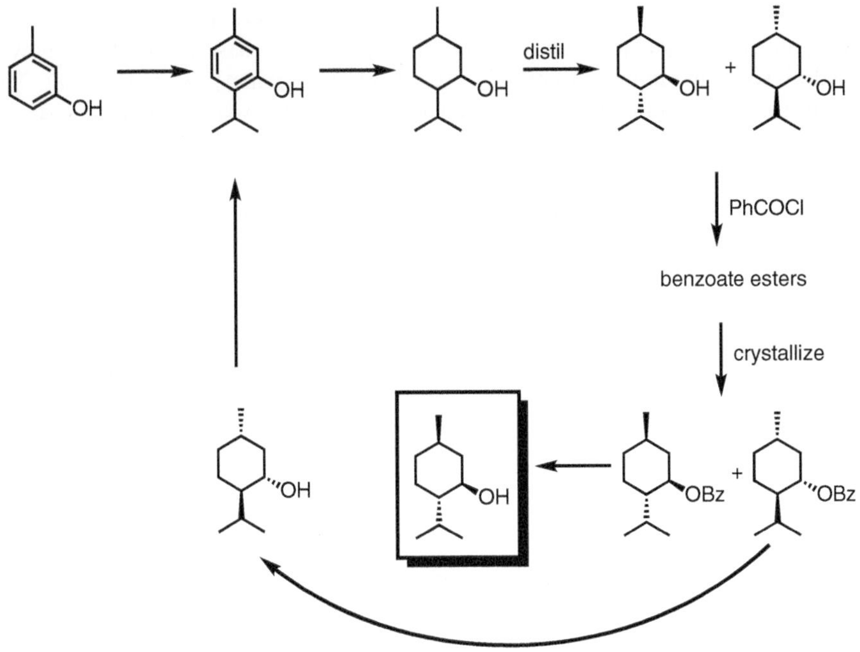

Scheme 2.12 *Haarmann and Reimer process for menthol production*

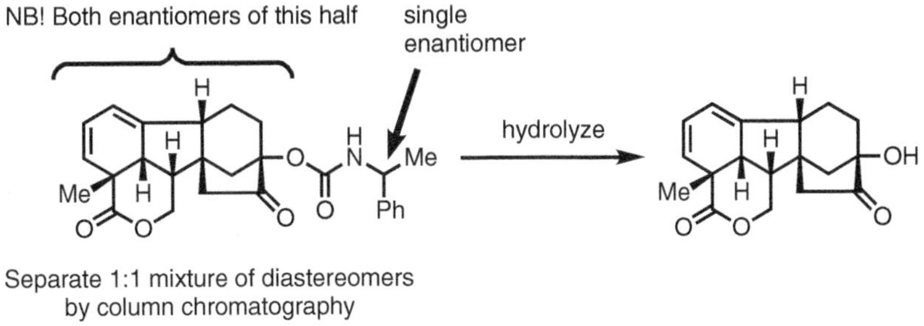

Scheme 2.13 *Separation of diastereomeric derivatives.*

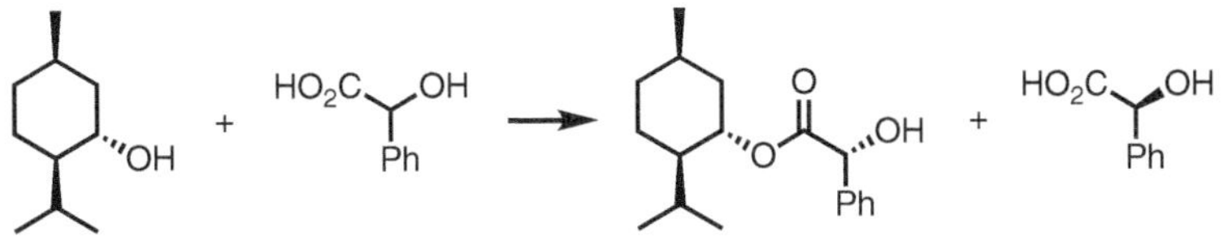

Scheme 2.14 *Kinetic resolution*

Asymmetric transformations, albeit still rarely predictable and thus of limited utility, could provide an interesting method for obtaining enantiopure compounds. Distinction must be made between *first-order* asymmetric transformations, in which the equilibrium is shifted to favor one enantiomer in solution, and *second-order* asymmetric transformations, in which one of the enantiomers crystallises from the solution and thus drives the equilibrium towards this side. Second-order asymmetric transformations have the unique property that the rate of crystal formation increases as the temperature is increased [39]. Second-order asymmetric transformations are also known as *dynamic kinetic resolution* (DKR), or dynamic kinetic asymmetric transformation (DYKAT) processes.

A specific example of second-order asymmetric transformations is provided by the transformations of phenylglycine derivatives [40]. Phenylglycines were converted in solution to their Schiff base derivatives followed by protonation (and concomitant hydrolysis of the imine) to precipitate the D-phenylglycinate (+)-tartrate salt. The optical purities approached 100% with practically all carbonyl compounds tested for the Schiff base formation. A similar process is used in the preparation of the following benzodiazepinone in nearly enantiopure form (Scheme 2.15) [41]. The benzodiazepinone is a useful intermediate for the synthesis of a number of cholecystokinin receptor ligands [42]. In this case, the crystallisation-induced second-order asymmetric transformation is brought about through the use of 3,5-dichlorobenzaldehyde and camphorsulfonic acid (CSA). The process affords over 90% yield of practically enantiopure material.

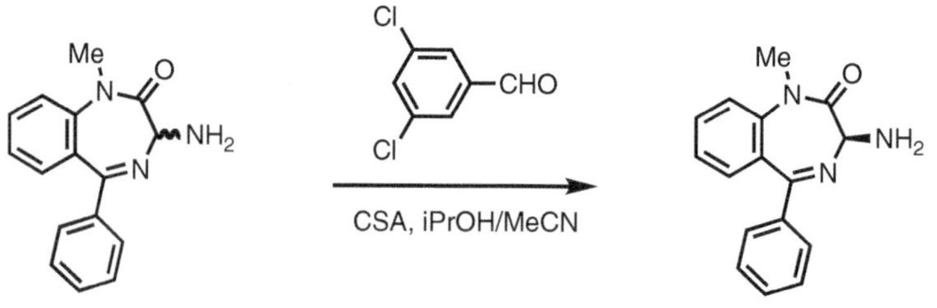

Scheme 2.15 *Second-order asymmetric transformation or dynamic kinetic resolution*

The underlying basis for second-order asymmetric transformations lies in the thermodynamics of the two interconverting species. One of the enantiomers must be removed from the equilibrium, thus giving the driving force to shift the equilibrium to that side. The interconversion rate must also be sufficiently fast so that equilibration is feasible. If the energy barrier of the path connecting the two (enantiomeric) compounds is low enough (60–75 kJmol^{-1} at room temperature), the two forms of the reactants and products readily interconvert. If the reactants and products are enantiomeric, their free energies are equal in an achiral environment, and no separation can be achieved. However, if the products are diastereomeric, with only one of the chiral centers being stereochemically labile, the ground state energies of the products will differ, and separation (enrichment) can be achieved. Scheme 2.16 presents an example where dynamic kinetic resolution has been achieved with high selectivity utilising a chiral catalyst to provide a chiral environment favoring the removal of one of the enantiomers from the equilibrium [43]. According to the Curtin–Hammett principle, the rapid equilibration between the enantiomeric ketones invokes a low activation barrier for this interconversion, and thereby the differences in the activation energies of the ketoesters with the Ru-BINAP catalyst will determine the reaction product distribution.

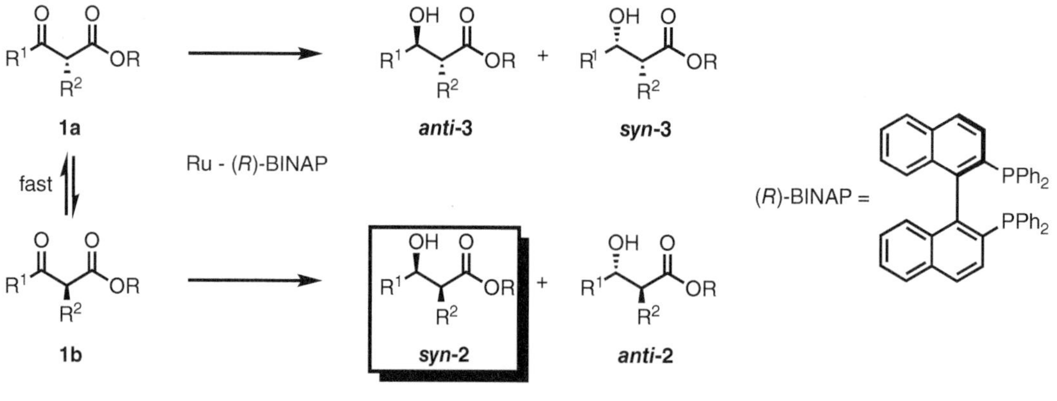

Scheme 2.16 *Dynamic kinetic resolution in catalysis*

In each case studied, the enantioselectivities were high (90–98% *ee*). With the (*R*)-BINAP catalyst, the *syn-2* products predominated when the R^2 group was an acylamino substituent. In the case of cyclopentan-1-one-2-carboxylate (R^1,R^2 = CH$_2$CH$_2$CH$_2$), the *anti-3* product was formed in 99:1 diastereomeric ratio with opposite absolute stereochemistry

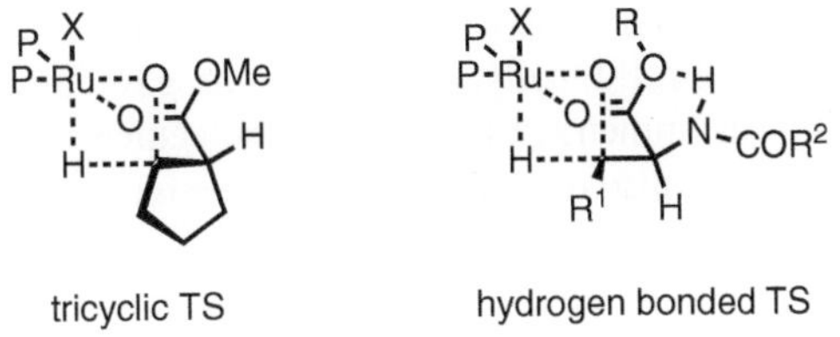

Figure 2.23 *Proposed transition states for hydrogenations of Scheme 2.16*

at the labile C2. These observations were rationalised as follows (Figure 2.23). The absolute stereochemistry at C3 is determined by the BINAP catalyst (*R* or *S*). The *anti* selectivity of the hydrogenation of the cycloalkyl substrate was rationalised in terms of a constrained tricyclic transition state model. In the case of the amide substrates, the amido NH is capable of participating in intramolecular hydrogen bond formation, which directs the delivery of the hydrogen with high *syn* selectivity. This rationalisation was further supported by the strong solvent dependence of the reduction: the diastereoselectivity dropped to 7:3 in changing the solvent from dichloromethane to methanol.

For second-order asymmetric transformations to be of practical utility, the reactants must be in rapid equilibrium, making the removal of one enantiomer the rate determining step. The structures of the starting materials must allow clear distinction between the two transition states in terms of the stabilities of the forming *syn* and *anti* products, and finally the catalyst must be strongly biased towards selecting one of the enantiomers of the starting materials.

Meso compounds potentially give access to chiral compounds simply if one has a way to break the internal mirror symmetry of the compound. This can be achieved by kinetic resolution in several cases, and the most common ones rely on enzymatic reactions (Scheme 2.17). Thus, one can for instance take the cyclic diester derived from a benzene 1,2-diol (microbial oxidation product of benzene), and achieve kinetic resolution by the action of porcine liver esterase (PLE). The optical yields are often quite high [44].

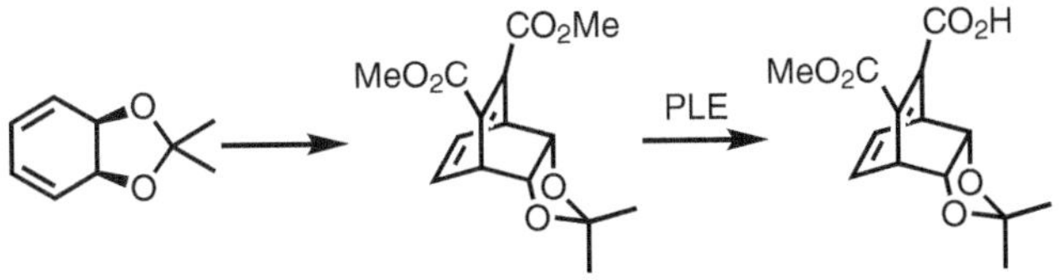

Scheme 2.17 *Enzymatic resolution*

A very useful example of kinetic resolution is based on the asymmetric epoxidation reaction developed by Sharpless (Scheme 2.18) [45]. In the kinetic resolution, a racemic allyl alcohol is converted to a mixture of the corresponding enantiomerically enhanced epoxy alcohol (glycidol) and the starting material. As we shall later see, this transformation is highly dictated by the reagent, and one can often achieve kinetic resolution with very high enantiomeric excess obtained for either product.

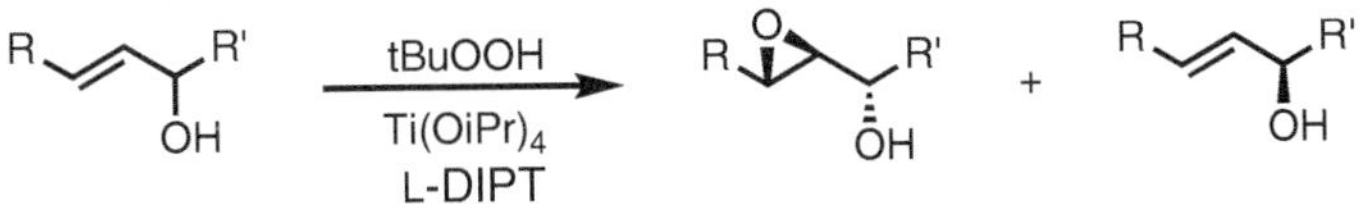

Scheme 2.18 *Kinetic resolution*

Asymmetric induction-based methods give the concluding category of transformation for obtaining optically pure products. The term "asymmetric induction" was initially introduced by (*Friedrich Gustav Carl*) *Emil Erlenmeyer* (1864–1921)[1] in 1914 in an intriguing paper, where he claimed to have induced asymmetry into benzaldehyde simply by heating it with tartaric acid [46]! Of course, today we understand why he was so convinced. Benzaldehyde was

---

[1] Son of Richard August Carl Emil Erlenmeyer or Emil Erlenmeyer (1825–1909), who formulated the Erlenmeyer rule, and whom we remember for the flask. The son is the inventor of the Erlenmeyer azlactone synthesis.

depicted as PhCHL.OL where L indicates an unoccupied position. The carbonyl carbon has four different substituents (Ph, H, L, and OL), and therefore must be chiral! Finally, Gustav Kortüm gave his definition of asymmetric induction in 1932 as "the action of a force exerted by asymmetric molecules on molecules capable of changing from a symmetrical into an asymmetrical configuration [47]." He also distinguished between intra- and intermolecular types of asymmetric induction.

Because of the tremendous developments in the level of asymmetric induction achieved over the past two or three decades, we shall place main emphasis on those processes which will give high levels of asymmetric induction (> 90% *ee*). We shall further divide the asymmetric processes in three major sub-classes: *internal*, *external*, and *relayed asymmetric induction* processes based on the origin of the chiral information as described in the following. The asymmetric information can originate from either the starting material or the reagent. Since the starting material can contain a chiral center as an integral part of the structure, or it can be temporarily introduced into it, we can conceptually think of three principal ways of introducing chirality in the emerging centers in the molecule.

One could think of a way, or route, where one uses chiral starting materials, whose intrinsic asymmetric information is transferred into the new chiral centers. In such cases we talk about *internal asymmetric induction*. The original stereocenter is thus an integral part of the final product, and this requires careful selection of starting materials in the planning stages of the synthesis. The utilisation of the chiral pool compounds [48] (amino acids, carbohydrates, and terpenes) as starting materials is a hallmark of this approach, whose extension to also include synthetic chiral materials is widely known as the Chiron approach [49]. The chiral center is bound to the reacting system irreversibly through a covalent bond, and remains connected throughout the sequence. The asymmetric induction in such cases heavily relies on diastereoselective processes, as we shall see later. The synthesis of (+)-vincamine from aspartic acid (Scheme 2.19) [50] utilised the chiral α-center of the starting amino acid as the sole source of chiral information in the construction of the three chiral centers in the target molecule. The α-carboxyl group, and thus also the original stereocenter, is lost later in the synthesis, but other atoms of the starting amino acid are incorporated in the final product as shown by the bold lines.

Scheme 2.19 *Internal asymmetric induction*

One can also think of a route where the chiral information is introduced into the molecule at a suitable stage before the chirality-forming reaction, then used as above, and the chiral originator then being removed (*relayed asymmetric induction*). One often speaks of chiral auxiliaries in such cases. An inherent drawback of such processes is that one usually needs two extra synthetic operations: one to introduce the chiral auxiliary, and another one to remove the chiral originator. Furthermore, in many cases the chiral auxiliary is completely lost in the following operations, including the chiral information, which is a wasteful process in terms of the economy of information.

The following scheme illustrates a synthesis of MeBmt, a non-proteinogenic amino acid in cyclosporine, utilising the chiral auxiliary technology developed by *David A. Evans* (1941–2022) of Harvard University (Scheme 2.20). It is noteworthy that in this synthesis the two oxazolidinone containing achiral auxiliaries are derived from D- and L-phenylalanine. All of the chiral information is thus derived from the chiral auxiliaries [51]. We shall return to this important class of chiral originators, and we will see that several non-destructive methods have been developed for the removal of the oxazolidinone moiety.

Finally, the third group of reactions involves processes where the chiral information is brought transiently into the reacting system, usually into the transition state, through reversible and/or weak bond formation, used to assemble the reacting partners appropriately, and finally released in its original form at the end of the reaction. Such *external asymmetric induction* processes are usually catalytic in the chiral moiety. External asymmetric induction thus corresponds to

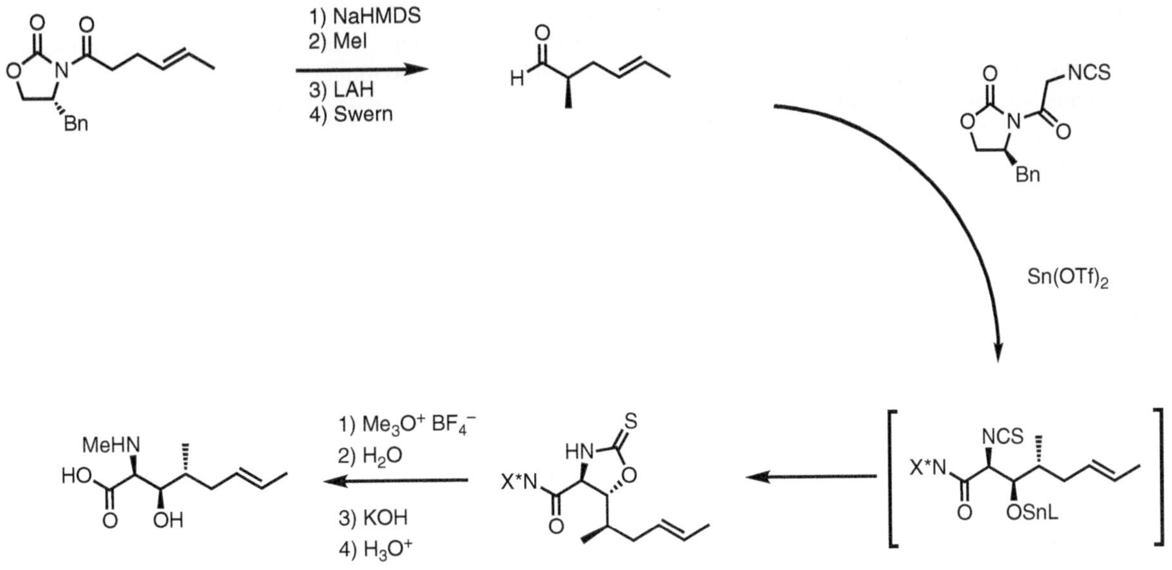

Scheme 2.20   *Relayed asymmetric induction*

Marckwald's classical definition of asymmetric synthesis. These processes are economically the most desirable ones, since we only need a catalytic quantity of the chiral originator, which is later recovered in its original form and thus reusable in other experiments. These are also the ideal goals for the development of asymmetric processes, and as we shall later see, many reactions have already succumbed to such catalyst development.

Takasago Chemicals (Japan) has developed the second major industrial menthol synthesis based on catalytic asymmetric synthesis. Whereas the H&R menthol synthesis (Scheme 2.12) represents very robust 'hard' chemistry, the Takasago synthesis relies on organometallic transformations (Scheme 2.21). Natural monoterpene myrcene is first hydroaminated under dissolving metal reduction conditions. The ensuing allylic amine is then isomerised under Rh-BINAP catalysis to give citronellal enamine in nearly complete enantioselectivity. Acid hydrolysis of the enamine gives (*R*)-citronellal, which undergoes a zinc-mediated *ene* reaction to give isopulegol. Final hydrogenation over a

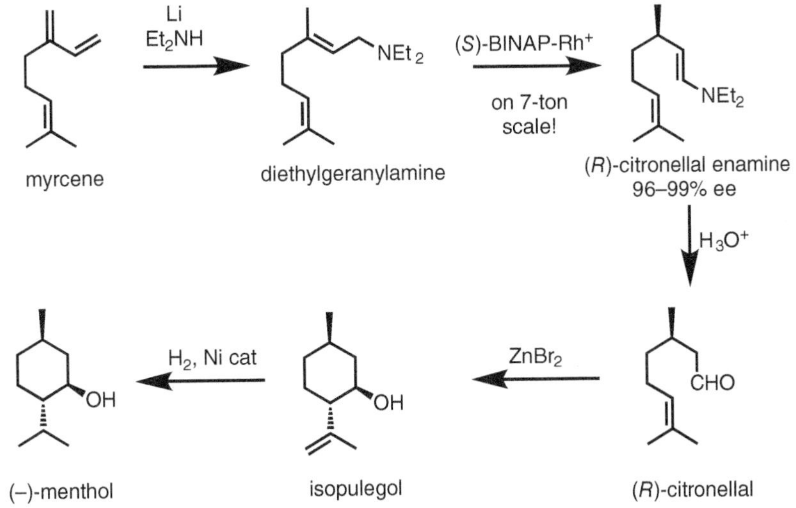

Scheme 2.21   *Takasago menthol synthesis*

nickel catalyst gives (−)-menthol. It is worth noting that the crucial stereochemistry inducing isomerisation step can be run on 7 ton scale, and the overall scheme consists of five chemical steps, four of which are metal-catalysed or mediated. Takasago production of (−)-menthol in 2009 was estimated at 1500 tonnes.

BASF launched the world's largest (−)-menthol production facility in 2012. The process starts with separation of neral from citral (which is a mixture of enal double bond isomers; neral and geranial) (Scheme 2.22). Neral is then hydrogenated over a chiral Rh-catalyst, and the synthesis is completed by cycloisomerisation and hydrogenation. Nearly simultaneously with the BASF announcement, both Symrise and Takasago retaliated by publicising that they will also double their production capacities.

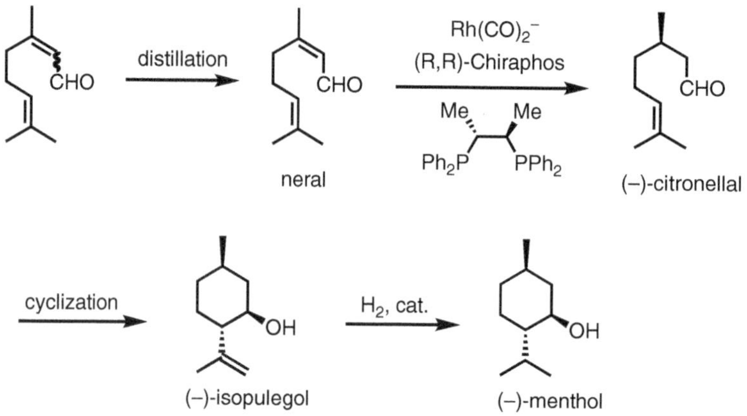

Scheme 2.22   BASF menthol synthesis

The following is a particularly intriguing example of using a catalyst for the enantioselective alkylation of aldehydes to produce optically active secondary alcohols (Scheme 2.23). We shall return to this process in more detail in Section 3.2.1.3, but it is noteworthy that this reaction proceeds with excellent enantioselectivity even when the chiral catalyst is only of 15% *ee* (i.e. 57.5:42.5 ratio of enantiomers) [52].

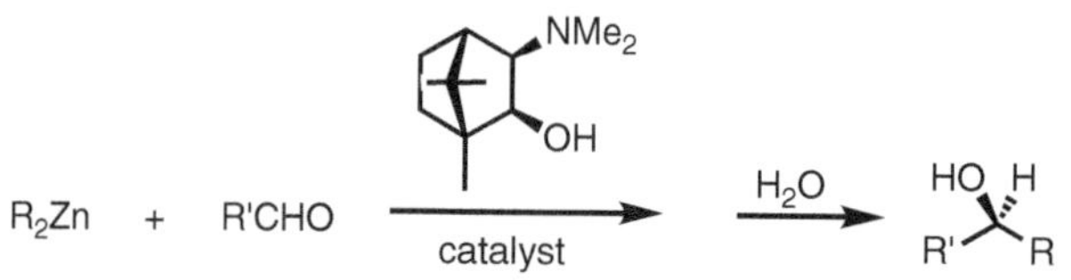

Scheme 2.23   External asymmetric induction

References

1. Moss, G.P. Basic Terminology of Stereochemistry, *Pure & Appl. Chem.* **1996**, *68*, 2193–2222.
2. Biot, J.B. *Mém. Acad. Sci.* **1838**, *15*, 93–279.
3. Horeau, A. *Tetrahedron Lett.* **1969**, 3121–3124.
4. Yamaguchi, S., Mosher, H.S. *J. Org. Chem.* **1973**, *38*, 1870–1877.
5. Gawley, R. *J. Org. Chem.* **2006**, *71*, 2411–2416.
6. a) Dale, J.A., Lull, D.L., Mosher, H.S. *J. Org. Chem.* **1969**, *34*, 2543–2549; b) Dale, J.A., Mosher, H.S. *J. Am. Chem. Soc.* **1973**, *95*, 512–519.
7. Pelšs, A., Kumpulainen, E.T.T., Koskinen, A.M.P. *J. Org. Chem.* **2009**, *74*, 7598–7601.
8. Huffman, J.W., Charles, J.T. *J. Am. Chem. Soc.* **1968**, *90*, 6486–6492.
9. Mislow, K., Raban, M. *Topics in Stereochemistry*, vol. *1*. (Allinger, N.L., Eliel, E.L., eds.), John Wiley & Sons: New York, **1967**, 1–38.

10. a) Hanson, K.R. *J. Am. Chem. Soc.* **1966**, *88*, 2731–2742; b) Hirschmann, H., Hanson, K.R. *Tetrahedron* **1974**, *30*, 3649–3956.
11. Hill, R.K., Chan, T.-H. *Tetrahedron* **1965**, *21*, 2015–2019.
12. Marckwald, W. *Ber. Dtsch. Chem. Ges.* **1904**, *37*, 349–354.
13. Whitesell, J.K., Felman, S.W. *J. Org. Chem.* **1980**, *45*, 755–756.
14. Shirai, R., Tanaka, M., Koga, K. *J. Am. Chem. Soc.* **1986**, *108*, 543–545.
15. Izawa, H., Shirai, R., Kawasaki, H., Kim, H., Koga, K. *Tetrahedron Lett.* **1989**, *30*, 7221–7224.
16. Rein, T., Pedersen, T.M. *Synthesis* **2002**, 579–594.
17. a) Hanessian S., Delorme, D., Beaudoin, S., Leblanc, Y. *J. Am. Chem. Soc.* **1984**, *106*, 5754–5756; b) Bennanni, Y.L.; Hanessian, S. *Chem. Rev.* **1997**, *97*, 3161–3195.
18. Whitesell, J.K., Allen, D.E. *J. Org. Chem.* **1985**, *50*, 3025–3026.
19. Partridge, J.J., Chadha, N.K., Uskokovic, M. *J. Am. Chem. Soc.* **1973**, *95*, 7171–7172.
20. Partridge, J.J., Chadha, N.K., Uskokovic, M. *J. Am. Chem. Soc.* **1973**, *95*, 532–540.
21. Brown, H.C., Jadhav, P.K. in *Asymmetric Synthesis* (Morrison, J.D., ed.) vol 2, Part A. Academic Press: New York, **1983**, pp. 1–43.
22. Cohen, N., Lopresti, R.J., Neukom, C., Saucy, G. *J. Org. Chem.* **1980**, *45*, 582–588.
23. Ihle, N.C., Correia, C.R.D., Wender, P.A. *J. Am. Chem. Soc.* **1988**, *110*, 5904–5906.
24. San Feliciano, A., Barrero, A.F., Medarde, M., de Corral, J.M.M., Aramburu, A., Perales, A., et al. *Tetrahedron Lett.* **1985**, *26*, 2369–2373.
25. Cahn, R.S., Ingold, C.K., Prelog, V. *Angew. Chem.* **1966**, *78*, 413–447.
26. Heathcock, C.H., Buse, C.T., Kleschick, W.A., Pirrung, M.C., Sohn, J.E., Lampe, J. *J. Org. Chem.* **1980**, *45*, 1066–1081.
27. Seebach, D., Prelog, V. *Angew. Chem., Int. Ed. Engl.* **1982**, *21*, 654–660.
28. a) Masamune, S., Ali, Sk. A., Anitman, D.L., Garvey, D.S. *Angew. Chem.* **1980**, *92*, 573–575; b) Masamune, S., Kaiho, T., Garvey, D.S. *J. Am. Chem. Soc.* **1982**, *104*, 5521–5523.
29. Piutti, A. *Compt. Rend. Hebd. Seances Acad. Sci.* **1886**, *103*, 134–137; *J. Chem. Soc. Abstr.* 1886, **50**, 1013.
30. Fabre, J.H. *Social Life in the Insect World*, T. Fisher Unwin, Ltd. **1911**.
31. a) Mori, K. *Chirality* **2011**, *23*, 449–462; b) Finefield, J.M., Sherman, D.H., Kreitman, M., Williams, R.M. Angew. *Chem. Int. Ed.* **2012**, *51*, 4802–4836.
32. Rasmussen, L.E.L., Greenwood, D.R. *Chem. Senses* **2003**, *28*, 433–446.
33. a) Parshall, G.W., Nugent, W.A. *CHEMTECH* **1988**, 184–190, 314–320, 376–383; b) Kagan, H.B. *Bull. Soc. Chim. Fr.* **1988**, 846–853; c) Scott, J.W. In *Topics in Stereochemistry*; vol *19*.; Eliel, E.L., Wilen, S.H., eds.; John Wiley & Sons: New York, **1990**; d) Crosby, J. *Tetrahedron* **1991**, *47*, 4789–4846; e) Farina, V., Reeves, J.T., Senanayake, C.H., Song, J.J. *Chem. Rev.* **2006**, *106*, 2734–2793; f) Carey, J.S., Laffan, D., Thomson, C., Williams, M.T. Org. *Biomol. Chem.* **2006**, *4*, 2337–2347.
34. Jacques, J., Collet, A., Wilen, S.H. *Enantiomers, Racemates and Resolution*, John Wiley & Sons: New York, **1981**.
35. Hopp, R., Lawrence, B.M. In *Mint – The Genus Mentha* (Lawrence, B.M. ed.) CRC Press: Boca Raton, FL, **2007**, 371–397.
36. a) Pirkle, W.H., Finn, J. In *Asymmetric Synthesis* (Morrison, J.D., ed.) vol *1*. Academic Press: New York, **1983**, 87–123; b) Allenmark, S.G. *Chromatographic Enantioseparation: Methods and Applications* John Wiley & Sons: New York, **1988**.
37. Corey, E.J., Narisada, M., Hiraoka, T., Elllison, R.A. *J. Am. Chem. Soc.* **1970**, *92*, 396–397.
38. Marckwald, W., McKenzie, A. *Ber. dtsch. Chem. Ges.* **1899**, *32*, 2130–2136.
39. Turner, E.E., Harris, M.M. *Quart. Rev.* **1947**, *1*, 299–330.
40. Clark, J.C., Phillips, G.H., Steer, M.R. *J. Chem. Soc., Perkin I*, **1976**, 475–481.
41. Reider, P.J., Davis, P., Hughes, D.L., Grabowski, E.J.J. *J. Org. Chem.* **1987**, *52*, 955–957.
42. Bock, M.G., DiPardo, R.M., Evans, B.E., Rittle, K.E., Whitter, W.L., Veber, D.F., et al. *J. Med. Chem.* **1989**, *32*, 13–16.
43. a) Noyori, R., Ikeda, T., Ohkuma, T., Widhalm, M., Kitamura, M., Takaya, H., et al. *J. Am. Chem. Soc.* **1989**, *111*, 9134–9135; b) Kitamura, M., Ohkuma, T., Tokunaga, M., Noyori, R. *Tetrahedron: Asymmetry* **1990**, *1*, 1–4.
44. a) Roberts, S.M. *Chem. Ind.* **1988**, *384*; b) Pratt, A.J. *Chem. Brit.* **1989**, 282–286.
45. a) Finn, M.G., Sharpless, K.B. In *Asymmetric Synthesis* (Morrison, J. D., ed.) Academic Press: New York, **1985**, vol. *5*, chapter 8; b) Woodward, S.S., Finn, M.G., Sharpless, K.B. *J. Am. Chem. Soc.* **1991**, *113*, 106–113; c) Pfenninger, A. *Synthesis* **1986**, 89–116.
46. Erlenmeyer, E., Landsberger, F., Hilgendorff, G. *Biochem. Z.* **1914**, *64*, 382–392.
47. Ebert, L., Kortum, G. *Ber. dtsch. Chem. Ges.* **1931**, *64*, 342–358.
48. a) Coppola, G.M., Schuster, H.F. *Asymmetric Synthesis: Construction of Chiral Molecules Using Amino Acids*, John Wiley & Sons: New York, **1987**; b) Ho, T.-L. *Enantioselective Synthesis: Natural Products from Chiral Terpenes*, John Wiley & Sons: New York, **1992**; c) Morrison, J.D. (ed.) *Asymmetric Synthesis*, vol. *5*, Academic Press: New York, **1985**.

49. Hanessian, S. *Total Synthesis of Natural Products: The Chiron Approach*, Pergamon Press: Oxford, **1986**.
50. Gmeiner, P., Feldman, P.L., Chu-Moyer, M.Y., Rapoport, H. *J. Org. Chem.* **1990**, *55*, 3068–3074.
51. Evans, D.A., Weber, A.E. *J. Am. Chem. Soc.* **1986**, *106*, 6757–6761.
52. Kitamura, M., Okada, S., Suga, S., Noyori, R. *J. Am. Chem. Soc.* **1989**, *111*, 4028–4036.

3

Asymmetric Synthesis

It is proposed that asymmetric syntheses be divided into enantioselective and diastereoselective syntheses.

Y. Izumi, 1971

In this chapter, we shall examine the asymmetric reactions that give access to functionalised chiral molecules utilising the principles introduced in Chapter 2. The field of asymmetric synthesis is developing rapidly, and a comprehensive survey of all the known reactions would be impossible. We therefore restrict our study to those representative reactions, which give a good idea of where we currently stand, and where the future research might take us. We shall study only two functional groups; the carbonyl group and the olefinic double bonds, as these are by far the best studied, and therefore also the most widely used functionalities in novel bond constructions.

To better understand the underlying reasons for stereoselective processes, we must appreciate the effects of even rather small energy differences. As we have already seen (Table 2.1), a difference between two alternative activation energies of only about 5 kJmol^{-1} is needed to bring about a 95:5 ratio, or 90% *ee* in the products. It is also important to be aware of the fact that the relationship between energy and ratio is logarithmic. This directly implies that prediction of large selectivities requires less accuracy!

We must also understand the origins of the subtle differences in the energy contents of the reacting species. Conformational analysis is the tool that gives us valuable insight, but we should remember that we are really interested in transition-state geometries, not those of ground states, because "although one conformation of a molecule is more stable than other possible conformations, this does not mean that the molecule is compelled to react as if it were in this conformation or that it is rigidly fixed in any way." [1] Direct observation of transition states is, of course, impossible, fortunately computational methods have advanced to the level that one can reliably model most of the reaction types using relative easily accessible computational facilities. [2]

Before commencing our studies on individual types of processes amenable to the creation of stereocenters, we shall review some aspects of conformational analysis central to understanding the rationalisations presented later.

3.1 Allylic Strain

It is well known that in cyclohexane the substituents tend to adopt equatorial position in preference over axial orientation (Scheme 3.1). This bias can be rationalised as arising from steric gauche-type repulsions with the axial protons located two C–C bonds away. Each substituent has a different tendency towards equatorial bias, and simple considerations of the various 1,3-interactions usually lead to qualitatively correct analyses of the cyclohexane conformations.

Asymmetric Synthesis of Natural Products, Third Edition. Ari M.P. Koskinen.
© 2023 John Wiley & Sons Ltd. Published 2023 by John Wiley & Sons Ltd.

Scheme 3.1   *Axial substituents experience gauche repulsions*

The situation in cyclohexenes is slightly different, especially when the olefinic carbon atoms are also substituted. Such systems were analysed by Francis Johnson, [3] and he introduced the concept of *allylic strain*, which has found widespread utility, not only in the chemistry of cyclohexene but also in rationalising acyclic stereocontrolled reactions. We shall briefly inspect the salient points of allylic strain.

We shall begin with the conformational behavior of allylically substituted cyclohexenes (Scheme 3.2). The energy barrier for ring inversion of cyclohexenes is *ca.* 22 kJmol^{-1}, considerably lower than that of cyclohexanes (44 kJmol^{-1}). [4] 2-Cyclohexenol prefers the conformation with the hydroxyl group pseudoaxial by *ca.* 4 kJmol^{-1}. [5] The preference for pseudoaxial hydroxyl group is in contrast to the pseudoequatorial preference in the case of a methyl substituent (2 kJmol^{-1}). [6] Taken together, this means that in simple cyclohexene derivatives, the equilibration between the conformers is fast, and the Curtin–Hammett principle applies.

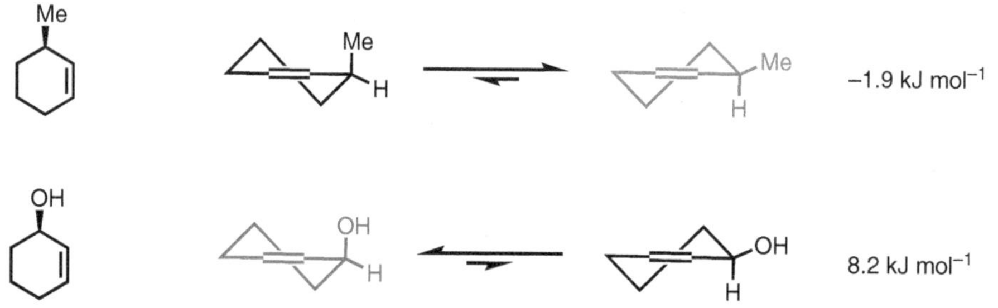

Scheme 3.2   *Conformations of substituted cyclohexenes*

When the olefin also carries a substituent, the energetics are affected accordingly. In 2,3-dialkylcyclohexenes, in which the alkyl groups are relatively large (Scheme 3.3), the dihedral angle between C_1–R$'$ and C_2–R is considerably smaller (roughly 35°) than the ideal value of 60°. This is manifested in steric interaction between the substituents R and R$'$ (conformer **B**) which forces the allylic substituent to adopt a pseudoaxial position (conformer **A**). This type of strain is called A^{1,2} strain to designate its allylic nature and more specifically that it arises from 1,2-interactions of substituents.

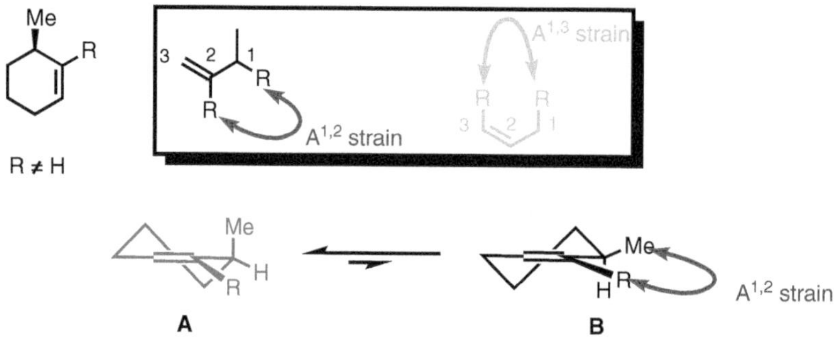

Scheme 3.3   A^{1,2}-*strain*

The facial selectivity (below or above the plane of the ring) on the attack of an electrophilic reagent (e.g. epoxidising agent) on the double bond is controlled by two factors: (i) the stereoelectronic effects would demand as much continuous

π-orbital overlap as possible in the transition state, favoring attack from the face opposite to the substituent R′; whereas (ii) the steric hindrance imposed by the substituent R' would force it to adopt a pseudoaxial orientation. In conformer **A**, the substituent R' causes little steric hindrance to the axial approach of the reagent from the lower face of the molecule. In conformer **B**, a large substituent R' effectively blocks the upper face (which would be favored by stereoelectronics). Attack from the lower face would necessitate the adoption of a boat-like conformation along the reaction coordinate, which implies a high energy of activation.

The second type of allylic strain, the $A^{1,3}$ strain, is concerned with alkyl substituents at the termini of an allyl system (Scheme 3.4). In methylenecyclohexane conformation **C**, the substituent R' lies nearly in the plane of the double bond (in cyclohexanone the dihedral angle O=C–C–H$_{eq}$ is only 4.3°), and can thus experience severe repulsion with a substituent R^2 at the opposite terminus of the double bond. Thus, in the case of large substituents R^2 and R', the conformational equilibrium would favor the conformation **D** with R' pseudoaxial.

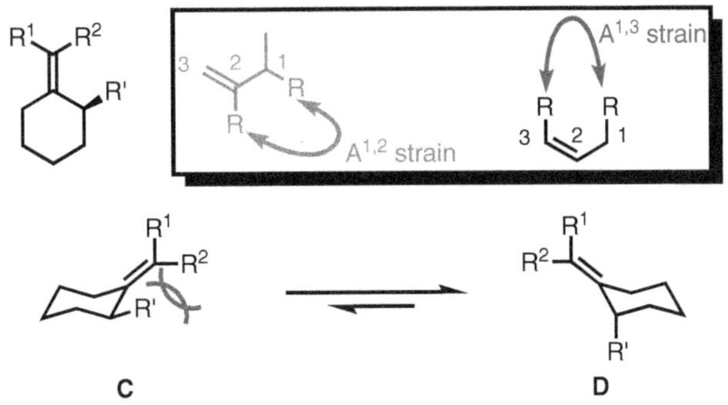

Scheme 3.4 $A^{1,3}$-strain

Allylic 1,3-strain has proven to be one of the most powerful tools in understanding acyclic stereocontrol. [7] Molecular modeling calculations show that the conformational bias in acyclic systems can be quite large indeed (Scheme 3.5). [8] In 3-methylbutene, rotation around the indicated allylic bond gives two possible minima (the staggered conformation is actually not an energy minimum), of which the one with the allylic hydrogen eclipsed is favored by 3.1 kJ/mol. Addition of a methyl group at the terminus of the alkene to form a Z alkene raises the energy of the methyl eclipsed conformation so much that it actually represents the maximum energy conformer at > 16 kJ/mol! The staggered conformation is, in this case, an energy minimum but 14.4 kJ/mol above the lowest energy conformer. Thus, for all practical purposes, the ground state is appropriately represented by the hydrogen-eclipsed conformation.

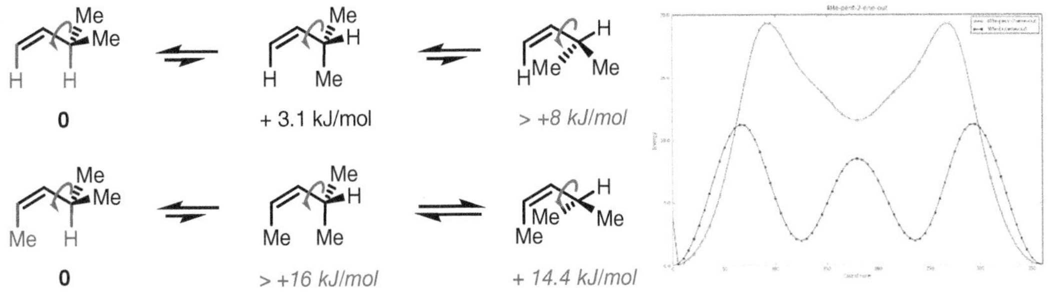

Scheme 3.5 *Conformations of butenes*

Although the energy differences are quite large in the pentene case, we should remember that these calculations represent the ground state conformations. Rotation of the dihedral angle is relatively free of penalty in energy up to *ca.* 30° rotation, and the associated small energy changes can be easily overcome.

When the reacting double bond (alkene or carbonyl) carries a substituent X in the allylic position, its reactions with electrophiles or nucleophiles will produce a new stereocenter and result in diastereomeric products. Electronegative and electropositive substituents X affect the reactivity of the double bond. The direction of electrophilic attack and nucleophilic attack are said to be under stereoelectronic control (Figures 3.1 and 3.2). [9] In order to facilitate the reaction by lowering the transition state energy, the substituent X must be aligned so as to maximise the overlap with either the HOMO (electrophilic attack) or LUMO (nucleophilic attack). This occurs best when the σ-bond to the substituent is parallel to the π-orbital of the double bond. The two alternative conformations can be distinguished by the developing allylic 1,3-strain, and generally good levels of asymmetric induction will be observed.

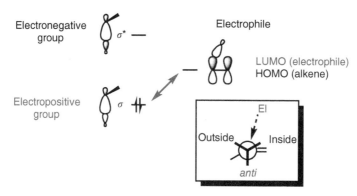

Figure 3.1   *Electrophilic attack*

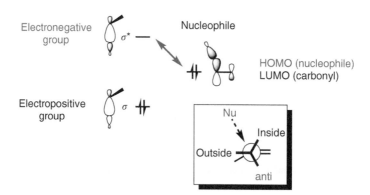

Figure 3.2   *Nucleophilic attack*

In the electrophilic attack (Figure 3.1), [9] the interaction of the LUMO of the electrophile with the HOMO of the alkene gives a transition state LUMO, which is stabilised by the arrangement of the proximal substituent in such a fashion that the most electropositive allylic group favors *anti* orientation to maximise donation from a high-lying σ orbital to the transition state LUMO.

In the nucleophilic attack (Figure 3.2), the most electronegative group favors *anti* conformation to maximise electron withdrawal from the reacting π system, and the most electropositive group (D) prefers outside position to minimise electron donation to the π system of the already electron-rich transition state. Electropositive groups prefer the outside or inside positions because the interaction of the σ_{C-D} orbital is destabilising. This destabilisation is maximal when the σ bond is *anti* to the carbonyl plane, and minimised when the σ lies in the plane of the carbonyl, which is best accommodated in the conformation where the electropositive substituent is outside.

3.2 Reactions of the Carbonyl Group

Carbonyl groups play a central role in synthesis due to their much developed chemistry. It is no surprise that asymmetric reactions of carbonyl compounds are perhaps the most important way of producing enantiopure compounds. Because of the electronegativity difference between carbon and oxygen, the carbonyl bond is intrinsically polarised and therefore reactive. The Lewis basicity of the oxygen atom allows for further polarisation of the carbonyl bond by complexation with Brønsted or Lewis acids, which opens the possibility for asymmetric induction via chiral catalysis. The multitude of reactions that carbonyl compounds undergo include direct attack at the carbonyl carbon, either via reduction (hydride addition), nucleophilic addition of alkyl, allyl and propargyl groups, etc. The α-center can be induced to function as a nucleophile, giving access to alkylation reactions and aldol-type addition reactions with a second carbonyl functionality. In α,β-unsaturated carbonyl compounds, the β-carbon is electrophilic, and 1,4- (Michael-) additions give rise to a further class of substituted carbonyl groups.

3.2.1 Nucleophilic Additions on the Carbonyl Carbon

The stereochemical aspects of the addition of a nucleophile onto a carbonyl group is one of the earliest examples of studies in stereoselectivity. In order to generate a (secondary) chiral center from a carbonyl carbon atom, one has, in principle, two alternative possibilities (Scheme 3.6). It would seem to be justified to expect that if reduction of a ketone with a suitable chiral hydride reducing agent gives access to one enantiomeric form of the product, C–C-bond formation through a chirally mediated delivery of the nucleophile R'$^-$ should give access to the other enantiomer. For most processes, this is exactly what is observed. [10]

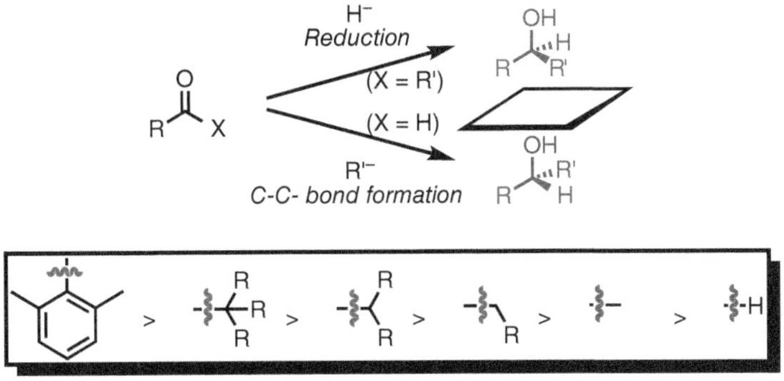

Scheme 3.6   Additions to carbonyl group

The difference between the sizes of R and X are the principal factors affecting the level of asymmetric induction; the larger the difference, the better the selectivity one would expect to obtain. Electronic factors do play a significant role, albeit their effects are not yet fully understood.

The addition of a hydride nucleophile onto the carbonyl carbon atom is a process whose geometrical features have been studied both experimentally and computationally. However, C–C-bond construction with either enantio- or diastereocontrol has lagged behind in development, save for cyclisation reactions and manipulations of carbocyclic compounds. It is only rather recently that important general progress has been made. We shall start our study of carbonyl group reactions by first inspecting some rationalisations on the general geometrical principles of addition reactions.

Starting from the presumption that the nucleophile adds along the plane containing the C=O bond and being at right angle to the plane of the carbonyl group (the normal plane), one can define an approach vector to define the steric trajectory the incoming nucleophile would follow. [11] Both qualitative considerations, [12] model calculations, [13] and X-ray crystallographic studies [14] led to the formulation that the nucleophile follows a trajectory along the normal plane, making an angle with the C=O line near the tetrahedral angle in magnitude (approximately 107°, however, subtle changes in the steric environment can lead to variations of this angle). [15] The ingenious use of crystallographic data enabled *Hans-Beat Bürgi* (1942–) and *Jack Dunitz* (1923–2021) to propose an estimate for this angle, which is known as the *Bürgi–Dunitz angle* (α_{BD}, Figure 3.3). [14]

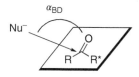

Figure 3.3   *Bürgi–Dunitz angle,* α_{BD}

The observed angle can be rationalised also in terms of frontier molecular orbital (FMO) theory. [16] The HOMO–HOMO interaction is destabilising since both orbitals are occupied. Destabilisation is maximised when the attack angle is acute, and correspondingly, destabilisation is minimised with an obtuse approach angle. The stabilising interaction between the LUMO of the electrophilic component (carbonyl acceptor) with the HOMO of the nucleophile (Nu⁻) is also maximised with an obtuse angle, since the overlap integrals between the HOMO and the *p*-orbitals on the carbonyl LUMO are of opposite sign (Figure 3.4). Attack angles on carbonyl groups are smaller than those on alkenes and additional polarisation of the carbonyl by coordination with a Lewis acid further reduces the magnitude of the angle.

Figure 3.4   *Nucleophilic attack*

If the group R in Scheme 3.6 is chiral, the reaction will proceed through diastereomeric transition states, and two products will be formed in unequal amounts (Scheme 3.7). The rationalisation for the effect of the neighboring stereocenter was first put forth by *Donald Cram* (1919–2001) of UCLA, [17] who suggested that nucleophilic attack on the asymmetric carbonyl compound takes place from the side of the smallest group attached to the asymmetric carbon atom. This model does not correspond to either the ground-state or the transition-state structure, and several alternative models have been advanced.

Scheme 3.7   *Nucleophilic attack on chiral carbonyl compounds produces diastereomers*

The different models are shown in Figure 3.5, grouped according to the level where they explain the diastereoselectivity. The original Cram model was purely a mnemonic model which explained the observations. Karabatsos was the first one to consider the conformations of the starting material, and this led to an improvement in the explanation. [18] *Hugh Felkin* (1922–2001) proposed in 1968, a model which took into account also the torsional effects in the transition state, [19] but it was only the so-called Felkin–Anh model, introduced in 1976, which first considered the approach angle

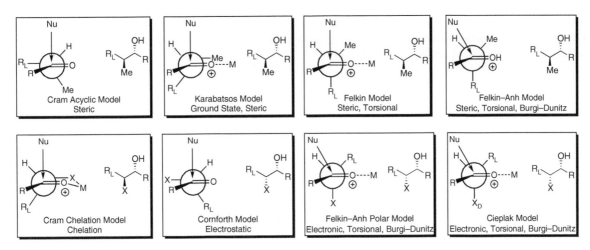

Figure 3.5 *Explanations for diastereoselectivity. X = OR, SR, NR₂, heteroatom substituent*

of the nucleophile (the Bürgi–Dunitz angle). [20] All these models explain the observed stereoselectivity correctly for cases where the stereocenter next to the reacting carbonyl carries electroneutral substituents.

For cases where the substituent is electronegative, the Cram rule does not predict the stereochemistry reliably. Approximately at the same time, Cram proposed his chelation-controlled model [21] and *John Warcup Cornforth* (1917–2013) his model, which is based on considerations of dipoles. [22] Again, both models are based on ground-state considerations, and as Barton had cautioned, [1] may not represent the true situation. The first *ab initio* calculations of Anh and Eisenstein provided the first theoretical evidence that the ground-state and transition-state arguments predict different results, and this resulted in the Felkin–Anh polar model, which correctly predicts the stereochemical outcome.[13a] Further refinements, mainly by more advanced computational methods, have been incorporated in the models by *Andrzej Cieplak*. [23] The addition reactions of enolates onto heteroatom-substituted aldehydes are important reactions generating e.g. polyketides, and interest in understanding the diastereoselectivity in detail remains of high priority. [24]

The steric interactions of an α-chiral aldehyde are not unlike the allylic strain effects. In the ground state conformation, the side chain tends to be oriented in such a manner that the largest group eclipses the carbonyl group. Solvation also affects the conformational equilibrium, and a generalisation for reactions where the transition state is electronegative can be put together (Figure 3.6). [25] In approaching the transition state, rotation of the C=O–C$_\alpha$ bond gives the arrangement with the large group perpendicular with the carbonyl group axis to maximise the σ^* stabilisation of the developing new bond. In case of electronegative substituent, the electronegative substituent prefers to be aligned *anti* to the C=O bond. Attack of the nucleophile occurs from the less hindered side, opposite to the large substituent. In this arrangement, the antibonding orbital σ^* of the C$_\alpha$–R$_L$ bond is *anti* to the forming C–Nu bond. The end result is the same as that described by all the models. The original (mainly correct) interpretation was given by Donald Cram, and the later investigations have given more detailed insights into the model. Therefore, it is appropriate to still refer to the Cram model.

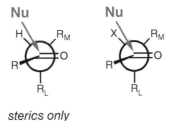

sterics only

Figure 3.6 *Cieplak generalisation; X = OR, SR, NR₂, heteroatom substituent*

In connection with the studies just reviewed, Cherest *et al.* also observed that lithium aluminum hydride reduction of a series of α-chiral dialkyl ketones gave increasing diastereoselectivities on increasing the size of the alkyl group on the side opposite to the chiral center (Table 3.1). [19]

Table 3.1 *Diastereoselectivity in reduction depends on steric effects*

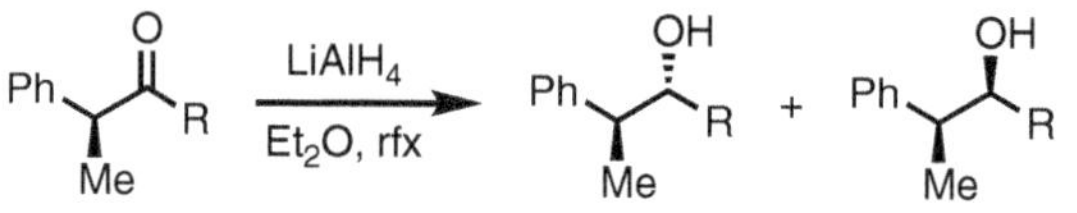

R	A	B
Me	3	1
Et	3	1
i-Pr	5	1
t-Bu	50	1

Clayton H. Heathcock (1936 –, University of California berkeley) and coworkers have formulated an explanation on the above and related observations. [26] If the two alkyl substituents on the carbonyl carbon are of similar size, the incoming nucleophile will follow the approach along the Bürgi–Dunitz angle. However, as the size difference between the two substituents increases, the incoming nucleophile will try to avoid the larger one, and the approach vector will thus be tilted away from the normal plane. This is called the Flippin–Lodge angle (α_{FL}, Figure 3.7), [27] which is related to the difference in steric bulk on the two sides of the normal plane. A similar explanation was also provided by *Samuel Danishefsky* (1936–) in his studies on Lewis acid–mediated cycloadditions of silyloxydienes to chiral aldehydes. [28]

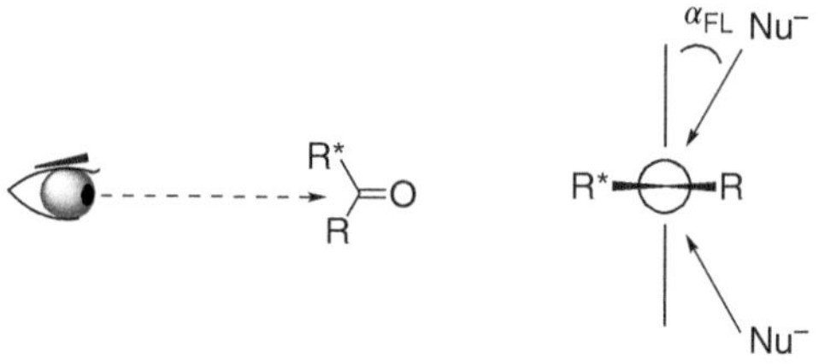

Figure 3.7   *Flippin–Lodge angle, α_{FL}*

In the reduction of the above series of ketones, one should expect that the larger the R group is, the further the Flippin–Lodge angle will be tilted towards the chiral center, and therefore the diastereoselectivity should also increase. The observed results are in full accordance with this rationalisation.

3.2.1.1   *Asymmetric reduction of carbonyl compounds*

The early developments of diastereoselective reductions of carbonyl groups were discussed in Chapter 1.6. The two faces of the plane of an unsymmetrically substituted carbonyl compound R–CO–R' are enantiotopic, and thereby the reaction of such compounds with a chiral reducing agent gives rise to asymmetric induction. The development of efficient chiral versions of the common hydride reducing agents based on boron and aluminum has occupied a pronounced position in the method development over the past few decades. [29]

The first claim to asymmetric reduction of a ketone with a chirally modified lithium aluminum hydride is based on simple alcohol modifiers. [30] Reduction of aliphatic ketones with a reagent prepared by mixing lithium aluminum hydride and camphor (which is presumably reduced to mostly isoborneol) was claimed to give optically active alcohols.

Nearly two decades later the observation could be reproduced, however, giving only 2% *ee*. [31] The strategy based on monohydroxy modifiers is hampered by facile disproportionation of the chiral alkoxyaluminumhydride species. Efforts to overcome this problem have led to significant developments of the reagents. Two of the most successful strategies are based on i) 1,2-diol modifiers; and ii) chiral 1,3-amino alcohol modifiers.

The first successful diol modifiers were based on monosaccharide-derived diols by *Austin R. Tatchell* (Figure 3.8). [32] It was found that one of the two active hydrogens in the aluminum hydride complex reacted much faster than the other one, resulting in *S* stereochemistry in the product alcohol. Treatment of the reagent with one equivalent of ethanol led to inversion of the observed sense of asymmetric induction to give predominantly the *R*-alcohol. This can be explained by preferential removal of the more reactive hydrogen H$_a$ by the alkoxide and the formation of a trialkoxy aluminum hydride complex.

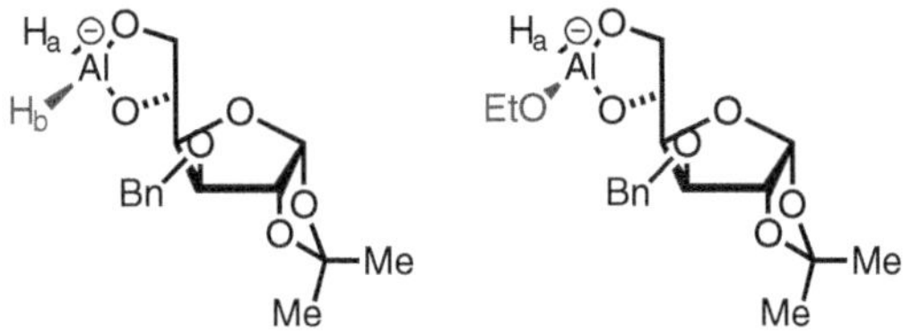

Figure 3.8   *Tatchell glucose auxiliaries*

Utilisation of aminoalcohols as chiral modifiers led to efficient asymmetric induction soon after the diol-base strategy, and especially the methods employing (2*S*,3*R*)-4-dimethylamino-3-methyl-1,2-diphenyl-2-butanol (Darvon alcohol) as the chiral modifier. [33] Cohen has further developed the methodology based on Darvon reduction. [34] The reduction is highly enantioselective especially for acetylenic ketones (typically 60–90% *ee*). The application in the synthesis of asteriscanolide by Wender was already mentioned (Scheme 2.10), and another highly selective example was provided by Trost in the synthesis of sterepolide (Scheme 3.8). [35]

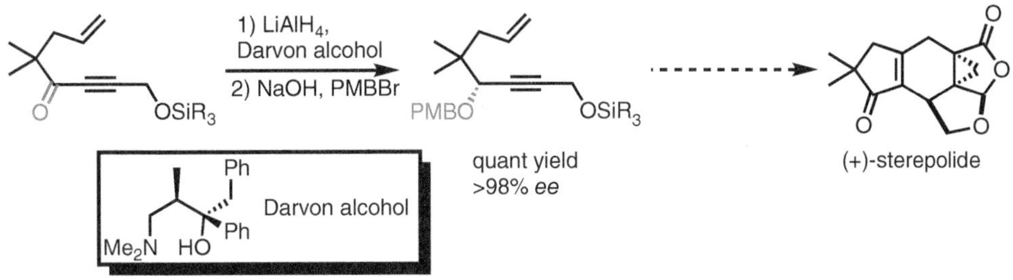

Scheme 3.8   *Synthesis of sterepolide*

By far, the best results with diol modifiers are obtained with the Noyori reagent, derived from axially chiral binaphthol, lithium aluminum hydride, and a simple alcohol (methanol or ethanol). [36] The reducing agent, 1,1′-bi-2,2′-naphthol–lithium aluminum hydride complex (BINAL-H, Figure 3.9), is capable of reducing α,β-unsaturated carbonyl compounds with exceptionally high enantioselectivities, very often with nearly complete enantioselectivity. The presence of the secondary modifier (methanol or ethanol) is crucial: in the reduction of acetophenone without the ethanol, only 2% *ee* was obtained, and utilisation of bulkier alcohols led to reversal of the sense of asymmetric induction.

The results were rationalised with the diastereomeric six-membered transition state structure shown in Figure 3.10 (top). The 1,3-diaxial interaction will favor the placement of the bulkier substituent equatorial and the less bulky one axial. The alternative model will be disfavored because of binaphthyl repulsion with the secondary alcohol modifier. For enones (Figure 3.10, bottom), a secondary repulsion arising from the interaction between a lone pair on the

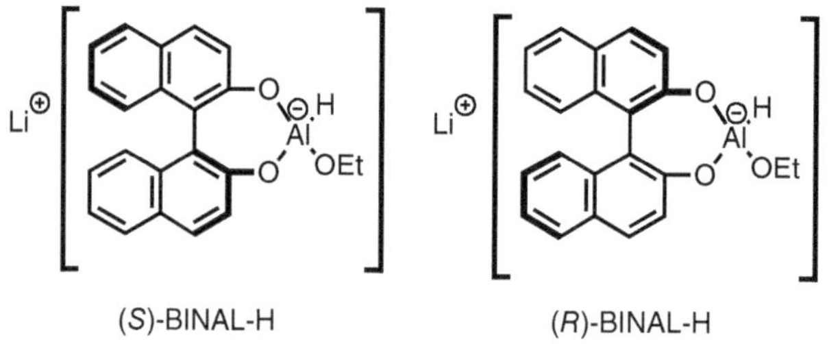

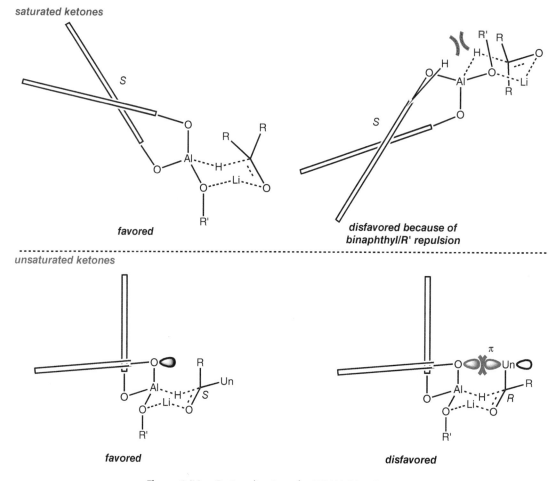

Figure 3.9   *Binaphthol-modified lithium aluminum hydride, BINAL-H*

Figure 3.10   *Rationalisations for BINAL-H reductions*

1,1′-bi-2,2′-naphthol (BINOL) and the π-orbitals of the electron-poor alkene are evoked to explain the high selectivity in enone and ynone reductions.

Alkyl acetylenic ketones were reduced in 84–94% *ee*, but increasing the size of the alkyl group has a deleterious effect on the enantiofacial differentiation. Thus, ethynyl isopropyl ketone gave only 57% *ee*. Enones were also reduced in high enantiomeric excess (see explanation in Figure 3.10). The dienone β-ionone gave nearly 100% *ee* upon reduction with (*S*)-BINAL-H. Reductions of the enones en route to prostaglandins with (*S*)-BINAL-H gave nearly quantitative asymmetric induction.[36c] Reduction of cyclopentenedione with the same reagent gave 4-hydroxy-2-cyclopenten-1-one in

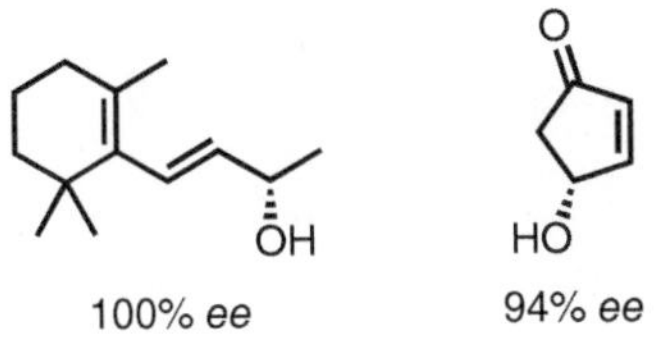

Figure 3.11   *BINAL-H reductions*

94% *ee* (Figure 3.11). [37] This compound is a key nucleus in the three-component synthesis of prostaglandins (see Chapter 8.2.1).

In an interesting application of chiral nucleophiles, a stannyl ketone was first reduced with (*S*)-BINAL in 93% *ee* (Scheme 3.9). After protection of the secondary carbinol as the methoxymethyl (MOM) ether, the stannane-derived lithio compound behaved as a chiral α-alkoxyalkyl anion. The final butanolide was obtained with undiminished enantiomeric purity. [38]

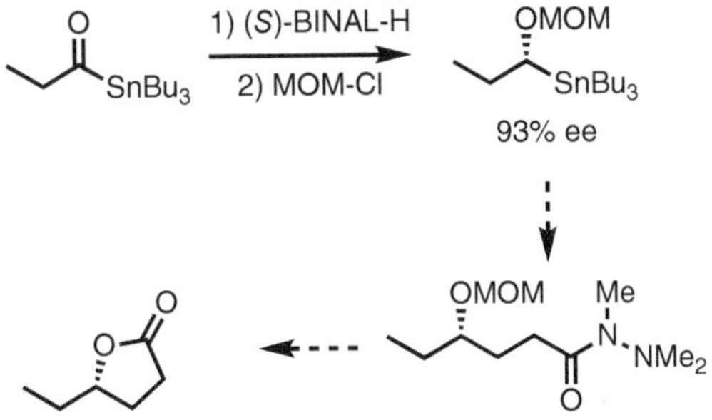

Scheme 3.9   *Synthesis of (S)-4-hexanolide*

A number of chiral terpene-derived borane reducing reagents have been developed, mainly by the groups of *Herbert C. Brown* (1912–2004) and *M. Mark Midland*. [39] The structures of the most widely used compounds, all derived from α-pinene, are shown in Figure 3.12. The proposed six-membered boat-like transition state for the reduction is also shown. [40] The large substituent R_L occupies the pseudoequatorial position to avoid steric interference with the R (methyl in alpine borane) group of the reagent. The B–C–C–H arrangement is nearly planar, which leads to enhancement of the reaction rate. [41]

Alpine borane (*B*-3-pinanyl-9-borabicyclo[3.3.1]nonane) reduces aldehydes, 1,2-dicarbonyl compounds and especially α,β-unsaturated ketones and acetylenic ketones rapidly and with high enantioselectivity. [42] A most remarkable distinction between the two faces of an alkynyl vinyl ketone was observed by Midland (Scheme 3.10). [43]

The triple bond provides a good functional handle for further elaborations (Scheme 3.11). It can be isomerised to the terminal position without affecting the chiral propargylic center with potassium 3-aminopropylamide (KAPA). [44] This strategy has been applied in the synthesis of a hexenolide in practically optically pure form. [45] The *R*-isomer, massoilactone, is the defense allomone of the formicine ant (*Camponotus*).

Reduction of alkynyl ketones is possible in the presence of other ketones, including aromatic ones, but not in the presence of aldehydes, as these are reduced much faster than the ynones. Without solvent, the reduction of α-keto esters and α-halo ketones proceeds with high enantioselectivity. [46] Aryl ketones can be reduced under high-pressure conditions (2000 atm). [43], [47] However, aliphatic ketones and enones give low enantioselectivities even under these conditions. Reduction of chiral ketones was also studied at high pressures, and of the ketones studied, carvone provided interesting results (Scheme 3.12). With *R*-Alpine borane, (*R*)-carvone was reduced in 4.6:1 ratio of diastereomers with the diequatorial isomer predominating. With the same reagent, (*S*)-carvone resisted reduction completely, even after 5 days at 6000 atm.

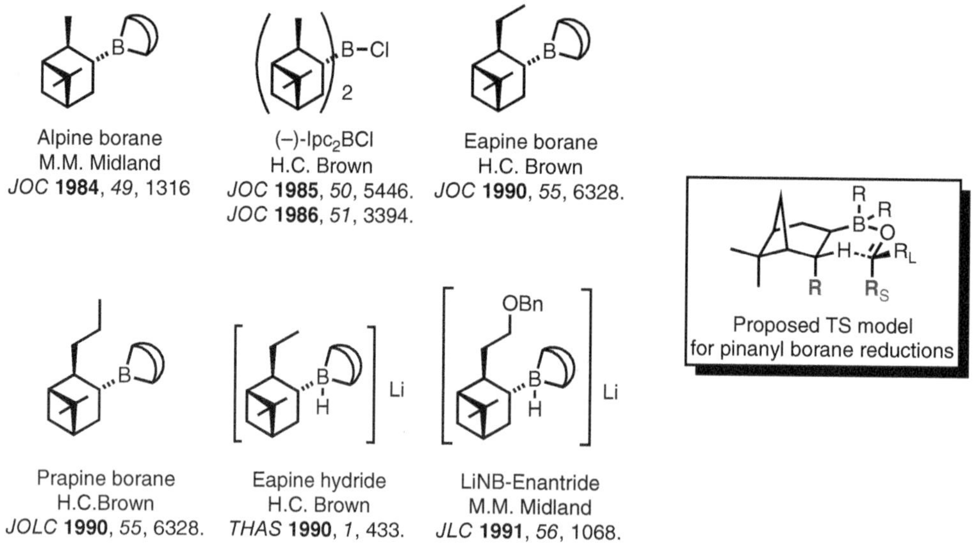

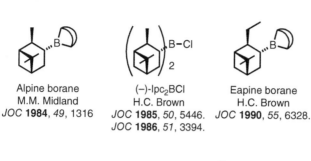

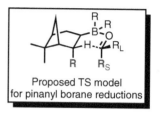

B
B−Cl
2
B
Alpine borane
M.M. Midland
JOC 1984, 49, 1316
(−)-Ipc2BCl
H.C. Brown
JOC 1985, 50, 5446.
JOC 1986, 51, 3394.
Eapine borane
H.C. Brown
JOC 1990, 55, 6328.
R
R
B
O
H
RL
R
RS
Proposed TS model
for pinanyl borane reductions
OBn
B
B
H
Li
B
H
Li
Prapine borane
H.C.Brown
JOLC 1990, 55, 6328.
Eapine hydride
H.C. Brown
THAS 1990, 1, 433.
LiNB-Enantride
M.M. Midland
JLC 1991, 56, 1068.

Figure 3.12 *Chiral borane reducing agents*

O
C8H17
CO2Et
Alpine borane
62%
OH
C8H17
98% ee
CO2Et

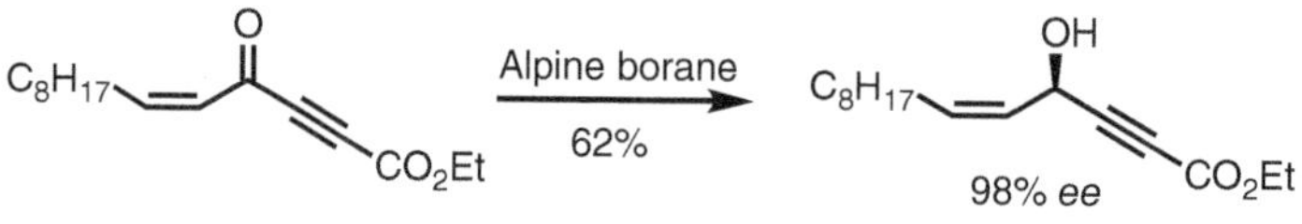

Scheme 3.10 *Alpine borane reduction*

O
C5H11n
Alpine borane
OH
C5H11n
NH2 NH−K+
KAPHA
O
O
C5H11n
massoilactone
1) n-BuLi
2) CO2
3) H2–Lindlar
OH
C5H11n

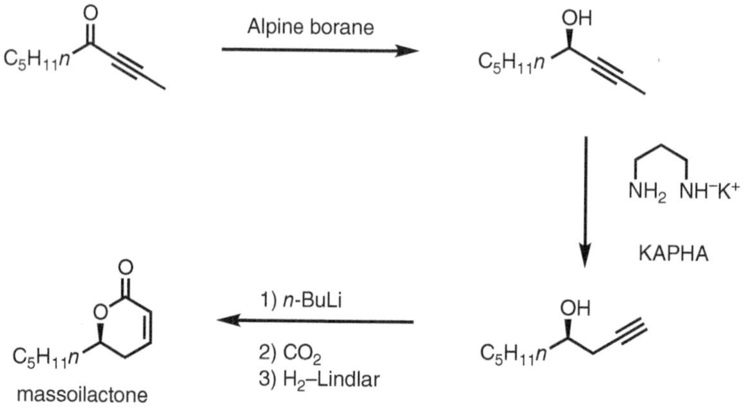

Scheme 3.11 *Synthesis of massoilactone*

O
BBN
6000 atm H2
> 90%
HO
HO
+
4.6
:
1

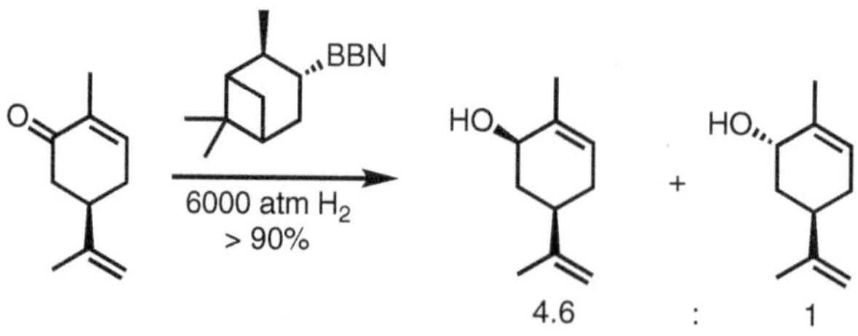

Scheme 3.12 *High-pressure Alpine borane reduction*

The low reactivity of Alpine borane towards ketones can be enhanced by increasing the Lewis acidity of the boron atom. This has been achieved through the development of chlorodiisopinocampheylborane (Ipc$_2$BCl). [48] Aromatic ketones and α-tertiary alkyl ketones can be reduced with high enantioselectivity, but simple aliphatic ketones still give poor results.

Satoru Masamune (1928–2003) of MIT developed the C$_2$-symmetric 2,5-dimethylborolane as a reducing agent capable of achieving the reduction of relatively symmetrical dialkyl ketones in high enantiomeric purity (Figure 3.13). [49] 2-Butanone and 2-octanone are reduced in *ca.* 80% *ee*, and larger difference between the two ketone substituents leads to practically complete facial discrimination by the reagent. The mechanism of the reduction is believed to be complex: The actual reacting species is formed from a dimer of the borolane and the corresponding 2,5-dimethylborolanyl methanesulfonate, which is formed during the preparation of the reagent. [50] Unfortunately, the preparation of the reagent (or more explicitly, its more stable dimeric precursor) is very tedious, involving multiple steps via air-sensitive intermediates at nearly every stage. Perhaps for that reason, the reagent has found very limited use in synthesis.

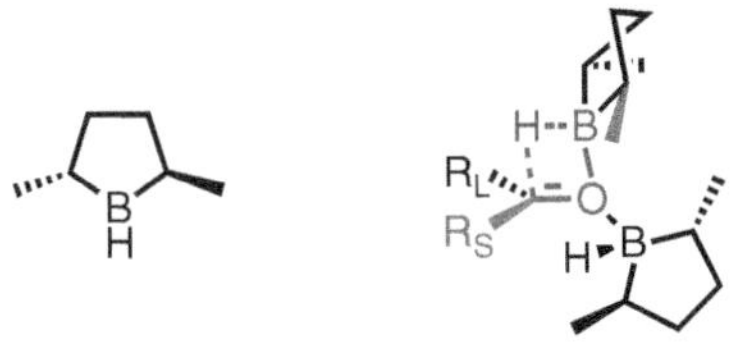

Figure 3.13 *Masamune borane and proposed model for stereoselectivity*

One of the singularly important developments in asymmetric reduction methods is based on Itsuno's initial observation that a reducing agent derived from (*S*)-2-amino-3-methyl-1,1-diphenylbutan-1-ol, itself derived from L-valine, and borane gives high enantioselectivity on reduction of ketones (95% *ee* for the reduction of acetophenone). [51] Corey et al. have extended this method and developed a more efficient reagent derived from proline (Figure 3.14). [52] They were able to isolate the active chiral reductant, the oxazaborolidine reagent. The higher selectivity observed with this so-called Corey–Bakshi–Shibata (CBS) reagent is presumably due to the higher bias exerted by the concave/convex discrimination of the bicyclic oxazaborolidine reagent.

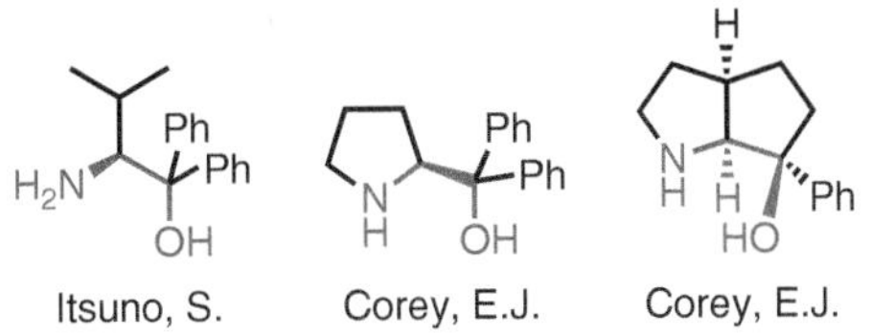

Figure 3.14 *Amino alcohol modifiers for borane*

The major breakthrough in the development of the Corey oxazaborolidine reagent came with the observation that the isolated oxazaborolidines function as efficient chiral catalysts in the reduction reactions. Usually only *ca.* 5–10 mol-% of the catalyst is needed to effect the reduction in high enantioselectivity (typically > 95% *ee*). Among the hydride donors, borane-tetrahydrofuran complex, borane-dimethyl sulfide complex (BMS), and catecholborane are the favored ones. The structure of the oxazaborolidine has been the subject of further development, and variations include changing the substituents (β-naphthyl instead of phenyl, [53] boron substituents, etc.). However, the original diphenylprolinol-derived compound is rather general and is a good choice as the starting point for more detailed optimisation studies. [54]

The B–H complex is air and moisture sensitive whereas the B-alkyl catalysts can be stored and handled in air. [55] This has obvious practical advantages, and the reagent is also more readily prepared than the hydrogen analogue. The observed enantioselectivities retained the high values.

Forskolin is a labdane diterpene isolated from the plant *Coleus forskohli*, a member of the mint family that grows natively in subtropical areas of India, Burma, and Thailand. Forskolin activates the enzyme adenylyl cyclase, which leads to raised levels of cyclic AMP (cAMP). It has been investigated for use in the treatment of allergies, respiratory problems, cardiovascular diseases, glaucoma, psoriasis, hypothyroidism, and weight loss. Recently, forskolin has made medical news as a natural remedy for urinary tract infections (UTI) by enhancing the ability of antibiotics to kill the bacteria that cause 90% of infections in the bladder.

In the synthesis of forskolin (Scheme 3.13), the CBS catalyst gave a 94% yield with 93% *ee* on reduction of the cyclohexenone. Further improvements of the route by modifying the reduction substrate and utilising catecholborane as the hydride source led to 98% *ee*. [56]

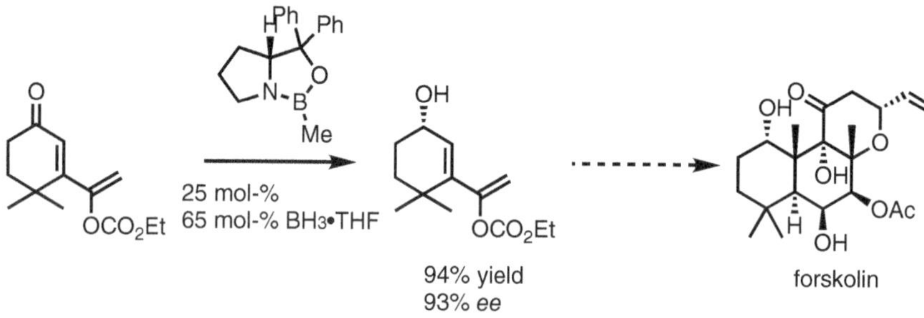

Scheme 3.13   Synthesis of forskolin

Fluoxetine is an antidepressant of the selective serotonin reuptake inhibitor (SSRI) class. In combination with olanzapine it is known as *symbyax*. Fluoxetine is approved for the treatment of major depression (including pediatric depression), obsessive-compulsive disorder (in both adult and pediatric populations), bulimia nervosa, panic disorder, and premenstrual dysphoric disorder. Another powerful synthetic example is given in Scheme 3.14, where the CBS reduction gives access to the correct stereochemistry of the antidepressant fluoxetine (Prozac®) with the oxazaborolidine-catalysed reduction of β-chloropropiophenone as a key step. [57]

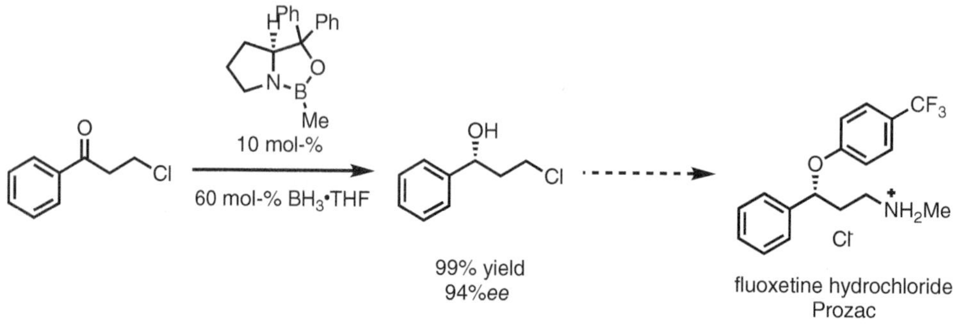

Scheme 3.14   Synthesis of fluoxetine

Whereas the B–H and B–Me catalysts work rapidly at 0 °C or room temperature but fail at lower temperatures, the B-butyl catalyst is capable of effecting the reduction at low temperatures. Especially with catecholborane as the stoichiometric hydride donor, the non-catalysed reaction is suppressed, and enones and α,α,α-trihalomethyl ketones undergo rapid reduction to give the corresponding alcohols in high enantiopurity. [58] In the case of the trifluoroacetyl mesitylene (Scheme 3.15), the reduction with the (S)-catalyst was observed to give 100% *ee* of the (R)-product. [59] The sense of enantioselectivity was inverted with the acetylmesitylene (99.7% *ee*). This result implies that the catalyst coordinates to the lone pair *anti* to the CF_3. The observed extremely high enantioselectivity cannot be simply explained by the mere steric size difference between the CF_3 and the mesityl groups, but substantial electronic repulsion between the electron-rich fluorine atoms and the negatively charged boron was evoked.

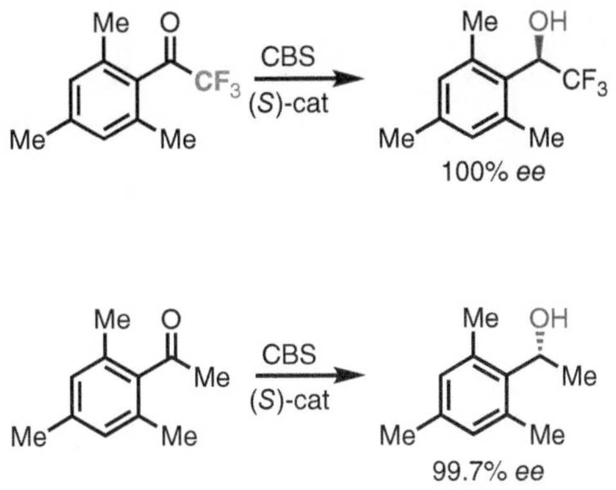

Scheme 3.15 Electronic effects in CBS reduction of acetophenones

A mechanistic rationale for the CBS reduction has been presented (Scheme 3.16), [54] and the catalysts have been modeled using quantum chemical calculations. [60] The reaction is described as involving a six-membered boat-like transition state with one molecule each of the catalyst, the borane, and the substrate. The borane and the carbonyl compound are complexed *cis* relative to each other on the oxazaborolidine, and hydrogen is transferred via a boat-form six-membered transition state.

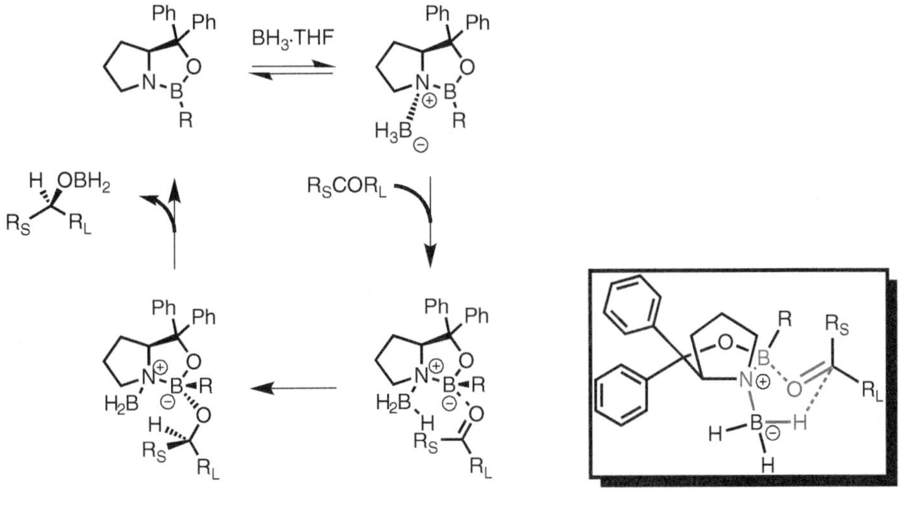

Scheme 3.16 Mechanistic rationalisation of the CBS reaction

BINAP (2,2'-bis(diphenylphosphino)-1,1'-binaphthyl) is a C_2-symmetric chiral bidentate diphosphine ligand widely used in asymmetric synthesis. When complexed with Ru, the complexes give high enantioselectivities in the reduction of a wide range of carbonyl compounds (Figure 3.15). [61] A number of BINAP-derivatives have been developed, and together with suitable chiral diamines, very high turnover numbers and turnover frequencies can be achieved. For example, hydrogenation of 3-(dimethylamino)propiophenone with an (*S*)-XylBINAP/(*S*)-DAIPEN ruthenium complex with an S/C ratio of 10 000/1 gave the corresponding *R* amino alcohol in 97.5 % *ee* (Scheme 3.17). [62]

3.2.1.2   1,3-Dioxygenated systems

The aldol reaction and its modifications provide a most powerful tool for setting up 3-hydroxycarbonyl systems usually with excellent stereocontrol. Stereospecific reduction of the carbonyl function would then provide a plausible access

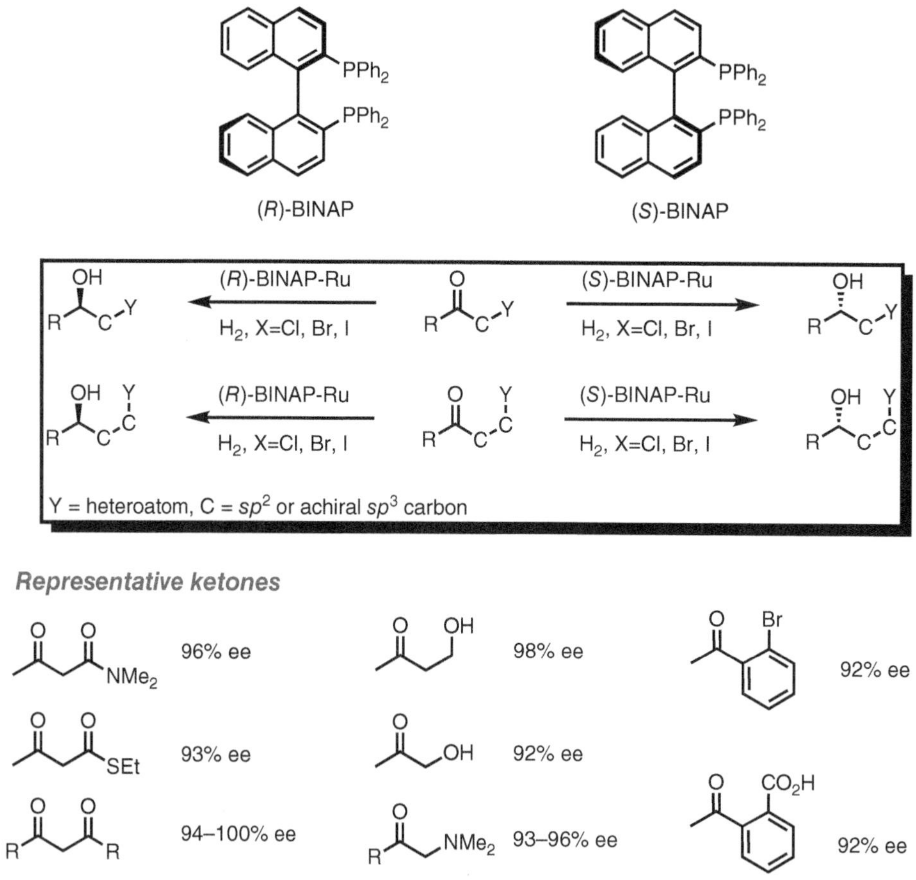

PPh2
PPh2
(R)-BINAP
(S)-BINAP
(R)-BINAP-Ru
H2, X=Cl, Br, I
(S)-BINAP-Ru
H2, X=Cl, Br, I
Y = heteroatom, C = sp2 or achiral sp3 carbon
Representative ketones
96% ee
98% ee
92% ee
93% ee
92% ee
94–100% ee
93–96% ee
92% ee

Figure 3.15 Ru-BINAP reductions

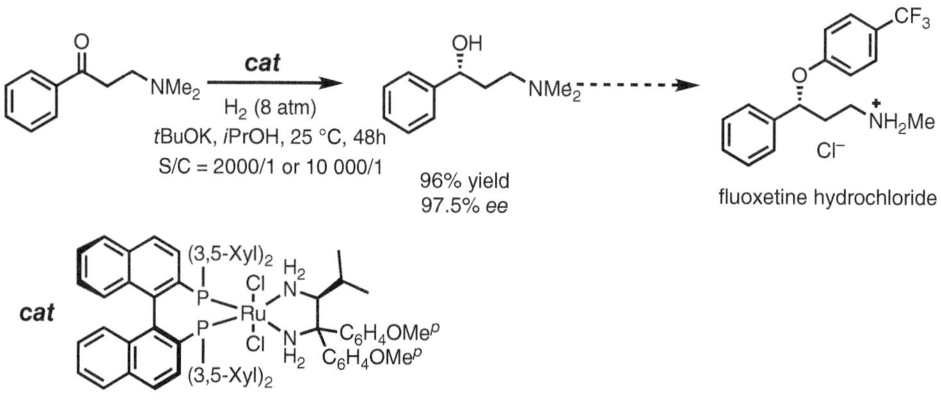

cat
H2 (8 atm)
tBuOK, iPrOH, 25 °C, 48h
S/C = 2000/1 or 10 000/1
96% yield
97.5% ee
fluoxetine hydrochloride
cat

Scheme 3.17 trans-[RuCl$_2${(S)-xylbinap}{(S)-daipen}] in the synthesis of fluoxetine

to *syn-* and *anti*-1,3-diols. [63] Methods have been devised for the utilisation of the existing chirality in directing the facial selectivity in hydride reductions.

In a study on the selective reduction of sugars, F. Smith had already observed in 1951, that ketoses may be reduced with sodium borohydride. [64] However, the diastereoselectivity was practically negligible. Some years later, *Olof Theander* (1924–) observed "Methyl α-3-oxo-glucoside (V), with an axial, bulky methoxyl group β to the oxo-group … gave a large proportion of the axial isomer on reduction with borohydride, as was expected" [65] (Scheme 3.18).

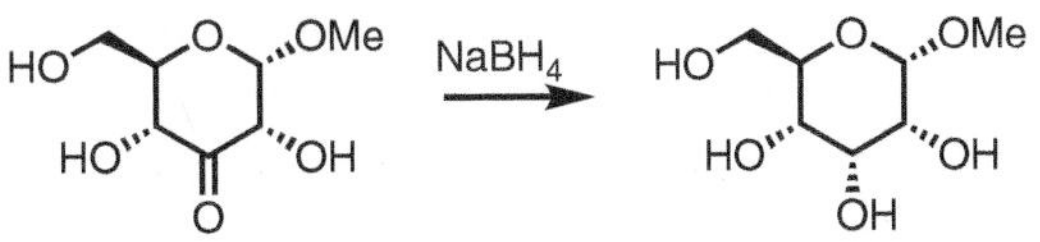

Scheme 3.18   *Early diastereoselective sodium borohydride reduction*

Koichi Narasaka realised in 1980 that the existing hydroxyl group in a 3-hydroxyketone can be used to direct the diastereoselectivity of the reduction through intramolecular coordination (Scheme 3.19). [66] Treatment of the hydroxyketone with a borane reagent led to the formation of a borinate whose chelation with the carbonyl oxygen gives a rigid intermediate. Hydride delivery is now under stereoelectronic control (axial hydride attack), and high diastereoselectivity is achieved. Chemists at Novartis optimised the method employing n-Bu$_2$BOMe as the chelating agent. [67] The hydride attacks on the six-membered chelate to give the more stable chair product according to the Fürst–Plattner rule. [68]

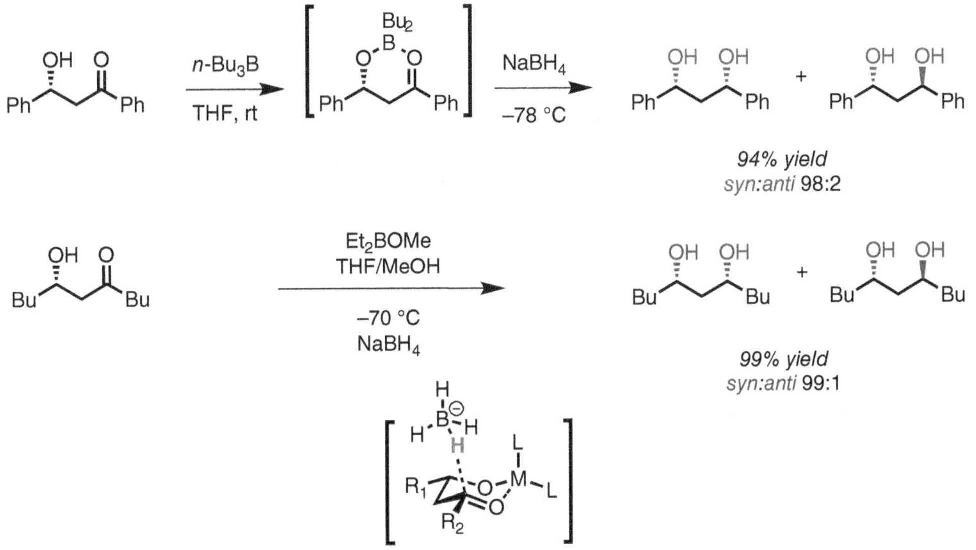

Scheme 3.19   *Narasaka–Prasad reduction*

Slight modification of the protocol allows diastereoselective reduction of 3-hydroxy ketones to anti 1,3-diols. Sodium triacetoxyborohydride, easily prepared from sodium borohydride and acetic acid, [69] functions also as a selective reducing agent. This possibility of modulating the reactivity of the borohydride reagent was realised by the Schering–Plough chemists in 1983. [70] In studying the diastereoselective reduction of veratrum alkaloids, *Anil Saksena* surmised that triacetoxyborohydride, being a weak reducing agent, is incapable of reducing the carbonyl function, whereas the derived alkoxydiacetoxyborohydride is more powerful as a reducing agent. Thus, the precomplexation is a precondition for the reduction, giving the desired distinction between intramolecular and intermolecular reductions, besides activating the reducing agent. Thus, sodium triacetoxyborohydride selectively coordinates with

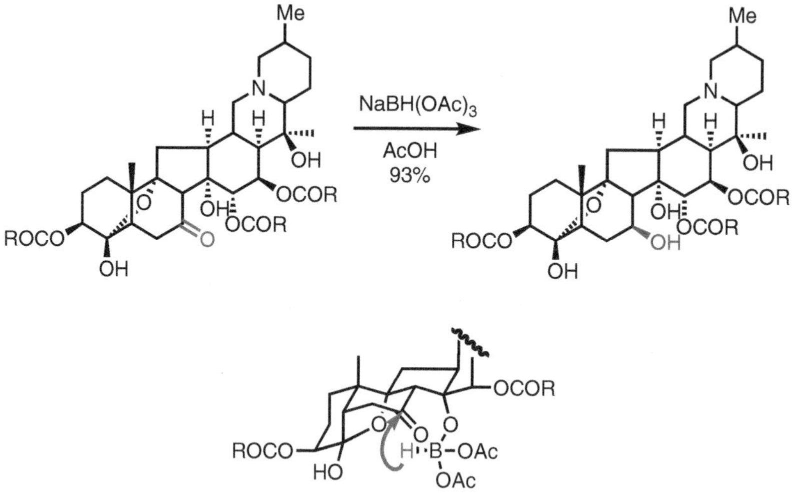

Scheme 3.20 *Saksena reduction*

a hydroxy group in the substrate (Scheme 3.20). A more reactive alkoxydiacetoxyborohydride is generated, and the intramolecular hydride delivery directs the formation of the equatorial alcohol.

This method has found numerous applications in the syntheses of several natural products and pharmacologically interesting compounds, as exemplified in Scheme 3.21 for a key step in the synthesis of a portion of milbemycin β_1. [71]

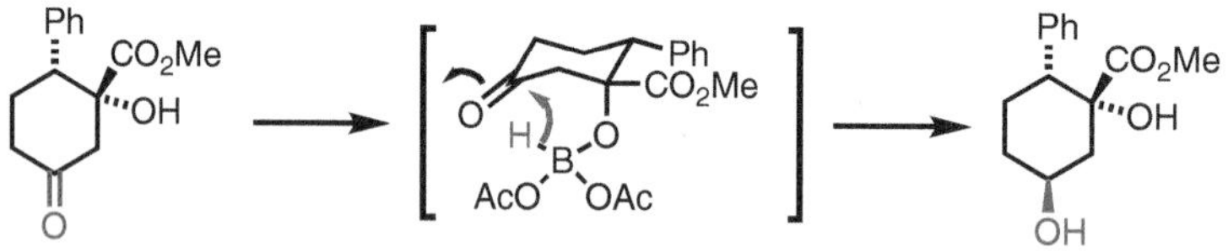

Scheme 3.21 *Diastereoselective synthesis of a fragment of milbemycin β_1*

Sodium triacetoxyborohydride suffers from limited solubility and sluggish reactions with hydroxy ketones. Evans has optimised the reagents and introduced tetramethylammonium triacetoxyborohydride as the favored reducing agent. [72] Scheme 3.22 displays the complementary diastereoselective reductions of acyclic 3-hydroxy ketones, the Saksena–Evans reduction giving rise to *anti* diols and the Narasaka–Prasad reduction to *syn* diols.

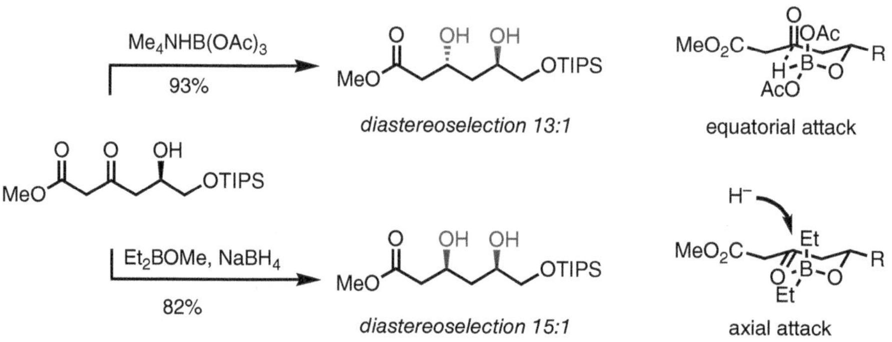

Scheme 3.22 *Complementary Saksena–Evans and Narasaka–Prasad reductions*

Cholesterol is mainly produced in the liver, where one of the first steps is the formation of mevalonic acid by reduction of 3-hydroxy-3-methylglutaryl-Coenzyme A to mevalonic acid by hydroxymethyl-CoA (HMG-CoA) reductase (see Section 9.1). Scientists at Sankyo Pharmaceuticals in Japan reasoned in early 1970s that inhibition of this enzyme would lead to inhibition of biosynthesis of cholesterol, a valuable target to lower blood cholesterol, and thereby aid prevention of cardiovascular disease. The first natural-product candidate identified was compactin (mevastatin), which was isolated from *Penicillium* strains independently by scientists at the Beecham Pharmaceuticals and Sankyo Laboratories. Shortly thereafter, Merck scientists isolated a second fungal metabolite, mevinolin, from *Aspergillus* strains. Again, the same compound was isolated by the Sankyo scientists from *Monascus ruber*, and they called it monacolin K, which was later renamed lovastatin. The development of compactin was terminated in 1980 due to safety concerns, but it served as a lead to the development of simvastatin by Merck & Co. Later investigations by Merck, Sharpe and Dohme Research Labs, and Parke–Davis Warner–Lambert Company (later Pfizer) scientists showed that the actual 'active' part of the molecule is the β-hydroxy-δ-lactone, and all commercially available statins contain this particular structural unit (Figure 3.16). [73] In the open-chain form of these lactones, the diol is in a 1,3-*syn*-configuration, and the Saksena reduction cannot be applied directly. Pfizer's atorvastatin eventually became the biggest blockbuster drug, which reached global sales in excess of 12 billion USD per annum.

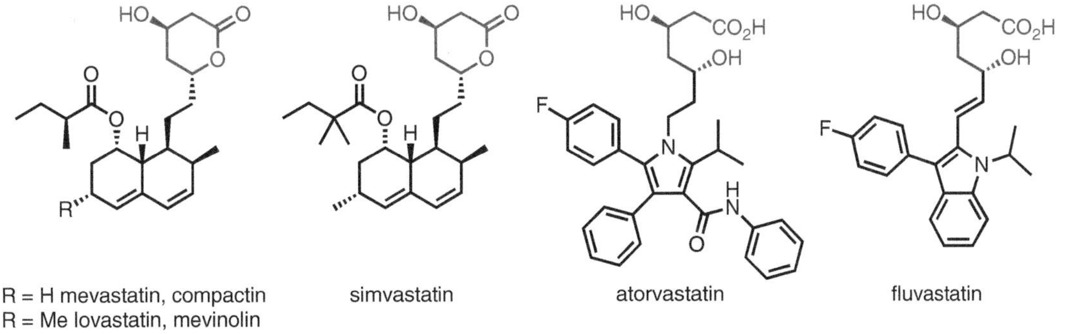

R = H mevastatin, compactin
R = Me lovastatin, mevinolin   simvastatin   atorvastatin   fluvastatin

Figure 3.16   *Structures of HMG-CoA inhibitors*

The development of cholesterol-lowering agents (hydroxymethyl CoA inhibitors), was a major focus area in the pharmaceutical chemistry in the 1980s, and the realisation of the correct stereochemistry of 1,3-diols was crucial. For *syn* selectivity in acyclic systems, the Narasaka reduction was implied. Precomplexation of the hydroxyketone is preferred with either alkoxydialkylborane [74] or lithium iodide. [75] Robust methods were needed for industrial production, and Scheme 3.23 shows one application from industrial laboratories. [76]

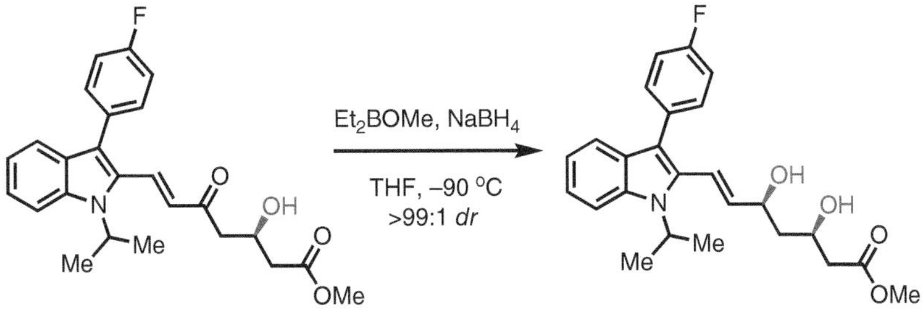

Scheme 3.23   *Modified Narasaka reduction in the synthesis of fluvastatin*

The Evans–Tishchenko reaction is a powerful method of converting a hydroxyketone stereo- and regioselectively to a protected 1,3-*anti*-diol (Scheme 3.24). [77] Treatment of the hydroxyketone with excess non-enolizable aldehyde and catalytic samarium diiode in THF gives the monoprotected 1,3-*anti*-diol usually in very high yield and selectivity. The

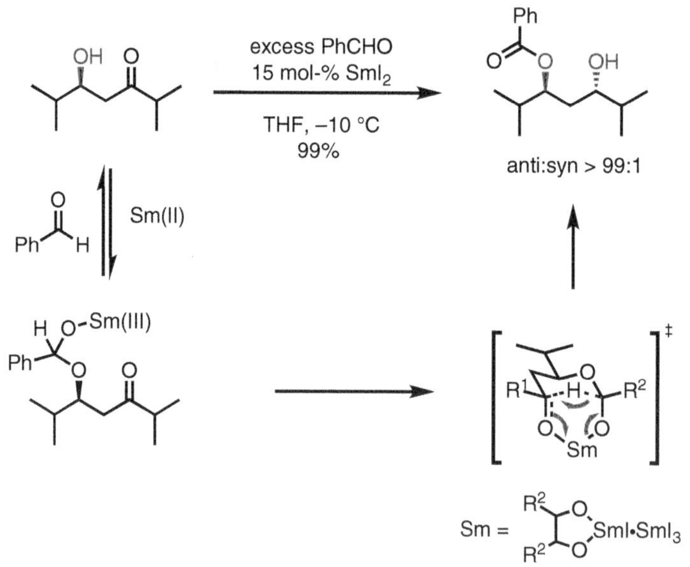

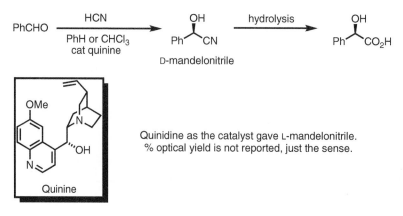

Scheme 3.24 *Evans–Tishchenko reduction*

mechanism for the Evans–Tishchenko reaction is similar to the aldol-Tishchenko reaction of two different aldehydes. The hydride-donor aldehyde forms a hemiacetal with the β-hydroxy ketone. The subsequent Tishchenko step proceeds through the metal-coordinated bicyclic [6,6] transition structure, leading to the 1,3-*anti*-diol structure with excellent stereoselectivity. The enantiopurity of the starting material is retained in the reaction. It has been proposed that the actual catalytic species is a complex of samarium triiodide and a pinacol species derived from the aldehyde via a radical coupling. [78] The Tishchenko reaction and its various modifications have been reviewed. [79]

3.2.1.3 *Asymmetric alkylation of the carbonyl group*

The earliest enantioselective carbon–carbon bond-forming reaction was actually of the type currently called organocatalytic. As you recall from earlier, after Emil Fischer's reports, the early development of new reactions involved modifications of known reactions with readily available chiral natural compounds. It is thus no surprise that *Georg Bredig* (1868–1944) attempted, in 1912, cyanohydrin formation of benzaldehyde catalysed with quinine and quinidine (Scheme 3.25). [80] The enantioselectivities were modest (only *ca.* 8% *ee*), but unlike the unfortunate

Scheme 3.25 *Bredig organocatalytic synthesis of mandelic acid*

experiments by Erlenmeyer, true catalytic asymmetric C–C bond formation was achieved. For reasons that have remained unclear, the field remained quite dormant for decades.

The next important success on chiral auxiliaries was achieved by Prelog in the synthesis of atrolactic acid from phenyl glyoxylate with methyl Grignard reagent. [30] This led to the formulation of the Prelog rule (Figure 1.28). In the same issue of the journal, the Prelog group also presented an intriguing case of using a steroidal directing group (Scheme 3.26). The enantiomeric excess was 69% *ee*, which must be considered extremely high in those days. [81]

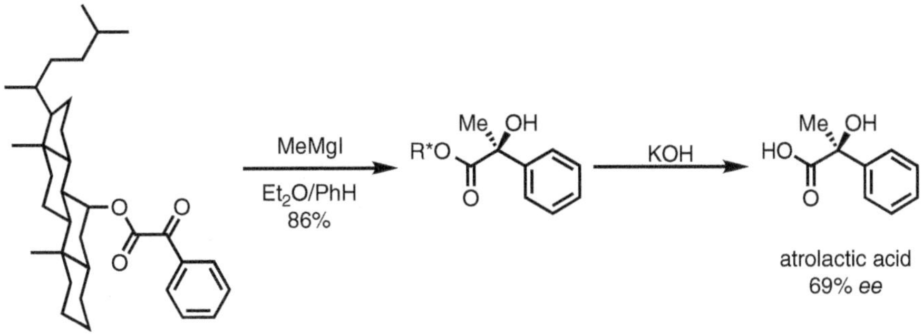

Scheme 3.26   *Prelog synthesis of atrolactic acid*

Some two decades later, Whitesell demonstrated the utility of 8-phenylmenthol as a powerful chiral auxiliary (Scheme 3.27). Grignard alkylations of the 8-phenylmenthyl esters of α-keto esters gave up to 98% *de*. [82]

Scheme 3.27   *Whitesell's use of phenylmenthol as chiral auxiliary*

As already discussed, addition of organometallic reagents to aldehydes provides a powerful tactic for the generation of secondary alcohols (Scheme 3.28). One can simultaneously construct a new chiral center and a new carbon–carbon bond. Although development of an asymmetric version of these reactions has met with limited success so far, the addition of unsaturated nucleophiles (most notably allyl and propargyl) has the added advantage of introducing a further

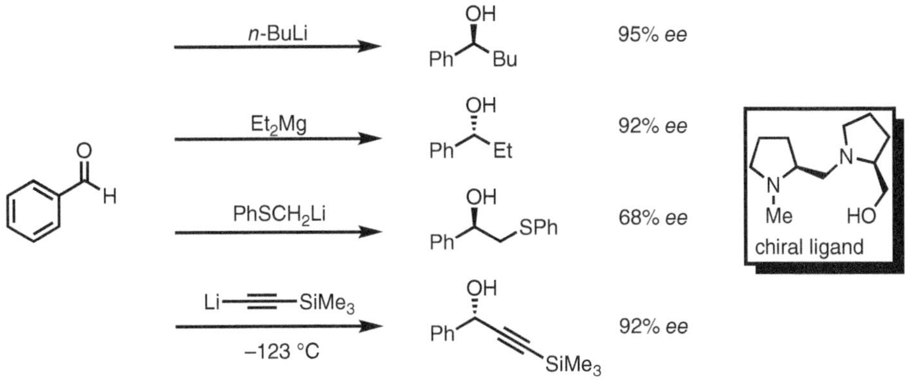

Scheme 3.28   *Mukaiyama proline catalyst for organolithium additions*

functional handle, which can provide access to more elaborate structures. Coordination of chiral diamine ligands with the organometallic species gives rise to chiral complexes capable of enantioface differentiation on the aldehyde targets. Both the nucleophilicity and basicity of the organometals are enhanced by coordination. [83] *Teruyaki Mukaiyama* (1927–2018) successfully coupled alkyl lithium [84] and Grignard reagents [85] with aromatic aldehydes in the presence of a proline derived diamine ligand to give access to enantiomerically enriched products. Although the ligands can usually be recovered and recycled, at least a stoichiometric amount of the ligand is needed because of the competing uncatalysed reactions, which produce racemic material.

Organozinc reagents are among the oldest known organometallic species, in fact diethylzinc was the first organometallic synthesised. [86] They react relatively sluggishly with aldehydes even at room temperature in nondonor solvents. This reactivity can be accelerated by coordination with Lewis bases such as amines. Using chiral amino alcohols provides for ligand-accelerated asymmetric catalysis. In Chapter 2, we have already seen the case of chirality multiplication in the alkylation reaction of benzaldehyde with dialkylzinc reagents in the presence of partially racemic camphor-derived amino alcohol. A number of chiral ligands have been devised for this purpose, including the ones shown in Figure 3.17 (enantioselectivities are given for the addition of diethylzinc to benzaldehyde). [87]

Corey, E.J.
Tetrahedron Lett. **1987**, *28*, 5233.
95% *ee* (*S*)

Soai, K.
J. Am. Chem. Soc. **1987**, *109*, 7111.
99% *ee* (*S*)

Tanaka, K.
Tetrahedron Lett. **1989**, *30*, 1700.
97% *ee* (*R*)

Brown, H.C.
Tetrahedron Lett. **1989**, *30*, 5551.
95% *ee* (*R*)

Ohno, M.
Tetrahedron Lett. **1989**, *30*, 7095.
98% *ee* (*S*)

(−)-DAIB
Noyori, R.
J. Am. Chem. Soc. **1989**, *111*, 4028.
99% *ee* (*S*)

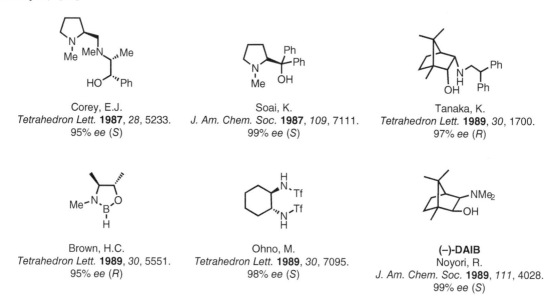

Figure 3.17 *Asymmetric ligands for organozinc reagents*

The Noyori (−)-3-*exo*-(dimethylamino)isoborneol (DAIB)-catalysed reaction has especially proven to be very powerful (Scheme 3.29). The aldehyde and the diethylzinc both coordinate on the same sterically less hindered face of the ligand, and the alkyl group is transferred to the *Si* face of the carbonyl group. A significant feature of this reaction is that it shows high levels of nonlinear amplification: even if the catalyst is only 15% *ee*, the product is formed in 95% *ee*!

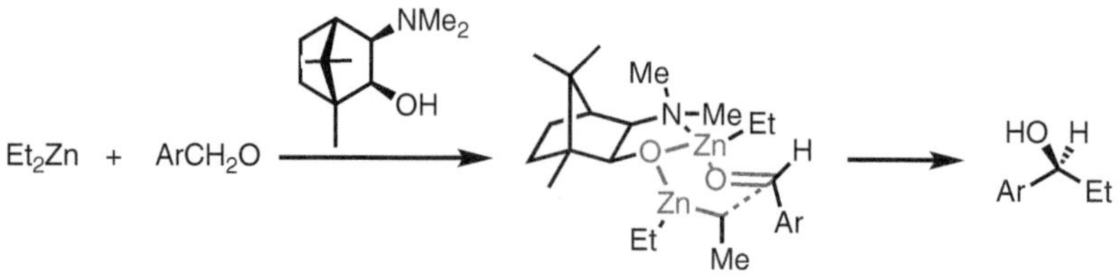

Scheme 3.29 *Noyori DAIB–catalysed alkylation of benzaldehydes*

This quite remarkable chirality amplification has its origins in the sterics-induced stability and instability of the dimeric complexes formed from the ligand and the organozinc reagent (Scheme 3.30). [88] The DAIB ligand accelerates the addition of the organozinc reagent to the carbonyl acceptor only as a monomeric species. However, the dimeric species are much more stable, and due to steric reasons, the heteromeric meso complex (i.e. from (+)- and (−)-DAIB) is the thermodynamically most stable one (its total dipole moment is smaller than that of the homochiral complexes). In solution, the chiral complex dissociates easier into the reacting monomeric species than the meso complex. Therefore, the minor enantiomer is captured in the stable meso complex, while the major enantiomer produces the more easily dissociable chiral dimer.

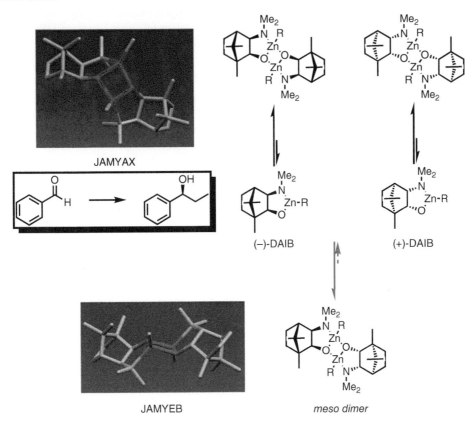

Scheme 3.30 *Chirality amplification*

Titanium complexes have also found wide applications as chiral catalysts for nucleophilic addition to carbonyl compounds (Figure 3.18). [89] These complexes can be readily prepared from inexpensive tartaric acid, which is commercially available in both enantiomeric forms. The aryl substituents are introduced by simple Grignard reaction so a broad

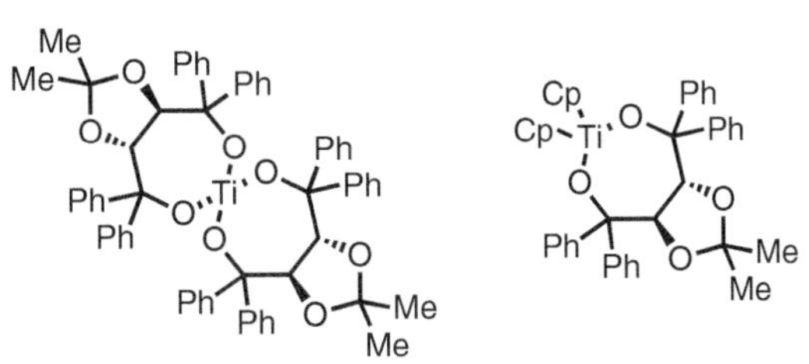

Figure 3.18 *Ti-TADDOL Lewis acids*

variety of TADDOL (α,α,α',α'-tetraaryl-2,2-disubstituted 1,3-dioxolane-4,5-dimethanol) ligands can be prepared. Several quite structurally varied catalyst systems have been developed, including the tartrate-derived seven-membered ring titanium complexes shown. These are efficient ligands for alkylzinc, lithium, and Grignard reagents giving enantioselectivities up to 98% *ee*.

The same TADDOL ligand has also been used for enantioselective Grignard alkylation with a wide variety of carbonyl substrates (Scheme 3.31). [90] Addition of the alkyl Grignard reagents to acetophenone occurs exclusively from the *Re* face. Aryl, heteroaryl, and α,β-unsaturated alkyl ketones give the highest enantioselectivities, whereas alkyl ketones give lower selectivities coupled with lower yields. The reaction is sensitive to the solvent: when the solvent is changed from THF to Et$_2$O, the selectivity is lost. The reaction mixture is heterogeneous, and the asymmetric induction may be mediated either by chiral aggregate formation [(metal-R)$_m$(metal-X*)$_n$], or by chiral Lewis acid (R*OMgBr) activation of the ketone substrate in the rate-determining step. However, the yields often remain modest.

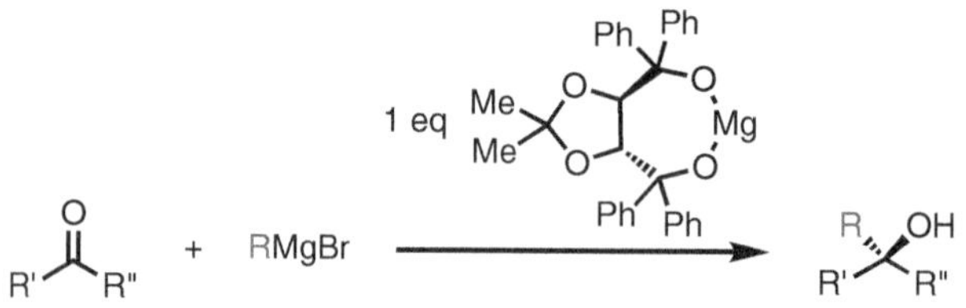

Scheme 3.31   TADDOL-mediated Grignard alkylation

Aryl ketones also react with aryl Grignard reagents in the presence of modified BINOL ligands (Scheme 3.32). [91] The yields and enantioselectivities are good to excellent, and the method benefits from the fact that the ligand is easily prepared and no other metals besides magnesium are required for the reaction. However, the substrate and reagent spaces are limited.

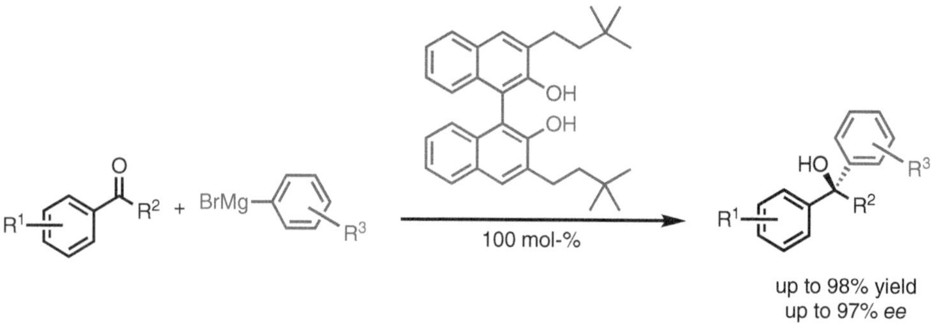

up to 98% yield
up to 97% *ee*

Scheme 3.32   Enantioselective Grignard additions with a BINOL-ligand

Through extensive ligand screening, Kavanagh and Gilheany were able to develop a 2,5-dimethylpyrrole-decorated diaminophenol ligand with improved enantioselectivities and broadened the scope of the methodology (Scheme 3.33). [92] High *ee*'s (mostly 87–95% *ee*) are obtained with aryl or heteroaryl ketones reacting with primary alkyl nucleophiles. The power of the methodology was showcased with the one-step synthesis of (*S*)-gossoronol in high yield and 93% *ee*. The fragrant diterpenoid gossonorol was first isolated in 1984 from cotton essential oil and has since then been extracted from other natural sources such as chamomile (*Chamomilla recutita*).

3.2.1.4   *Asymmetric allylation/propargylation of the carbonyl group*

Many of the above catalyst systems handle allyl and propargyl transfers quite satisfactorily, but some more specific systems have also been developed. These reactions bear a striking resemblance to the aldol reaction, and many of

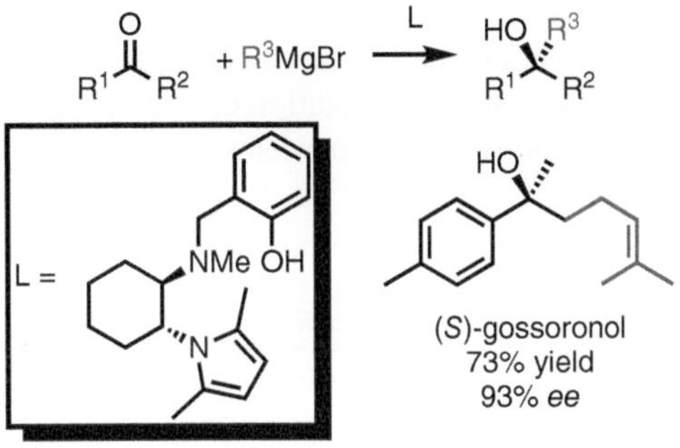
+ R3MgBr
L
L =
NMe OH
(S)-gossoronol
73% yield
93% ee

Scheme 3.33 Improved diaminophenol ligand for enantioselective Grignard

the generalisations presented in that context will hold true for the allylmetal reactions with aldehydes. In the allylations, the metal in the allyl metal species can vary tremendously, giving the chemist a possibility to fine-tune the nucleophilicity/basicity of the reagent (Scheme 3.34).

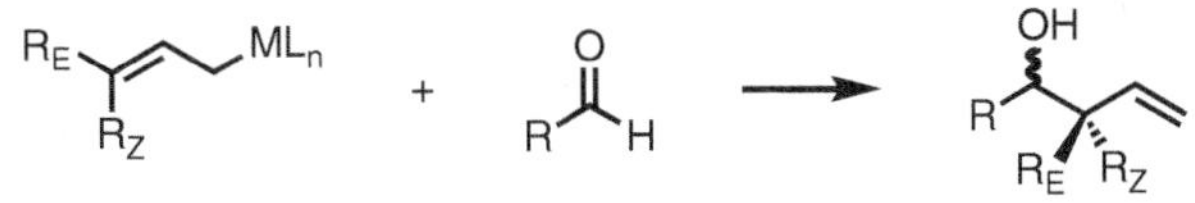
M = B, Al, Si, Ti, Cr, Zr, Sn

Scheme 3.34 General allylation/crotylation of aldehydes

The reaction is successful with a wide variety of metals (B, Al, Si, Ti, Cr, Zr, Sn). [93] The diastereochemical outcome can be classified into three groups: (i) type 1, in which the *syn*/*anti* ratio of the products reflects the *Z*- or *E*-geometry of the allyl moiety (B, Al, Sn); (ii) type 2 reactions which are *syn* selective irrespective of the olefin geometry (Sn, Si); and (iii) type 3 reactions which are *anti* selective irrespective of the olefin geometry (Ti, Cr, Zr). [94]

Figure 3.19 lists some commonly used chiral boron reagents for carbonyl group allylation, along with typical enantioselectivities for benzaldehyde. [95]

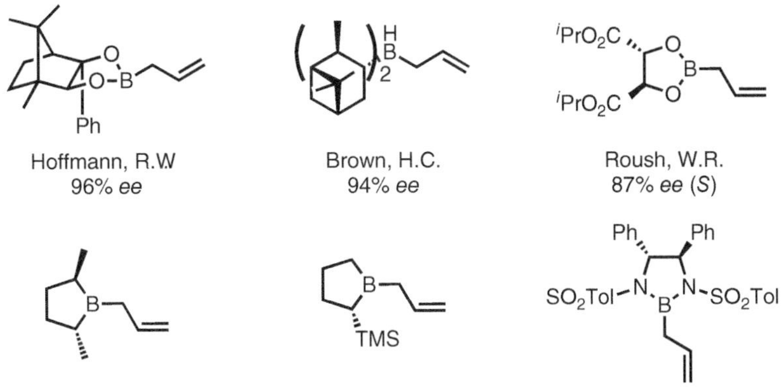
Hoffmann, R.W
96% ee
Brown, H.C.
94% ee
Roush, W.R.
87% ee (S)
Masamune, S.
for aliphatic aldehydes:
86–97% ee (S)
Masamune, S.
96% ee
Corey, E.J.
94% ee (R)

Figure 3.19 Chiral allylboron reagents

Crotylation with stereochemically defined crotylmetal reagents (R_E or R_Z = Me) gives rise to masked aldol products which can be converted to 3-hydroxyaldehydes simply by ozonolysis. Powerful asymmetric reagents, based on crotylboranes [96] and crotylboronates, [97] have been described by *William R. Roush* (1952–).

The proposed origin of the asymmetric induction in the case of allyl boronates is shown in Scheme 3.35. Rotation of the B—O bond clockwise moves the aldehyde nonbonding lone pair away from the proximate ester, which is pseudoaxial in the five-membered dioxaborolane ring. The alternative rotation in a counterclockwise sense would give an assembly with the ester carbonyl and the aldehyde nonbonding lone pair close to each other.

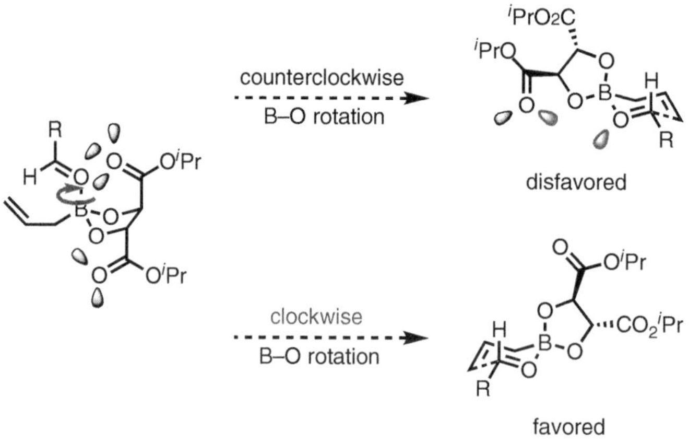

Scheme 3.35   Allylation with allyl boronates

Chiral allyl titanates have also been successfully utilised in crotylation reactions. [98] Reactions with various chiral and achiral aldehydes give excellent enantio- and diastereoselectivities, as exemplified by the allylation of the serine-derived Garner's aldehyde (Scheme 3.36). In a control reaction, allyl magnesiumbromide gave a nearly 1:1 mixture of *syn* and *anti* diastereomers. The allylations with the Ti-TADDOL allylating reagents were practically completely under reagent control, both the *R,R*- and *S,S*-reagents giving > 96% *de*, a fine example of external asymmetric induction.

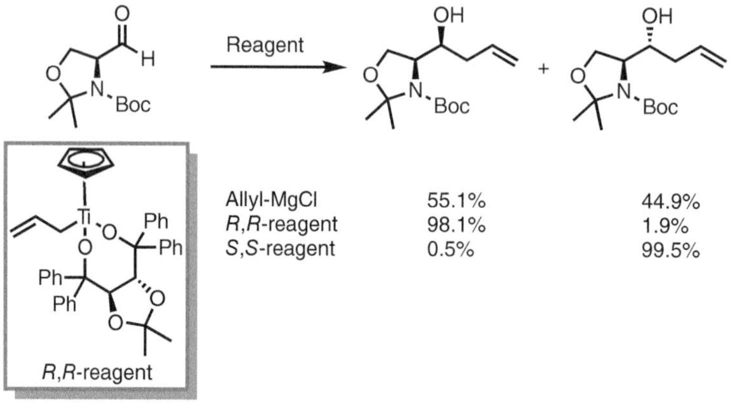

Reagent		
Allyl-MgCl	55.1%	44.9%
R,R-reagent	98.1%	1.9%
S,S-reagent	0.5%	99.5%

Scheme 3.36   Ti-TADDOL, reagent-controlled allylation

The Duthaler reagent was employed also in the synthesis of (+)-dodoneine, a naturally occurring 5,6-dihydro-2*H*-pyran-2-one isolated from *Tapinanthus dodoneifolius*, which is a parasitic medicinal plant that grows on the Sheanut trees in West Africa (Scheme 3.37). [99] The phenylpropanal derivative [two steps from commercially available methyl 3-(4-hydroxyphenyl)propanoate] was allylated with the (*S,S*)-Duthaler reagent, giving the homoallyl alcohol in excellent yield and excellent enantioselectivity. The terminal alkene was then subjected to cross metathesis with ethyl acrylate using the second generation Hoveyda–Grubbs catalyst to give the unsaturated hydroxy ester. Reaction

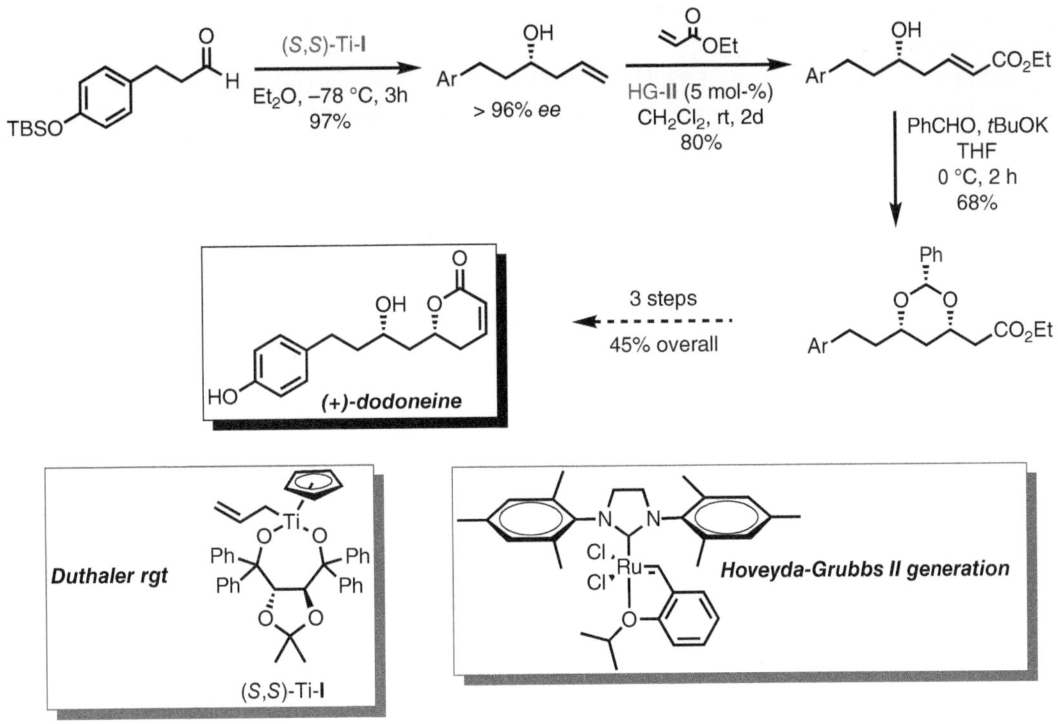

Scheme 3.37 *Enantioselective allylation in the synthesis of (+)-dodoneine*

with benzaldehyde under basic conditions triggered a cascade of reactions involving hemiacetal formation followed by conjugate addition of the alcoholate of the hemiacetal to form the six-membered 1,3-dioxane ring with all substituents equatorial. Three straightforward operations led to (+)-dodoneine.

Transfer-hydrogenative coupling reactions provide an interesting and economically feasible way to form C–C bonds. Krische has developed iridium-catalysed carbonyl allylation reactions from the alcohol or the aldehyde oxidation level using allyl acetate as the allyl donor. [100] Aliphatic, allylic, and benzylic alcohols are converted to the corresponding homoallylic alcohols with high levels of enantioselectivity. This reaction was recently applied in a short synthesis of bryostatin 7 (Scheme 3.38). [101]

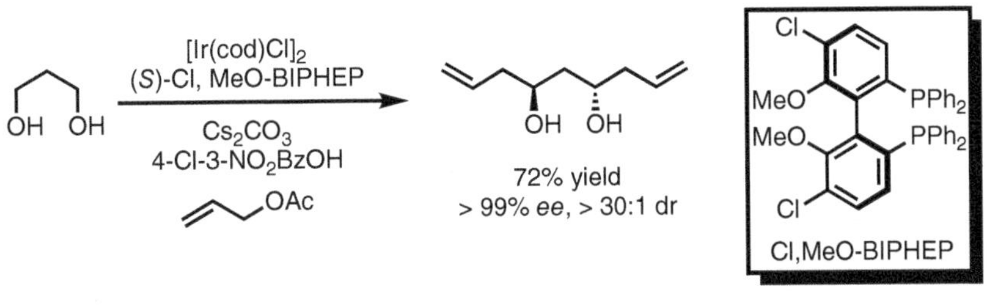

Scheme 3.38 In situ *oxidative asymmetric allylation*

Enantioenriched propargylic alcohols and amines are versatile building blocks in asymmetric synthesis. For example, activation of the alcohol and direct displacement with a nucleophile or in an S_N2' fashion can lead to quarternary carbon stereogenic centers or optically active allenes. Selective reduction of the alkyne affords the alkane or alkene. Benzylic alcohols are available through metal-catalysed [2 + 2 + 2] cycloadditions, and hydrosilylation can furnish vinylsilanes which can be further functionalised to trisubstituted olefins or oxidised to hydroxy ketones. Several enantioselective routes to propargylic alcohols have been developed, and here we shall highlight two based on direct addition of alkyne anions to aldehydes.

An early example is shown in Scheme 3.39. [102] Thanks to the acidity of acetylene protons, the inexpensive commodity chemical 2-methyl-3-butyn-2-ol functions as a masked acetylene unit and can be activated with zinc triflate and a (−)-*N*-methylephedrine as the chiral base to form the nuclophilic alkynyl species. Addition reactions to aliphatic, *α*,*β*-unsaturated and benzylic aldehydes proceed in high yields and excellent enantioselectivities. The terminal alkyne can be liberated from the acetone adduct simply by first silyl protecting the secondary alcohol and then inducing a base-catalysed fragmentation reaction.

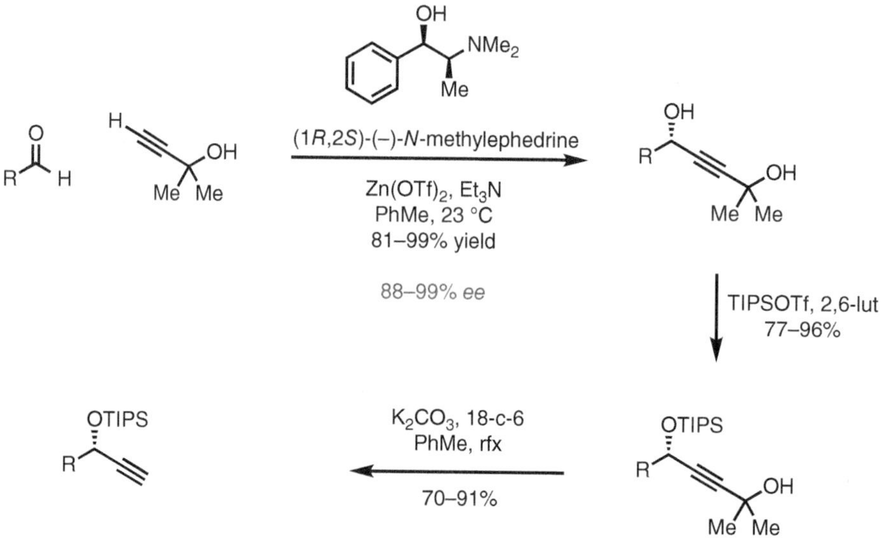

Scheme 3.39 *Enantioselective synthesis of terminal propargylic alcohols*

Trost has developed a versatile chiral ligand, ProPhenol, useful as an enantioselective catalyst for base-mediated nucleophilic addition reactions. General conditions for the asymmetric alkynylation of aldehydes through Zn−ProPhenol-catalysed alkyne addition have been developed (Scheme 3.40). [103] This synthetically efficient methodology operates with relatively low catalyst loading and can avoid the use of excess alkyne and dialkylzinc reagents. The ProPhenol ligand is readily prepared in both enantiomeric forms from L- or D-proline and the ligands are commercially available. A wide variety of aliphatic, *α*,*β*-unsaturated and benzylic aldehydes as well as different terminal alkynes are tolerated giving high yields and enantioselectivities. Scheme 3.40 also shows a mechanistic explanation of the reaction, which has been applied to the syntheses of several natural products.

3.2.2   Reactions at the α-Carbon (Enolate Chemistry)

The *α*-center of carbonyl compounds provides an efficient entry point for the introduction of a new stereocenter in a molecule. As enolate chemistry is one of the most extensively studied fields in organic synthesis, it is no surprise that methods for asymmetric *α*-alkylation of carbonyl compounds have also been well developed. Two factors need to be considered in discussing asymmetric induction in enolate alkylations, namely the *E*/*Z*-geometry of the enolate and the origin of asymmetric information, i.e. enantiofacial selectivity.

To define an unambiguous system of nomenclature for the enolate geometry, we shall in this text distinguish the two stereoisomeric forms of the enolate (or enol ether) as *E* and *Z* with the descriptors for the atoms defining the relationship in parenthesis (Figure 3.20). We shall use the main carbon chain (R) and the enolized oxygen (O) as the reference groups. With this agreement, the encircled R' and O⁻ determine whether one is an *E*(O,R)- or *Z*(O,R)-enolate regardless of the nature of the other groups (e.g. ketone, ester, thioester, and alkoxyketone). We can thus simply speak of *E*- and *Z*-enolates.

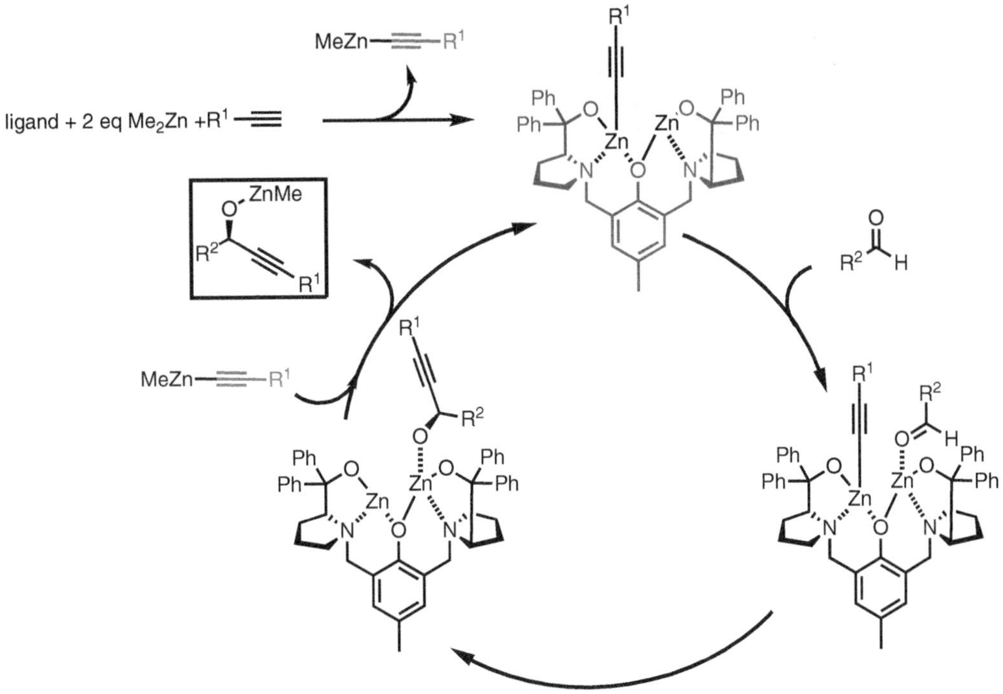

Scheme 3.40 *ProPhenol-mediated alkynyl zinc addition to aldehydes*

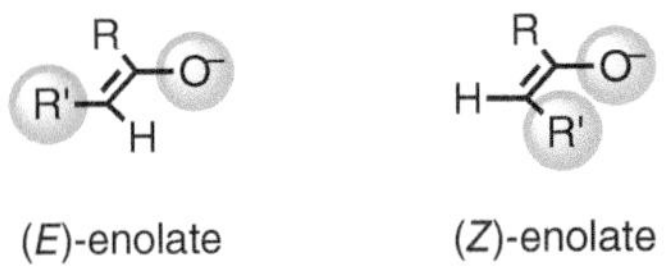

(*E*)-enolate   (*Z*)-enolate

Figure 3.20 *Enolate geometries*

The geometry of the enolate is of prime importance for the stereochemical outcome of any reaction occurring at the α-carbon. It is therefore important to recap the factors determining the stereochemistry of enolization. We shall describe the process in terms of a simplified molecular orbital theory and conformational analysis (Figure 3.21). From conformational analysis, we know that the most stable (ground state) geometry of a ketone is the one having the α-hydrogen eclipsed with the carbonyl oxygen (structure **A**). As this is the ground state conformation, and as the ground state can hardly be expected to be reactive enough to drive the reaction, we must consider what happens as we rotate the C_{α}–C=O bond. Rotation of the bond by 30° clockwise (structure **B**), or by 90° counterclockwise (structure **C**) brings the C_{α}–H bond perpendicular to the plane of the carbonyl group, or expressed in other words, parallel to the plane of the π-orbital of the carbonyl group. This leads to maximal overlap of the σ-orbital of the C_{α}–H bond with the (electron-attracting) π-orbital. Conformation **B** gives the *kinetic* (*E*)-enolate. Under equilibrating conditions, conformation **B** suffers from developing strain between the R' and R groups compared to conformation **C**, which is thus the favored one leading to the *thermodynamic* (*Z*)-enolate product.

The practical value of this analysis is the notion that the hydrogen, which is (nearly) perpendicular to the plane of the carbonyl group, is the most acidic one. [104] This is clearly manifested in the acidities of the α-protons in cyclohexanone. The equatorial proton is nearly coplanar with the carbonyl group and thus the $\sigma_{C–H}$-orbital is nearly perpendicular to the π-system, whereas the axial proton places the $\sigma_{C–H}$-orbital almost perfectly aligned with the $\pi^*_{C=O}$. In acyclic ketones, this leads to stereoselectivity of enolate formation as follows. Conformation **B** is the favored pathway leading to the *E* enolate, which is the kinetic product. However, as one increases the size of the R' group, the steric repulsion between the two alkyl substituents (R and R') becomes dominant (allylic A^{1,2} strain) and the order of

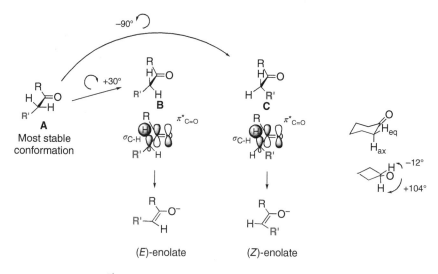

Figure 3.21   *Enolization of a carbonyl group*

the stabilities of the two pathways leading to *E* and *Z* enolates will be reversed. Similar arguments also hold for the protonation of enolate anions. [105]

According to the *Ireland mechanism* an (*E*)-enolate is formed via a chair-form transition state (R = large alkyl group) (Figure 3.22). If the R' group is also large, a (*Z*)-enolate is formed! The nitrogen lone pair is the actual proton abstractor and the role of the metal is to coordinate the base to the carbonyl compound as a Lewis acid. [106] The Ireland model

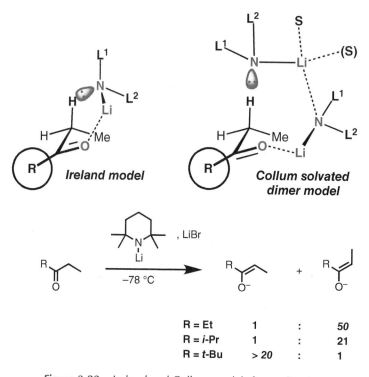

R = Et	1	:	50
R = *i*-Pr	1	:	21
R = *t*-Bu	> 20	:	1

Figure 3.22   *Ireland and Collum models for enolization*

can be used to explain *Z:E* selectivities in enolization. The transition state is necessarily loose: note large Li–N and Li–O distances (2.0 Å), and aggregation and solvation effects are not taken into account. David Collum has studied the enolization process intimately and has presented a model where the dimeric nature of the base, as well as solvation, are taken into account. This model is stereoelectronically much more sensible. With $L_1 = L_2 = Me_3Si$, enolization rates are accelerated by tertiary amines. [107]

A remarkable example of the stereoelectronically controlled enolization is provided by an attempted alkylation of a late intermediate in the synthesis of taxol (Scheme 3.41). [108] The eight-membered ring ketone is held rigidly in a conformation where the methyl group is axially disposed. When the ketone was subjected to equilibrating enolization conditions and quenched with deuterium oxide, a single diastereomer of deuterated product was obtained.

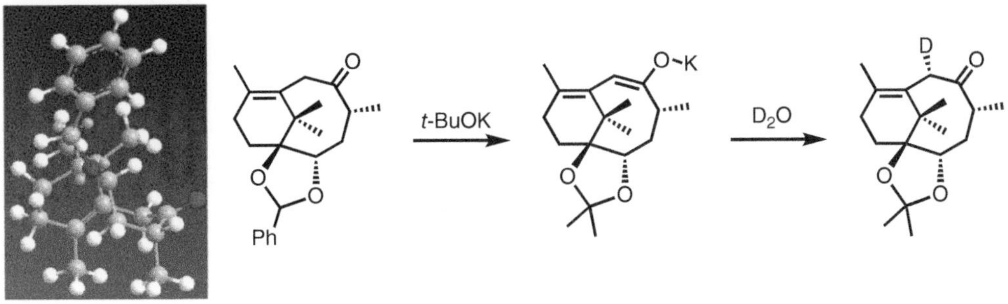

Scheme 3.41   *Stereoelectronically controlled regiospecific enolization*

3.2.2.1   α-*Alkylation of Enolate Anions*

In the asymmetric α-alkylation of the enolate anions, the methods developed thus far utilise either internal or relayed asymmetric induction methods. Thus, the chiral information is either covalently connected to the starting material, leading to diastereotopic selection in the transition states, or brought into the transition state with the aid of a chiral auxiliary, which must be cleaved later. We shall not discuss the first case in this chapter, but a few comments are sufficient for the latter case.

The early alkylations of carbonyl compounds by Pincock and Rolston relied on the application of 1,4-asymmetric induction, in a fashion related to the principles of the Prelog rule. [109] Thus, an acid was converted to a chiral ester with either isoborneol or menthol, and the derived enolate was treated with an electrophile. However, the optical rotation of the derived alcohol was negligible, and the work remained unnoticed!

The first successful asymmetric alkylation of a carbonyl compound was published in 1969 by *Shinji Yamada* (Scheme 3.42). [110] Asymmetric induction was attempted through proline-derived enamine alkylation, but the enantioselectivity remained moderate at best.

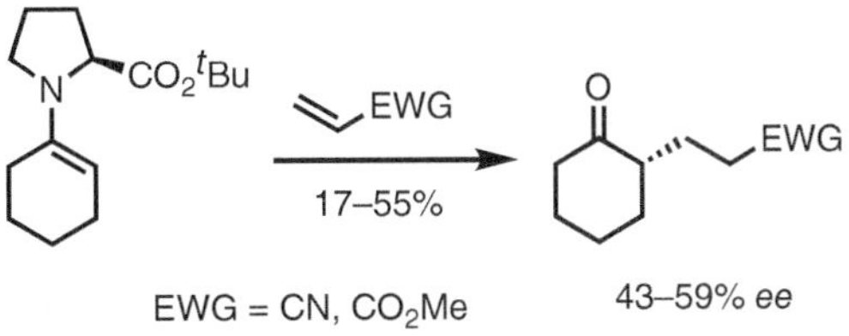

Scheme 3.42   *Early asymmetric α-alkylation*

In the 1980s, Koga produced quite a bit of interesting chemistry on enantioselective alkylation. An early example of inducing asymmetry in the generation of quaternary centers by an enolate is shown in Scheme 3.43. [111] The cyclohexanonecarboxylic acid was first converted to an imine with a valine ester. Treatment with LDA gave a chelated

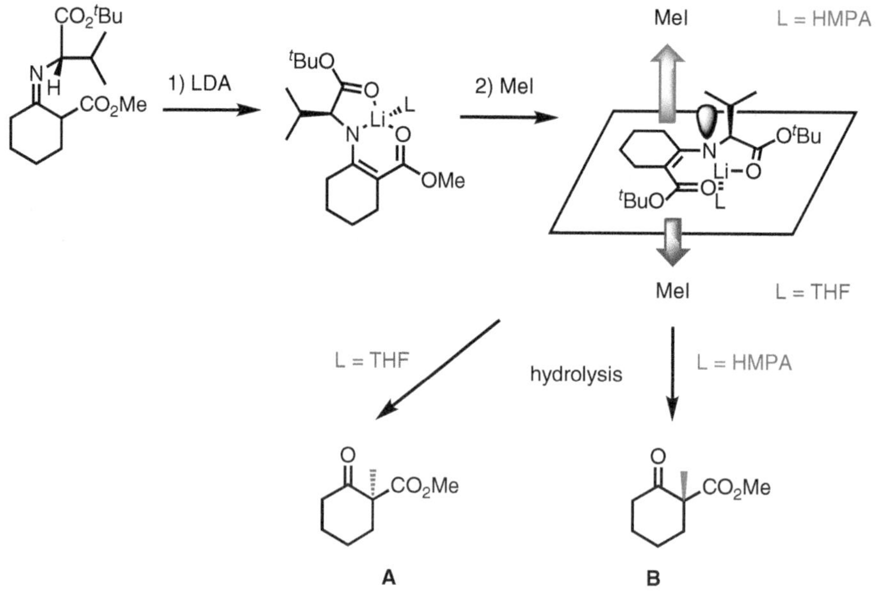

Scheme 3.43 *Koga asymmetric alkylation*

enolate, which is essentially flat, save for the stereochemistry-inducing isopropyl group. The lithium atom, in its quest for tetrahedral coordination, coordinates the solvent (L) from the face opposite to the isopropyl group. When the enolate is treated with an electrophile (methyl iodide), the electrophile displaces the solvent and the approach will be directed by coordination to the lithium giving product **A**. Simple addition of a better-coordinating co-solvent like HMPA displaces the solvent THF, and prevents exchange with methyl iodide. In this case, the electrophile approach from the bottom face is hindered by HMPA and alkylation results in the opposite enantiomer **B**.

David A. Evans developed a highly successful general protocol for α-alkylation of carbonyl compounds, with particular applications in aldol reactions (Section 3.2.2.2). Carboxylic acids can be readily derivatised with chiral oxazolidinones. These, in turn, are readily available in high enantiopurity from natural amino acids (Scheme 3.44). The acyloxazolidinones can then be enolized, and the enolates react with high diastereoselectivity. The facial selectivity is dependent on the enolization method: if the metal is not coordinated with the oxazolidone carbonyl oxygen, the attack of an electrophile occurs from the *Re* face of the enolate double bond. [112] This is explained to be the result of the

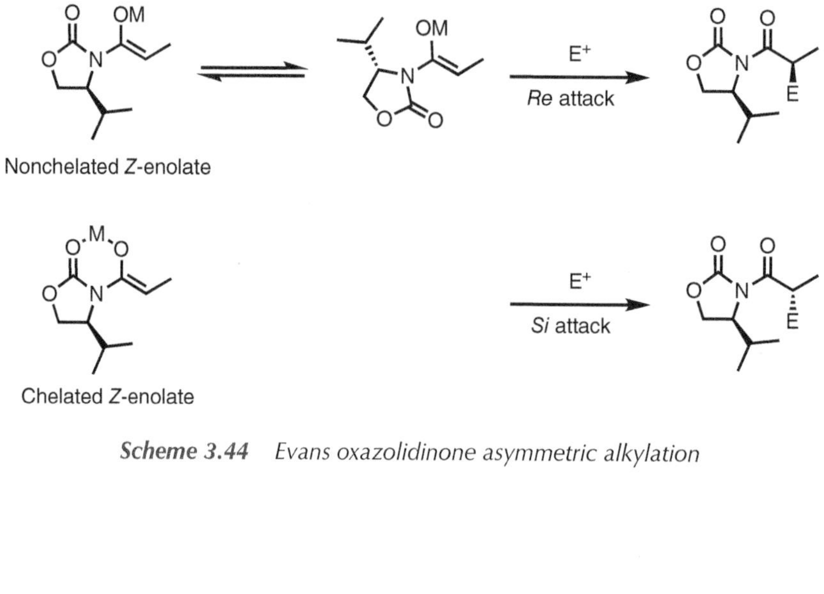

Scheme 3.44 *Evans oxazolidinone asymmetric alkylation*

equilibration of the oxazolidinyl enolate to the conformation, which places the dipoles opposite to each other. With a coordinating metal (e.g. Ti), the attack occurs from the *Si* face to give the diastereomeric product. [113] The Evans synthesis of MeBmt-amino acid of cyclosporine, shown in Scheme 2.19, relies on this methodology.

Several chiral auxiliaries have been developed, [114] and some of the most general ones are shown in Figure 3.23. The Evans oxazolidinones will be discussed again with the aldol reaction. *Dieter Enders* (1946–2017) developed the proline-derived SAMP- and RAMP-hydrazones as highly general chiral auxiliaries. [115] In alkylations of, e.g. aldehyde enolates, electrophiles such as alkyl halides, oxiranes, aziridines, carbonyl compounds, and Michael acceptors can be used with typical *dr*'s of 2:1 to 20:1. Pseudoephedrine amides, developed by Andrew Myers, [116] have been widely used as an auxiliary for alkylations because of high diastereoselectivities (usually in the range of 13:1 to 99:1).

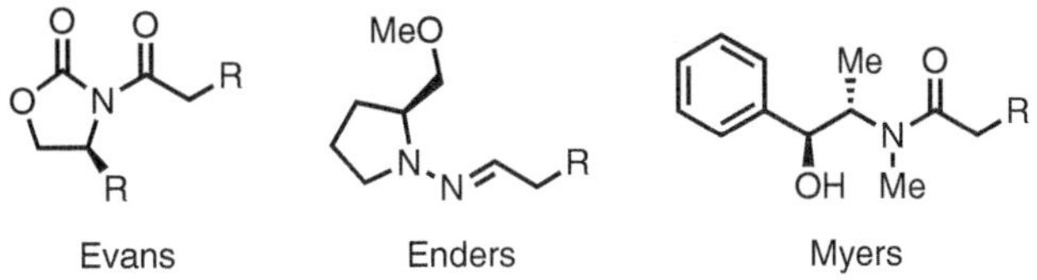

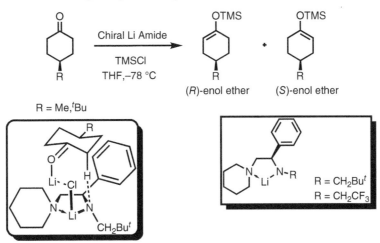

Figure 3.23 *Chiral auxiliaries for carbonyl alkylation*

In contrast to direct α-alkylation of aldehydes and carboxylic acid derivatives, direct asymmetric alkylation (including allylation) of ketones remains elusive. An early approach relied on enolization of prochiral ketones employing chiral lithium bases, pioneered independently by Koga and Simpkins. [117] An early example is shown in Scheme 3.45, where the enantioselective deprotonation reaction of prochiral 4-substituted cyclohexanones by chiral bidentate lithium amides in the presence of excess trimethylsilyl chloride (TMSCl) (Corey's internal quench method) gave the corresponding (*R*)-enol ethers in good chemical and optical yields of up to 90%. [118]

Scheme 3.45 *Early enantioselective deprotonation by chiral lithium base*

Despite extensive investigations involving the use of chiral auxiliaries, metal catalysis, phase transfer catalysis, or more recently, organocatalysis, no general solution has been found to the problem of asymmetric α-alkylation of simple ketones. [119]

3.2.2.2 *Aldol reaction*

Aldol reaction is, undoubtedly, one of the most extensively studied reactions in synthetic organic chemistry. The first aldol condensation reaction, the acid-catalysed dimerisation of acetone to give mesityl oxide, was reported by Kane, as early as 1838. [120] Methods to avoid elimination of water were independently developed by *Alexandr Borodin* (1833–1887) and *Charles Adolphe Wurtz* (1817–1884). [121] Borodin graduated from MedicalSurgical Academy in St. Petersburg and spent post-doctoral time in Heidelberg with Emil Erlenmeyer. After returning to St. Petersburg, he also befriended himself with the composer Mily Balakirev, and eventually gave up chemistry for musical composition in 1875. Wurtz's career was much more orthodox: he studied with Liebig in Giessen and became the first professor of organic chemistry at Sorbonne in 1875. Today, his name can be seen on the Tour Eiffel in Paris as one of the 72 prominent French persons of culture, together with other scientists such as Antoine Lavoisier, Michel Eugène Chevreul, Ètienne Louis Malus, Louis Le Chatelier, and Joseph Louis Gay-Lussac.

Two- and three-carbon units; acetate and propionate, play a central role in the biosynthesis of most natural products. These (related) relatively simple structural units are used in a most versatile way to construct structures of tremendous stereochemical diversity, employing only a limited number of distinctly different reaction types. The reaction of an enolate with an aldehyde, the aldol reaction (Scheme 3.46), provides nature with a rapid means of setting up several stereochemical relationships in a single step. The important natural-product class of polyketides will be covered in Chapter 8. Recent advances in understanding the stereochemical aspects of the aldol reaction have given us means to correctly predict the outcome of such a process, and design a specific reaction for a given stereochemical problem. Early on, it was discovered that with sterically bulky alkyl ethyl ketones, the *Z*-enolates favored the formation of *syn* products, and the *E*-enolates gave predominantly *anti* aldol products. [122]

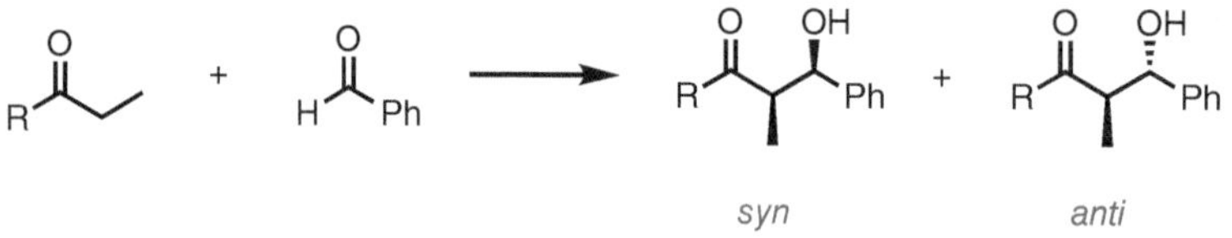

R	Enolate	*syn*	*anti*
H	Z	50	50
	E	65	35
iPr	Z	90	10
	E	45	55
tBu	Z	98	2
	E	8	92

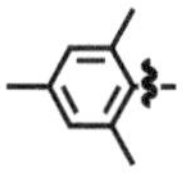

Bulky R - high selectivity

Z-enolate ⇒ *syn*
E-enolate ⇒ *anti*

Scheme 3.46   *Aldol reaction diastereoselectivity depends on enolate geometry*

In this section, we will inspect the factors affecting product stereochemistry in light of the currently accepted rationalisations on the mechanism of aldol additions. The reaction has been studied by several groups, and numerous reviews have been written to highlight different aspects of the reaction. [123]

A number of transition-state geometries have been suggested to explain the general trends in the diastereoselectivity of aldol reactions. The earliest model was the pericyclic model proposed by Zimmerman and Traxler. [124] This model describes the reaction proceeding through a six-membered chelated chair-like transition state (Scheme 3.47), and is called a type 1 aldol reaction. Chelation forces the reacting enolate double bond and the carbonyl-acceptor double bond in a *gauche* arrangement. To avoid steric repulsion with the enolate R^1 substituent, the aldehyde substituent R^3 prefers to assume an equatorial position.

When this model is applied to the aldol reaction, one can explain the general conclusion that *E*-enolates will preferentially give *anti*-aldol products, whereas *Z*-enolates would lead to *syn* products. This generalisation holds true in broad terms, but several further refinements need to be done. Inspection of the transition structures in the scheme leads to the following conclusions. The correlation is dependent on the relative sizes of the substituents R. Bulky R^3 groups

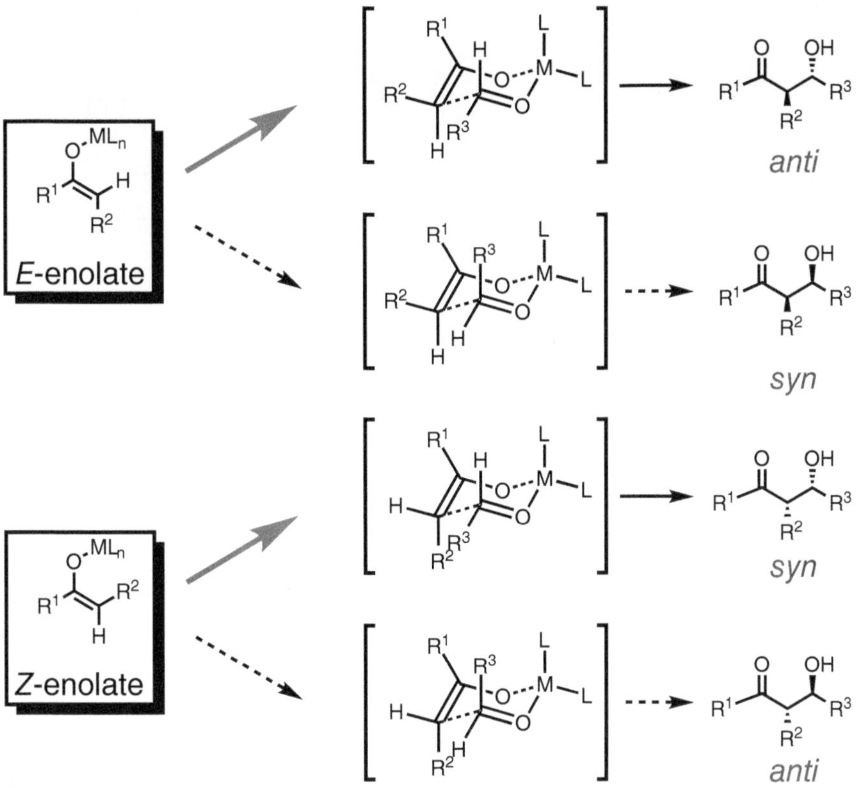

Scheme 3.47 *Zimmermann–Traxler transition states for type 1 aldol reactions*

will lead to higher selectivity. Especially in the case of Z-enolates, the correlation is enhanced if both R^1 and R^3 are sterically demanding. If R^2 is very large, the normal trends ($Z \rightarrow syn$, $E \rightarrow anti$) are reversed. Steric congestion in the transition structure also enhances the selectivities, and boron enolates usually give higher selectivities than the more loose lithium enolates (typical bond lengths are collected in Table 3.2). [125] Similarly, increase in bond length [126] and decrease in solvation enthalpy [127] lead to looser transition structures and breakdown of the selectivity suggested by the Zimmerman–Traxler model.

Table 3.2 *Crystallographic bond distances between some metals and oxygen*

Bond	Bond length, Å
Li O	1.92–2.00
Mg O	2.01–2.03
Zn O	1.92–2.16
Al O	1.92
B O	1.36–1.47
Ti O	1.62–1.73
Zr O	2.15

Not all aldol reactions follow the generalisations given above, and thus, further classification of the reactions has emerged (Scheme 3.48). The reactions following the Zimmerman–Traxler cyclic transition state yielding mainly Z-*syn*,

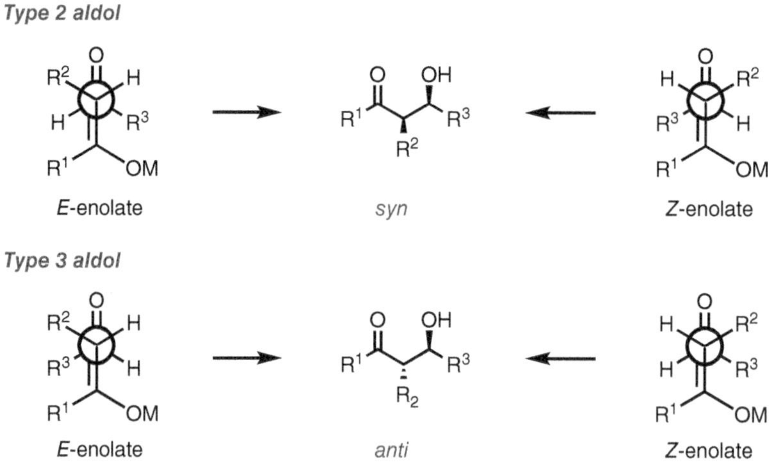

Type 2 aldol

E-enolate   *syn*   Z-enolate

Type 3 aldol

E-enolate   *anti*   Z-enolate

Scheme 3.48 *Transition states for type 2 and 3 aldol reactions*

E-anti selectivity are known as type 1 aldol processes. Type 2 reactions are *syn* selective regardless of the enolate geometry and are common to a wide range of enol derivatives, including enol silanes (the Mukaiyama aldol reaction), enol stannanes, zirconium enolates, and enol borates. Both open and cyclic transition state geometries have been proposed for these reactions. Ketene acetals and ketene thioacetals tend to give *anti* products regardless of the enolate geometry. These *anti*-selective aldol reactions are known as type 3 processes.

The open transition state was proposed by *Yoshinori Yamamoto* (1942–) to account for the *syn* selective addition of tin or zirconium enolates to aldehydes (Scheme 3.42, top). [128] The enolate and carbonyl moieties are aligned in an anti-periplanar fashion. In both, the *E*- and *Z*-enolates, the transition state leading to the *anti*-aldol products is destabilised through steric interaction between the R^2 and R^3 substituents.

Ab initio calculations suggest that (i) the open transition state is favored for metal-free enolates; (ii) cyclic transition states best account for the case of lithium enolates with distorted Bürgi–Dunitz attack angles due to short bonds to lithium; and (iii) a close balance of chair- and twist-boat cyclic transition structures exists for enol borinates and enol borates depending on the substitution pattern. [129]

Aldol reactions show increasing preference for the participation of the cyclic transition state and thereby *syn*-product preference (synclinal orientation of the enolate and the aldehyde) with increasing cation-coordinating ability ($K^+ <$ $Na^+ < Li^+ < MgBr^+$). Reactions with a "naked" enolate, where the cation has been sequestered, show a strong preference for an open transition state in which the dipole moments of the enolate and aldehyde tend to cancel each other (anti-periplanar orientation). Solvent effects are negligible, especially for the strongly coordinating cations (e.g. Li). The use of additives, such as HMPA, effectively leads to attenuation of the coordinating power of the cation. As a result, the tendency towards an open transition state is increased. [130]

Introduction of asymmetry into the aldol reaction can be achieved using any of the general methods discussed in Chapter 2. Here we shall examine two of the generally useful strategies, namely the utilisation of a chiral auxiliary (relayed a.i.), and the use of chiral reagents (or catalysts, external a.i.).

Of the chiral auxiliaries, the Evans chiral oxazolidinones have gained widespread use in many laboratories, thanks to their generally high and predictable level and sense of asymmetric induction. The normal mode for the Evans aldol process produces *syn* products as the favored products. This is due to the chelated six-membered transition state, which favors the observed products (Scheme 3.49). [102] Of the two possible *syn* aldol products, the one forming through β-attack is favored on steric grounds. Rotation of the enolate carbonyl–nitrogen bond can give rise to two different enolates, whose reactions from either α- or β-face will result in diastereomeric *syn* aldol products. In the addition, approach of the electrophile from the β-face is favored, and this is explained as being the result of the stabilisation of the enolate geometry to the one depicted through dipolar organisation.

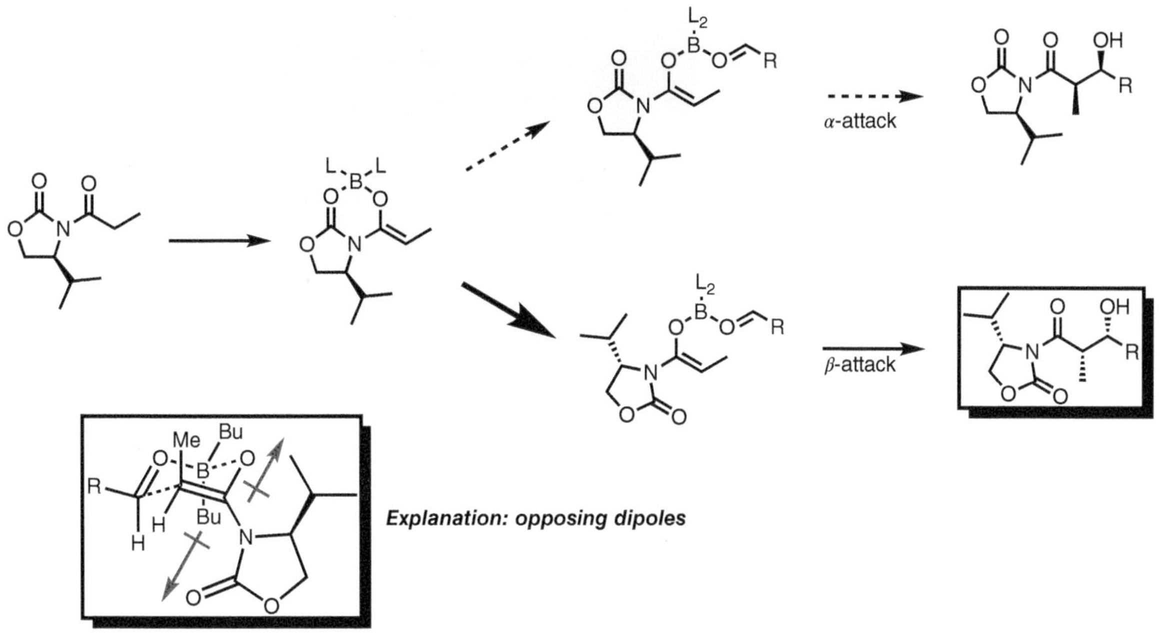

Scheme 3.49 *Evans syn selective aldol*

The chelated oxazolidinyl enolates allow control of the stereochemistry of the aldol products at the α-carbon if one can control whether or not the metal is chelated at the time of the aldol reaction. Titanium enolates can be employed to obtain the non-Evans *syn* aldol products (Scheme 3.50). [131] The chelated complex is slightly looser (Ti–O bond length 1.62–1.73 Å), and the titanium is hexacoordinate, allowing an approximately octahedral coordination around titanium, whereas boron is approximately tetrahedrally coordinated.

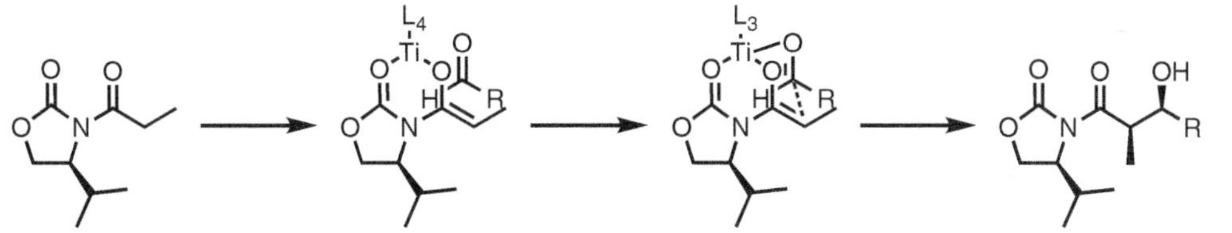

Scheme 3.50 *Thornton non-Evans syn-selective aldol*

The X-ray structure of the titanium tetrachloride complex of 4-benzyl-*N*-propionyloxazolidin-2-one is shown in Figure 3.24. [132] The phenyl group efficiently shields one face of the ketone, and this is used to explain the stereochemistry.

Simple modification of the Evans auxiliary by *Dieter Seebach* (1937–) by incorporating a 5,5-diphenyl moiety (Figure 3.25) led to a useful alternative to the Evans oxazolidinones. [119] The Seebach auxiliaries displayed high diastereoselectivities in both of the *syn*-selective aldols. A model for the diastereoselectivity was advanced on the basis of the X-ray structure in Figure 3.24, and one of the phenyl groups was noted to hinder the approach of the electrophile from the bottom face (Figure 3.25, right panel). The new auxiliary presented some operational advantages over the classical Evans auxiliaries, including: the acyloxazolidinones are readily crystalline; the acylation of the auxiliary can be done at higher temperatures; generation of the enolates can be done with BuLi; and the acyloxazolidinones can be cleaved with NaOH without detectable attack on the oxazolidinone ring.

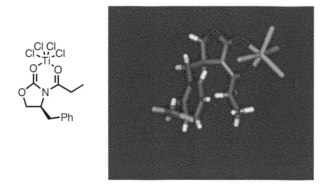

Figure 3.24   *X-ray structure of 4-benzyl-N-propionyloxazolidin-2-one (GOMLUP)*

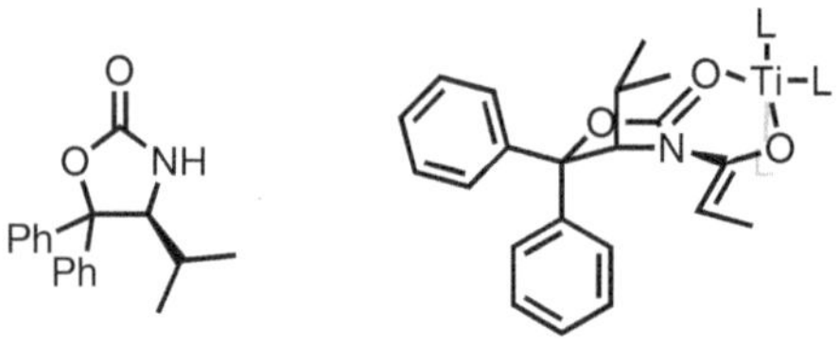

Figure 3.25   *Seebach oxazolidinone*

The *syn/anti* ratio in the Evans aldol reaction can be affected very easily through the choice of the Lewis acid employed to activate the aldehyde partner (Scheme 3.51). [133] Assuming an open transition state, the form **A** is favored if the Lewis acid employed is small (e.g. $SnCl_4$, $TiCl_4$), because it minimises the gauche interactions about the forming bond. The product will therefore be the non-Evans *syn* aldol. However, if one uses a large Lewis acid (e.g. Et_2AlCl), transition state **B** becomes more favorable because of the methyl–Lewis acid interaction. The product will therefore be the *anti* aldol. The observed *anti/syn* selectivities range from 74:26 (PhCHO) to 95:5 (*i*-PrCHO and *t*-BuCHO).

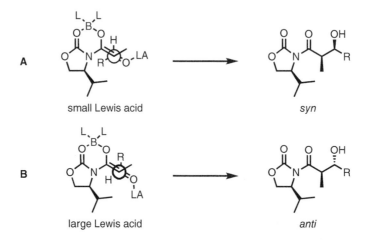

Scheme 3.51   *Lewis acid effect on* syn/anti *selectivity*

Mild enolization methods have been developed recently, and magnesium halides in the presence of an amine base and chlorotrimethylsilane catalyse the direct aldol reaction with high *anti* diastereoselectivity. When one combines the mild enolization to the use of the Nagao *N*-acylthiazolidinethiones, one has access to the 'missing' Evans *anti* aldol

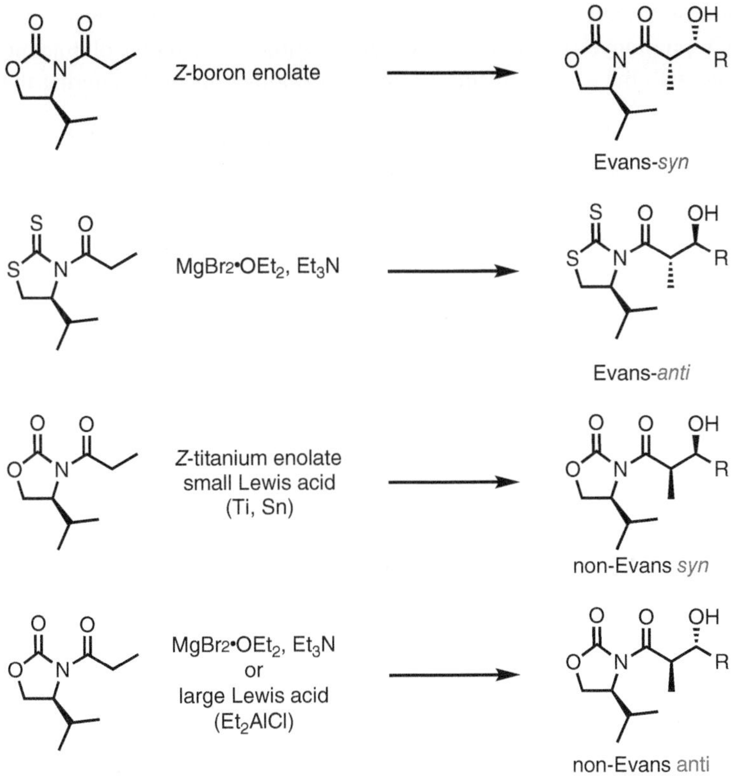

Scheme 3.52   *All aldols from acyloxazolidinone derivatives*

(Scheme 3.52). [134] The mild enolization works also with the *N*-acyloxazolidinones, giving an alternative access to the non-Evans *anti* products in high selectivity. [135] A comprehensive review details other selective aldol variants. [136]

Michael Crimmins has developed a protocol to obtain the Evans *syn* and non-Evans *syn* aldol products from the same starting materials employing soft enolization together with different amounts of (−)-sparteine as a chiral modifier (Scheme 3.53). [137] With 1 eq. (equivalent) of titanium tetrachloride and 2 eq. of (−)-sparteine as the base, or 1 eq.

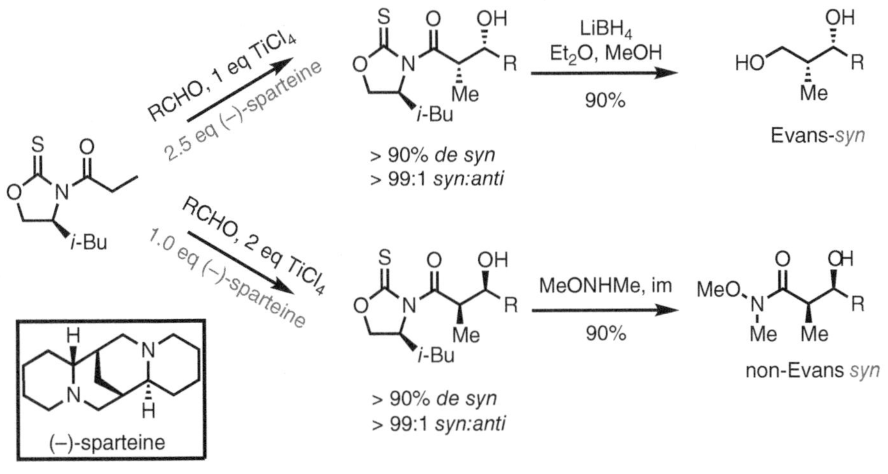

Scheme 3.53   *Soft enolization and (−)-sparteine to both* syn *aldols*

of (−)-sparteine and 1 eq. of N-methyl-2-pyrrolidinone, selectivities of 97:3 to >99:1 were obtained for the Evans *syn* aldol products using *N*-propionyl oxazolidinones, oxazolidinethiones, and thiazolidinethiones. The non-Evans *syn* aldol adducts are available with the oxazolidinethione and thiazolidinethiones by altering the Lewis acid/amine base ratios.

When enolates react with α-chiral aldehydes as electrophiles, the situation follows the Felkin–Anh rules as well as the Zimmerman–Traxler transition state (Scheme 3.54). [138] For *E*- and *Z*-enolates, there are two alternative transition states, one of which is disfavoured by the *syn* pentane interactions. Thus the *E*-enolates give the *anti*, *syn* (Felkin) product whereas the *Z*-enolates prefer the formation of the *syn*, *anti* (*anti*-Felkin) products.

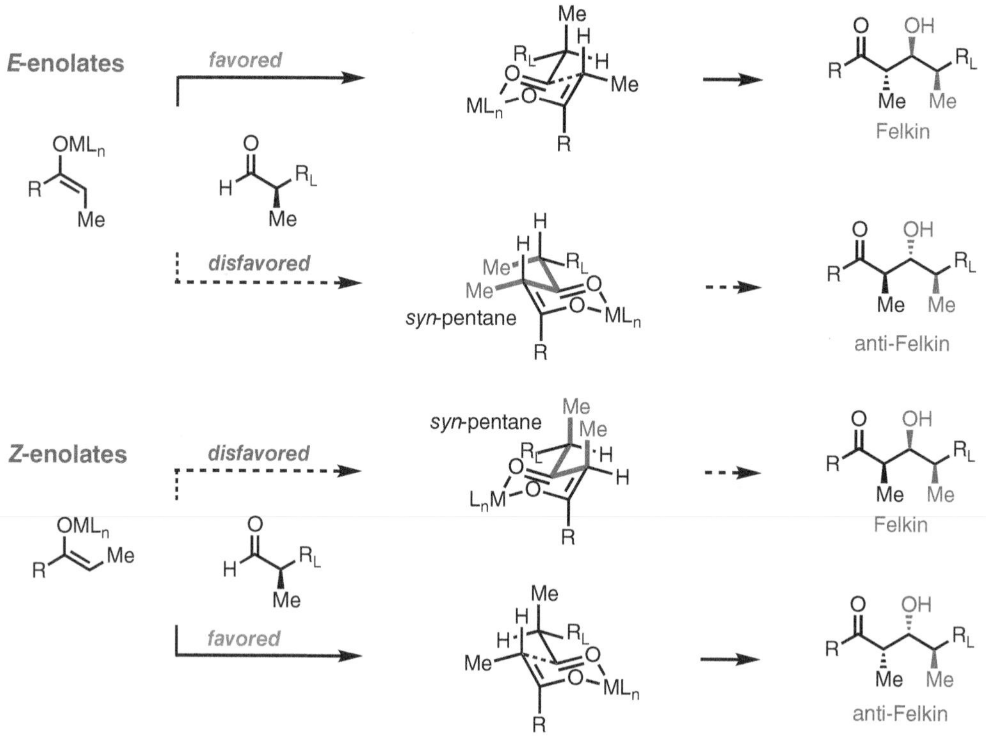

Scheme 3.54   Diastereofacial selectivity of the aldol reactions of α-chiral aldehydes

The *anti*-Felkin–Anh selective aldol reaction has found use in the total synthesis of the polyether ionomycin (Scheme 3.55). [139] Generation of the Z-boryl enolate from the chiral crotonoyl oxazolidone and allowing this to react with (*R*)-3-(benzyloxy)-2-methylpropanal gave the required *anti*-Felkin–Anh product in high selectivity and high yield.

A version of aldol reactions catalytic in the chiral source has been much sought after. The first example involved the use of gold catalysis together with a chiral ferrocenyl phosphine ligand to assemble the isocyanatoacetate and aldehyde (Scheme 3.56). [140] The resulting oxazolines were obtained with high to excellent *trans:cis* ratio and the *trans* product was obtained in moderate to high enantioselectivity.

An early example of catalytic asymmetric aldol reactions was described by Kobayashi and Mukaiyama. A chiral proline-derived diamino ligand effected the Mukaiyama aldol of thioacetate and thiopropionate ketene thioacetals with a number of aldehydes in good to excellent enantioselectivities and high diastereoselectivities (Scheme 3.57). [141]

Chiral catalysts based on diazaborolidines [142] have been developed for aldol reactions. In the example in Scheme 3.58, the *syn/anti* selectivity can be controlled simply by the choice of the aldol substrate. Ester enolates give predominantly *anti* products whereas thioester enolates give rise to *syn* aldol products. [125]

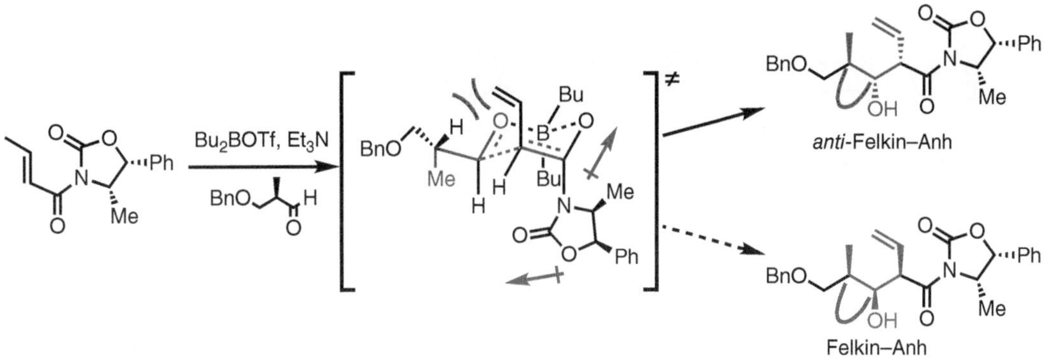

Scheme 3.55 Anti-*Felkin–Anh* selective aldol reaction

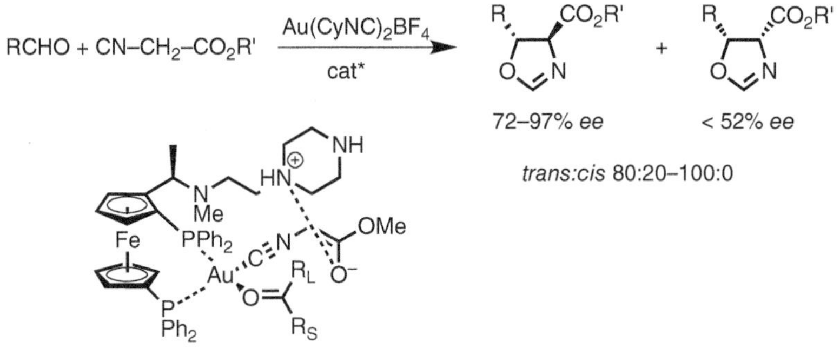

Scheme 3.56 *Au-catalysed catalytic asymmetric aldol*

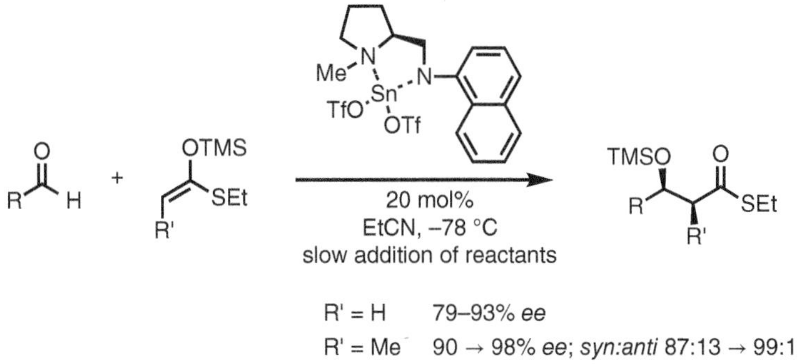

R' = H   79–93% ee
R' = Me   90 → 98% ee; syn:anti 87:13 → 99:1

Scheme 3.57 *Early catalytic aldols*

Chiral acyloxyboranes (CAB) can be used in less than stoichiometric quantities in Mukaiyama-type aldol reactions (Scheme 3.59). [143] Thus, treatment of a silyl enol ether and an aldehyde with 20 mol-% of the tartaric acid-derived CAB gives the *syn* aldol products (regardless of the geometry of the starting enol ether) in high % ee. The details of the transition state have been investigated using a combination of dispersion corrected DFT calculations and transition-state force fields (TSFF) developed using the quantum-guided molecular mechanics (Q2MM) method. [144] A closed transition state involving a nontraditional hydrogen bond is favored over the corresponding open transition state.

Acetate aldols provide a challenge for stereoselective reactions, because the application of the Zimmermann–Traxler transition state (Scheme 3.47) is not straightforward due to the missing substituent R^2. Early attempts by Manfred Braun in the early 1980s with mandelic acid-derived diol acetates were successful and gave enantioselectivities up to 94%

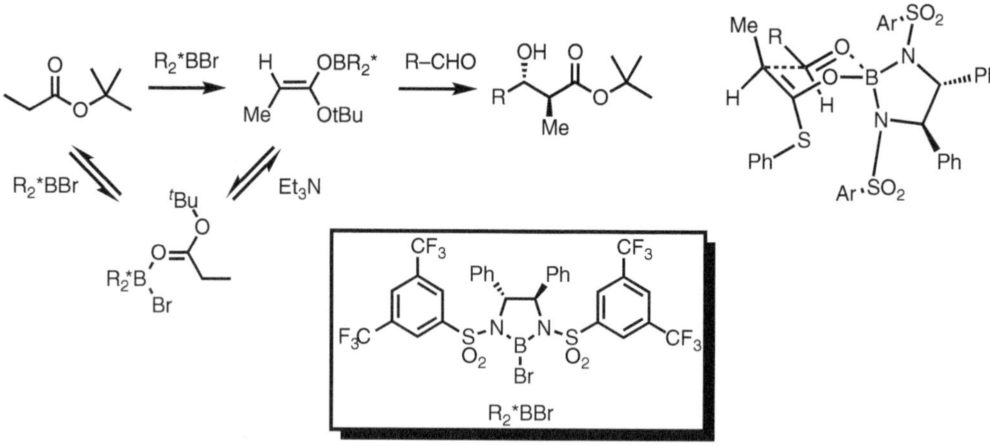

Scheme 3.58 *Chiral catalysis in aldol reactions*

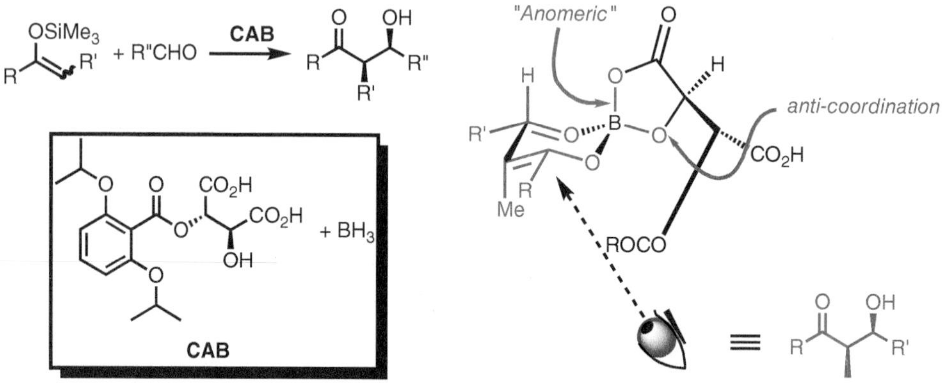

Scheme 3.59 *Chiral acyloxyborolidines as catalysts for Mukaiyama aldol*

ee. [145] The Braun aldol was popular during the development of the cholesterol-lowering statins, and Scheme 3.60 presents the application in a synthesis of fluvastatin (Lescol, by Sandoz, later Novartis). [146] The model shown explains the sense of absolute stereoselection, and is in line with the observations by the Sandoz group that all phenyl groups play a role in the enantioselectivity. The original phenyl group of mandelic acid is crucial; if it is replaced with a methyl group, the enantioselectivity collapses. [147]

Erick Carreira (1963–) developed a titanium BINAM catalyst that efficiently gives enantioselectivities of 88–97% *ee* in the Mukaiyama aldol additions of silyl ketene acetals onto aldehydes. [148] This reaction was applied by Smith during studies on the synthesis of phorboxazole (Scheme 3.61). [149]

Evans has applied copper bisoxazoline catalysis in the addition of acetate ketene thioacetals onto pyruvic acid derivatives. The catalyst loading is relatively low, and the reaction proceeds with excellent enantioselectivities at −10 °C (Scheme 3.62). [150]

The vinylogous extension of the Mukaiyama aldol reaction (VMAR) allows the synthesis of larger fragments generating a 1,5-dioxygenated system with a double bond. Scheme 3.63 shows an example where an initial VMAR is followed by a Lewis acid-mediated Prins-type cyclisation with a second aldehyde to give rise to 2,3,5-trisubstituted tetrahydropyrans with three new stereocenters in a one-pot process. [151] The VMA reaction has been reviewed. [152]

In an important series of developments related to the aldol reaction, one must note that in 1909 *Henry Drysdale Dakin* (1880–1952) had already observed that the Knoevenagel condensation between an aldehyde and an active methylene

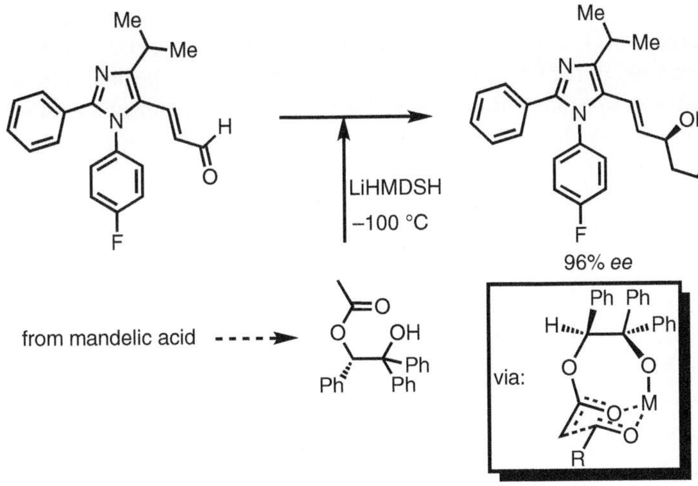

Scheme 3.60   *Braun acetate aldol in the synthesis of fluvastatin*

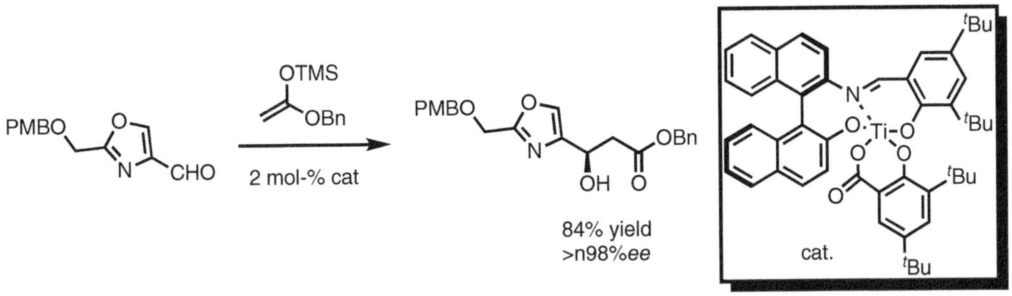

84% yield
>n98%ee

Scheme 3.61   *Carreira catalysis in Smith's phorboxazole synthesis*

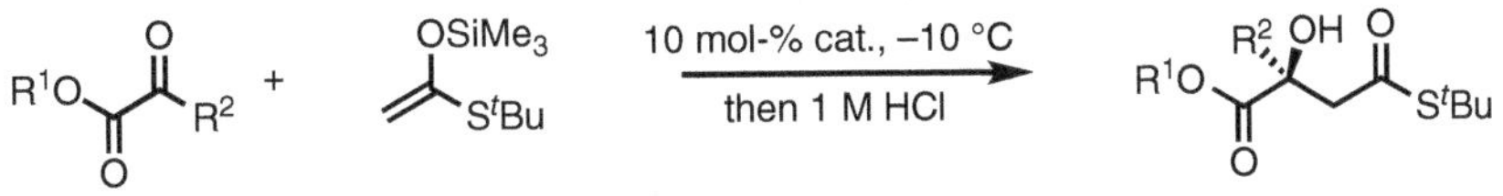

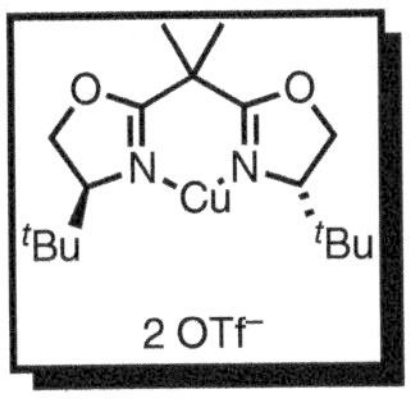

R^1	R^2	% ee
Me	Me	99
Bn	Me	99
^{t}Bu	Me	99
Me	Et	94
Me	^{i}Bu	94
Et	^{i}Bu	36

Scheme 3.62   *Evans acetate aldol*

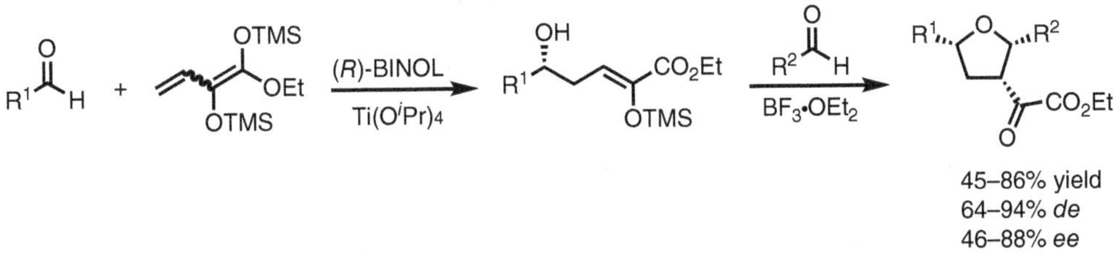

45–86% yield
64–94% de
46–88% ee

Scheme 3.63   *Enantioselective vinylogous Mukaiyama aldol route to tetrahydrofurans*

compound can be catalysed with amino acids. [153] Achiral enamine catalysis was developed mainly in the 1950s and 1960s, pioneered by Gilbert Stork. [154] During the 1970s, synthetic access to steroids became of very high economic interest, and the asymmetric formation of the CD ring system of steroids was an especially pressing problem (Chapter 9.3.2). The most straightforward way to achieve this is to utilise the classical Robinson annulation, provided that the reaction could be harnessed in an enantioselective form. Catalysis by enamines is known for these reactions, and utilisation of a chiral amine for the formation of the enamine nucleophile accomplishes the desired function (Scheme 3.64). [155] Thus, proline catalysis of the so-called Hajos–Parrish–Eder–Sauer–Wiechert reaction gives access to the nearly enantiopure product aldol, a result of enantiofacial selection of the two enantiotopic carbonyl groups. It is noteworthy that natural L-proline leads to the natural configuration of steroids. The mechanistic aspects of the reaction have been studied rather recently, after the renaissance of organocatalysis in the 2000s. [156]

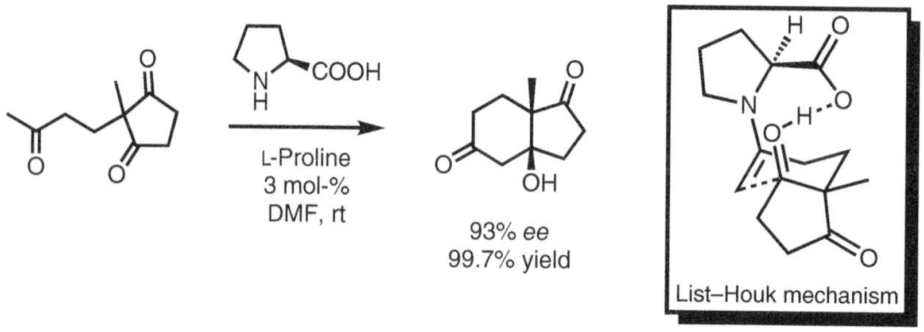

Scheme 3.64 *Hajos–Parrish–Eder–Sauer–Wiechert reaction*

Major advances in the field of organocatalytic aldol reactions have been achieved during the last two decades, and although major questions concerned with, e.g. efficient direction and control of *syn/anti* selectivity still remain, the whole field of organocatalysis has taken its major role as the new enantioselective-catalysis mode together with organometallic catalysis and enzyme (bio-) catalysis. The broad topic of organocatalysis in general is beyond the scope of this treatise. [157]

3.2.3 Reactions at the β-Carbon of an Enone

In 1953, Terentjev had already reported that cyclohexanone and 2-methyl cyclohexanone added to acrylonitrile will yield products with non-zero optical rotation when the reaction was carried out in basic ethanol in the presence of quartz (Scheme 3.65). [158]

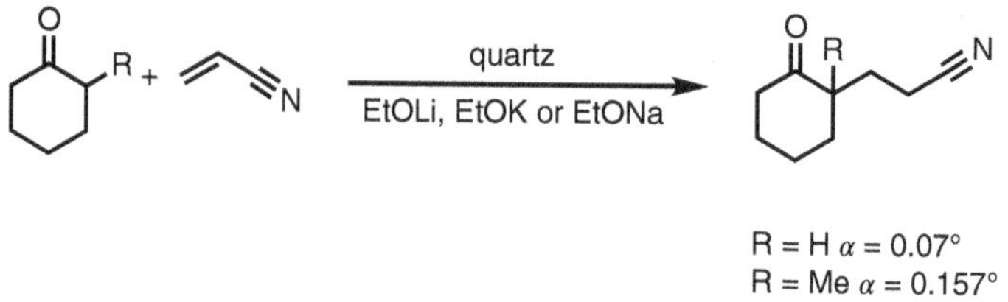

R = H α = 0.07°
R = Me α = 0.157°

Scheme 3.65 *First asymmetric Michael addition*

The first asymmetric conjugate addition was reported as late as 1962, by Inouye and *Harry M. Walborsky* (1923–2002) (Scheme 3.66). [159] Addition of phenylmagnesium bromide to (–)-menthyl crotonate gave (S)-3-phenylbutyric acid in 5–7% optical yield, and when the reaction was catalysed with cuprous chloride, the opposite (R)-enantiomer was formed. Walborsky's work went practically unnoticed, and a few years later Kawana and Emoto reported similar findings in the reaction of crotonyl esters of sugar derivatives and menthol with PhMgBr.

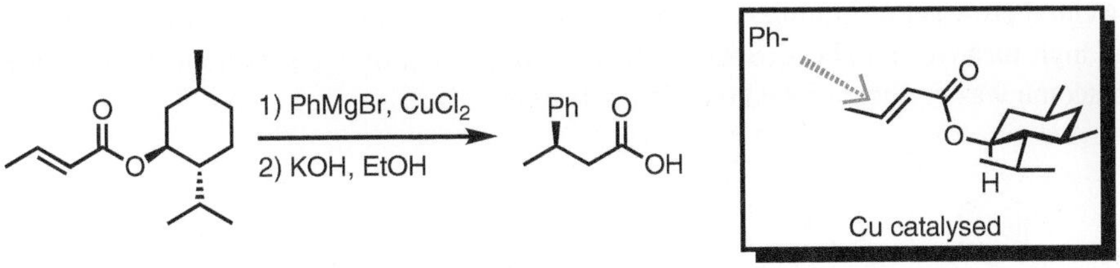

Scheme 3.66   Early asymmetric conjugate addition

[160] The best results were obtained with 3-*O*-crotonyl-1,2;5,6-di-*O*-isopropylidene-α-D-glucofuranose, when (*R*)-3-phenylbutyric acid was formed in 33% optical yield. In the presence of cuprous chloride, the enantioselectivity increased to 74%. The authors also describe a model to explain the sense of the enantioselectivity. The reversal of the stereochemistry upon addition of copper salts was explained by change of mechanism to involve an *s-cisoid* configuration with the magnesium reagent, and an *s-transoid* configuration with phenyl copper.

In the first reported catalytic asymmetric Michael addition, Långström and Bergson used partially resolved (57% *ee*) (*R*)-(+)-2-hydromethylquinuclidine as the chiral catalyst in the addition of methyl 2-carboxy-1-indanone to acrolein or α-isopropylacrolein (Scheme 3.67). [161] Equimolar amounts of the nuclephile and Michael acceptor were used, and the catalyst loading was as low as 1:3500. The products exhibited optical rotation, but the authors conclude by saying "As yet it is not possible to say anything regarding the degree of asymmetric induction in these reactions."

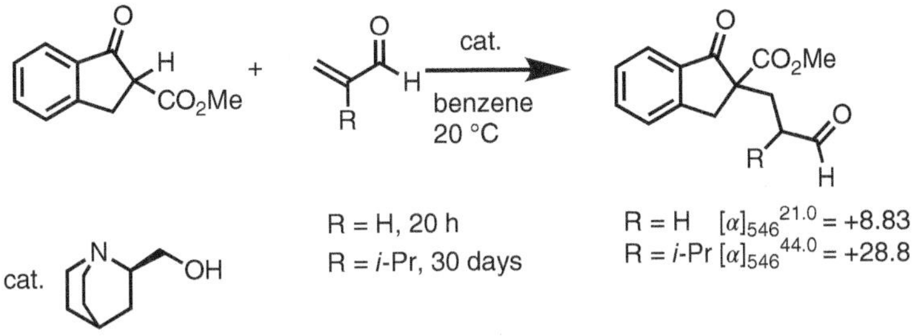

Scheme 3.67   First catalytic asymmetric Michael addition

Soon thereafter, *Hans Wynberg* (1922–2011) published his seminal work on cinchona alkaloid-catalysed conjugate additions (Scheme 3.68). [162] The enantiomeric excess was determined to be 56% *ee*. Interestingly, the solvent effect on yield was barely noticeable, whereas asymmetric induction heavily depended on the solvent. Cinchonine, the pseudo enantiomer of quinine, gave the enantiomeric product.

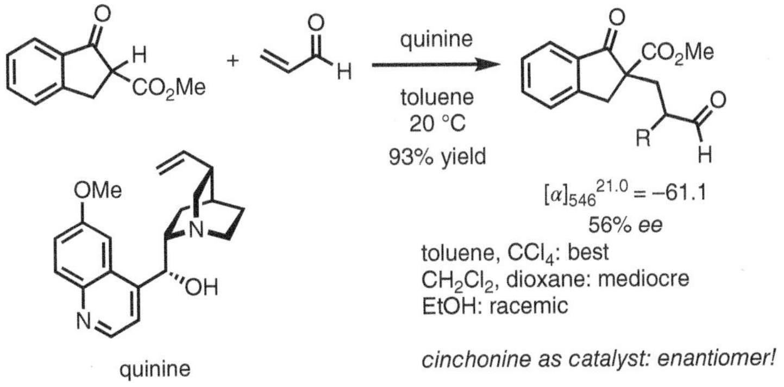

Scheme 3.68   Quinine catalysis in Michael addition

Cram utilised a chiral crown ether complexed to potassium to catalyse the Michael addition of β-ketoesters to methyl vinyl ketone with high turnover numbers (Scheme 3.69). [163] Enantioselectives up to 99% *ee* were achieved. The stereochemical outcome was predicted based on steric effects.

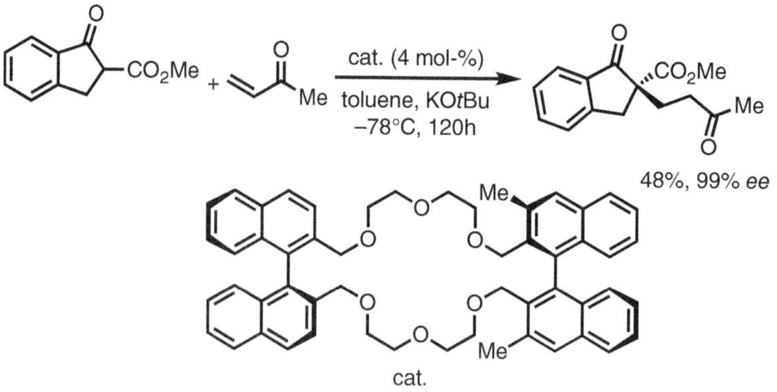

Scheme 3.69   *First phase-transfer-catalysed asymmetric Michael addition*

The mechanism of asymmetric induction in the case of chiral auxiliary–mediated conjugate additions was much studied by *Wolfgang Oppolzer* (1937–1996). The camphor-derived bicyclic dicyclohexyl sulfonamide derivative functions as an efficient chiral auxiliary in the alkylation of enoates (Scheme 3.70). [164] The sense of asymmetric induction is rationalised by steric (and electronic) reasons, the enoate adopts an *s-transoid* conformation preferentially. The *s-transoid* conformer is only *ca.* 1.3 kJmol^{-1} more stable in the ground state, but coordination with a Lewis acid further stabilises this conformation. The nucleophile now attacks the double bond from the *Si* face.

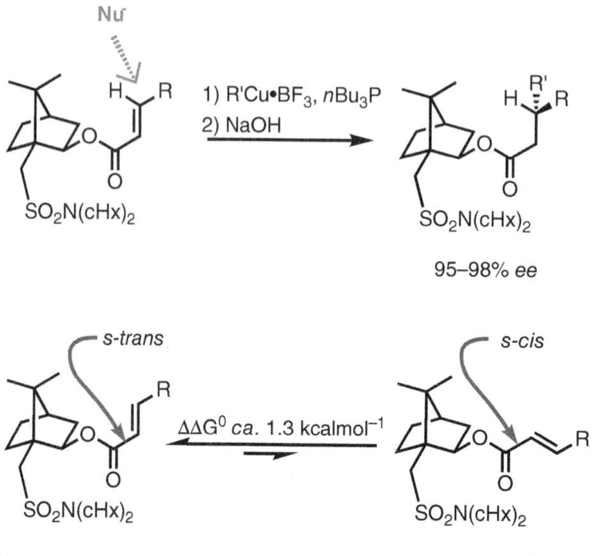

Scheme 3.70   *Oppolzer rationalisation for diastereoselectivities*

The related camphor-derived *N*-enoyl sultams undergo a highly diastereoselective addition with a wide range of organometallic reagents. Phosphine-stabilised alkyl and alkenyl cuprates give diastereoselectivities typically higher than 85% *de*, but as the products are highly crystalline, they can be crystallised to practically 100% diastereopurity (Scheme 3.71). [165] Note that the sense of asymmetric induction is opposite to that observed for the bicyclic camphorsulfonamides.

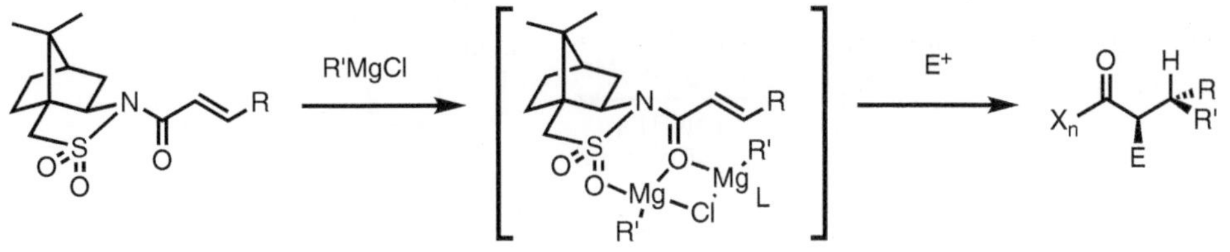

Scheme 3.71   Oppolzer camphor sultam auxiliary

Other chiral auxiliaries for carboxylic acid derivatives have also been used with moderate to good diastereoselectivity. The *E*-crotonate esters derived from 8-phenylmenthol have given good diastereoselectivities (up to 99% *de*) on reaction with Mukaiyama cuprates (RCu•BF$_3$). The corresponding *Z*-isomers as well as tri- and tetrasubstituted enoates gave only modest selectivity. [166]

Fumaric acid semialdehydes have been converted to the corresponding hemiaminals and aminals with a number of chiral aminols and diamines, respectively. Particularly promising results have been obtained by the Mukaiyama and Scolastico groups. Mukaiyama has utilised the aminals derived from (*S*)-prolinamine whose reaction with Grignard reagents under copper(I) catalysis gives rise to the absolute stereochemistry *R* at the new stereocenter (Scheme 3.72). [167] Complexation of the magnesium to the more basic bridgehead nitrogen is suggested to be the directing factor.

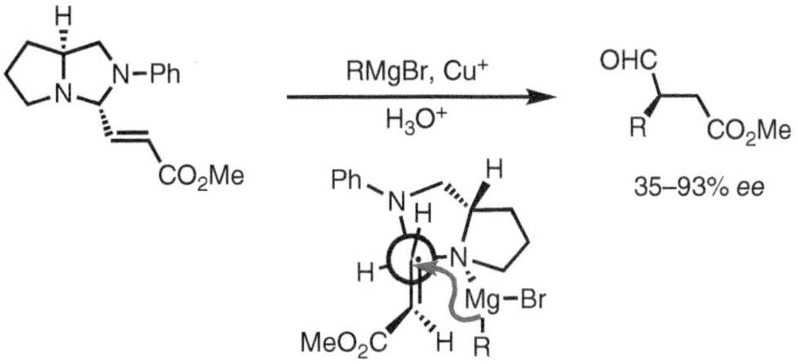

Scheme 3.72   Mukaiyama chiral auxiliary

Carlo Scolastico of Milan employed hemiaminals derived from norephedrine (Figure 3.26). α,β-Unsaturated aldehydes, ketones and esters with the oxazolidine auxiliary reacted with lithium dialkylcuprates to give the alkylation products with 80–90% *de*. [168]

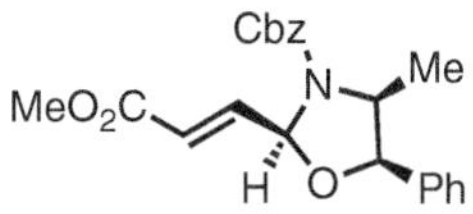

Figure 3.26   Scolastico chiral auxiliary

Masakatsu Shibasaki (1947–) has developed lanthanum-based chiral catalysts for the conjugate addition of β-ketoesters onto cyclic enones using an *N*-linked BINOL-type ligand (Scheme 3.73). [169] The reaction works also with α-substituted β-ketoesters to provide vicinal quaternary and tertiary centers with a diastereomeric ratio up to 86/14 and an *ee* up to 86%.

During synthetic studies towards brefeldin C, *Stuart L. Schreiber* (1956–) investigated an asymmetric enamine–enal cycloaddition, a reaction that can also be considered a sequential (organocatalytic) enamine-catalysed conjugate addition followed by ring closure (Scheme 3.74). [170] The starting aldehyde–enal was conveniently obtained by careful ozonolysis of 1,3-cyclooctadiene. When this was treated with the oxazolidone derived from condensation of pivalaldehyde and norephedrin (a 3:2 mixture of diastereomers!), a 17:1 mixture of the cycloadducts was generated. The

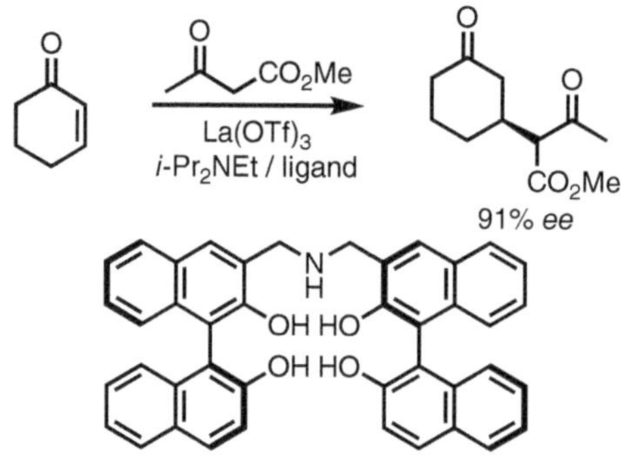

Scheme 3.73 *Shibasaki catalytic conjugate addition*

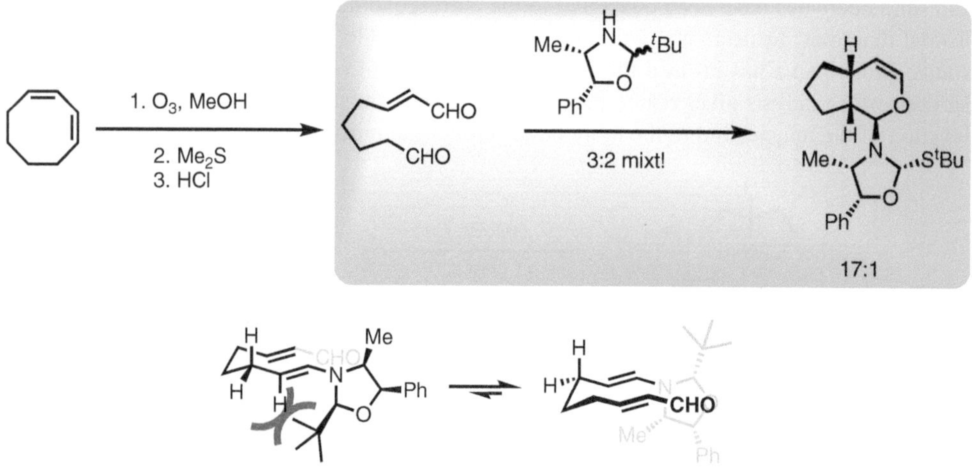

Scheme 3.74 *Schreiber enamine enal condensation*

stereoselectivity was explained to arise from the facial selectivity of the enamine, the disfavored enamine suffering from steric interference of the *tert*-butyl group and the alkene (a strain somewhat reminiscent of allylic 1,3-strain; remember that the enamine nitrogen is planarised, thus positioning the nitrogen substituents nearly in plane with the alkene).

The field of organocatalysis is being developed rapidly, and conjugate additions have been subjected to reviews. [138] Here we showcase two examples involving short practical syntheses of pharmaceuticals. *Karl Anker Jørgensen* (1955–) developed the first organocatalytic conjugate addition of malonates to cinnamaldehyde derivatives (Scheme 3.75). [171] The prolinol-catalysed conjugate addition gave the desired adducts generally in 86–95% ee. The *p*-fluorocinnamaldehyde adduct was converted to the lactam by reductive amination followed by lactam formation. Another three literature steps remain to convert the intermediate to the antidepressant paroxetine (paxil).

(+)-*cis*-2-Methyl-4-propyl-1,3-oxathiane has an important role in a natural flavor, showing a strong tropical-fruity odor with a green note; it is the main component responsible for the passion fruit aroma. By contrast, its enantiomer possesses a typical sulphurous, linseed oil-like odor. A simple organocatalytic Michael addition of benzyl mercaptan to *trans*-2-hexenal catalysed by the Jørgensen–Hayashi catalyst gave the Michael adduct in 84% ee (Scheme 3.76). [172] Cleavage of the benzyl-protecting group followed by conversion to the thioacetal gave the aroma compound in just four simple operations.

The final example comes from a conjugate addition of a Meldrum's acid derivative onto a nitro olefin (Scheme 3.77). [173] Despite prior art of using malonates as soft nucleophiles, we chose Meldrum's acid as a convenient choice, since the conjugate addition product can be converted, in two straightforward steps, to pregabalin, an anticonvulsant drug used to treat epilepsy and neuropathic pain marketed by Pfizer under the brand name Lyrica®. The cinchona alkaloid-derived

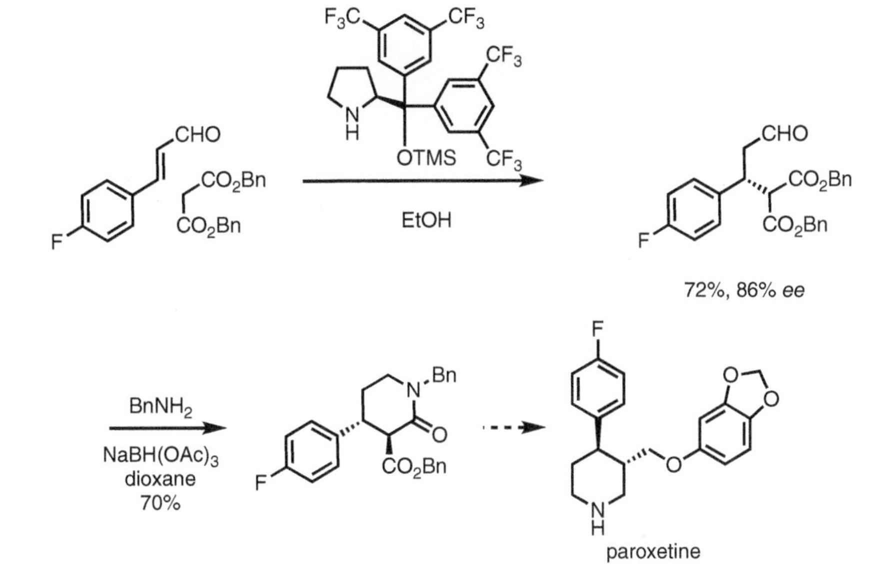

Scheme 3.75 *Jørgensen organocatalytic synthesis of paroxetine*

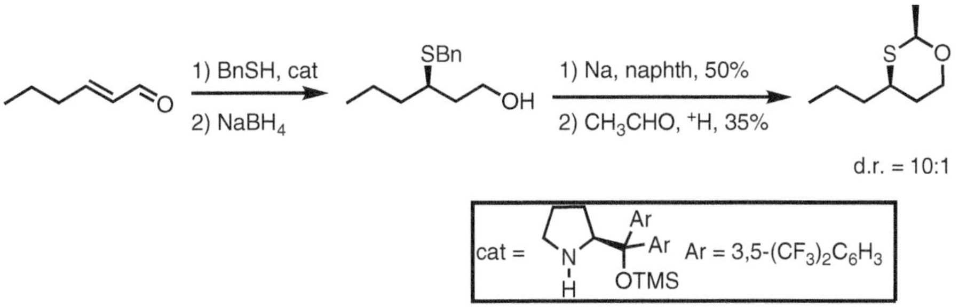

Scheme 3.76 *Organocatalytic Michael addition to passion fruit aroma compound*

thiourea activates the Michael acceptor through double NH hydrogen bonding, and the role of the quinuclidine is to orient the nucleophile properly, as shown in Scheme 3.77.

3.3 Reactions of Olefins

The enantiotopic faces of double bonds provide a means of converting prochiral sp^2 centers to chiral sp^3 centers either through hydrogenation (reduction) or the introduction of oxygen atoms at both ends of the double bond. In this Section, we will briefly examine enantioselective methods for both types of reactions. More comprehensive discussion concerning reduction or oxidation methods and their mechanisms, as well as asymmetric carbometallation reactions, is beyond the scope of this textbook.

3.3.1 Reduction

The first reported asymmetric reduction of a double bond was based on relayed asymmetric induction, the use of chiral auxiliaries (Scheme 3.78). [174] Vavon esterified β-methyl cinnamic acid with a number of optically active alcohols and subjected the ester to hydrogenation over platinum black. The resulting esters were hydrolysed to the optically

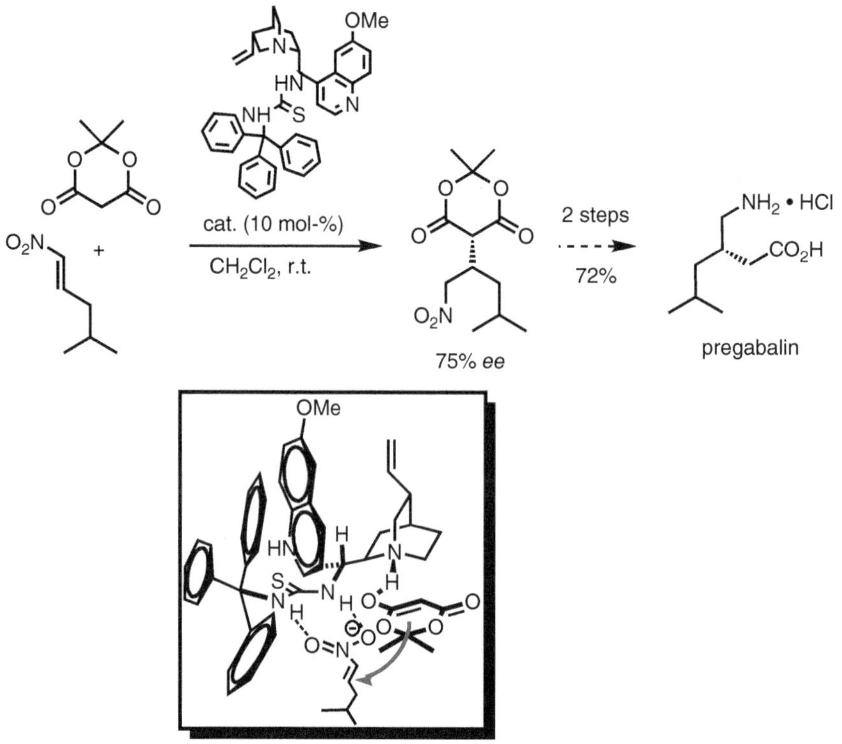

Scheme 3.77 *Koskinen organocatalytic synthesis of pregabalin*

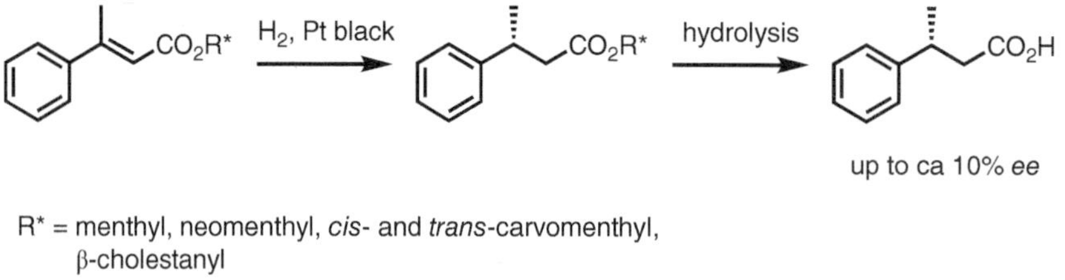

R* = menthyl, neomenthyl, *cis*- and *trans*-carvomenthyl,
β-cholestanyl

Scheme 3.78 *First asymmetric reduction of a double bond*

active acids. The observed asymmetric inductions where low, and the authors conclude that the reasons for asymmetric induction cannot be ascertained.

William Standish Knowles (1917–2012, Nobel Prize in chemistry in 2001) of the Monsanto Company first reported catalytic enantioselective hydrogenation of olefins using as catalyst (or correctly as catalyst precursor) a soluble rhodium complex which contains optically active tertiary phosphine ligands (Scheme 3.79). [175] Mixing (−)-methylpropylphenylphosphine of 69% optical purity with rhodium trichloride hydrate gave the solid catalyst complex. Hydrogenating α-phenylacrylic acid or itaconic acid in the presence of 0.15 mol-% of this catalyst gave hydratropic acid and methylsuccinic acid of modest optical purity.

This discovery soon led to the development of the so-called Monsanto process for the production of amino acids from inexpensive unsaturated acids (see Section 5.1.2). Further extensive improvement of the phosphine ligands has led to several successful, mostly C_2-symmetric bisphosphine ligands, as exemplified in Figure 3.27. [176]

Chiral Crabtree-like Ir-catalysts have also been developed intensively for olefin hydrogenation. [177] Some of these catalysts achieve phenomenal turnover numbers as exemplified by the Ir-SpiroPAP catalyst used in the synthesis of the NK-1 receptor antagonist aprepitant (Scheme 3.80). [178]

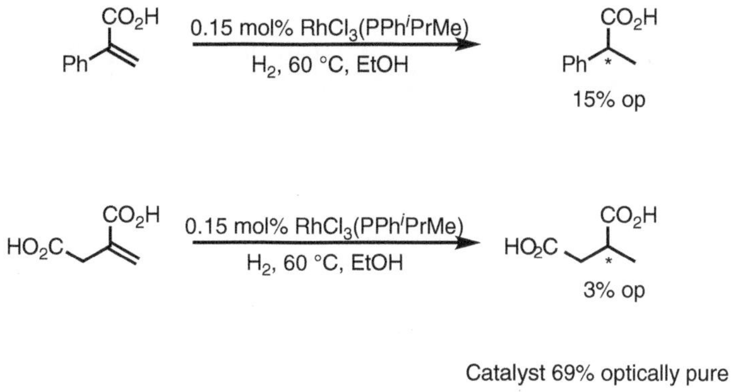

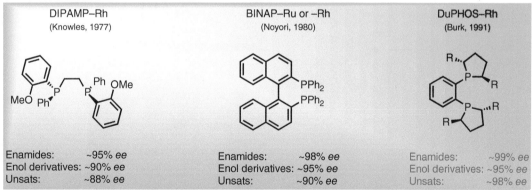

Scheme 3.79   *First chiral phosphine ligand for Rh-catalysed asymmetric hydrogenation*

Figure 3.27   *Chiral phosphine ligands for asymmetric olefin hydrogenation*

The Ir-Xyliphos catalyst is employed in the industrial synthesis for the production of (*S*)-metolachlor, the active ingredient in Dual Magnum®, one of the most important grass herbicides for use in maize. The industrial synthesis is only three steps long (Scheme 3.81), involving imine formation, the asymmetric reduction of the imine, and final acylation of the chiral secondary amine. [179] This process is today's largest application of asymmetric catalysis, reaching 10 000 tons/year. The Ir-Xyliphos hydrogenation catalyst achieves an unprecedented number of 2 million turnovers with a turnover frequency of 600 000 h^{-1}. The development started in 1982, and the first production batch was run in November 1996.

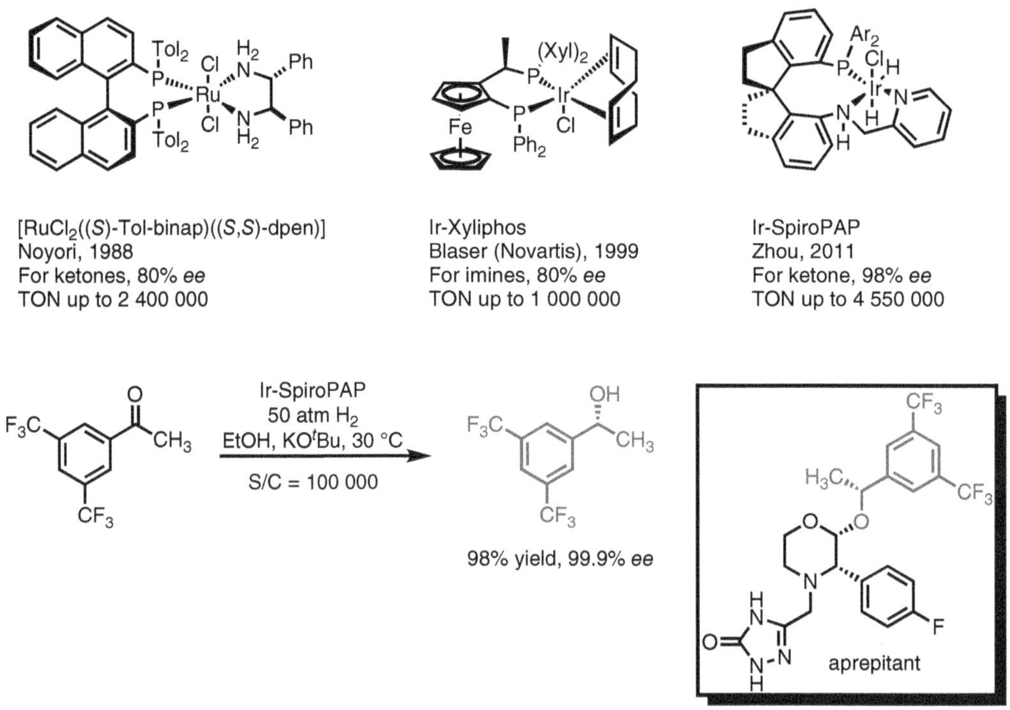

[RuCl2((S)-Tol-binap)((S,S)-dpen)]
Noyori, 1988
For ketones, 80% ee
TON up to 2 400 000
Ir-Xyliphos
Blaser (Novartis), 1999
For imines, 80% ee
TON up to 1 000 000
Ir-SpiroPAP
Zhou, 2011
For ketone, 98% ee
TON up to 4 550 000
Ir-SpiroPAP
50 atm H2
EtOH, KOtBu, 30 °C
S/C = 100 000
98% yield, 99.9% ee
aprepitant

Scheme 3.80 *High turn-over numbers can be achieved with Ir catalysts*

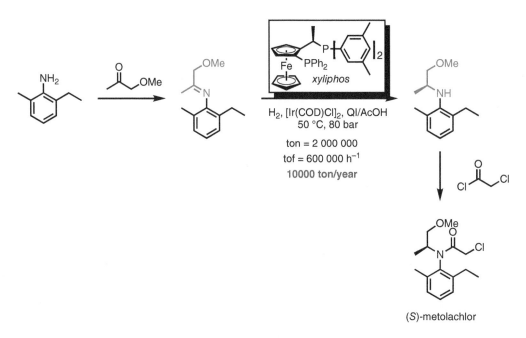

xyliphos
H2, [Ir(COD)Cl]2, QI/AcOH
50 °C, 80 bar
ton = 2 000 000
tof = 600 000 h−1
10000 ton/year
(S)-metolachlor

Scheme 3.81 *Largest industrial-scale catalytic asymmetric process*

3.3.2   Oxidation

The 1,2-dihydroxylated unit is a very common structural unit in natural products, and these functional systems have become widely studied targets. Since readily accessible alkenes are potential starting materials for the target compound types, this section will cover the stereocontrolled synthesis of deoxygenated species through direct oxidative methods.

Setting up a 1,2-dihydroxy system is most often carried out without formal change of the oxidation level of the two participating carbon atoms. Starting with an olefin, one can simply add the equivalent of hydroperoxide onto it to obtain the diol (Scheme 3.82). An alternative protocol is to first epoxidise the alkene and then open the epoxide to give the diol. If the alkene contains a stereocenter at the allylic position, the stereochemistry (diastereoselectivity) of the addition will be affected by the existing chiral center.

Scheme 3.82   Dioxygenation of allylically substituted alkenes

Depending on whether one wants to introduce the two hydroxyl groups *syn* or *anti* with respect to the emerging single bond, one has several alternative routes to follow. The simplest one is to rely on *cis* hydroxylation, usually carried out through osmylation (the so-called Upjohn dihydroxylation with catalytic OsO_4, using *N*-methyl morpholine *N*-oxide, NMMO, or trimethylamine *N*-oxide, TMANO as the stoichiometric oxidant). However, this requires the pre-existence of the desired *cis* or *trans* olefin, which in many cases becomes the limiting factor. The alternative epoxidation/epoxide-opening strategy requires access to stereochemically homogeneous olefins, and introduces another factor: regioselective opening of the epoxide. Powerful methodologies have been developed for both direct asymmetric dihydroxylation (AD) and asymmetric epoxidation (AE) for all of the six structurally distinct classes of olefins (Figure 3.28). As a rule of thumb, the most nucleophilic (electron-rich) olefin reacts fastest.

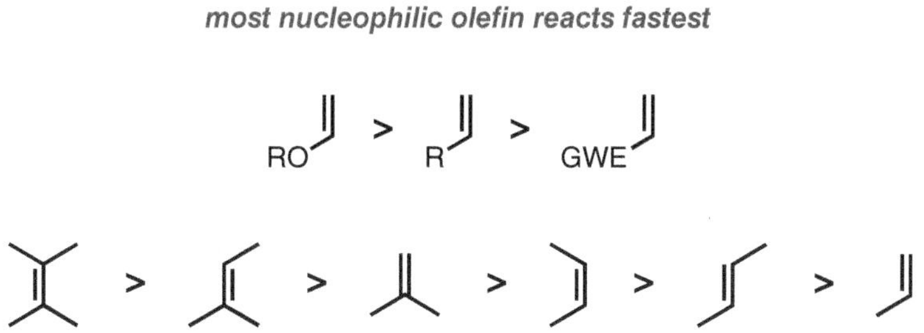

most nucleophilic olefin reacts fastest

Figure 3.28   Alkene types and reactivities

3.3.2.1   *Asymmetric dihydroxylation*

The stereochemical questions regarding the introduction of asymmetry from the existing chiral centers in the molecule have been studied by experimental and theoretical methods for allyl alcohol substrates. Allylic $A^{1,3}$-strain has been suggested to be the key factor governing the stereoselectivity (Figure 3.29). [3], [7] The lowest-energy ground-state conformation of allyl alcohols places the carbinol hydrogen in plane with the olefin to avoid steric repulsion of the alkyl or oxygen groups with the group R_Z on the distal atom of the olefin.

Figure 3.29   Allylic strain in allyl alcohol derivatives

In dihydroxylation, the delivery of the two oxygens would be either from the same face where the hydroxyl group resides (by its participation through a chelation-controlled process), or the face selectivity would be simply governed by steric effects, i.e. the differences in the sizes of the RO and R groups. *Yoshito Kishi* (1937–) of Harvard University has contributed significantly on the study of osmium-mediated *cis*-hydroxylations, and his group has shown that allyl alcohols and allyl esters give poorer diastereoselectivity than the corresponding allyl ethers. [180] Chelation plays a minimal role, and it was originally suggested that the selectivity is size controlled. *Kendall Houk* (1943–) further developed the model for asymmetric induction, and according to his model, the ethereal oxygen does participate in the direction of the hydroxylation by donating electron density from the σ^* orbital to the olefin's π orbital, effectively increasing the electron density on the face opposite to the oxygen, and thus making it more nucleophilic towards the (electrophilic) oxidising agent (inside alkoxy effect, see Section 3.1).[9] This would also explain why allyl esters are poorer in directing the stereoselectivity, as they are also poorer donors of electron density.

The earliest asymmetric dihydroxylation was reported by Alexander McKenzie in 1908 (Scheme 3.83). [181] He treated L-bornyl fumarate with potassium permanganate and observed that the reaction produced "a slight excess of L-bornyl-L-tartrate. Similar results were obtained in the oxidation of L-menthyl fumarate."

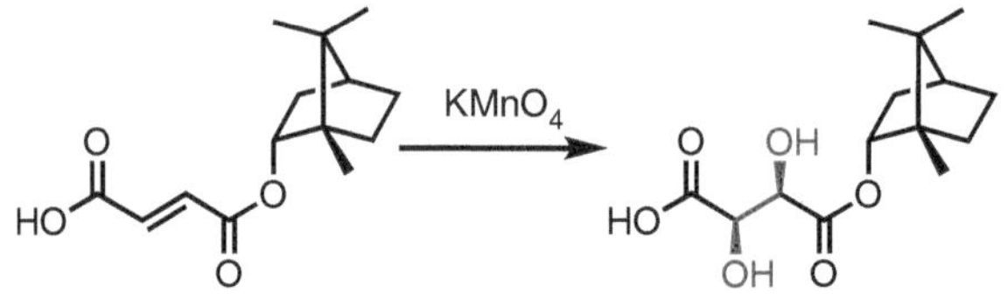

Scheme 3.83   *Early asymmetric dihydroxylation by McKenzie*

Dihydroxylation necessarily leads to the introduction of two new hydroxyl groups on the same face of the existing olefin. Dihydroxylation of olefins with osmium tetroxide is accelerated with (Lewis basic) donor ligands to osmium (ligand-acceleration effect), which makes the system especially responsive for asymmetric catalysis. External asymmetric induction is thus possible, and a number of efficient catalytic ligands have been designed and tested (Figure 3.30). [182]

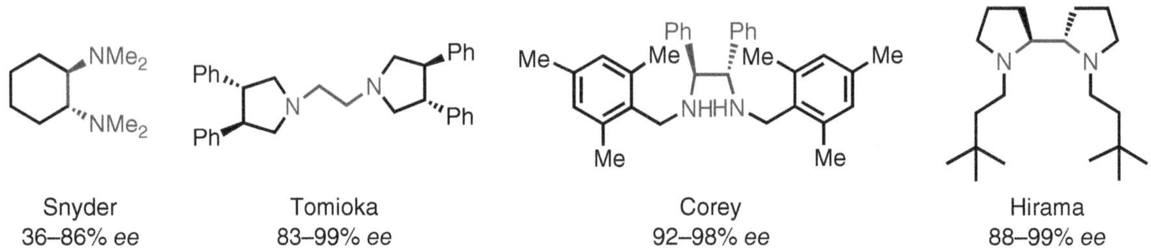

Snyder	Tomioka	Corey	Hirama
36–86% *ee*	83–99% *ee*	92–98% *ee*	88–99% *ee*

Figure 3.30   *Chiral ligands for asymmetric dihydroxylation*

K. Barry Sharpless (1941–) has been active in developing oxidation reactions of olefins. His group has devised catalytic systems capable of delivering asymmetric information very powerfully in both dihydroxylation (the Sharpless asymmetric dihydroxylation, SAD) and epoxidation reactions. In dihydroxylations, the chirality can be conveniently induced from derivatives of quinine. [183] Several improvements have been made over the years, and the method is also used industrially. A simple mnemonic model is shown in Figure 3.31.

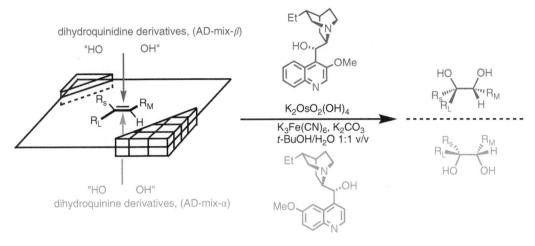

Figure 3.31 *Sharpless asymmetric dihydroxylation mnemonic model*

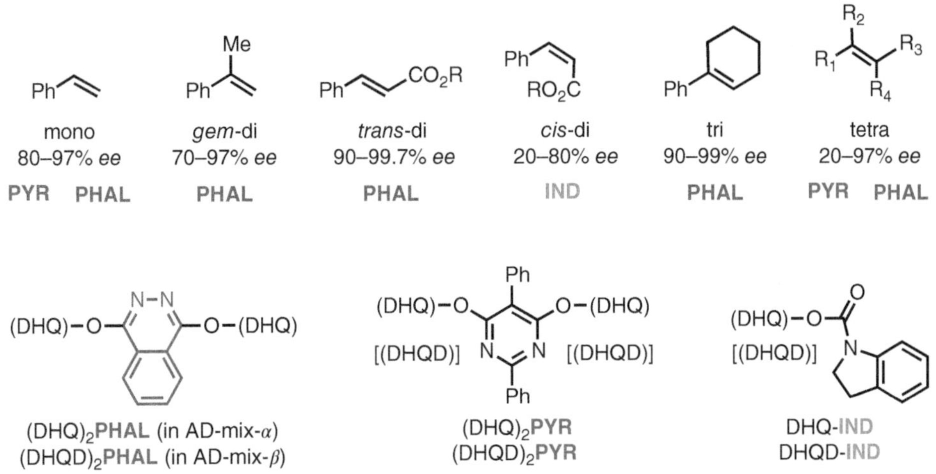

mono	*gem*-di	*trans*-di	*cis*-di	tri	tetra
80–97% *ee*	70–97% *ee*	90–99.7% *ee*	20–80% *ee*	90–99% *ee*	20–97% *ee*
PYR PHAL	PHAL	PHAL	IND	PHAL	PYR PHAL

(DHQ)$_2$PHAL (in AD-mix-α)
(DHQD)$_2$PHAL (in AD-mix-β)

(DHQ)$_2$PYR
(DHQD)$_2$PYR

DHQ-IND
DHQD-IND

Figure 3.32 *Substrate types and ligands for SAD*

Sharpless asymmetric dihydroxylation with commercial phthalazine ligands (AD-mix) works for most olefin types (Figure 3.32). However, some olefin types give poor enantioselectivities, and the Sharpless group has developed further ligands to broaden the scope of SAD. Notably, ligands derived from pyrazine (PYR) and dihydroindole (IND) practically cover the cases difficult for PHAL ligands.

Although the Sharpless asymmetric dihydroxylation is a reliable, high-yielding way of accessing the *cis*-diols, *trans*-dihydroxylation of olefins is very much undeveloped, and usually one has to adopt detours. Jørgensen has developed an organocatalytic method to achieve this. [184] The method is basically an organocatalytic epoxidation variation of the Juliá–Colonna oxidation (Scheme 3.84). The enal undergoes a nucleophilic epoxidation reaction, and under the reaction conditions, basic methanol-assisted epoxide opening initiates a series of ring-opening/-closing steps. The eventual product is the acetal of the initial aldehyde. In a similar way, using TsNHOTs in place of hydrogen peroxide gives aminohydroxylation (the product is the corresponding azetidine) with equally high enantio- and diastereoselectivities. Due to the mechanism of the reaction, the nitrogen atom ends up selectively in the β-position of the original aldehyde.

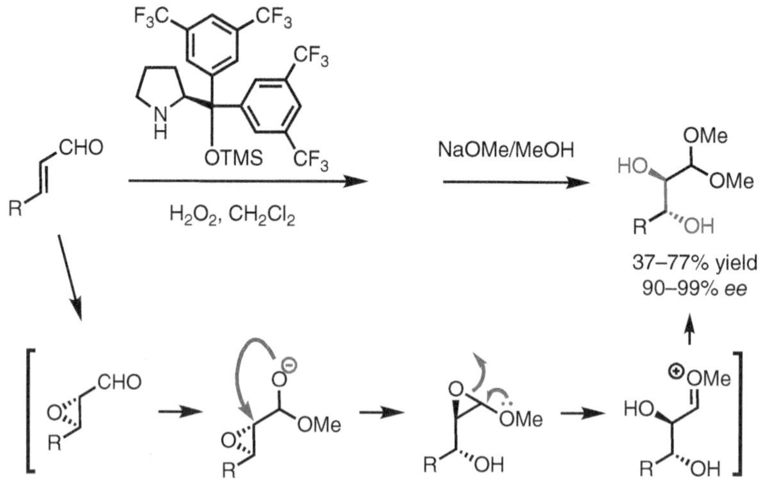

Scheme 3.84 *Jørgensen anti-selective dihydroxylation*

3.3.2.2 Asymmetric epoxidation

Epoxidation of an olefin leads to the addition of the epoxy oxygen on one face of the molecule. In the Prilezhaev epoxidation, an alkene is treated with a peroxy carboxylic acid to give the epoxide. [185] The lower the pK_a of the peracid, the higher it's reactivity. An early attempt of asymmetric Prilezhaev epoxidation dates back to 1960s, when, during a study on the reaction rates' dependence on solvents, they Ewins, Henbest, and McKervey encountered enantioselectivity in the oxidation of styrene with (+)-*cis*-monopercamphoric acid (Scheme 3.85). [186] Although the asymmetric induction was very low, the method has been applied in the synthesis of the quinoline alkaloid balfourodine and its congeners with 2–10% optical induction. [187]

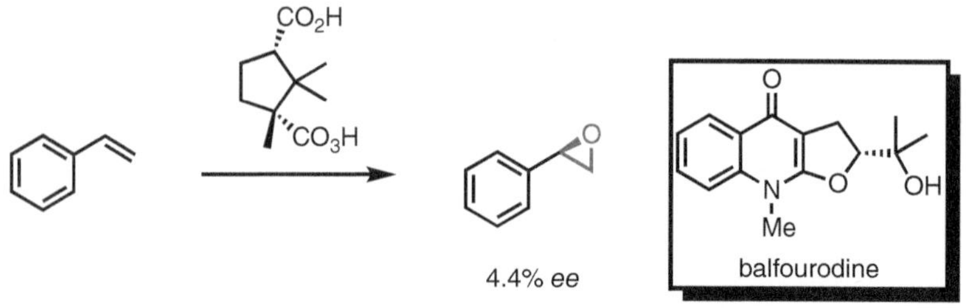

Scheme 3.85 *First asymmetric epoxidation*

If the olefin carries a substituent on the adjacent carbon atom, one can utilise the existing chirality either as a directing group (active volume) or as a blocking group (inactive volume) (Scheme 3.86). [188] Careful choice of the oxidant gives one the option to choose either an intramolecular delivery of the oxygen (Henbest oxidation) [189] to give the *syn*-addition product, or alternatively by pre-blocking the allylic (e.g. hydroxyl) with a suitable (bulky) protecting group to protect this same face from the attack of the oxidant, to give the *anti* product.

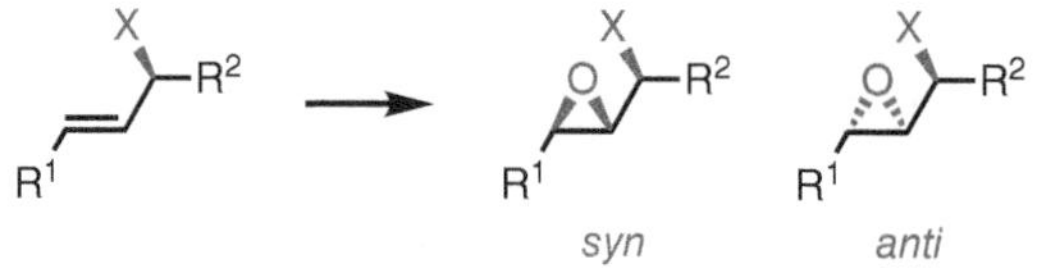

Scheme 3.86 *Diastereoselectivity in epoxidation*

The rationale is based on the application of the allylic $A^{1,3}$-strain, and accordingly, the selectivity is most pronounced with Z-olefins. The minimum-energy conformation corresponds to the conformer where the carbinol hydrogen is eclipsed with the olefinic linkage. [190] In this conformer, the two faces of the olefin are clearly distinct, and in the case of an oxidant capable of complexation with the hydroxyl function, the faces are amenable to specific means of epoxidation.

The formation of epoxides from cyclic allylic alcohols with peracid epoxidation occurs on the side *cis* to the alcohol group, and both diastereoselectivity and rate of the reaction are increased by unprotected allylic alcohol groups (Scheme 3.87). [159] The reaction follows the 'butterfly mechanism' of Bartlett, involving the interaction of the nucleophilic alkene with the electrophilic peracid. [191] The overall transition state of the synchronous reaction has been postulated using *ab initio* calculations and experimental kinetic isotope effects. The mechanism involves a *trans* antiperiplanar arrangement of the peracid O–O bond and the reacting alkene (*n-p** stabilisation). [192]

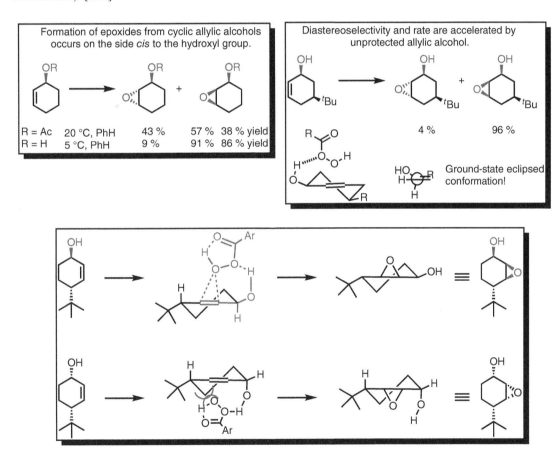

Scheme 3.87　*Allyl alcohols are favored substrates for Henbest epoxidation*

Oxyphilic transition metals can be used to enhance the diastereoselective delivery of the oxygen. [193] The reactions, applications, and rationalisations have been covered in an instructive review. [194]

With the development of metal-directed epoxidations, the stage was set for the discovery of the first catalytic asymmetric epoxidation reactions. Nearly simultaneously, the groups of *Shunichi Yamada* from Japan and Barry Sharpless from USA reported the first successful asymmetric epoxidations, using Mo or V as the directing metals, respectively (Figure 3.33). [195]

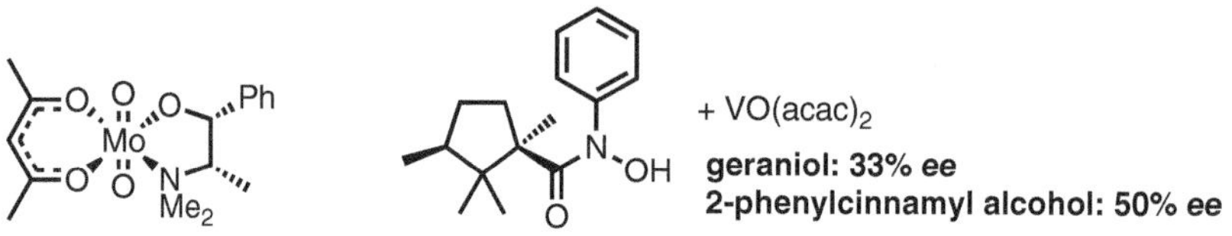

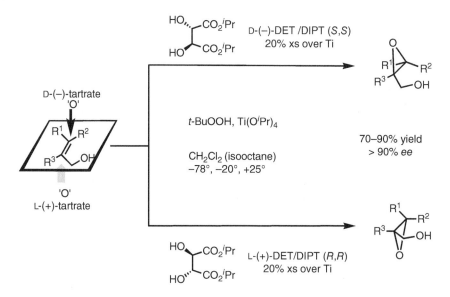

Figure 3.33 *Early asymmetric ligands for epoxidation*

A major breakthrough came in 1980 when Tsutomu Katsuki and Barry Sharpless published the seminal paper on asymmetric epoxidation (Scheme 3.88). [196] The new oxidation "gave uniformly high asymmetric inductions throughout a range of substitution patterns in the allylic alcohol substrate … and upon use of a given tartrate enantiomer, the system seems obliged to deliver the epoxide oxygen from the same enantioface of the olefin regardless of the substitution pattern." To top it all, the reagents (diethyl tartrate, titanium tetraisopropoxide, and *tert*-butyl hydroperoxide) are all commercially available at low cost.

Scheme 3.88 *Sharpless asymmetric epoxidation mnemonic model*

No wonder this reaction became known as the Sharpless asymmetric epoxidation (SAE). [197] SAE has become one of the reactions that have had the widest impact on asymmetric synthesis over the last three decades. This reaction has been developed to allow industrial-scale operations, and it has found vast applications in the syntheses of innumerable natural products and medicinally important agents. The Sharpless epoxidation is suitable to practically any allylic alcohol, which themselves are easily produced by classical means from simpler starting materials.

A complex formed from a tartrate ester (usually diisopropyl or diethyl tartrate) and titanium tetraisopropoxide is used as the chiral catalyst. The catalyst combines the olefin to be oxidised and the oxidant (*tert*-butyl hydroperoxide or cumene hydroperoxide) in such a manner that the delivery of the oxygen occurs principally from one face of the alkene. The detailed mechanism of the reaction has been studied, [198] and a reasonable model has been advanced to explain all the experimental observations (Scheme 3.89).

Although the Sharpless epoxidation has proven to be of wide utility, it still has some limitations. Only allylic alcohols seem to work well with regard to asymmetric induction. In fact, adding just one carbon, to give a homoallylic alcohol,

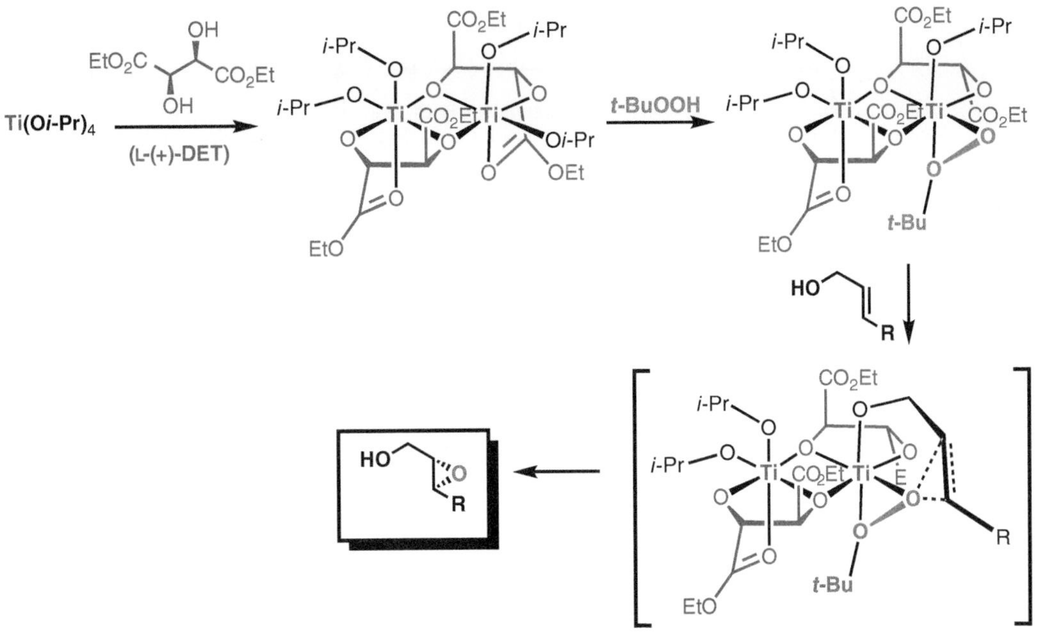

Scheme 3.89   Sharpless asymmetric epoxidation mechanistic rationalisation

degrades the induction down to *ca.* 50% *ee* levels, hardly meeting the desired criteria any more. This seems to be due to the structure of the catalyst and the transition state, as inversion of the sense of absolute stereochemistry at the emerging chiral center has also been observed. The reaction also uses toxic organic alkyl peroxides.

In order to avoid at least some of the problems associated with SAE (and other epoxidation reactions), *Hisashi Yamamoto* (1943–) has developed a tungsten-catalysed asymmetric epoxidation (Scheme 3.90). [199] It is the first highly enantioselective epoxidation protocol using a tungsten catalyst, and it uses environmentally benign aqueous H_2O_2 as oxidant instead of toxic organic alkyl peroxides. The reactions are simply performed under air, and in most cases, at RT, require no anhydrous solvent or preparation of metal–catalyst complex prior to the catalytic process. The substrate scope is broad, and high *ee*'s and good chemoselectivities for primary alcohols over secondary and tertiary alcohols are observed.

The second major problem lies in the fact that the substrate needs to be an alcohol. Without the hydroxyl group, the olefin does not bind to the titanium species and is thus not amenable to asymmetric induction. Much effort has been directed at overcoming this shortfall, and some new catalytic systems can ameliorate this problem, at least in part, as illustrated by the manganese-catalysed epoxidation by *Eric Jacobsen* (1960–) (Scheme 3.91). [200]

A related (salen) manganese(III)-complex catalyst system was also developed by *Tsutomu Katsuki* (1946–2014) (Scheme 3.92). [201] The chiral Mn-salen complex is formed from 1,2-diphenylethane-1,2-diamine (DPEN) and two axially chiral binaphthol derivatives. The BINOL-parts efficiently shield one quadrant of the complex from attacking the alkene, as illustrated by the X-ray structure of the diaqua complex of the catalyst precursor (MEBCUR).

The potassium salt of Caro's acid (potassium monoperoxysulfate, $KHSO_5$) readily oxidises ketones to the corresponding endoperoxides (dioxiranes), and in an early attempt at using ketone endoperoxides as asymmetric epoxidising reagents, Curci demonstrated that, e.g. isopinocamphone-derived ketone endoperoxide does give asymmetric induction, albeit in low levels. [202] Eventually, a robust, reproducible, and readily available ketone was found by Shi, and a practical organocatalyst for the epoxidation of a wide range of alkenes has been developed (Scheme 3.93). [203]

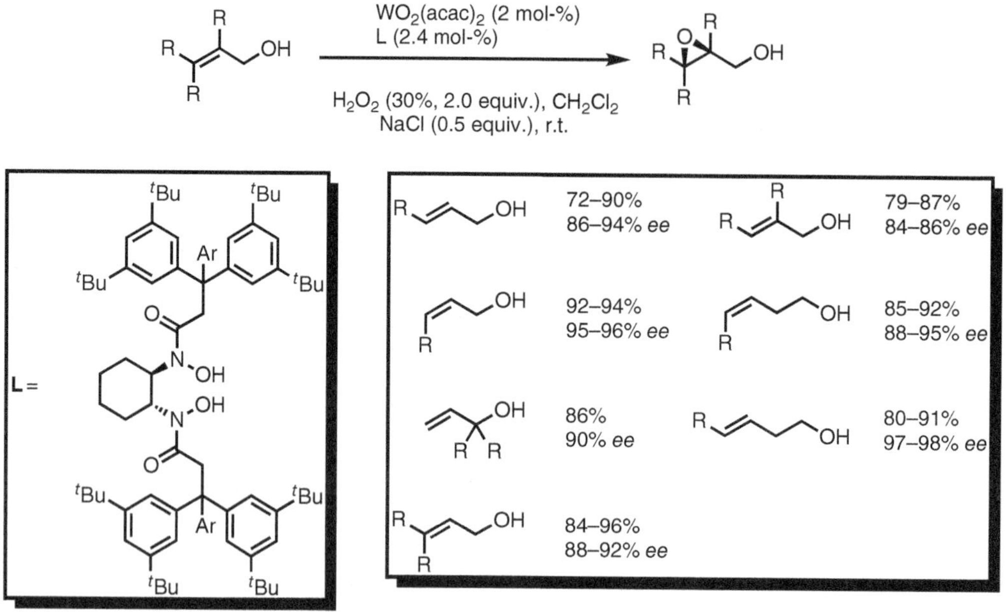

Scheme 3.90 *Tungsten-catalysed asymmetric epoxidation of allylic and homoallylic alcohols*

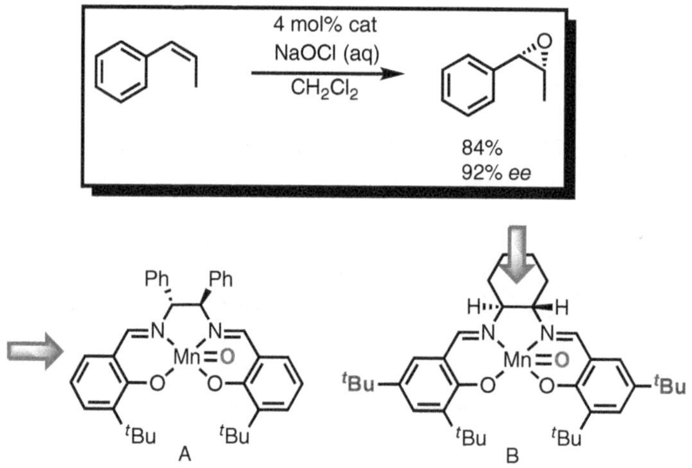

Scheme 3.91 *Jacobsen epoxidation*

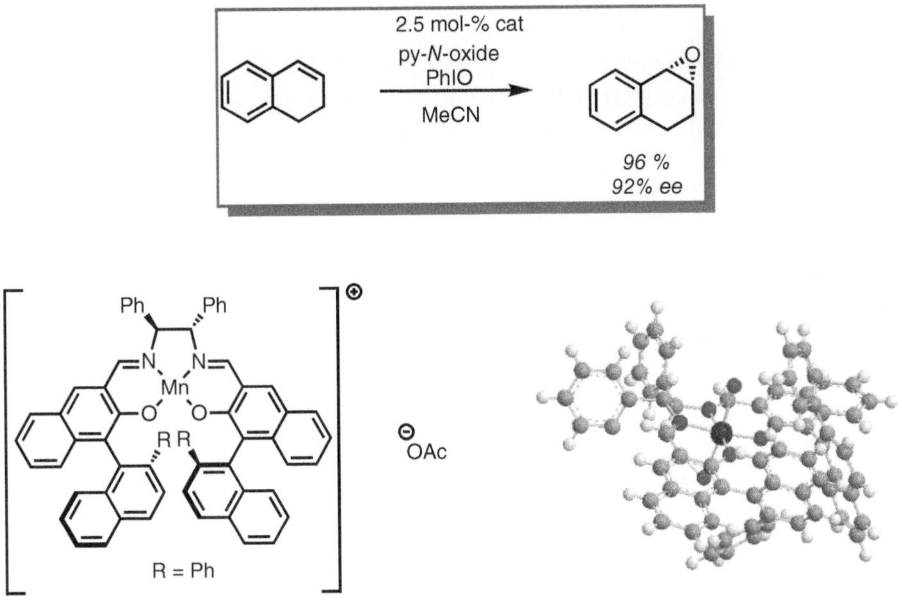

2.5 mol-% cat
py-*N*-oxide
PhIO

MeCN

96 %
92% ee

R = Ph

Scheme 3.92   *Katsuki epoxidation*

KHSO$_5$

Trifluoroacetone gives a more reactive dioxirane!

Curci *Chem. ommun.* **1984**, 155.

Shi catalysts:

from D-fructose   from L-fructose

Baumstark & McCloskey,
THL **1987**, *28*, 3311.

HSO$_5^-$

EtO$_2$C

cHx

Me

30 mol-% cat

138 mol-% Oxone
0.05 M Na$_2$B$_4$O$_7$
MeCN:DMM, 0 °C

EtO$_2$C

cHx

Me

89%, 94% ee

Shi, Y. *J. Am. Chem. Soc.* **1997**, *119*, 11224–11235.

Scheme 3.93   *Shi organocatalytic epoxidation*

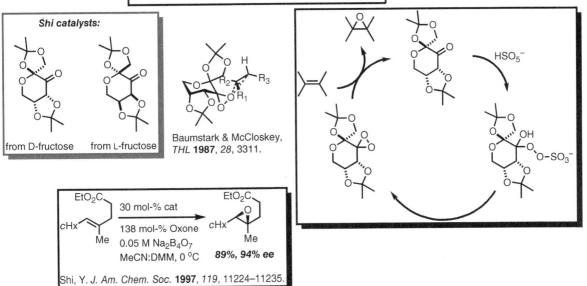

References

1. Barton, D.H.R. *Experientia* **1950**, *6*, 316–329.
2. Some representative textbooks: a) Höltje, H.-D., Folkers, G. *Molecular Modeling: Basic Priciples and Applications*, Wiley-VCh: Weinheim, **1997**, 187 pp; b) Lewars, E. *Computational Chemistry Introduction to the Theory and Applications of Molecular and Quantum Mechanics*, Kluwer Academic: New York, **2004**, 471 pp; c) Leach, A. *Molecular Modelling: Principles and Applications*, 2nd ed., Prentice-Hall: **2001**, 774 pp; d) Hinchliffe, A. *Molecular Modelling for Beginners*, John Wiley & Sons: Chichester, **2003**, 415 pp; e) Heine, T., Joswig, J.-O., Gelessus, A. *Computational Chemistry Workbook: Learning Through Examples*, Wiley-VCh, Weinheim, **2009**, 208 pp; f) Jensen, J.H. *Molecular Modeling Basics*, Taylor & Francis: Boca Raton, FL, **2010**, 166 pp.
3. Johnson, F. *Chem. Rev.* **1968**, *68*, 375–413.
4. Anet, F.A.L., Haq, M.Z. *J. Am. Chem. Soc.* **1965**, *87*, 3147–3150.
5. Senda, Y., Imazumi, S. *Tetrahedron* **1974**, *30*, 3813–3815.
6. Rickborn, B., Lwo, S.-Y. *J. Org. Chem.* **1965**, *30*, 2212–2216.
7. Hoffmann, R.W. *Chem. Rev.* **1989**, *89*, 1841–1860.
8. The calculations have been conducted with MacroModel, version 9.9, Schrödinger, LLC, New York, NY, **2011**.
9. Houk, K.N., Paddon-Row, M.N., Rondan, N.G., Wu, Y.-D., Brown, F.K., Spellmeyer, D.C., et al. *Science* **1986**, *231*, 1108–1117.
10. Whitesell, J.K., Deyo, D., Bhattacharya, A. *J. Chem. Soc., Chem. Commun.* **1983**, *802*.
11. Baldwin, J.E. *J. Chem. Soc., Chem. Comm.* **1976**, 738–741.
12. a) Menger, F.M. *Tetrahedron* **1983**, *39*, 1013–1040; b) Liotta, C.L., Burgess, E.M., Eberhardt, W.H. *J. Am. Chem. Soc.* **1984**, *106*, 4849–4852.
13. a) Anh, N.T., Eisenstein, O. *Nouv. J. Chim.* **1977**, *1*, 61–70; b) Scheiner, S., Lipscomb, W.M., Kleier, D.A. *J. Am. Chem. Soc.* **1976**, *98*, 4770–4777.
14. Bürgi, H.B., Dunitz, J.D. *Accts. Chem. Res.* **1983**, *16*, 153–161.
15. a) Bürgi, H.B., Dunitz, J.D., Shefter, E. *J. Am. Chem. Soc.* **1973**, *95*, 5065–5067; b) Bürgi, H.B., Dunitz, J.D., Lehn, J.M., Wipff, G. *Tetrahedron* **1974**, *30*, 1563–1572.
16. a) Fukui, K. *Accts. Chem. Res.* **1971**, *4*, 57–64; b) Houk, K.N. *Accts. Chem. Res.* **1975**, *8*, 361–369; c) Fleming, I. *Frontier Orbitals and Organic Chemical Reactions*, John Wiley & Sons, London, **1976**.
17. Cram, D.J., Abd Elhafes, F.A. *J. Am. Chem. Soc.* **1952**, *74*, 5828–5835.
18. Karabatsos, G.J. *J. Am. Chem. Soc.* **1967**, *89*, 1367–1371.
19. a) Cherest, M., Felkin, H., Prudent, N. *Tetrahedron Lett.* **1968**, 2199–2204; b) Cherest, M., Felkin, H. *Tetrahedron Lett* **1968**, 2205–2208.
20. Anh, N.T., Eisenstein, O. *Tetrahedron Lett.* **1976**, *17*, 155–158.
21. Cram, D.J., Kopecky, K.R. *J. Am. Chem. Soc.* **1959**, *81*, 2748–2755.
22. Cornforth J.W., Cornforth R.H., Mathew, K.K. *J. Am.Chem.Soc.* **1959**, *81*, 112–127.
23. a) Cieplak, A.S. *J. Am. Chem. Soc.* **1981**, *103*, 4540–4552; b) Cieplak, A.S. *Chem. Rev.* **1999**, *99*, 1265–1336.
24. Evans, D.A., Siska, S.J., Cee, V.J. *Angew. Chem. Int. Ed.* **2003**, *42*, 1761–1765.
25. a) Cieplak, A.S., Wiberg, K.S. *J. Am. Chem. Soc.* **1992**, *114*, 9226–9227; b) Cieplak, A.S. *Chem. Rev.* **1999**, *99*, 1265–1336.
26. Heathcock, C.H. *Aldrichimica Acta* **1990**, *23*, 99–111.
27. a) Heathcock, C.H., Flippin, L.A. *J. Am. Chem. Soc.* **1983**, *105*, 1667–1668; Lodge, E.P., Heathcock, C.H. *J. Am. Chem. Soc.* **1987**, *109*, 3353–3361.
28. Danishefsky, S., Kato, N., Askin, D., Kerwin, Jr., J.F. *J. Am. Chem. Soc.* **1982**, *104*, 360–362.
29. a) *Asymmetric Synthesis*; Morrison, J.D., ed.; Academic Press: New York, **1985**; Vol. 2; b) Singh, V.K. *Synthesis* **1992**, 605–617; c) Itsuno, S. *Org. React.* **1998**, *52*, 395–576; d) Cho, B.T. *Chem. Soc. Rev.* **2009**, *38*, 443–452.
30. a) Bothner-By, A.A. *J. Am. Chem. Soc.* **1951**, *73*, 846; b) Portoghese, P.S. *J. Org. Chem.* **1962**, *27*, 3359–3360.
31. Minoura, Y., Yamaguchi, H. *J. Polym. Sci. A-1* **1968**, *6*, 2013–2021.
32. a) Landor, S.R., Miller, B.J., Tatchell, A.R. *Proc. Chem. Soc.* **1964**, *227*; b) Landor, S.R., Miller, B.J., Tatchell, A.R. *J. Chem. Soc.* **1967**, 197–201.
33. a) Yamaguchi, S., Mosher, H.S. *J. Org. Chem.* **1973**, *38*, 1870–1877; b) Reich, C.J., Sullivan, G.R., Mosher, H.S. *Tetrahedron Lett.* **1973**, 1505–1508.
34. Cohen, N., Lopresti, R.J., Neukom, C., Saucy, G. *J. Org. Chem.* **1980**, *45*, 582–588.
35. Trost, B.M., Hipskind, P.A., Chung, J.Y.L., Chan, C. *Angew. Chem. Int. Ed.* **1989**, *28*, 1502–1504.

36. a) Noyori, R., Tomino, I.;,Tanimoto, Y. *J. Am. Chem. Soc.* **1979**, *101*, 3129–3131; b) Noyori, R., Tomino, I., Tanimoto, Y., Nishizawa, M. *J. Am. Chem. Soc.* **1984**, *106*, 6709–6716; c) Noyori, R., Tomino, I., Yamada, M., Nishizawa, M. *J. Am. Chem. Soc.* **1984**, *106*, 6717–6725.

37. Suzuki, M., Morita, Y., Koyano, H., Koga, M., Noyori, R. *Tetrahedron* **1990**, *46*, 4809–4822.

38. Chong, J.M., Mar, E.K. *Tetrahedron Lett.* **1990**, *31*, 1981–1984.

39. Midland, M.M. *Chem. Rev.* **1989**, *89*, 1553–1561.

40. Midland, M.M., McLoughlin, J.I. *J. Org. Chem.* **1984**, *49*, 1316–1317.

41. a) Midland, M.M., Tramontano, A., Zderic, S.A. *J. Organomet. Chem.* **1978**, *156*, 203–211; b) Midland, M.M., Zderic, S.A. *J. Am. Chem. Soc.* **1982**, *104*, 525–528.

42. a) Midland, M.M., Greer, S., Tramontano, A., Zderic, S.A. *J. Am. Chem. Soc.* **1979**, *101*, 2352–2355; b) Midland, M.M., McDowell, D.C., Hatch, R.L., Tramontano, A. *J. Am. Chem. Soc.* **1980**, *102*, 867–869.

43. Midland, M.M., Graham, R.S. *Org. Synth.* **1984**, *63*, 57–65.

44. Midland, M.M., Halterman, R.L., Brown, C.A., Yamaichi, A. *Tetrahedron Lett.* **1981**, *22*, 4171–4172.

45. Midland, M.M., Tramontano, A., Kazubski, A., Graham, R.S., Tsai, D.J.S., Cardin, D.B. *Tetrahedron* **1984**, *40*, 1371–1380.

46. Brown, H.C., Pai, G.G. *J. Org. Chem.* **1985**, *50*, 1384–1394.

47. Midland, M.M., McLoughlin, J.I., Gabriel, J. *J. Org. Chem.* **1989**, *54*, 159–165.

48. Brown, H.C., Chandrasekharan, J., Ramachandran P.V. *J. Am. Chem. Soc.* **1988**, *110*, 1539–1546.

49. Imai, T., Tamura, T., Yamamuro, A., Sato, T., Wollmann, T.A., Kennedy, R.M., et al. *J. Am. Chem. Soc.* **1986**, *108*, 7202–7204.

50. a) Masamune, S., Kennedy, R.M., Petersen, J.S., Houk, K.N., Wu, Y.-D. *J. Am. Chem. Soc.* **1986**, *108*, 7404–7405; b) Rarig, R.-A.F., Nelson, J.M., Vedejs, E. *J. Org. Chem.* **2017**, *82*, 12757–12762.

51. a) Itsuno, S., Ito, K., Hirao, A., Nakahama, S. *J. Chem. Soc., Chem. Commun.* **1983**, 469–470; b) Itsuno, S., Ito, K., Hirao, A., Nakahama, S. *J. Org. Chem.* **1984**, *49*, 555–557; c) Itsuno, S., Nakano, M., Miyazaki, K., Masuda, H., Ito, K., Hirao, A., et al. *J. Chem. Soc., Perkin 1*, **1985**, 2039–2044; d) Itsuno, S., Nakano, M., Ito, K., Hirao, A., Owa, M., Kanda, N., Nakahama, S. *J. Chem. Soc., Perkin 1*, **1985**, 2615–2619; e) Itsuno, S., Sakurai, Y., Ito, K., Hirao, A., Nakahama, S. *Bull. Chem. Soc. Jpn* **1987**, *60*, 395–396.

52. Corey, E.J., Bakshi, R.K., Shibata, S. *J. Am. Chem. Soc.* **1987**, *109*, 5551–5553.

53. a) Corey, E.J., Link, J.O. *Tetrahedron Lett.* **1989**, *30*, 6275–6278; b) Corey, E.J., Link, J.O. *J. Org. Chem.* **1991**, *56*, 442–444.

54. a) Corey, E.J., Shibata, S., Bakshi, R.K. *J. Org. Chem.* **1988**, *63*, 2862–2864; b) Mathre, D.J., Jones, T.K., Xavier, L.C., Blacklock, T.J., Reamer, R.A., Mohan, J.J., et al. *J. Org. Chem.* **1991**, *56*, 751–762.

55. Corey, E.J., Bakshi, R.K., Shibata, S., Chen, C.-P., Singh, V.K. *J. Am. Chem. Soc.* **1987**, *109*, 7925–7926.

56. a) Corey, E.J., Da Silva Jardine, P., Mohri, T. *Tetrahedron Lett.* **1988**, *29*, 6409–6412; b) Corey, E.J., Da Silva Jardine, P. *Tetrahedron Lett.* **1989**, *30*, 7297–7300; c) Corey, E.J., Helal, C.J. *Angew. Chem., In. Ed.* **1998**, *37*, 1986–2012.

57. Corey, E.J., Reichard, G.A. *Tetrahedron Lett.* **1989**, *30*, 5207–5210.

58. Corey, E.J., Bakshi, R.K. *Tetrahedron Lett.* **1990**, *31*, 611–614.

59. Corey, E.J., Cheng, X.-M., Cimprich, K.A., Sarshar, S. *Tetrahedron Lett.* **1991**, *32*, 6835–6838.

60. a) Nevalainen, V. *Tetrahedron: Asymmetry* **1991**, *2*, 63–74; b) Nevalainen, V. *Tetrahedron: Asymmetry* **1991**, *2*, 1133–1155; c) Corey, E.J., Lok, T.-P. *J. Am. Chem. Soc.* **1991**, *113*, 8966–8967.

61. Noyori, R., Ohkuma, T. *Angew. Chem. Int. Ed.* **2001**, *40*, 40–73.

62. Ohkuma, T., Ishii, D., T:akeno, H., Noyori, R. *J. Am. Chem. Soc.* **2000**, *122*, 6510–6511.

63. Oishi, T., Nakata, T. *Synthesis* **1990**, 635–645.

64. Abdel-Akher, M., Hamilton, J.K., Smith, F. *J. Am. Chem. Soc.* **1951**, *73*, 4691–4692.

65. Theander, O. *Acta Chem. Scand.* **1958**, *12*, 1883–1885.

66. Narasaka, K., Pai, F.C. *Chem. Lett.* **1980**, 1415–1418.

67. Chen, K.-M., Hardtmann, G.E., Prasad, K., Repic, O., Shapiro, M.J. *Tetrahedron Lett.* **1987**, *28*, 155–158.

68. Fürst, A., Plattner, P.A. *Helv. Chim. Acta* **1949**, *32*, 275–283.

69. a) Gribble, G.W., Ferguson, D.C. *J. Chem. Soc., Chem. Commun.* **1975**, 535–536; b) Gribble, G.W., Nutaitis, C.F. *Org. Proc. Prep. Int.* **1975**, *17*, 317–384.

70. Saksena, A.K., Mangiaracina, P. *Tetrahedron Lett.* **1983**, *24*, 273–276.

71. Turnbull, M.D., Hatter, G., Ledgerwood, D.E. *Tetrahedron Lett.* **1984**, *25*, 5449–5452.

72. a) Evans, D.A., Chapman, K.T. *Tetrahedron Lett.* **1986**, *27*, 5939–5942; b) Evans, D.A., Chapman, K.T., Carreira, E.M. *J. Am. Chem. Soc.* **1988**, *110*, 3560–3578; c) Evans, D.A., Gauchet-Prunet, J.A., Carreira, E.M., Charette, A.B. *J. Org. Chem.* **1991**, *56*, 741–750.

73. Roth, B.D. *Progr. Med. Chem.* **2002**, *40*, 1–22.

74. a) Chen, K.-M., Hardtmann, G.E., Prasad, K., Repic, O., Shapiro, M.J. *Tetrahedron Lett.* **1987**, *28*, 155–158; b) Sletzinger, M., Verhoeven, T.R., Volante, R.P., McNamara, J.M., Corley, E.G., Liu, T.M.H. *Tetrahedron Lett.* **1985**, *26*, 2951–2924.

75. Mori, Y., Kuhara, M., Takeuchi, A., Suzuki, M. *Tetrahedron Lett.* **1988**, *29*, 5419–5422.

76. For review: Repic, O., Prasad, K., Lee, G.T. *Org. Proc. Res. Dev.* **2001**, *5*, 519–527.

77. Evans, D.A., Hoveyda, A.H. *J. Am. Chem. Soc.*, **1990**, *112*, 6447–6449.

78. Romo, D., Meyer, S.D., Johnson, D.D., Schreiber, S.L. *J. Am. Chem. Soc.* **1993**, *115*, 7906–7907.

79. Koskinen, A.M.P., Kataja, A.O. *Org. React.* **2015**, *86*, 105–409.

80. a) Bredig, G. *Chem. Ztg.* **1912**, *35*, 324–325; b) Bredig, G., Fiske, P.S. *Biochem. Z.* **1913**, *46*, 7–23.

81. Dauben, W.G., Dickel, D.F., Jeger, O., Prelog, V. *Helv. Chim. Acta.* **1953**, *36*, 325–336.

82. a) Whitesell, J.K., Bhattacharya, A., Henke, K. *J. Chem. Soc., Chem. Commun.* **1982**, 988–989; b) Whitesell, J.K., Deyo, D., Bhattacharya, A. *J. Chem. Soc., Chem. Commun.* **1983**, 802.

83. Mazaleyrat, J.-P., Cram, D.J. *J. Am. Chem. Soc.* **1981**, *103*, 4585–4586.

84. a) Soai, K., Mukaiyama, T. *Chem. Lett.* **1978**, 491–492; b) Mukaiyama, T., Soai, K., Sato, T., Shimizu, H., Suzuki, K. *J. Am. Chem. Soc.* **1979**, *101*, 1455–1460.

85. Soai, K., Mukaiyama, T. *Bull. Chem. Soc. Jpn* **1979**, *52*, 3371–3376.

86. Frankland, E. *Justus Liebigs Ann. Chem.* **1849**, *79*, 171.

87. a) Soai, K., Niwa, S. *Chem. Rev.* **1992**, *92*, 833–856; b) Takahashi, H., Kawakita, T., Yoshioka, M., Kobayashi, S., Ohno, M. *Tetrahedron Lett.* **1989**, *30*, 7095–7098; c) Tanaka, K., Ushio, H., Suzuki, H. *J. Chem. Soc., Chem. Commun.* **1989**, 1700–1701; d) Joshi, N.N., Srebnik, M., Brown, H.C. *Tetrahedron Lett.* **1989**, *30*, 5551–5554; e) Kitamura, M., Okada, S., Suga, S., Noyori, R. *J. Am. Chem. Soc.* **1989**, *111*, 4028–4036; f) Corey, E. J., Hannon, F.J. *Tetrahedron Lett.* **1987**, *28*, 5233–5236; g) Corey, E.J., Hannon, F.J. *Tetrahedron Lett.* **1987**, *28*, 5237–5240; h) Corey, E.J., Yuen, P.-W., Hannon, F.J., Wierda, D.A. *J. Org. Chem.* **1990**, *55*, 784–786.

88. a) Noyori, R., Kitamura, M. *Angew. Chem., Int. Ed. Engl.* **1991**, *30*, 49–69; b) Noyori, R. *Science* **1990**, *248*, 1194–1199.

89. a) Duthaler, R.O., Hafner, A. *Chem. Rev.* **1992**, *92*, 807–832; b) Seebach, D., Beck, A.K., Heckel, A. *Angew. Chem. Int. Ed.* **2001**, *40*, 92–138.

90. Weber, B., Seebach, D. *Angew. Chem., Int. Ed. Engl.* **1992**, *31*, 84–86.

91. Osakama, K., Nakajima, M. *Org. Lett.* **2016**, *18*, 236–239.

92. Kavanagh, S.E., Gilheany, D.G. *Org. Lett.* **2020**, *22*, 8198–8203.

93. Hoffmann, R.W. *Angew. Chem.* **1982**, *94*, 569–590.

94. Denmark, S.E., Weber, E.J. *Helv. Chim. Acta* **1983**, *66*, 1655–1660.

95. a) Hoffmann, R.W., Herold, T. *Chem. Ber.* **1981**, *114*, 375–383; b) Brown, H.C.;,Jadhav, P.K. *J. Am. Chem. Soc.* **1983**, *105*, 2092–2093; c) Roush, W.R., Walts, A.E., Hoong, L.K. *J. Am. Chem. Soc.* **1985**, *107*, 8186–8190; d) Garcisa, J., Kim, B.M., Masamnue, S. *J. Org. Chem.* **1987**, *52*, 4831–4832; e) Short, R.P., Masamune, S. *J. Am. Chem. Soc.* **1989**, *111*, 1892–1984; f) Corey, E.J., Yu, C.-M., Kim, S.S. *J. Am. Chem. Soc.* **1989**, *111*, 5495–5496; g) Lachance, H., Hall, D.G. *Org. Reactions* **2008**, *73*, 1–573.

96. Brown, H.C., Bhat, K.S., Randad, R.S. *J. Org. Chem.* **1989**, *54*, 1570–1576.

97. a) Roush, W.R., Hoong, L.K., Palmer, M.A.J., Park, J.C. *J. Org. Chem.* **1990**, *55*, 4109–4117; b) Roush, W.R., Hoong, L.K., Palmer, M.A.J., Straub, J.A., Palkowitz, A.D. *J. Org. Chem.* **1990**, *55*, 4117–4126.

98. Hafner, A., Duthaler, R.O., Marti, R., Rihs, G., Rothe-Streit, P., Schwarzenbach, F. *J. Am. Chem. Soc.* **1992**, *114*, 2321–2336.

99. Dittoo, A., Bellosta, V., Cossy, J. *Synlett* **2008**, 2459–2460.

100. a) Patman, R.L., Bower, J.F., Kim, I.S., Krische, M.J. *Aldrichim. Acta* **2008**, *41*, 95–104; b) Hassan, A., Krische, M.J. *Org. Process Res. Dev.* **2011**, *15*, 1236–1242.

101. Lu, Y., Woo, K., Krische M.J. *J. Am. Chem. Soc.* **2011**, *113*, 13876–13879.

102. Boyall, D., López, F., Sasaki, H., Frantz, D., Carreira, E J. *Org. Lett.* **2000**, *2*, 4233–4236.

103. a) Trost, B.M., Weiss, A.H. *Adv. Synth. Catal.* **2009**, *351*, 963–983; b) Trost, B.M., Bartlett, M.J., Weiss, A.H., von Wangelin, J.A., Chan, V.S. *Chem. Eur. J.* **2012**, *18*, 16498–16509.

104. Corey, E.J., Sneen, R.A. *J. Am. Chem. Soc.* **1956**, *78*, 6269–6278.

105. Zimmerman, H.E. *Accts. Chem. Res.* **1987**, *20*, 263–268.

106. a) Ireland, R.E., Mueller, R.H., Willard, A.K. *J. Am. Chem. Soc.* **1976**, *98*, 2868–2877; b) Ireland, R.E., Wipf, P., Armstrong, J.D. *J. Org. Chem.* **1991**, *56*, 650–657.

107. Hall, P.L., Gilchrist, J.H., Collum, D.B. *J. Am. Chem. Soc.* **1991**, *113*, 9571–9574; b) Woltornist, R.A., Collum, D.B. *J. Am. Chem. Soc.* **2021**, *143*, 17452–17464.

108. Stork, G., Manabe, K., Liu, L. *J. Am. Chem. Soc.* **1998**, *120*, 1337–1338.

109. Pincock, R.E., Rolston, J.H. *J. Org. Chem.* **1964**, *29*, 2990–2992.
110. Yamada, S., Hiroi, K., Achiwa, K. *Tetrahedron Lett.* **1969**, *48*, 4233–4236.
111. a) Tomioka, K., Ando, K., Takemasa, Y., Koga, K. *J. Am. Chem. Soc.* **1984**, *106*, 2718–2719; b) Tomioka, K., Ando, K., Takemasa, Y., Koga, K. *Tetrahedron Lett.* **1984**, *25*, 5677 5680; c) Koga, K. *Pure Appl. Chem.* **1994**, *66*, 1487–1492.
112. Evans, D.A., Bartroli, J., Shih, T.L. *J. Am. Chem. Soc.* **1981**, *103*, 2127–2129.
113. Evans, D.A., Ennis, M.D., Mathre, D.J. *J. Am. Chem. Soc.* **1982**, *104*, 1737–1739.
114. Gnas, Y., Glorius, F. *Synthesis* **2006**, 1899–1930.
115. Job, A., Janeck, C.F., Bettray, W., Peters, R., Enders, D. *Tetrahedron* **2002**, *58*, 2253–2329.
116. Myers, A.G., Yang, B.H., Chen, H., Gleason, J.L. *J. Am. Chem. Soc.* **1994**, *116*, 9361–9362.
117. a) Shirai, R., Tanaka, M., Koga, K. *J. Am. Chem. Soc.* **1986**, *108*, 543–545; b) Cain, C.M., Cousins, R.P.C., Coumbarides, G., Simpkins, N.S. *Tetrahedron* **1990**, *46*, 523–544.
118. Toriyama, M., Sugasawa, K., Shindo, M., Tokutake, N., Koga, K. *Tetrahedron Lett.* **1997**, *38*, 567–570.
119. Cano, R., Zakarian, A., McGlacken, G.P. *Angew. Chem. Int. Ed.* **2017**, *56*, 9278–9290.
120. a) Kane, R. *Ann. Phys.* **1838**, *120*, 473–494; b) Kane, R. *J. Prakt. Chem.* **1838**, *15*, 129–155.
121. a) Borodin, A. *J. Prakt. Chem.* **1864**, *93*, 413–425; b) Wurtz, C.A. *Bull. Soc. Chim. Fr.* **1872**, *17*, 436–442; c) Wurtz, C.A. *J. Prakt. Chem.* **1872**, *5*, 457–464; d) Wurtz, C.A. *Compt. rend. Acad. sci.* **1872**, *74*, 1361; e) Wurtz, C.A. *Ber.* **1872**, *5*, 326.
122. Heathcock, C.H., Buse, C.T., Kleschick, W.A., Pirrung, M.C., Sohn, J.E., Lampe, J. *J. Org. Chem.* **1980**, *45*, 1066–1081.
123. a) Evans, D.A., Nelson, J.V., Taber, T.R. In *Topics in Stereochemistry* (Eliel, E.L., Wilen, S.H., eds.) Wiley Interscience: New York, **1983**; Vol. 13, Chapter 1; b) Heathcock, C.H. In *Comprehensive Carbanion Chemistry* (Buncel, E., Durst, T., eds.) Elsevier: New York, **1984**; Vol. 5B, p. 177; c) Heathcock, C.H. In *Asymmetric Synthesis* (Morrison, J.D., ed.) Academic Press: New York, **1984**; Vol. 3, 213–274; d) Mukaiyama, T. *Org. React.* **1982**, *28*, 203–331; e) Heathcock, C.H. *Science* **1981**, *214*, 395–400.
124. Zimmerman, H.E., Traxler, M.D. *J. Am. Chem. Soc.* **1957**, *79*, 1920–1923.
125. a) Seebach, D. *Angew. Chem.* **1988**, *100*, 1685–1715; b) Nevalainen, V. *Tetrahedron: Asymmetry* **1991**, *2*, 63–74.
126. a) Amstutz, R., Schweizer, W.B., Seebach, D., Dunitz, J.E. *Helv. Chim. Acta* **1981**, *64*, 2617–2621; b) Willard, P.G., Carpenter, G.B. *J. Am. Chem. Soc.* **1986**, *108*, 462–468; c) Willard, P.G., Carpenter, G.B. *J. Am. Chem. Soc.* **1985**, *107*, 3345–3346.
127. House, H.O., Prabhu, A.V., Phillips, W.V. *J. Org. Chem.* **1976**, *41*, 1209–1214.
128. Yamamoto, Y., Maruyama, K. *Tetrahedron Lett.* **1980**, *21*, 4607–4610.
129. a) Li, Y., Paddon-Row, M.N., Houk, K.N. *J. Am. Chem. Soc.* **1988**, *110*, 3684–3686; b) Li, Y., Paddon-Row, M.N., Houk, K.N. *J. Org. Chem.* **1990**, *55*, 481–493.
130. Denmark, S.E., Henke, B.D. *J. Am. Chem. Soc.* **1991**, *113*, 2177–2194.
131. Nerz-Stormes, M., Thornton, E.R. *J. Org. Chem.* **1991**, *56*, 2489–2498.
132. Hintermann, T., Seebach, D. Helv. *Chim. Acta* **1998**, *81*, 2093–2126. CSD refcode GOMLUP.
133. a) Danda, H., Hansen, M.M., Heathcock, C.H. *J. Org. Chem.* **1990**, *55*, 173–181; b) Walker, M.A., Heathcock, C.H. *J. Org. Chem.* **1991**, *56*, 5747–5750; c) Hayashi, K., Hamada, Y., Shioiri, T. *Tetrahedron Lett.* **1991**, *32*, 7287–7290.
134. Evans, D.A., Downey, C.W., Shaw, J.T., Tedrow, J.S. *Org. Lett.* **2002**, *4*, 1127–1130.
135. Evans, D.A., Tedrow, J.S., Shaw, J.T., Downey, C.W. *J. Am. Chem. Soc.* **2002**, *124*, 392–393.
136. Geary, L.M., Hultin, P.G. *Tetrahedron: Asymmetry* **2009**, *20*, 131–173.
137. Crimmins, M.T., King, B.W., Tabet, E.A., Chaudhary, K. *J. Org. Chem.* **2001**, *66*, 894–902.
138. Roush, W. *J. Org. Chem.* **1991**, *56*, 4151–4157.
139. Evans, D.A., Dow, R.L., Shih, T.L., Takacs, J.M., Zahler, R. *J. Am. Chem. Soc.* **1990**, *112*, 5290–5313.
140. Ito, Y., Sawamura, M., Hayashi, T. *J. Am. Chem. Soc.* **1986**, *108*, 6405–6406.
141. a) Kobayashi, S., Fujishita, Y., Mukaiyama, T. *Chem. Lett.* **1990**, 1455–1458; b) Kobayashi, S., Furuya, M., Ohtsubo, A., Mukaiyama, T. *Tetrahedron: Asymmetry* **1991**, *2*, 635–638.
142. Corey, E.J., Kim, S.S. *J. Am. Chem. Soc.* **1990**, *112*, 4976–4977.
143. Furuta, K., Maruyama, T., Yamamoto, H. *J. Am. Chem. Soc.* **1991**, *113*, 1041–1042.
144. Lee, J.M., Zhang, X., Norrby, P.-O., Helquist, P., Wiest, O. *J. Org. Chem.* **2016**, *81*, 5314–5321.
145. Braun, M., Devant, R. *Tetrahedron Lett.* **1984**, *25*, 5031–5034.
146. Prasad, K., Chen, K.-M., Repič, O. *Tetrahedron: Asymm.* **1990**, *1*, 703–706.
147. Repič, O. *Principles of Process Research and Chemical Development in the Pharmaceutical Industry*, John Wiley & Sons, New York: **1998**, pp. 147–149.
148. Carreira, E.M., Singer, R.A., Lee, W. *J. Am. Chem. Soc.* **1994**, *116*, 8837–8838.
149. Smith III, A.B., Verhoest, P.R., Minbiole, K.P., Lim, J.J. *Org. Lett.* **1999**, *1*, 909–912.

150. Evans, D.A., Kozlowski, M.C., Burgey, C.S., MacMillan, D.W.C. *J. Am. Chem. Soc.* **1997**, *119*, 7893–7894.
151. Hoffmeyer, P., Schneider, C. *J. Org. Chem.* **2019**, *84*, 1079–1084.
152. Cordes, M., Kalesse, M. *Molecules* **2019**, *24*, 3040–3058.
153. Dakin, H.D. *J. Biol. Chem.* **1910**, *7*, 49–55.
154. Stork, G., Terrell, R., Szmuszkovicz, J. *J. Am. Chem. Soc.* **1954**, *76*, 2029–2030.
155. a) Hajos, Z.G., Parrish, D.R. *J. Org. Chem.* **1974**, *39*, 1615–1621; b) Eder, U., Sauer, G., Wiechert, R. *Angew. Chem. Int. Ed. Engl.* **1971**, *10*, 496–497.
156. a) Bahmanyar, S., Houk, K.N. *J. Am. Chem. Soc.* **2001**, *123*, 12911–12912; b) Hoang, L., Bahmanyar, S., Houk, K.N., List, B. *J. Am. Chem. Soc.* **2003**, *125*, 16–17.
157. a) Berkessel, A., Gröger, H. *Asymmetric Organocatalysis*, Wiley-VCh Weinheim: **2005**, 440 pp; b) *Enantioselective Organocatalysis*, Dalko P. I. (ed.), Wiley-VCh Weinheim: **2007**, 536 pp; c) Xiang, S.-H., Tan, B. *Nature Comm.* **2020**, *11*, 3786.
158. Terentjev, A.P., Klabunovskii, E.I., Budovskii, E.I. *Sbornik Statei po Obshchei Khimii* **1953**, *2*, 1612–1616.
159. Inouye, Y., Walborsky, H.M. *J. Org. Chem.* **1962**, *27*, 2706–2707.
160. Kawana, M., Emoto, S. *Bull. Chem. Soc. Jpn* **1966**, *39*, 910–916.
161. Långström, B., Bergson, G. *Acta Chem. Scand.* **1973**, *27*, 3118–3119.
162. Wynberg, H., Helder, R. *Tetrahedron Lett.* **1975**, 4057–4060.
163. Cram, D.J., Sogah, G.D.Y. *J.C.S. Chem. Comm.* **1981**, *13*, 625–628.
164. a) Oppolzer, W. *Angew. Chem.* **1984**, *96*, 840–854; b) Oppolzer, W., Dudfield, P., Stevenson, T., Godel, T. *Helv. Chim. Acta* **1985**, *68*, 212–215; c) Oppolzer, W. *Tetrahedron* **1987**, *43*, 1969–2004.
165. Oppolzer, W., Mills, R.J., Pachinger, W., Stevenson, T. *Helv. Chim. Acta* **1986**, *69*, 1542–1545.
166. Oppolzer, W., Moretti, R., Godel, T., Meunier, A., Löher, H. *Tetrahedron Lett.* **1983**, *24*, 4971–4974.
167. Asami, M., Mukaiyama, T. *Chem. Lett.* **1979**, 569–572.
168. a) Scolastico, C. *Pure Appl. Chem.* **1988**, *60*, 1689–1698; b) Bernardi, A., Cardani, S., Pilati, T., Poli, G., Scolastico, C., Villa, R. *J. Org. Chem.* **1988**, *53*, 1600–1607.
169. a) Sasai, H., Arai, T., Satow, T., Houk, K.N., Shibasaki, M. *J. Am. Chem. Soc.* **1995**, *117*, 6194–6198; b) Majima, K., Tosaki, S., Ohshima, T., Shibasaki, M. *Tetrahedon Lett.* **2005**, *46*, 5377–5381.
170. Schreiber, S.L., Meyers, H.V. *J. Am. Chem. Soc.* **1988**, *110*, 5198–5200.
171. Brandau, S., Landa, A., Franzén, J., Marigo, M., Jørgensen, K.A. *Angew. Chem. Int. Ed.* **2006**, *45*, 4305–4309.
172. Scafato, P., Colangelo, A., Rosini, C. *Chirality* **2009**, *21*, 176–182.
173. Bassas, O., Huuskonen, J., Rissanen, K., Koskinen, A.M.P. *Eur. J. Org. Chem.* **2009**, 1340–1351.
174. Vavon, G., Jakubowicz, B. *C. R. Acad. Sci.* **1933**, *196*, 1614–1617.
175. Knowles, W.S., Sabacky, M.J. *Chem. Comm.* **1968**, 1445–1446.
176. Burk, M.J. *J. Am. Chem. Soc.* **1991**, *113*, 8518–8519.
177. a) Brown, J.M. *Organometallics* **2014**, *33*, 5912–5923; b) Verendel, J.J., Pàmies, O., Diéguez, M., Andersson, P.G. *Chem. Rev.* **2014**, *114*, 2130–2169; c) Massaro, L., Zheng, J., Margarita, C., Andersson, P.G. *Chem. Soc. Rev.* **2020**, *49*, 2504–2522.
178. Xia, J.-H., Liu, X.-Y., Xie, J.-B., Wang, L.X., Zhou, Q.-L. *Angew. Chem. Int. Ed.* **2011**, *50*, 7329–7332.
179. Blaser, H.-U., Pugin, B., Splinder, F., Thommen, M. *Acc. Chem. Res.* **2007**, *40*, 1240–1250.
180. Cha, J.K., Christ, W.J., Kishi, Y. *Tetrahedron* **1974**, *40*, 2247–2255.
181. McKenzie, A., Wren, H. *Proc. Chem. Soc.* **1908**, *23*, 188.
182. a) Tokles, M., Snyder, J.K. *Tetrahedron Lett.* **1986**, *27*, 3951–3954; b) Tomioka, K., Nakajima, M., Koga, K. *J. Am. Chem. Soc.* **1987**, *109*, 6213–6215; c) Tomioka, K., Nakajima, M., Iitaka, Y., Koga, K. *Tetrahedron Lett.* **1988**, *29*, 573–576; d) Corey, E.J., DaSilva Jardine, P., Virgil, S., Yuen, P.-W., Connell, R.D. *J. Am. Chem. Soc.* **1989**, *111*, 9243–9244; e) Oishi, T., Hirama, M. *J. Org. Chem.* **1989**, *54*, 5834–5835; f) Hirama, M., Oishi, T., Ito, S. *J. Chem. Soc., Chem. Commun.* **1989**, 665–666.
183. a) Kolb, H.C., VanNieuwenhze, M.S., Sharpless, K.B. *Chem. Rev.* **1994**, *94*, 2483–2547; b) Noe, M.C., Letavic, M.A., Snow, S.L. *Org. Reactions* **2005**, *66*, 109–625.
184. Albrecht, L., Jiang, H., Dickmeiss, G., Gschwend, B., Hansen, S.G., Jørgensen, K.S. *J. Am. Chem. Soc.* **2010**, *132*, 9188–9196.
185. Prileschajew, N. *Ber. Dtsch. Chem. Ges.* **1909**, *42*, 4811–4815.
186. Ewins, R.C., Henbest, H.B., McKervey, M.A. *J. Chem. Soc., Chem. Commun.* **1967**, 1085–1086.
187. Bowman, R.M., Collins, J.F., Grundon, M.F. *J. Chem. Soc., Perkin Trans. 1*, **1973**, 626–632.
188. Winterfeldt, E. *Prinzipien und Methoden der Stereoselektive Synthese*, Vieweg & Sohn: Braunschweig, **1988**.
189. Henbest H.B., Wilson, R.A. *J. Chem. Soc.* **1957**, 1958–1965.

190. Hasan, I., Kishi, Y. *Tetrahedron Lett.* **1980**, *21*, 4229–4232.

191. Bartlett, P.D. *Rec. Prog. Chem.* **1950**, *11*, 47–51.

192. Singleton, D.A., Merrigan, S.R., Liu, J., Houk, K.N. *J. Am. Chem. Soc.* **1997**, *119*, 3385–3386.

193. Sharpless, K.B., Michaelson, R.C. *J. Am. Chem. Soc.* **1973**, *95*, 6136–6137.

194. Hoveyda, A.H., Evans, D.A. Fu, G.C. *Chem. Rev.* **1993**, *93*, 1307–1370.

195. a) Yamada, S., Mashiko, T., Terashima, S. *J. Am. Chem. Soc.* **1977**, *99*, 1988–1990; b) Michaelson, R.C., Palermo, R.E., Sharpless, K.B. *J. Am. Chem. Soc.* **1977**, *99*, 1990–1992.

196. Katsuki, T., Sharpless, K.B. *J. Am. Chem. Soc.* **1980**, *102*, 5974–5976.

197. a) Pfenninger, A. *Synthesis* **1986**, 89–116; b) Katsuki, T., Martin, V.V. *Org. Reactions* **1996**, *48*, 1–299; c) Riera, A., Sharpless, K.B. *Angew. Chem. Int. Ed.* **2002**, *41*, 2024–2032; d) Moreno, M. *Molecules* **2010**, *15*, 1041–1073.

198. a) Corey, E.J. *J. Org. Chem.* **1990**, *55*, 1693–1694; b) Woodard, S.S., Finn, M.G., Sharpless, K.B. *J. Am. Chem. Soc.* **1991**, *113*, 106–113; c) Finn, M.G., Sharpless, K.B. *J. Am. Chem. Soc.* **1991**, *113*, 113–126; d) Jorgensen, K.A. *Tetrahedron: Asymmetry* **1991**, *2*, 515–532.

199. Wang, C., Yamamoto, H. *J. Am. Chem. Soc.* **2014**, *136*, 1222–1225.

200. Jacobsen, E.N., Zhang, W., Muci, A.R., Exker, J.R., Deng, L. *J. Am. Chem. Soc.* **1991**, *113*, 7063–7064.

201. Hosoya, N., Hatayama, A., Irie, R., Sasaki, H., Katsuki, T. *Tetrahedron* **1994**, *50*, 4311–4322.

202. Curci, R., Fiorentino, M., Serio, M.R. *J. Chem. Soc., Chem. Commun.* **1984**, 155–156.

203. Tu, Y., Wang, Z.-X., Shi, Y. *J. Am. Chem. Soc.* **1996**, *118*, 9806–9807.

4

Sugars

New conceptual approaches to glycosylation and novel strategies for the construction of complex oligosaccharides and glycoconjugates are … welcome to meet the intrinsic structural diversity of carbohydrates.

R.R. Schmidt, 2009

In the early nineteenth century, individual sugars were often named after their source, e.g. grape sugar (Traubenzucker) for glucose and cane sugar (Rohrzucker) for saccharose (the name sucrose was coined much later). Cellulose (from French 'cellule' for cell and ending '-*ose*' to refer to sugars) was isolated and its overall composition elucidated in 1838 by the French chemist *Anselme Payen* (1795–1878). Its chemical formula was confirmed to be the same as that of dextrin (starch) [1]. The term 'carbohydrate' (French '*hydrate de carbone*') was applied originally to monosaccharides, in recognition of the fact that their empirical composition can be expressed as $C_n(H_2O)_n$. Although misleading, the term persists in general use in a wider sense, including monosaccharides, oligosaccharides (oligomers with a few monosaccharides), and polysaccharides (glycans consisting of a large number of monosaccharide units), as well as substances derived from monosaccharides by reduction, oxidation, or by replacement of one or more hydroxy group(s) by heteroatomic groups. We prefer the term 'sugar,' which is frequently applied to monosaccharides and lower oligosaccharides. Strictly speaking, cyclitols (Section 4.6) are generally not regarded as carbohydrates, but are sugars.

In a common parlance, sugars are often associated with a sweet taste. If we compare the sweetness of sugars to that of sucrose, the common sugar, which is a disaccharide formed of glucose and fructose, we observe that most of the sugars are only fairly sweet. Lactose, the milk sugar, a disaccharide of galactose and glucose, and maltose, the malt sugar, a disaccharide of two glucoses, are really not sweet at all (Table 4.1). Since glucose and other sugars are broken down in the body to produce energy, naturally occurring non-energy-producing sweeteners as well as artificial sweeteners have been developed. Steviol and rebaudoside (a steviol glycoside) are small terpenoid molecules considerably sweeter than sucrose (Figure 4.1). Sodium cyclamate (Hermesetas), acesulfame K (AceK, Sunett), aspartame (Nutrasweet), saccharine (Sweet'n'Low, discovered in 1879 at Johns Hopkins), and sucralose (Splenda) are artificial sweeteners. Natural proteins thaumatin, from the katemfe fruit (*Thaumatococcus daniellii*), and monellin, from serendipity berry (*Dioscoreophyllum cumminsii*), are among the sweetest compounds known. Perhaps the world record, however, is held by lugduname, a guanidino acetic acid derivative synthesised in Lyons (*Lat. Lugdunum*) in 1996.

In Chapter 1, we learned that in photosynthetic plants, atmospheric carbon is fixated into sugars, especially into glucose (Scheme 1.2). The metabolism of glucose forms the cornerstone of constructing the building blocks for higher-order organic natural compounds. In this chapter, we will begin our expedition into looking at the journey of carbon from glucose to more advanced building blocks of the classes of natural products, i.e. amino acid derivatives

Table 4.1   *Natural and artificial sweeteners*

Name	E number	Type of compound	Sweetness
Lactose		Disaccharide	0.16
Maltose		Disaccharide	0.33–0.45
Glucose		Monosaccharide	0.74–0.8
Sucrose		Disaccharide	1.00 (reference)
Fructose		Monosaccharide	1.17–1.75
Na cyclamate	E952	Sulfonate	26
Steviol glycosides	E960	Glycoside	40–300
Acesulfame K	E950	Sulfonate	200
Aspartame	E951	Dipeptide methyl ester	180–250
Saccharin	E954	Sulfonate	300–400
Sucralose	E955	Synthetic chlorinated sucrose	320–1000
Thaumatin	E957	Protein	2000
Monellin		Protein	3000
Lugduname		Guanidine	225 000

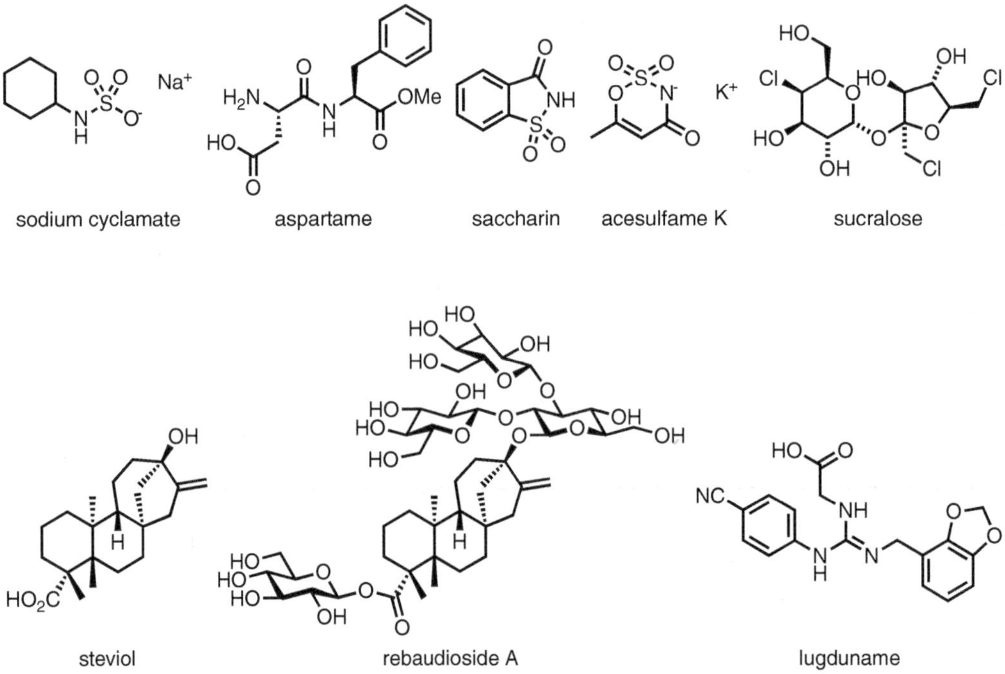

sodium cyclamate   aspartame   saccharin   acesulfame K   sucralose

steviol   rebaudioside A   lugduname

Figure 4.1   *Artificial and natural sweeteners*

(Chapter 5), phenylpropanoids (Chapter 6), nucleobases (Chapter 7), acetic acid-derived polyketides (Chapter 8), mevalonic acid-derived terpenes (Chapter 9), and finally alkaloids (Chapter 10).

4.1   Glucose Metabolism

Carbohydrates are formed as the result of the photosynthetic function of plants, algae, and bacteria. These organisms can utilise atmospheric carbon dioxide, which by the action of photosynthetic enzymes is converted to a chemically useful form, carbohydrates. Green-leafed plants and blue-green algae of the oceans are typical examples of efficient light-harvesting systems. Carbon fixation occurs in the dark reactions of the photosynthetic cycle, where the energy from light stored in ATP and NADPH is used to convert carbon dioxide and water to organic compounds. This dark reaction is known as the Calvin cycle (Scheme 1.1).

4.1.1 Glycolysis

Glycolysis is an oxygen-independent pathway and occurs in most organisms in the cytosol. Glycolysis is the initial phase of breakdown of sugars to produce not only energy, but for our purposes, also intermediate building blocks that can be used to synthesise primary and secondary metabolites. In this Section, we will discuss the breakdown of glucose, but other monosaccharides such as galactose and fructose are equally well substrates for glycolysis. The most common form of glycolysis is known as the Embden–Meyerhof–Parnas pathway. The Entner–Doudoroff pathway is another metabolic pathway that occurs most notably in Gram-negative bacteria, certain Gram-positive bacteria, and archaea. A third important metabolic pathway for glucose is the pentose phosphate pathway, which is important for the production of ribulose phosphate, a building block of nucleotides (Chapter 7). Here we will describe the Embden–Meyerhof–Parnas pathway, where glucose is broken down through the action of ten enzymes (Scheme 4.1).

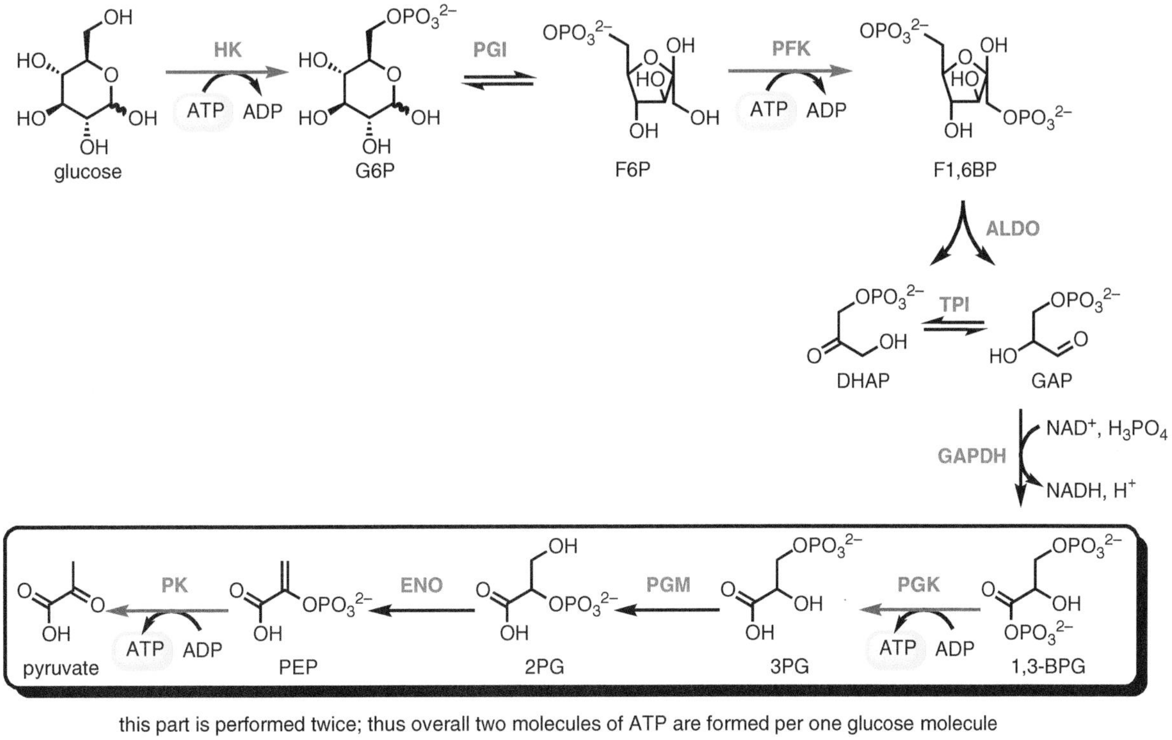

Scheme 4.1 *Glycolysis*

Glucose is first phosphorylated by hexokinase (HK) to glucose-6-phosphate (G6P) with consumption of one molecule of adenosine triphosphate (ATP) and formation of one molecule of adenosine diphosphate (ADP). Phospho-glucose isomerase (PGI) rearranges G6P to fructose-6-phosphate (F6P), which undergoes a second phosphorylation by phosphofructokinase (PFK) to form fructose-1,6-bisphosphate (F1,6BP), which is cleaved by aldolase (ALDO) to two three-carbon units at the same oxidation level (glyceraldehyde phosphate, GAP, and dihydroxyacetone phosphate, DHAP). Triosephosphate isomerase (TPI) converts DHAP to GAP, which is oxidatively phosphorylated to 1,3-bisphosphoglycerate (1,3-BPG) by glyceraldehyde phosphate dehydrogenase (GAPDH) in a process where nicotinamide adenine dinucleotide (NAD^+) is reduced to NADH with simultaneous liberation of a proton. Phosphoglycerate kinase (PGK) hydrolyses one of the phosphates to form 3-phosphoglycerate (3PG) and one molecule of ATP. Phosphoglyceromutase (PGM) isomerises 3PG to 2-phosphoglycerate (2PG), which loses water by the action of enolase (ENO) to form phosphoenolpyruvate (PEP). Finally, the phosphate is hydrolysed by pyruvate kinase (PK) to form pyruvate and one molecule of ATP. Thus, in the first five steps (from glucose to 1,3BPG), processing of one molecule of glucose utilises two molecules of ATP and two molecules of inorganic phosphate, and is thus energy consuming (the preparatory phase or the investment phase). The remaining five steps are performed twice, and thus four molecules of ATP and two molecules of NADH are produced during these stages. The overall balance thus is production of two molecules of ATP per one glucose, corresponding to a large amount of energy (62 kJ/mol) being liberated in chemical form in adenosine triphosphate (ATP). The initial product from glucose is pyruvic acid, which can be further broken down into acetic acid (acetyl coenzyme A, see Section 7.1).

4.1.2 Citric Acid Cycle

The breakdown of pyruvate is continued in the citric acid cycle (or tricarboxylic acid cycle, or Krebs cycle, or Szent–György–Krebs cycle, Scheme 4.2), where pyruvic acid (and thereby glucose) is effectively oxidised to carbon dioxide.

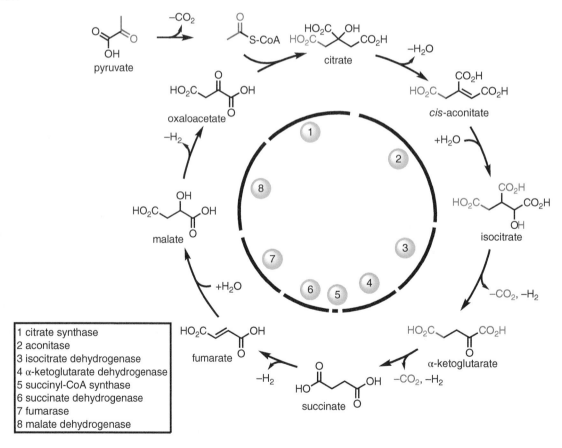

1 citrate synthase
2 aconitase
3 isocitrate dehydrogenase
4 α-ketoglutarate dehydrogenase
5 succinyl-CoA synthase
6 succinate dehydrogenase
7 fumarase
8 malate dehydrogenase

Scheme 4.2 Krebs cycle or citric acid cycle

The cycle begins with the enzyme citrate synthase. In preparation for the cycle, pyruvate dehydrogenase connects an acetyl group from pyruvate to the carrier coenzyme A. Citrate synthase transfers the acetyl group to oxaloacetate, forming citric acid. In the second step, the oxygen atom of citrate needs to be moved one carbon further to create a more reactive isocitrate molecule. Aconitase performs this isomerisation reaction, formally through elimination/readdition of water. In the third step, isocitrate dehydrogenase removes one of the carbon atoms as carbon dioxide, and transfers electrons to NAD$^+$ to form NADH. In the fourth step, a huge complex of three types of enzymes connected by flexible linkers performs a sequence of events: a second carbon atom is released as carbon dioxide; electrons are transferred to NAD$^+$ to give a second molecule of NADH; and the remaining part of the molecule is again connected to coenzyme A. The fifth enzymatic reaction is the only step in the cycle where ATP is made directly. The bond between succinate and coenzyme A is particularly unstable and provides the energy needed to build a molecule of ATP. In the sixth step, membrane-bound succinate dehydrogenase extracts hydrogen atoms from succinate and transfers them, first to the carrier FAD, and finally to the mobile electron carrier ubiquinone. In the seventh step, fumarase adds a water molecule to get everything ready for the final step of the cycle. In the final step, malate dehydrogenase recreates oxaloacetate, transferring electrons to NADH in the process. As the acetyl group is broken down, two molecules of carbon dioxide are formed and three molecules of NAD$^+$ are reduced to NADH. The electrons thus stored are delivered to the large protein complexes that generate the proton gradient to power ATP synthase.

The combined action of glycolysis and citric acid cycle can recover 1160 kJ/mol of the energy bound in glucose, which corresponds to *ca.* 40% efficiency (the heat of formation of glucose is 2870 kJ/mol).

4.2   Monosaccharides

Higher carbohydrates are formed by oligomerisation of monosaccharides, which are polyhydroxyaldehydes (aldoses) or polyhydroxyketones (ketoses) with three or more carbon atoms. The generic term monosaccharide refers to a single unit without glycosidic bond to another carbohydrate unit. Monosaccharides are classified according to their chemical properties and the structural classes are discussed separately in the next few paragraphs.

There are many ways of presenting the structures of sugars. Figure 4.2 shows the commonly used presentations for D-glucose. The classical representation is based on the Tollens–Fischer projection. In this presentation, at each stereocenter, the carbon chain going up and down is considered to be pointing behind the plane of the paper and the substituents pointing left or right. Notice the difference to the modern usage, the so-called Natta projection, shown on the left where a zigzag form of carbon chain is shown. In cyclic forms, a new stereocenter is formed and the formed six-membered ring can be presented in the currently obsolete Haworth projection, the more familiar chair conformation, or preferably the simple stereochemical projection.

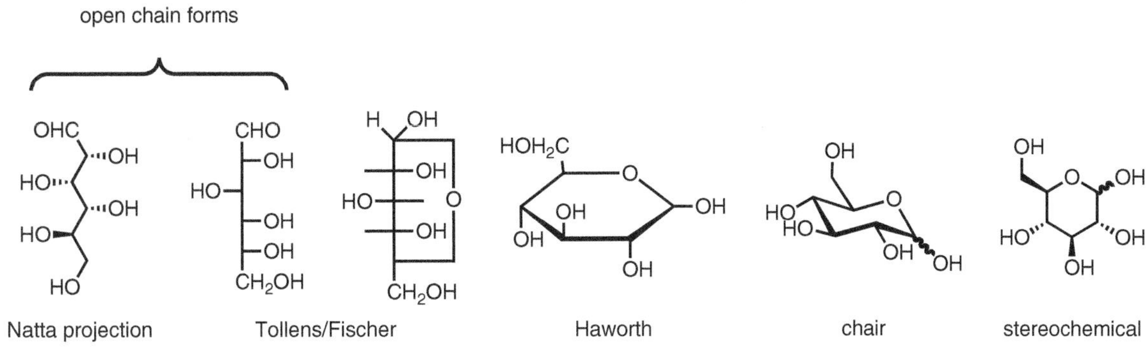

Figure 4.2   *Commonly used presentations for D-glucose*

4.2.1   Anomeric Effect

In carbohydrates, the formation of the six-membered ring in the aldohexose series leads to the generation of a new stereocenter. The two stereoisomeric forms are known as anomers, and they are conventionally distinguished as the α- and β-anomers (Figure 4.3). The early definition of the nomenclature was based on optical rotation: the α-anomer was the one (in the D-series) which showed higher optical rotation. This was based initially on the optical rotatory properties of the crystalline anomers of glucose. A more rigorous definition was based on structures: in the D-series, the β-form is the one having the hemiacetal hydroxyl group on the left in the Fischer projection.

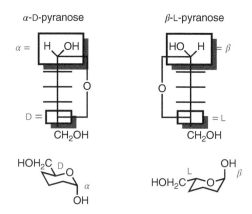

Figure 4.3   *Definitions of α/β- and D/L-designations*

The anomeric effect is a phenomenon typical, but not restricted to carbohydrate chemistry. Originally, the anomeric effect was related to the propensity of the C1 substituent in a pyranose ring to adopt an axial positioning, although the steric effects are not favorable. The phenomenon is not restricted to pyranoses but is common to all cyclic acetals, as well as to any compound with two electronegative (electron withdrawing) substituents bound to the same carbon atom. The anomeric effect is manifested structurally in the fact that, for instance, in a *cis*-2,3-dichloro-1,4-dioxane the C–Cl bond distances to the equatorial and axial chlorine atoms are different (CDCDOX, Figure 4.4) [2]. The equatorial chlorine is significantly closer to the ring carbon atom than the axial one.

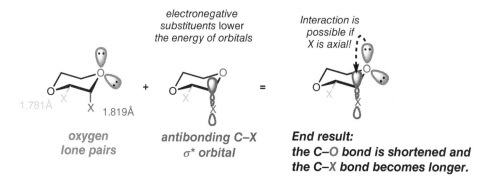

Figure 4.4   *Equatorial and axial substituents are different*

The anomeric effect has its origins in the orbital interactions. Since the antibonding σ^* orbital for a C–X bond is lower in energy than the corresponding antibonding orbital for a C–H bond, the overlap with the lone pair is more effective for the more electronegative substituent. When the axial C–X bond lies in the same plane as the axial lone pair on the ring ether oxygen, donation of electron density from the oxygen *n*-type (green) orbital to the antibonding σ^*_{C-X} orbital (violet) is possible, stabilising this arrangement (Figure 4.5) [3]. The $(n_O-\sigma^*_{C-X})$ interaction is bonding between carbon and oxygen, which strengthens and shortens the ring C–O bond. Charge transfer to the σ^*_{C-X} orbital also weakens and lengthens the axial C–X bond. In the corresponding equatorial C–X anomer, such hyperconjugative effect is not possible. The anomeric $n-\sigma^*$ interactions weaken the C–X bond thus making it more reactive. We will return to this in Section 4.3.1 Glycoside bond formation.

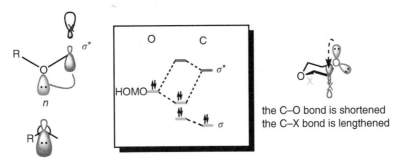

Figure 4.5 *Anomeric effect, orbital representation*

Another term, *exo*-anomeric effect, refers to the favored conformation of an alkoxy substituent O–R (Figure 4.6). The *exo*-anomeric effect has similar origins as the anomeric effect; however, this relates to the preferred orientation of the (axial, or α) O–R bond. Three possible staggered conformations can be considered (**A**–**C**, Figure 4.6) [4]. In conformations **A** and **C**, one of the lone pairs of the exocyclic oxygen (green in Figure 4.6) is disposed in an antiperiplanar fashion to the ring C–O bond, thus making a similar interaction possible as discussed for the anomeric effect above. Conformation **B** lacks such a possibility, and is thus disfavored against these two alternatives. Distinction between conformations **A** and **C** arises from unfavorable interactions of the R group with the axial hydrogens (axial 1,3-strain) in conformer **C**. Conformer **A** is thus the most favored one. However, the energy differences between the various conformations are rather small, and equilibration of the anomers is thus a facile process.

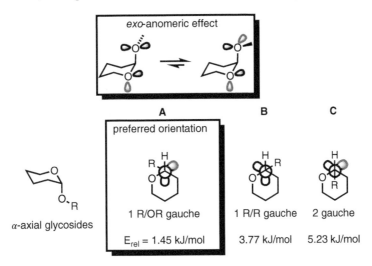

Figure 4.6 *exo-Anomeric effect*

4.2.2 Aldoses and Ketoses

In the 1880s, Emil Fischer and his contemporaries had already realised that sugars needed to be distinguished from one another, and a nomenclature based on the number of carbon atoms in the monosaccharide was adopted. Structures of sugars of three (trioses), four (tetroses), five (pentoses), and six (hexoses) carbon atoms are shown in their Fischer-projection forms in Figure 4.7, along with the recommended three-letter abbreviations in parentheses. Emil Fischer assigned the projection with the C5 OH group pointing to the right to the dextrorotatory glucose [5].

The absolute configuration was confirmed by correlation only in 1951 by the Dutch *Johannes Bijvoet* (1892–1980) when he succeeded in determining the absolute configuration of L-tartaric acid by X-ray crystallography [6]. In 1906, the Russian–American chemist *Martin André Rosanoff* (1874–1951) selected the enantiomeric glyceraldehydes as the point of reference to absolute stereochemistry [7]. According to this Rosanoff convention, the natural carbohydrates all belong to the D-series, i.e. they are conceptually derived from D-glyceraldehyde.

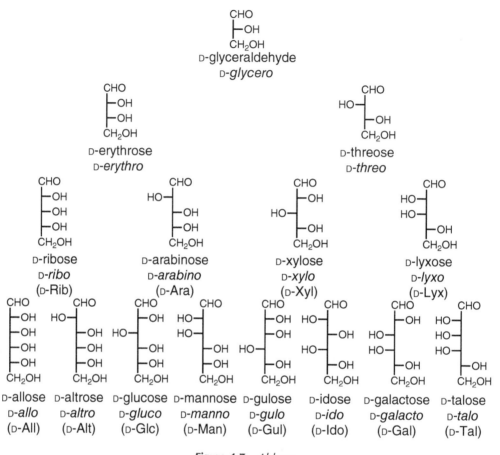

Figure 4.7   *Aldoses*

In ketoses, the carbonyl group is situated not at the terminus of the carbon chain, but at carbon atom 2 (Figure 4.8). Thus, the ketoses have one less stereocenter than the corresponding aldose; hence, there are only one ketotetrose, two ketopentoses, and four ketohexoses.

Carbohydrates with five or more carbon atoms seldom occur in the open chain form; rather, they cyclise to cyclic hemiacetals. The five-membered cyclic forms are called furanoses and the six-membered ones pyranoses. The cyclic forms of ribose (a furanose) and glucose (a pyranose) are shown in different projections in Figure 4.9.

The structural chemistry of carbohydrates is very diverse. Glucose itself can be dimerised in 25 different ways; similarly, a glucose trimer can be formed in 176 different ways! (And the anomeric isomers are not yet included.) The more complex carbohydrates are usually represented in a shorthand form: for instance, fructofuranose is Fruf; glucopyranose is Glcp; and galactopyranose Galp.

The hexopyranoses are shown in the conventional chair representation in Figure 4.10. In all cases, the anomeric hydroxyl group at C1 is drawn out with a wavy line to indicate that both α- and β-anomers are possible. The α-anomer has the substituent below the ring (bond down, axial), the β-anomer correspondingly above the ring (bond up,

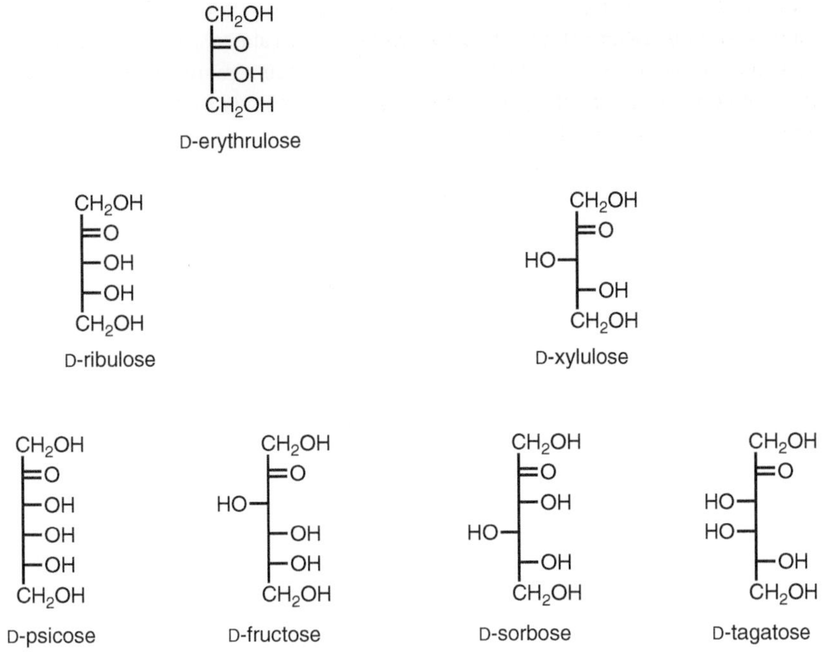

Figure 4.8 *Ketoses*

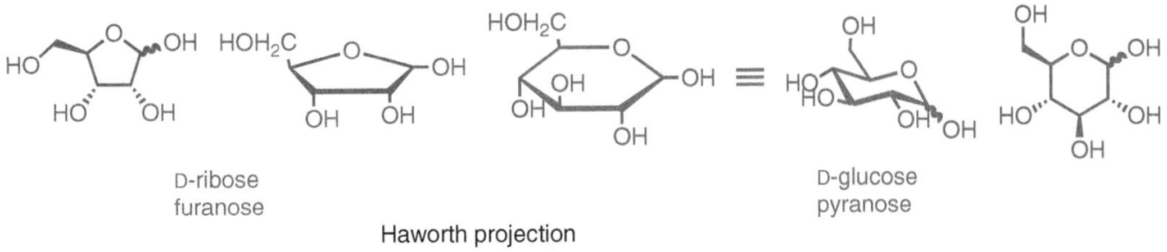

Figure 4.9 *Cyclic hemiacetal forms*

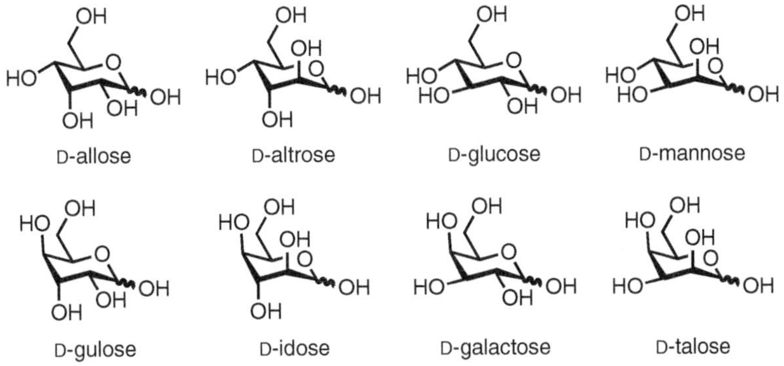

Figure 4.10 *Hexopyranoses*

equatorial). Of the remaining substituents in the tetrahydropyran ring, glucose has all its substituents equatorial, and is thus thermodynamically the most stable one. Mannose, allose, and galactose each have one axial substituent; and altrose, talose, and gulose have two axial substituents. Idose has all three remaining hydroxyl substituents in the axial position. Note that epimerisation of the C5 hydroxymethyl appendage leads to the enantiomeric series: for example, isomerisation of D-glucose leads to the L-antipode of idose.

All the sugars discussed so far contain an aldehyde function. The aldoses react with the Tollens' reagent (ammoniacal silver solution) to form a silver mirror and the corresponding carboxylic acid, aldonic acid. Because of this reactivity, the terminal aldoses are also called reducing sugars.

4.2.3 Deoxy Sugars

Monosaccharides in which an alcoholic hydroxyl group is replaced with a hydrogen atom are called deoxy sugars (Figure 4.11). Of the deoxy sugars, deoxyribose is a constituent of the deoxyribonucleic acid (DNA). L-Fucose (6-deoxy-L-galactose), L-rhamnose (6-deoxy-L-mannose), and D-quinovose (6-deoxy-D-glucose) are important components of bacterial cell walls. Recent information on protein glycosylation also indicates an important role for fucose in the signal transduction processes in mammalian cells as well as allergic reactions.

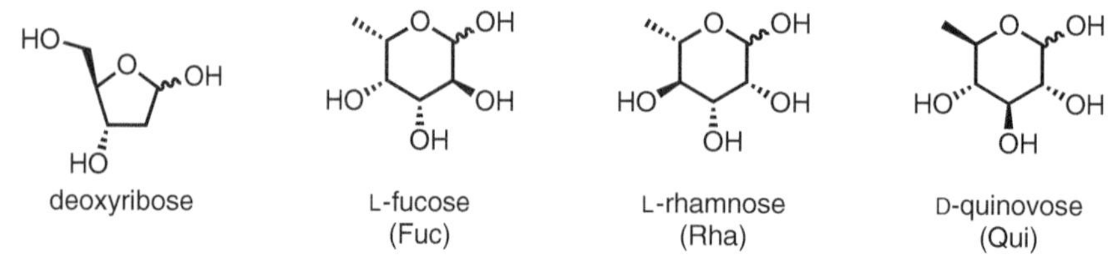

Figure 4.11 *Deoxy sugar*

4.2.4 Amino Sugars

Monosaccharides having one alcoholic hydroxyl group replaced by an amino group are called amino sugars, in contrast to iminosugars, where the endocyclic oxygen atom has been replaced by a nitrogen atom. The latter ones are called iminosugars and are the subject of Section 4.8.

The first polysaccharide identified was chitin, isolated from mushrooms in 1811 by *Henry Braconnot* (1780–1855), approximately 30 years before the isolation of cellulose [8]. Subsequent studies showed that chitin is a polymer of *N*-acetylglucosamine. Amino sugars occur in plants, animals, invertebrates, and microorganisms. They play many important physiological roles as components of antibiotics, chitin, glycolipids, and serum mucoproteins (blood group antigen determinants). They also play a role in cell recognition, differentiation, and protection, and they have been implicated in the pathogenesis of leukemia, liver diseases, and bacterial sepsis.

A number of commonly encountered amino sugars are shown in Figure 4.12. The most common amino sugar, glucosamine (2-amino-2-deoxy-D-glucose), occurs in its *N*-acetylated form in polysaccharides, glycoproteins, and proteoglycans as well as chitin. Glucosamine is a constituent of lobster shells, and galactosamine

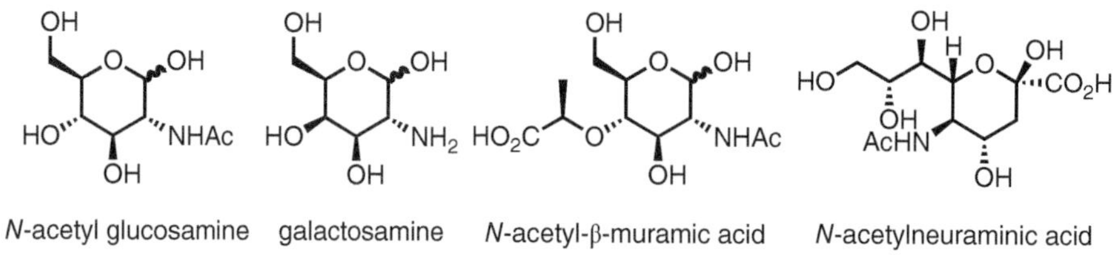

N-acetyl glucosamine   galactosamine   *N*-acetyl-β-muramic acid   *N*-acetylneuraminic acid

Figure 4.12 *Amino sugars*

(2-amino-2-deoxy-D-galactose) a component of cartilage. It was the extensive studies in the antibiotic field which started to reveal the multitude of amino sugars present in nature. Only few amino sugars occur in their free form, they are usually components of complex antibiotics and oligo- and polysaccharides.

N-Acetyl-*β*-muramic acid [2-acetamido-3-*O*-(1-carboxyethyl)-2-deoxy-D-glucose, NAM or MurNAc] is the product of addition of phosphoenolpyruvate and *N*-acetylglucosamine. It is a key building block of peptidoglycan in the bacterial cell wall.

Sialic acids are nine-membered *α*-keto sugar acids derived biosynthetically from glucosamine and phosphoenolpyruvate. *N*-Acetylneuraminic acid (5-acetamido-3,5-dideoxy-D-glycero-D-galacto-2-nonulopyranic acid, Neu5Ac) and the corresponding *N*-glycolylneuraminic acid are found in many glycoproteins. Many viruses such as adenoviruses (*Adenoviridae*), rotaviruses (*Reoviridae*), and influenza viruses (*Orthomyxoviridae*) can use host-sialylated structures for binding to their target host cell, and thus sialic acids play an important role in viral infections, especially on the surface of human erythrocytes and on the cell membranes of the upper respiratory tract. Widely used anti-influenza drugs (oseltamivir and zanamivir) are sialic acid analogs that inhibit the viral enzyme neuraminidase (see Section 4.2.4.1 Case study: oseltamivir).

Many rare amino sugars, such as 3-amino-, 4-amino-, and diamino sugars, are present as part of a number of antibiotics (Figure 4.13). Kanamycins and gentamicins belong to the 2,4-substituted 2-deoxystreptamine antibiotics, whose central part is a carbacyclic core decorated with e.g. kanosamine (3-amino-3-deoxy-D-glucose, in kanamycins) or gentosamine (3-methylamino-3-deoxy-xylose) [9]. Pyrrolo[1,4]benzodiazepine antibiotics mediate their antitumor activity by binding to DNA through the minor groove to form a covalent link with N^2 of guanosine. Sibiromycin, a glycosylated pyrrolo[1,4]benzodiazepine antibiotic, exhibits potent antitumor activity. It contains the rare aminosugar sibirosamine (4,6-dideoxy-3-methyl-4-(methylamino)-L-mannose) [10]. Several polyene antibiotics, such as amphotericin B and

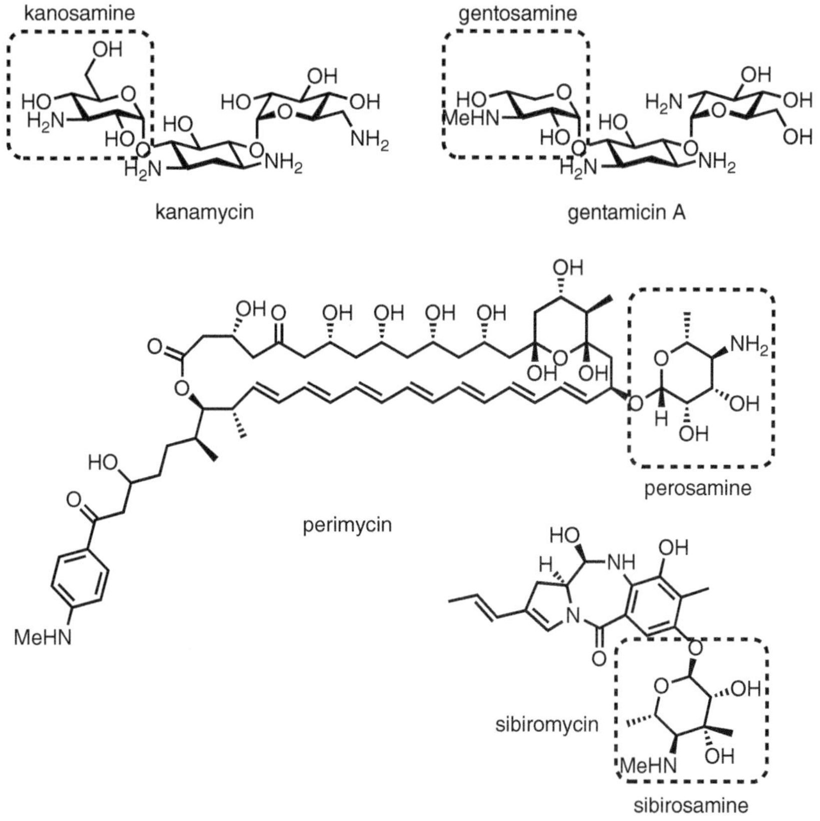

Figure 4.13 *Antibiotics with rare amino sugars*

nystatin (Chapter 7), usually contain mycosamine (3,6-dideoxy-3-amino-D-mannose) units, but the aromatic heptaene antibiotic perimycin contains perosamine (4,6-dideoxy-4-amino-D-mannose) [11], which also occurs in the lipopolysaccharides of several Gram-negative pathogens. The syntheses of the most common structural type, the 6-deoxysugars, have been reviewed [12].

2,3,4,6-Tetradeoxy-4-aminoglycosides are very rare structural units in natural products. Both stereoisomeric series with the 4-amino and 5-methyl groups *cis* or *trans* to each other are known (Figure 4.14). The first one characterised was the forosamine/tolyposamine series, in which the relationship is *trans*. Spiramycins [13], and the related chimeramycins [14] and shengjimycins [15] contain such structures. Even rarer is the series with the epimeric glycoside subunit ossamine, in which the dimethylamino/methyl relationship is *cis*. Ossamine was initially reported as the amino sugar fragment of the cytotoxic agent ossamycin [16]. This unit has only later been reported as part of dunaimycins [17], spinosyn G [18], and grecocyclines [19].

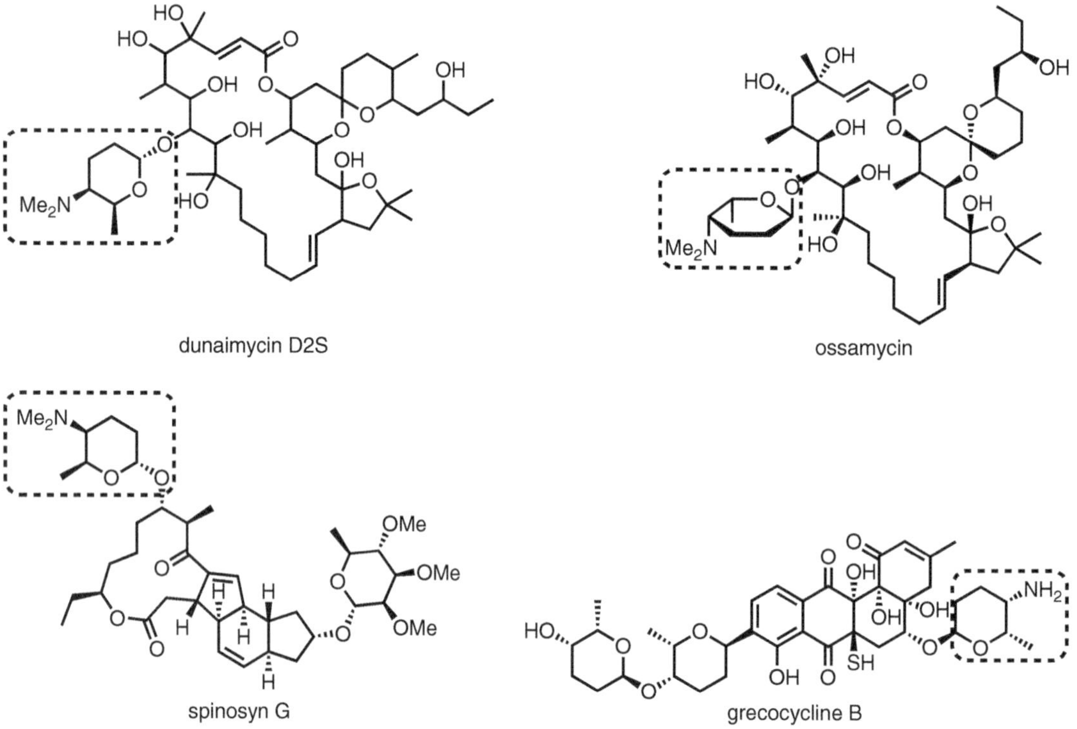

dunaimycin D2S

ossamycin

spinosyn G

grecocycline B

Figure 4.14   *Tetradeoxy amino sugars*

Synthesis of the tetradeoxy amino sugars from commercial sugars requires the removal of several existing oxygenated stereocenters, and eventual replacement of one hydroxyl group with an amino group. An expedient synthesis of the epitolyposamine/ossamine series has been readily achieved from L-threonine (Scheme 4.3) [20]. The fully protected L-threonine derivative is available in mol quantities in high yield. DIBAL-H reduction followed by a modified Horner–Wadsworth–Emmons olefination gave the enoate, which was reduced catalytically and subsequent deprotection with glacial acetic acid effected the ring closure to the lactone. Straightforward operations from the lactone serve to deliver either epi-tolyposamine derivatives, or through literature N-alkylation procedures, ossamine derivatives. The route is efficient also in terms of stereochemical information: the starting amino acid is safe from racemisation; and the intermediate amino aldehyde can be processed rapidly to avoid epimerisation.

2-Amino-2-deoxysugar derivatives can be synthesised from glycals through [4 + 2] cycloaddition with azodicarboxylate (Scheme 4.4) [21]. Photocatalysed cycloaddition of azodicarboxylate with the glycal gives the cycloadduct,

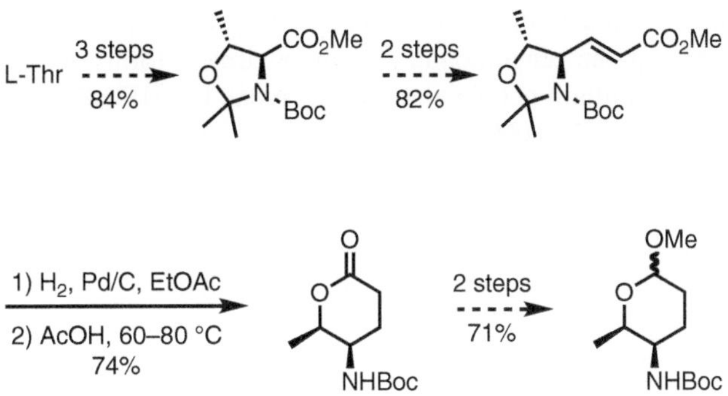

Scheme 4.3 *Synthesis of ossamine/epi-tolyposamine*

which can be decomposed to the 2-amino-2-deoxy sugar by treatment with acidic methanol followed by reduction of the hydrazine (Raney nickel) and acetylation. Silyl protecting groups are essential, as e.g. acetylated glycals give poorer yields and greatly eroded diastereoselectivity in the cycloaddition step.

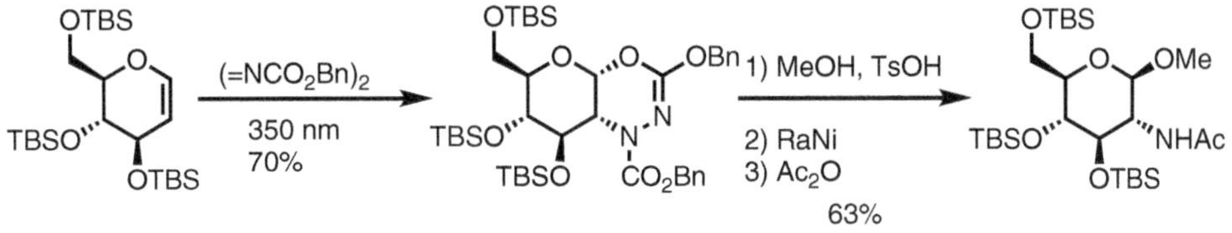

Scheme 4.4 *Synthesis of amino sugars from glucals*

Glycosylated 2-amino-2-deoxysugars can also be synthesised directly from glucals using a modification of the oxidation technology developed by *Samuel J. Danishefsky* (1936–) for oligosaccharide synthesis from glucals (Scheme 4.5, see also Section 4.2) [22]. Treatment of the glucal with iodonium di-*sym*-collidine perchlorate and benzenesulfonamide gives the *trans*-diaxial iodosulfonamide, which, upon treatment with a carbohydrate and a base, gives the 2-sulfonamido-*β*-glycoside *via* the aziridine [23].

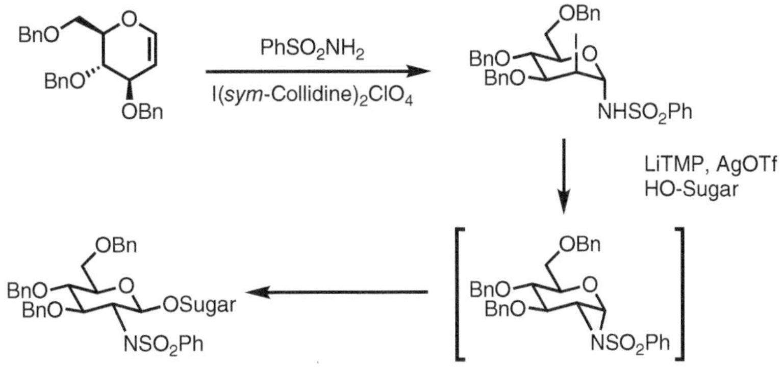

Scheme 4.5 *Synthesis of amino sugars from glucals using oxidative amination*

4.2.4.1 Case synthesis: Oseltamivir

Seasonal influenza is a severe viral infection of the respiratory system, annually affecting some 20% of the population worldwide and resulting in 250 000–500 000 deaths [24]. The infectivity, morbidity, and mortality rates of influenza are high because the virus can easily mutate to more virulent types. The viral surface glycoprotein antigens are composed of two main types: sixteen subtypes of hemagglutinin (H) and nine different subtypes of neuraminidases (N). Occasionally, major pandemics have arisen through the generation of extremely virulent forms of the virus, such as the Spanish Flu (H1N1, 1918–1920, killed 20–100 million people worldwide), Asian Flu (H2N2, 1958–1959, 1 million casualties), and Hong Kong Flu (H3N2, 1968–1969, 750 000 casualties). Recently, the 2007 Avian Flu (H5N1) and the 2009 Swine Flu (H1N1) were largely evaded thanks to rational drug design, which has yielded chemical agents to combat the virus neuraminidase.

The life cycle of the influenza virion begins with the formation of a bud on the surface of the infected cell. The bud is covered with a complex polysaccharide sialic acid, which contains a large proportion of neuraminic acid. The bud is released and opened by the action of neuraminidases (sometimes called sialidases); enzymes which specifically cleave the glycosidic bonds to terminal neuraminic acids. Structural analogues of neuraminic acid have therefore gained momentum as potential influenza cures (Figure 4.15).

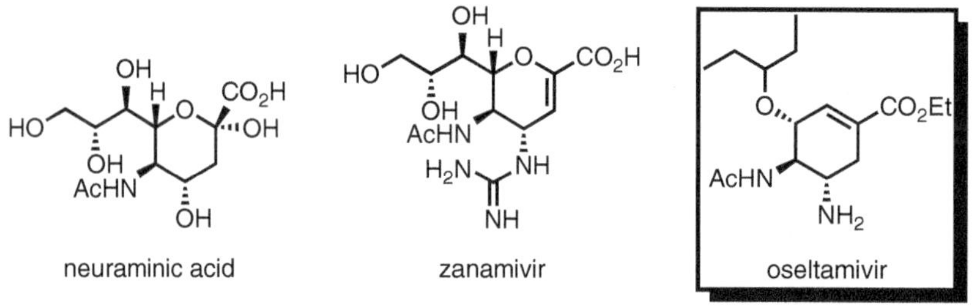

Figure 4.15   *Neuraminic acid and neuraminidase inhibitors*

Zanamivir was the first neuraminidase inhibitor developed at the Monash University and commercialised by GlaxoSmithKline. Oseltamivir, a structurally simpler analogue was discovered by Gilead Sciences and commercialised by F. Hoffmann–La Roche as the highly soluble phosphate salt (Tamiflu), which is hydrolysed to the active oseltamivir carboxylate by hepatic esterases. The synthetic routes to oseltamivir illustrate several different strategies for the source of asymmetric information, and we will look at a few of them [25].

The Hoffmann–La Roche commercial synthesis utilises the chiral pool approach, employing shikimic acid as the starting material (Scheme 4.6) [26]. Although shikimic acid was initially scarcely available, it has recently become widely available in multi-hundred-ton amounts by extraction of star anise, and by a fermentation process using a genetically engineered *Escherichia coli* strain. In the commercial synthesis, shikimic acid is converted to a trimesylate ester, which is then reacted with sodium azide. This reacts selectively with the allylic mesylate and, on reduction with triethyl phosphite, the azide spontaneously cyclises to an aziridine intermediate.

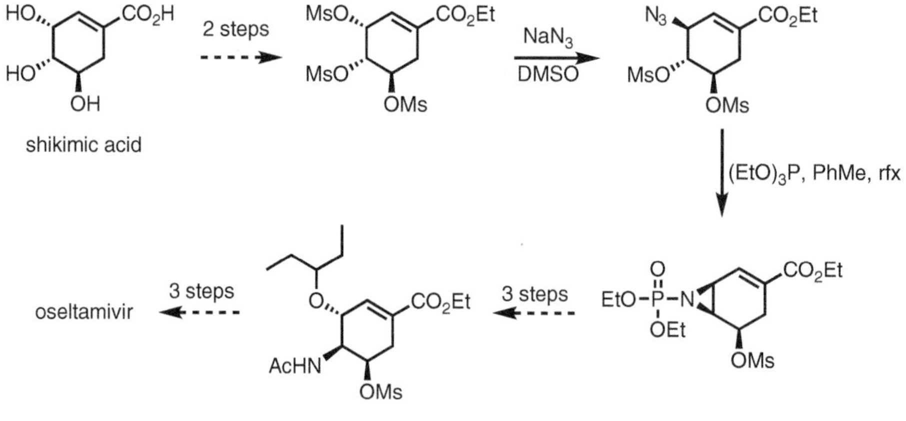

Scheme 4.6   *Hoffmann–La Roche synthesis of oseltamivir*

Regioselective ring opening with 3-pentanol followed by straightforward functional group manipulations lead to oseltamivir.

Corey has employed an oxazaborolidine catalysed Diels–Alder reaction to generate a chiral cyclohexene carboxylate (Scheme 4.7) [27]. This was reacted in a series of ten reactions to provide oseltamivir phosphate.

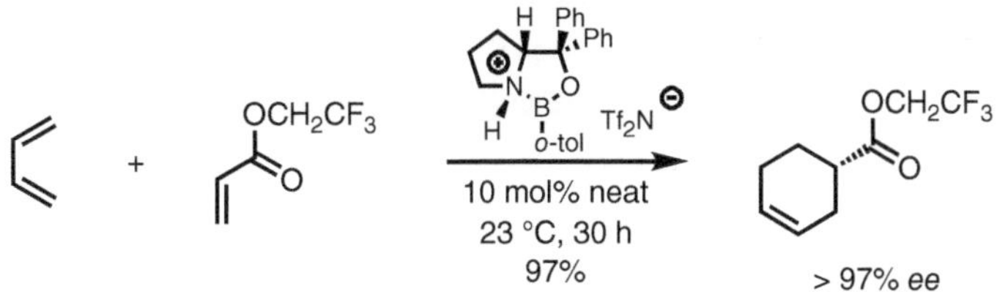

Scheme 4.7 *Asymmetric Diels–Alder reaction in the synthesis of oseltamivir*

Tomáš Hudlický (1947–2022, Brock University, Canada) developed approaches utilising aromatic dihydroxylation reactions in total syntheses, and a genetically modified *E. coli* strain accepts ethyl benzoate as its substrate [28]. This was utilised in the synthesis of oseltamivir (Scheme 4.8), and it was pointed out that with this invention one can avoid the use of bromobenzene, thus also evading a later use of Pd chemistry.

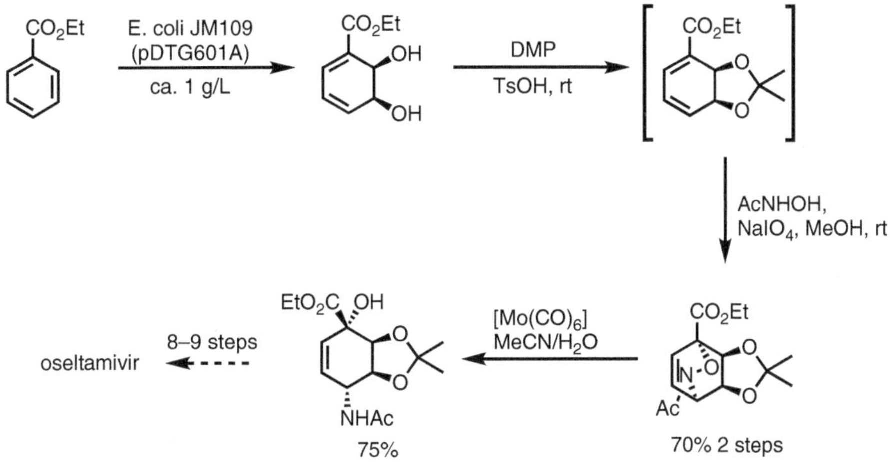

Scheme 4.8 *Hudlicky chemoenzymatic synthesis of oseltamivir*

Tamio Hayashi (1948–) has developed an organocatalytic cascade sequence, which effectively delivers the oseltamivir core in a one-pot sequence (Scheme 4.9). The transformation consists of a diphenylprolinol silyl ether-catalysed Michael addition of the 3-pentyloxyacetaldehyde to the nitro acrylate, followed by a second conjugate addition to a Henry-type Michael addition to a vinyl phosphonate. The product then undergoes an intramolecular Horner–Wadsworth–Emmons reaction to give the advanced cyclohexene core. Although the stereochemistry at C5 is incorrect, this could be corrected, and a further two one-pot operations gave Tamiflu in an impressive 57% overall yield [29].

Trost has utilised Pd-catalysed asymmetric allylic amination to convert the racemic bicyclic lactone to a nearly enantiopure intermediate for the synthesis of oseltamivir (Scheme 4.10) [30]. The synthesis involved only eight steps from the lactone and proceeded in 30% overall yield.

An aldol approach utilising the chiral Evans oxazolidone auxiliary and an L-glutamic hemialdehyde was developed by Salcik (Scheme 4.11) [31]. The boron enolate of the Evans oxazolidone predictably gave the *syn* aldol product with high Felkin control (see Section 3.2.2.2). The intermediate aldol product was converted to a dialdehyde intermediate,

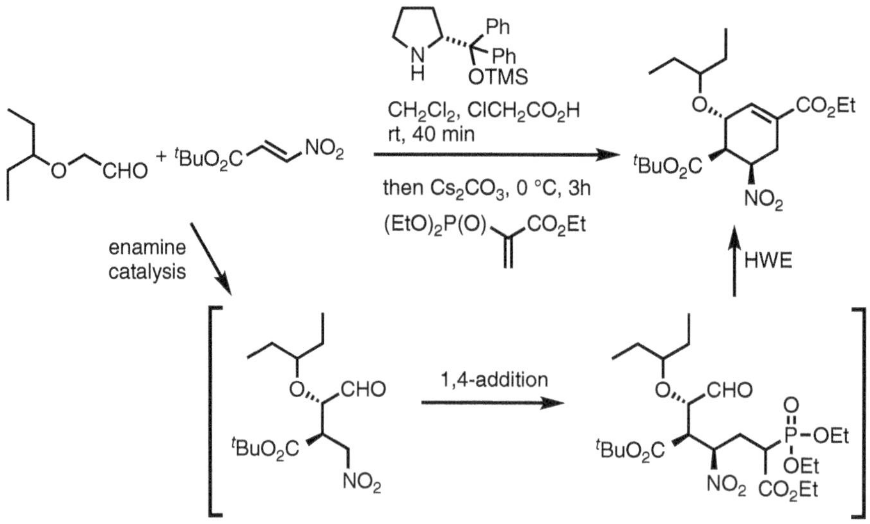

Scheme 4.9   *Hayashi organocatalytic synthesis of oseltamivir*

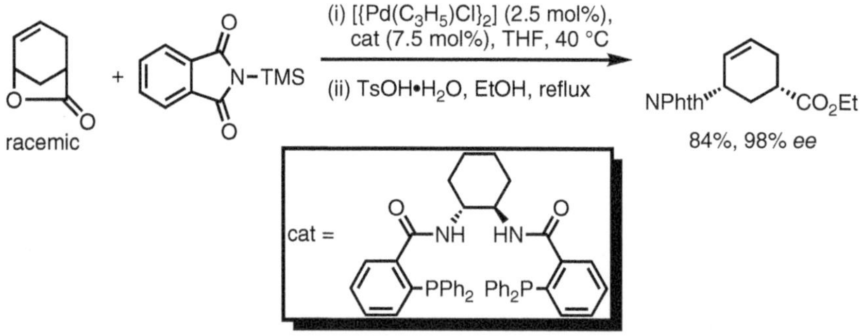

Scheme 4.10   *Trost allylic amination route to oseltamivir*

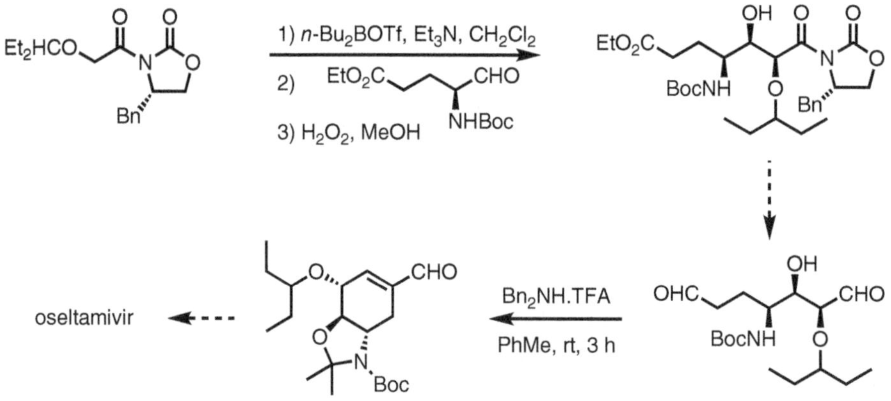

Scheme 4.11   *Aldol approach to oseltamivir*

which was subjected to a dibenzylamine-catalysed intramolecular aldol condensation process developed by Woodward [32]. Because of the high *syn* selectivity of the Evans boron aldol process, the configuration of the 4-hydroxyl group in the cyclohexenone ring needed to be inverted employing literature precedents.

The recent outbreaks of COVID-19 (SARS-CoV-2), the Middle East Syndrome (MERS-CoV), Severe Acute Respiratory Syndrome (SARS-CoV), influenza, and the Ebola hemorrhagic fever have emphasised the need for rapid response to synthesise therapeutic agents rapidly to combat these potential threats to pandemics. Time of synthesis is of essence then, and a recent review on "time economy" nicely compares, among others, the syntheses of oseltamivir from this perspective [33].

4.2.5 Sugar Alcohols

Sugar alcohols, or alditols, arise formally by reduction of the aldehyde or ketone function of an aldose or ketose to an alcohol (Figure 4.16). The sweetener xylitol (wood sugar alcohol) is perhaps the best-known example of these. D-Mannitol is present in seaweed, D-glucitol (D-sorbitol) is used as a sweetener and is found naturally in apples, pears, peaches, and prunes.

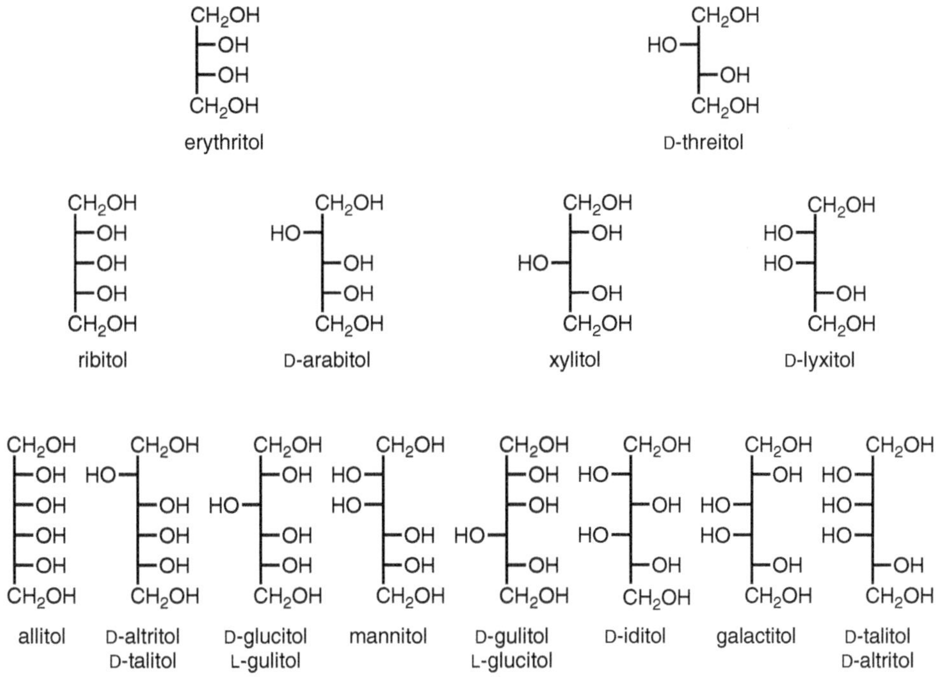

Figure 4.16 *Sugar alcohols*

4.2.6 Acidic Sugars

Acidic sugars fall into three main types, depending on which carbon atom has been oxidised (Figure 4.17). In aldonic acids, the aldehydic carbonyl group is oxidised to a carboxylic acid; in the uronic acids, the terminal hydroxymethyl; and in the aldaric acids both terminal groups are oxidised to the corresponding carboxylic acid oxidation state.

By far, the most important group of acidic sugars is the uronic acids, which commonly occur as the hexuronic acids. These are intermediates in the biosynthesis of pentoses from hexoses. Many of these occur in the gums of plants and as building blocks for the formation of bacterial cell walls, as well as incorporated into proteoglycans (copolymers of proteins and carbohydrates). Vitamin C (ascorbic acid) is also a member of this structural class.

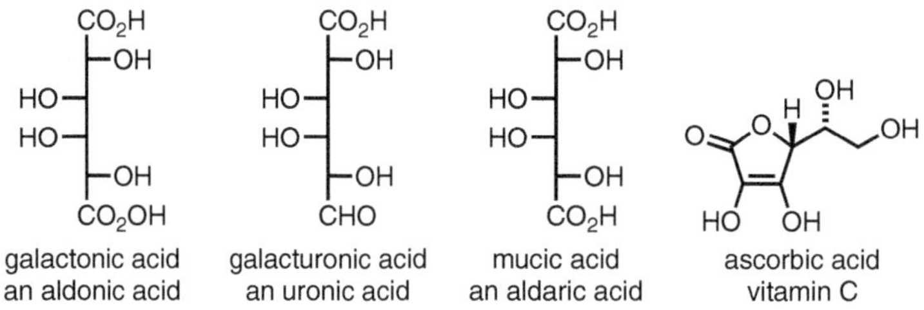

Figure 4.17 *Acidic sugars*

4.3 Oligo- and Polysaccharides

Monosaccharides can condense with each other to form oligomeric structures. As we will soon see, the structural variety of oligo- and polysaccharides can be truly overwhelming. Table 4.2 and Figure 4.18 show a few common di- and trisaccharides that we encounter in our daily lives. Cellobiose can be readily obtained by enzymatic or acidic hydrolysis of cellulose, maltose is malt sugar, lactose is milk sugar, and sucrose is the common table sugar. Fructooligosaccharides (FOS) such as 1-kestose (1-kestotriose; GF2), and nystose (1,1-kestotetraose; GF3) occur in many plants, including banana, onion, wheat, barley, asparagus, and Jerusalem artichokes. The human small intestine has no enzyme to hydrolyse the glycosidic linkages; therefore, FOS's are considered indigestible in the human small intestine. Fructooligosaccharides stimulate the growth of bifidobacteria in the human colon, suppress putrefactive pathogens, and reduce serum cholesterol concentrations. Due to their physiochemical properties, sweetening power, and low caloric value, FOS has been added to pastry, confectionery, and dairy products. Their energy value is theoretically lower than that of sucrose, since the energy value depends on the extent of absorption in the small intestine and fermentation in the colon. Vegetables (e.g. beans, sprouts, cabbage) are particularly rich in raffinose, which is also used as a starting material for the synthetic production of sucralose, a sweetener some 600 times sweeter than sucrose.

Table 4.2 *Common di- and trisaccharides*

Name	Structure
Disaccharides	
cellobiose	β-D-Glcp-(1-4)-D-Glc
lactose	β-D-Galp-(1-4)-D-Glc
maltose	α-D-Glcp-(1-4)-D-Glc
sucrose (saccharose)	α-D-Glcp-(1-2)-β-D-Fruf
Trisaccharides	
kestose	β-D-Fruf-(2-6)-β-D-Fruf-(2-1)-α-D-Glcp
raffinose	α-D-Galp-(1-6)-α-D-Glcp-(1-2)-β-D-Fruf

Carbohydrates provide a fascinatingly complex starting material for biopolymers. If one considers the possibilities for structural isomers, the existence of several sites for coupling can give rise to much larger number of polysaccharide isomers than is available for either polypeptides or polynucleic acids, as shown in Table 4.3 [34].

The structural versatility is an added bonus if one considers the possibility for information storage, and carbohydrate units connected to proteins play a central role in cellular recognition processes. Simultaneously, the biosynthesis of polysaccharides is much more complicated than that of polypeptides or polynucleotides – typically, the enzymatic

sucrose   lactose   maltose   cellobiose   chitobiose

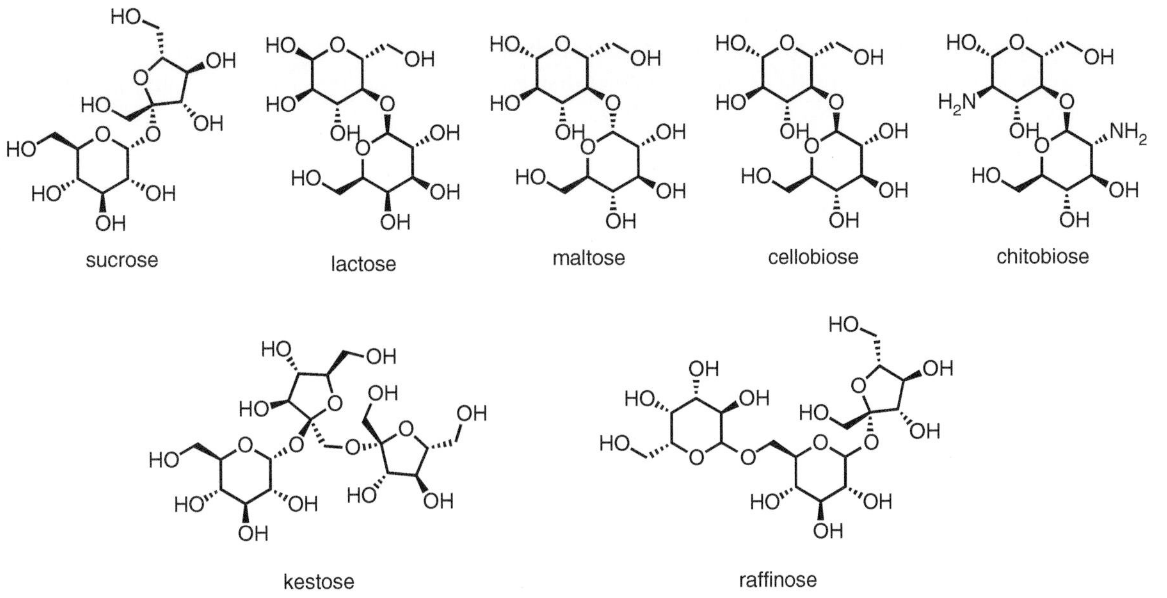

kestose   raffinose

Figure 4.18   *Di- and trisaccharides*

Table 4.3   *Isomers of biopolymers*

Product	Structure	Number of Isomers	
		Peptides, Nucleic acids	Saccharides
Monomer	Z	1	1
Dimer	Z_2	1	11
Trimer	Z_3	1	120
Tetramer	Z_4	1	1424
Pentamer	Z_5	1	17 872
Monomer	Z	1	1
Dimer	YZ	2	20
Trimer	XYZ	6	720
Tetramer	WXYZ	24	34 560
Pentamer	VWXYZ	120	2 144 640

processes used for connecting two carbohydrate units require a highly specific enzyme, and therefore polysaccharides with a defined cellular function are still unknown. However, we shall see in the next section that glycosylation of proteins is much better understood.

Starch, cellulose, and pectins are the most common polysaccharides from the plants. Teichoic acid (sugar phosphates) and mureins (branched copolymers of amino sugars and peptides) occur widely in the cell walls of Gram-positive bacteria.

The cell wall of Gram-negative bacteria is structurally considerably more complex. The cell wall peptidoglycan is surrounded by a cytoplasmic membrane, which is further surrounded by an outer membrane-like structure. This is composed of proteins, phospholipids, lipoproteins, and lipopolysaccharides (LPS, or endotoxin), which are not covalently bound to peptidoglycan. These LPS's are the basis for the species-specific immune reactions. The part of the LPS that is not responsible for the antigenic activity is called the core polysaccharide. This is bound covalently to the lipid material. The core polysaccharide is structurally much simpler than the antigenic regions, and it typically contains lipid A (Figure 4.19).

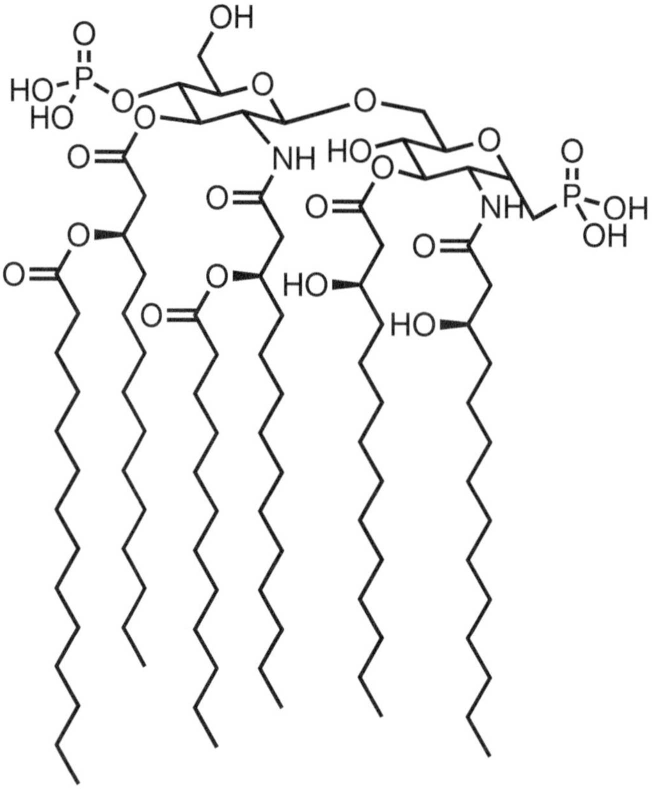

Figure 4.19   *Lipid A of* Escherichia coli

4.3.1   Glycoside bond formation

In the synthesis of oligo- and polysaccharides, formation of the glycosidic bond is traditionally conducted with a free alcohol (glycosyl donor) by converting the hemiacetal hydroxyl group to a good leaving group with any one of a number of activating reagents [32], and allowing this to react with a carbohydrate component with a free hydroxyl group (glycosyl acceptor) (Scheme 4.12).

Acid activation of the glycosyl donor gives rise to the protonated intermediate which can be converted to the glycosyl halide (the Koenigs–Knorr procedure, route A). This can then be coupled with the alcohol ROH in the presence of e.g. silver salts to give the disaccharide. Alternatively, the protonated intermediate can also be reacted directly with the alcohol ROH (Fischer–Helferich coupling, route B). The latter route is hampered by its reversibility, making it less attractive for polysaccharide synthesis.

Base activation of the glycosyl donor gives an alkoxide, which can participate in ring-chain tautomerism. However, this intermediate can be trapped with trichloroacetonitrile to give the imidate, which reacts under acid catalysis with alcohols to give the disaccharide.

The stereochemistry of glycosidation is dictated by anomeric effect (Scheme 4.13). When the glycosyl donor is activated, an intermediate oxonium ion is formed. The attack of the nucleophilic glycosyl acceptor is now under anomeric

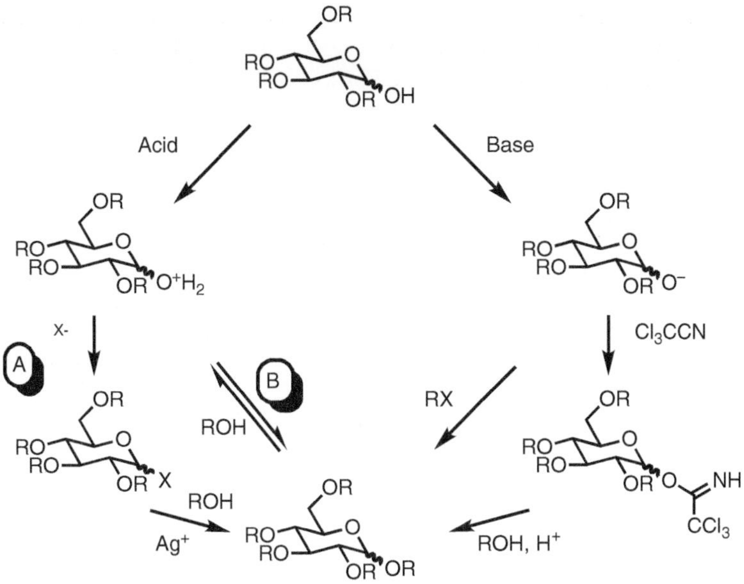

Scheme 4.12 *Glycosyl bond formation*

control. Attack from the top face (β-attack, red) will lead to a twist boat confirmation and after ring flip to a β-glycoside. However, direct axial attack from the α-face (blue) will give directly the stable a-glycoside.

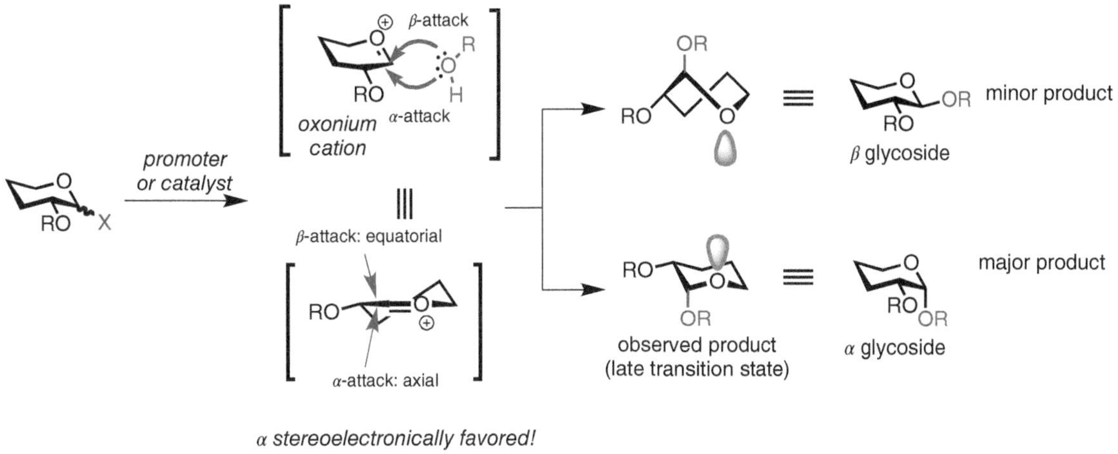

Scheme 4.13 *Stereochemistry of glycosidation*

The situation changes if the neighboring carbon atom C2 contains an equatorial nucleophilic substituent such as an acyl group (Scheme 4.14). The acyl oxygen can attack the oxonium ion to form a new fused 5-6-membered oxonium ion. This can preferably be attacked from the β-face to give the β-glycoside as the major product.

The Koenigs–Knorr coupling suffers from several shortcomings: the formation of the halogenoses requires rather drastic reaction conditions; the halogenoses are thermally unstable and prone to hydrolysis; and the need for heavy metal salts is an obvious problem, especially in large-scale work. Several new alternatives have been sought to overcome these problems, including the use of thiol or fluorine activation. An efficient method for glycosidation is based on the use of glycosyl fluorides (Scheme 4.15) [35]. Thus, 2,3,4,6-tetra-O-benzyl-β-D-glucopyranosyl fluoride reacts with various

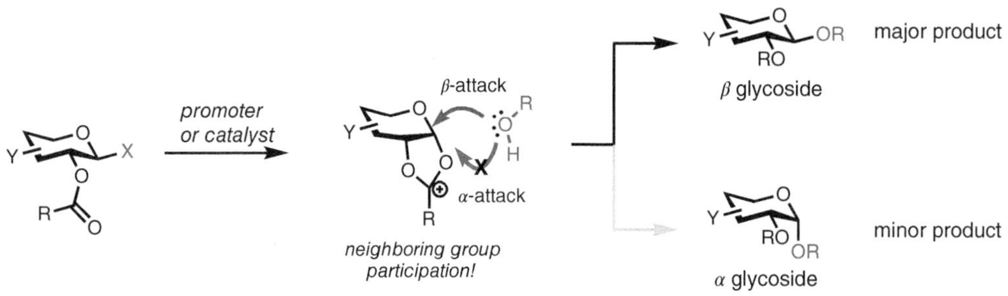

Scheme 4.14   *Neighboring group participation*

hydroxy compounds, including sterically hindered ones, to give the α-anomer with diastereomer ratios ranging from 4:1 to 12:1.

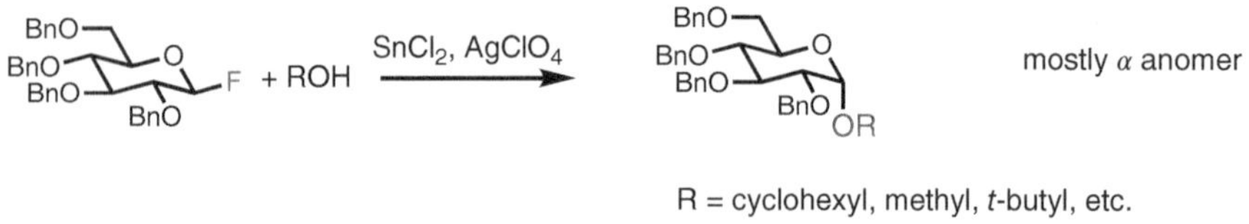

R = cyclohexyl, methyl, *t*-butyl, etc.

Scheme 4.15   *Glucosyl fluorides*

In connection with the synthesis of avermectin B$_{1a}$, a practical method for glycosidation was developed by K.C. Nicolaou (1946–) (Scheme 4.16) [36]. The fluorosugars can be conveniently synthesised from the corresponding phenylthiosugars by treatment with NBS and diethylaminosulfur trifluoride (DAST), or directly from the free alcohol by treatment with DAST [37]. Coupling is effected with the Mukaiyama stannous chloride–silver perchlorate activation. The method is mild enough so that most protecting groups and glycosidic linkages survive intact.

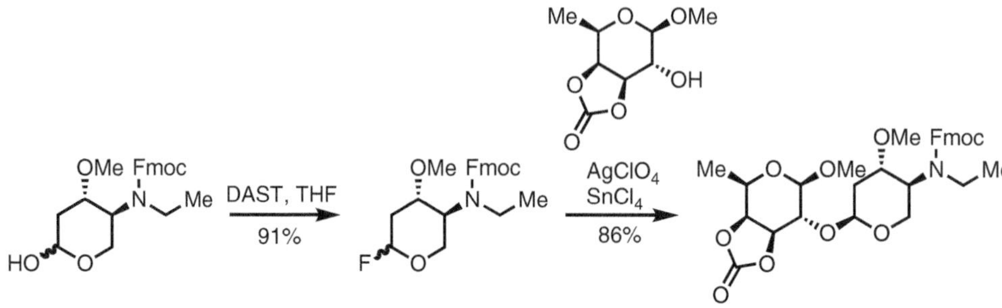

Scheme 4.16   *Glycosyl fluorides in the synthesis of avermectin B$_{1a}$*

Later, Suzuki reported more reactive and widely applicable systems employing cationic zirconocene and hafnocene complexes (the Suzuki modification) [38]. Scheme 4.17 shows an example of the Mukaiyama–Suzuki glycosylation during the synthesis of oligosaccharide selectin ligands [39], involving neighboring group participation by the acetate-protecting group to give the β-glycoside. Interestingly, the *N*-phthaloyl-protecting group acted similarly as a neighboring group in a subsequent reaction to direct the glycosylation to give the β-anomer product. Glycosyl fluorides can be activated by other reagents too, including LiClO$_4$, and strong Brønsted acids like TfOH.

Direct oxidative coupling of glycals is also possible, as shown by Danishefsky, by utilising one of two alternative powerful methods (Scheme 4.18). Oxidative coupling can be effected with iodonium di(*sym*-collidinium) perchlorate (IDCP) [19]. Careful choice of the protecting groups directs the bond formation: the acyl-protected glycal will donate its

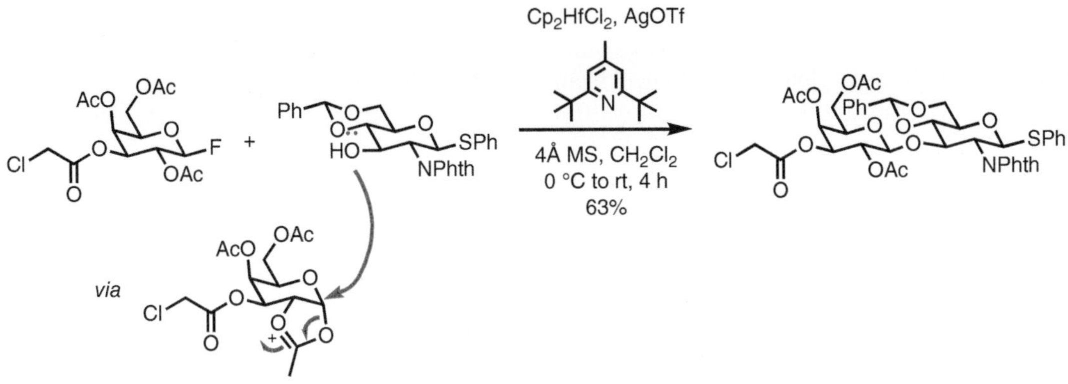

Scheme 4.17 *Mukaiyama–Suzuki glycosylation.*

free hydroxyl groups for bond formation with the ether-protected glycal but will not react with itself. This coupling strategy produces the diaxial α-linked 2-iodoglycosides, and the formation of β-linked glycosides requires a different strategy.

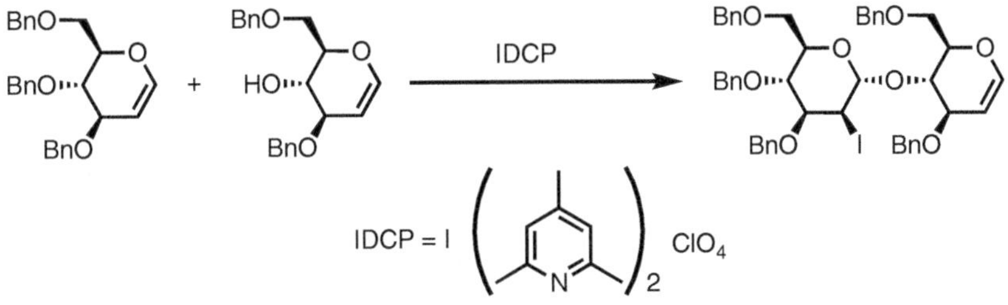

Scheme 4.18 *Direct oxidative glycosidation*

Epoxidation of glycals incorporating nonparticipating protecting groups with dimethyloxirane (DMDO, Scheme 4.19) leads to highly stereoselective epoxide formation ($\alpha:\beta$ ratio 20:1 with benzyl protection; with *tert*-butyl dimethylsilyl protection, only α-epoxide was observed). The epoxides were then coupled with the alcohols in the presence of anhydrous zinc chloride to give the β-glycoside as the sole product. The yields are low (50–58%) due to the instability of the epoxide towards the Lewis acid.

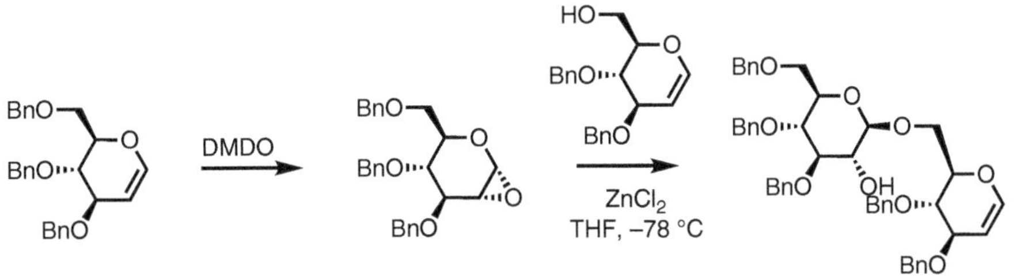

Scheme 4.19 *Glycosidation through epoxidation*

A powerful carbohydrate-coupling strategy has been developed, based on the initial observation on the reactivity of glycosyl halides by Paulsen [40]. He observed that halide donors with electron-withdrawing ester protecting groups were considerably more stable than the analogous benzyl ether-protected compounds. Nearly simultaneously *Bert Fraser-Reid* (1934–2020) discovered the use of pentenyl groups as activatable-protecting groups at the anomeric center [41]. Combination of these observations led to the powerful concept of armed/disarmed glycosyl donors and acceptors. This strategy is illustrated by the short synthesis of the trisaccharide glycan segment of the nephritogenic glucopeptide isolated from rat glomerular basement membrane (Scheme 4.20).

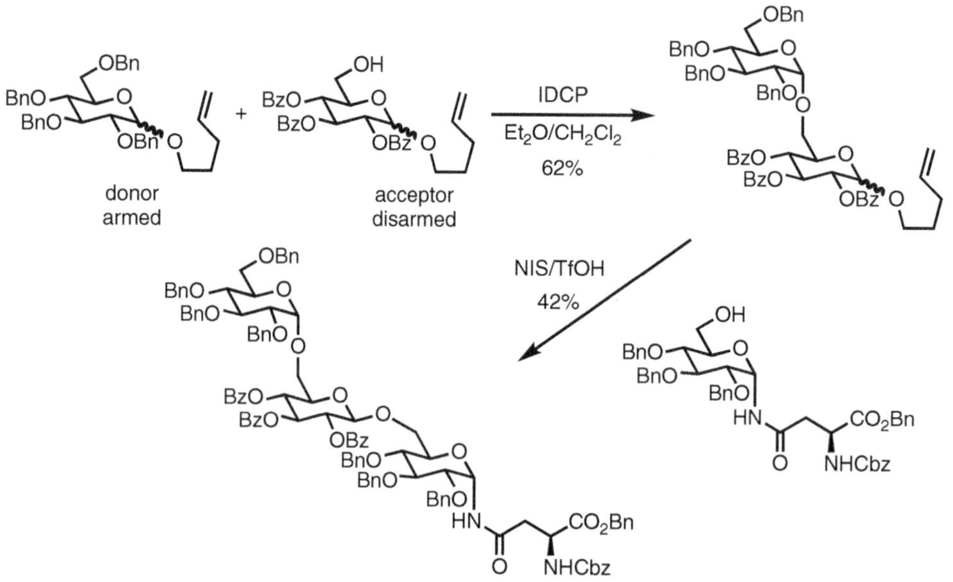

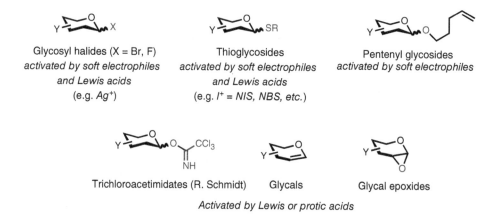

Scheme 4.20   *Armed/disarmed strategy for glycosidation*

To summarise, Figure 4.20 shows the common glycosyl donors. Glycosyl halides (bromides and fluorides) are classical examples of Koenigs–Knorr partners. Thioglycosides and pentenyl glycosides are both activated by soft electrophiles, the latter one leading to the possibility of armed and disarmed glycosyl donors. Trichloroacetimidates, glycals, and glycosyl epoxides are all activated by Lewis or protic acids.

Glycosyl halides (X = Br, F)
*activated by soft electrophiles
and Lewis acids*
(e.g. *Ag*⁺)

Thioglycosides
*activated by soft electrophiles
and Lewis acids*
(e.g. *I*⁺ = NIS, NBS, etc.)

Pentenyl glycosides
activated by soft electrophiles

Trichloroacetimidates (R. Schmidt)   Glycals   Glycal epoxides
Activated by Lewis or protic acids

Figure 4.20   *Summary of glycosyl donors*

Later developments of glycosylation have led to the realisation that the donor activation can, in fact, be considered as a continuum [42], which has eventually led to the realisation of automated oligosaccharide synthesis [43].

4.4 Glycoproteins and Proteoglycans

Glycopeptides comprise a group of molecules which contain both a carbohydrate domain and an amino acid domain. The term glycoprotein is often used to refer to all the macromolecular complexes of proteins and carbohydrates. Such a broad definition, however, mixes the true glycoproteins with proteoglycans and carbohydrate–protein complexes. In true glycoproteins, the protein chain is connected to a branched polysaccharide; in proteoglycans, the oligosaccharide is unbranched. Some glycoproteins are classified according to their functions in Table 4.4.

Table 4.4   Some glycoproteins with their functions

Suggested function	Glycoprotein
Enzyme	Cholinesterase
	Bromelain
	Ficin
	Ribonuclease
	Yeast invertase
Nutrient storage	Casein
	Ovalbumin
Hormone	Erytropoietin
	FSH (follicle stimulating hormone)
	LH (luteinising hormone)
	Thyroglobulin
Plasma and serum	α-, β-, and γ-glycoproteins
Protective mechanism	Fibrinogen
	Immunoglobulins
	Interferon
Structural proteins	Bacterial cell walls
	Collagen
	Extensin (plant cell wall)
Toxin	Fungal mycotoxins
	Ricin
Transport	Ceruloplasmin
	Haptoglobin
	Transferrin
Unknown	Avidin (egg white)
	Blood group antigens

Proteins usually perform their physiological function only in specific areas of the cell. The newly synthesised protein is transported from the rough endoplasmic reticulum to its site of function, guided by highly selective recognition processes, which, at least in eukaryotic cells, are dependent on glycoproteins. The carbohydrate part of the glycoprotein plays a key role in this process. On one hand, the carbohydrate moiety can protect the protein from degradation: the copper-transport protein ceruloplasmin has a biological half-life of 54 h; whereas its deglycosylated analogue asialoceruloplasmin has a half-life of less than 5 min. Another major function of the carbohydrate portion is within recognition and control processes such as cell growth and differentiation. Tumor cell membranes have altered glycoprotein structures, and these conjugates are, in part, tumor associated antigens. The blood group antigens are glycoproteins as well (Figure 4.21) [44]. In the ABO blood group system, the red blood cells are decorated with an oligosaccharide antigen, which reacts with Immunoglobulin M (IgM) antibodies. By chemical structure, the simplest type is the type O antigen, which contains a linear pentasaccharide. Type A and B antigens differ from type O in having a branch at the fucose bound nearest to the red blood cell surface: there is a single sugar, either a galactose (type B) or N-acetylgalactosamine (type A). Type AB red blood cells display both A and B antigens on their surface.

Figure 4.21   Blood group antigens

Proteoglycans are building blocks for connective tissue. Their core protein is often heavily glycosylated with covalently attached glycosaminoglycans (GAGs) (Figure 4.22). The glycosaminoglycans are usually of four structural types: chondroitin sulphate (composed of D-glucuronic acid and N-acetylgalactosamine); and dermatan sulphate (L-iduronic acid and N-acetylgalactosamine) occur largely in cartilage tissues. Heparan sulphate is present in all animals in the form of proteoglycans, where two or more heparin sulphate chains are bound to the cell surface or extracellular matrix proteins. Heparan sulphate is composed of dimers of D-glucuronic acid or L-iduronic acid with N-acetylglucosamine, or either O- or N-sulfated glucosamine units. Keratan sulphate is the glycosaminoglycan that occurs mainly in cornea and bone. It is composed of a repeating dimer of galactose and N-acetylglucosamine, which is then multiply sulfated. Disturbances of proteoglycan metabolism lead to a broad range of diseases known as mucopolysaccharidosis. Typical to these diseases is accumulation of proteoglycans inside the cell. Clinical symptoms vary from loss of sight due to opacity of the cornea (Hurler syndrome) to premature death (Sanfilippo syndrome).

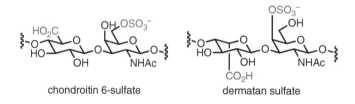

chondroitin 6-sulfate   dermatan sulfate

Figure 4.22   Chondroitin and dermatan sulfates

The glycosidic bonds are typically formed through either asparagine (amide bond formation with glucosamine), serine (β-glucosyl ether), or serine/threonine α-glycosidic bond with N-acetylgalactosamine. Typical structural types are shown in Figure 4.23. The syntheses of N- and O-glycosyl amino acids have been reviewed [45].

Glycosylation of proteins is a post-translational modification for the protein chain already constructed [46]. The core carbohydrate portion is first assembled on a polyprenol, dolichol (Figure 4.24), and then transferred to the nascent polypeptide chain in the rough endoplasmic reticulum. Final maturation occurs primarily in the Golgi apparatus of the cell by highly specific glycosyltransferases, which selectively transfer the peripheral sugars (N-acetylglucosamine, galactose, fucose, and N-acetylneuraminic acid) to the pentaglycosyl core structure.

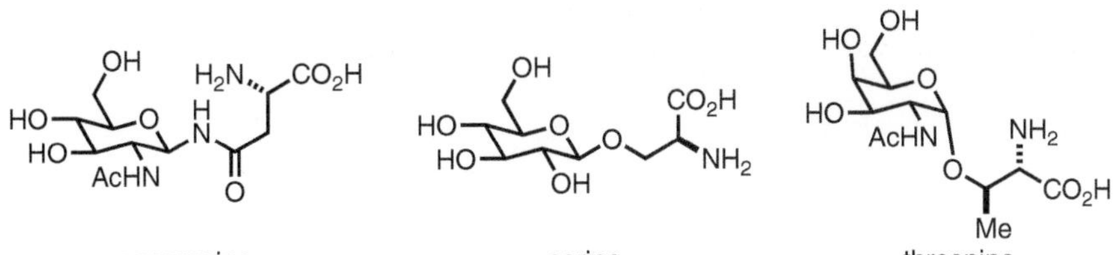

Figure 4.23   *Proteoglycan bonds*

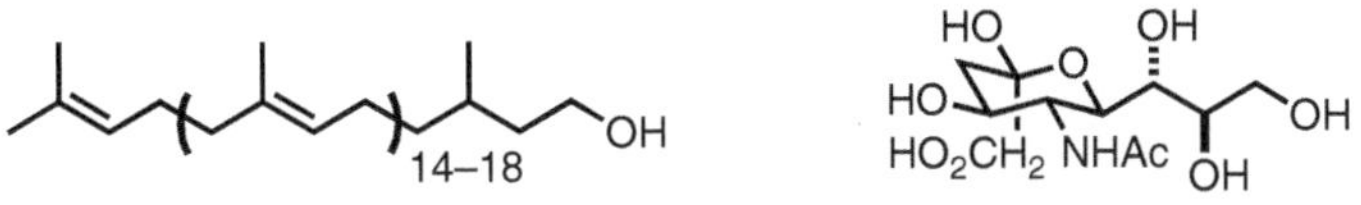

Figure 4.24   *Dolichol and N-acetylneuraminic acid*

4.5   Glycolipids

Glycolipids occur widely in nature, but they still represent only a small fraction of the total lipids. Glycolipids form parts of membranes and their actual role is only partly understood. They are known to participate in the biosynthesis of glycoproteins and complex polysaccharides: the polysaccharide is built on a glycolipid, from where it is transferred to a protein or carbohydrate at the end of biosynthesis. Glycolipids also have a regulatory role in the synthesis of proteoglycans, and they also inhibit the effects of toxic and antiviral agents. Toxins such as tetanus and cholera toxins bind to the carbohydrate part of glycolipids. Gangliosides inhibit the action of these toxins. The antiviral effectivity of interferon is increased by interaction with gangliosides.

The simplest glycolipids to occur in mammalian tissues are monoglycosyl ceramides, or cerebrosides. These compounds are sphingosine derivatives, and their structures are discussed in detail in Section 7.2.2. In the cerebrosides of the brain, sphingosine (or sphinganine) is usually glycosylated with D-galactose, in serum with D-glucose. The lipids are typically long-chain saturated (behenic acid, $C_{21}H_{43}COOH$, lignoceric acid, $C_{23}H_{47}COOH$), unsaturated (nervonic acid, $\Delta^{15}-C_{24:1}$), or hydroxy acids (cerebronic acid, $C_{22}H_{45}CHOHCOOH$).

Gangliosides are neuraminic acid-containing glycosphingolipids. They are composed of a hydrophilic oligosaccharide attached to a double lipophilic tail named ceramide, and to one or more neuraminic acid residues. The number of the neuraminic acid residues determines the polarity of the ganglioside. About fifty types of gangliosides are known, and they are typically constituents of all mammalian somatic cell membranes. Gangliosides also play an important role in nerve growth, nerve regeneration, and in brain functions: the mammalian brain cortex has the highest relative amount of gangliosides, *ca.* ten times that found in extraneural organs. Gangliosides also occur in pancreas, spleen, liver, kidneys, and most prevalently, in the grey matter of the brains. Accumulation of the lipids, lipidosis, is a group of serious genetic diseases, of which dozen or more distinct disorders are known. Tay-Sachs disease (affects brains), and Gaucher's disease (affects spleen and liver) are two examples for which the enzymatic malfunctions are known: for the former, the defective enzyme activity is hexosaminidase A, an enzyme that normally hydrolyses galactose residues from ganglioside GM_2 (Figure 4.25). In the case of Gaucher's disease, glucocerebrosidase activity is deficient.

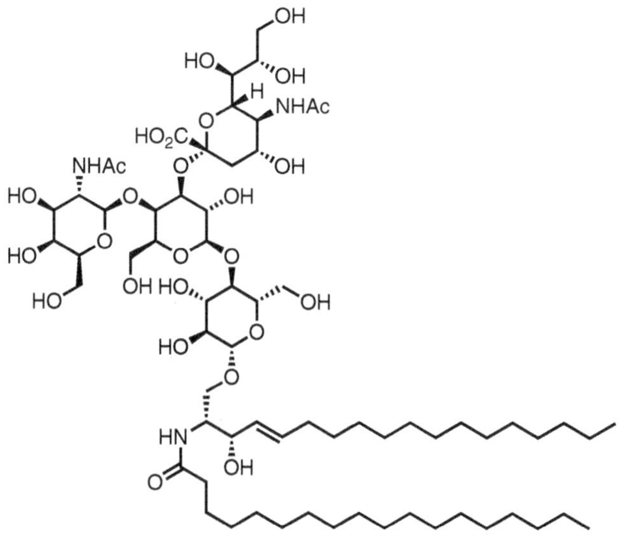

Figure 4.25 *Ganglioside GM$_2$*

4.6 Sugar Antibiotics

Many antibiotics contain rare carbohydrate units. Nucleoside antibiotics usually interfere with the DNA/RNA synthesis. Due to this activity, they are also very toxic, which is reflected in their restricted clinical applicability. Puromycin is a purine analogue, blasticidin is a pyrimidine analogue, and showdomycin, coformycin, and pentostatin are modified nucleoside antibiotics (Figure 4.26).

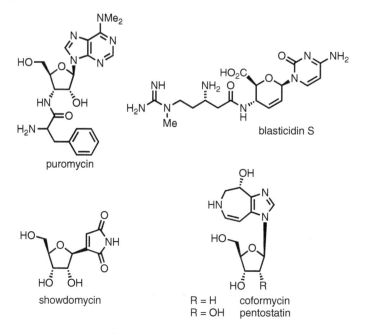

Figure 4.26 *Sugar antibiotics*

Sugars can also be connected to several aromatic aglycons which exhibit antitumoral activity. Daunomycin (also known as daunorubicin) and adriamycin (doxorubicin) are members of the antitumoral anthracycline antibiotics

(Figure 4.27). Their development began in the 1950s when *Federico Arcamone* and his team at Farmitalia laboratories isolated a red pigment from *Streptomyces peucitius* (allegedly referring to the Peucetia region in Italy). Eventually, pure daunomycin was isolated, and the same compound was also isolated by chemists at the French company Rhône-Poulenc, only to be called rubidomycin. The compound was not only antibiotic, but also highly successful against leukemias. Chemical modifications of the natural product leads have led to the development of a number of less toxic anti-leukemic drugs.

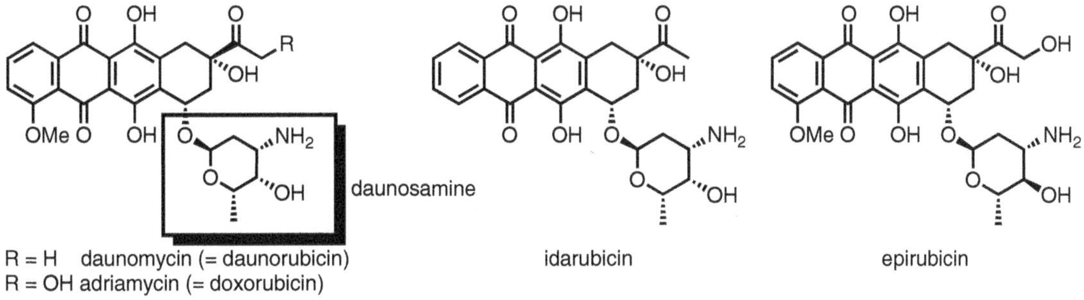

R = H daunomycin (= daunorubicin)
R = OH adriamycin (= doxorubicin)

daunosamine

idarubicin

epirubicin

Figure 4.27 Anthracycline antibiotics

In macrolide antibiotics, a macrocyclic lactone, formed via the polyketide pathway, is usually glycosylated, connected to a carbohydrate unit, typically with an amino sugar. These are effective against Gram-positive bacteria, but their effects on Gram-negative ones are usually weak. Erythromycin, leucomycins (see Chapter 7 for structures), and avermectins (Figure 4.28) are typical examples of such structures. Half of the 2015 Nobel Prize in Physiology or Medicine was awarded to *William C. Campbell* and *Satoshi Ōmura* for the discovery of avermectin, "the derivatives of which have radically lowered the incidence of river blindness and lymphatic filariasis, as well as showing efficacy against an expanding number of other parasitic diseases."

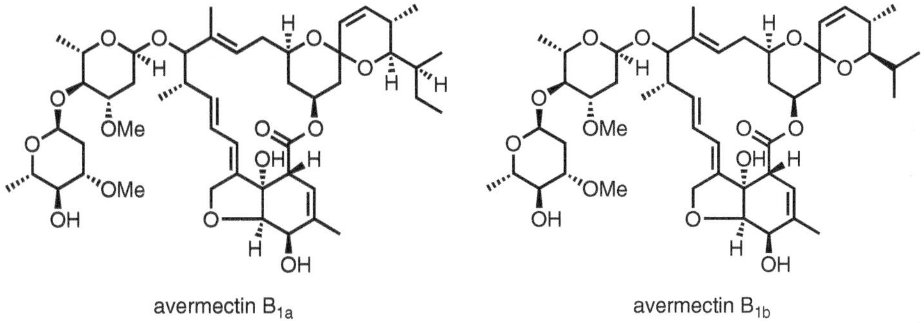

avermectin B_{1a}

avermectin B_{1b}

Figure 4.28 Avermectins B_{1a} and B_{1b}

Figure 4.29 shows some other antibiotics containing a carbohydrate unit. The discovery of penicillin by Alexander Fleming in 1928 and the first attempts of treatments of patients in 1941, although with varying success [47], pointed to the large number of fungi as a potentially rich source of antibiotics. In 1943, Waksman and his students discovered that *Streptomyces griseus* produces an antibiotic compound different to penicillin and chemically more stable. The compound, streptomycin, was to become the first of a growing class of aminoglycoside antibiotics, which still are indispensable to combat tuberculosis. Since 1948, streptomycin has been considered the drug of choice for bubonic, septicemic, and pneumonic plague [48]. Due to issues in supply (streptomycin is manufactured by one pharmaceutical company and is available only in modest supplies and by request) and serious side effects, other drugs, in particular gentamicin (Figure 4.13) have gained attention because it can be given in a single intravenous dose and appears to be as effective as streptomycin [49].

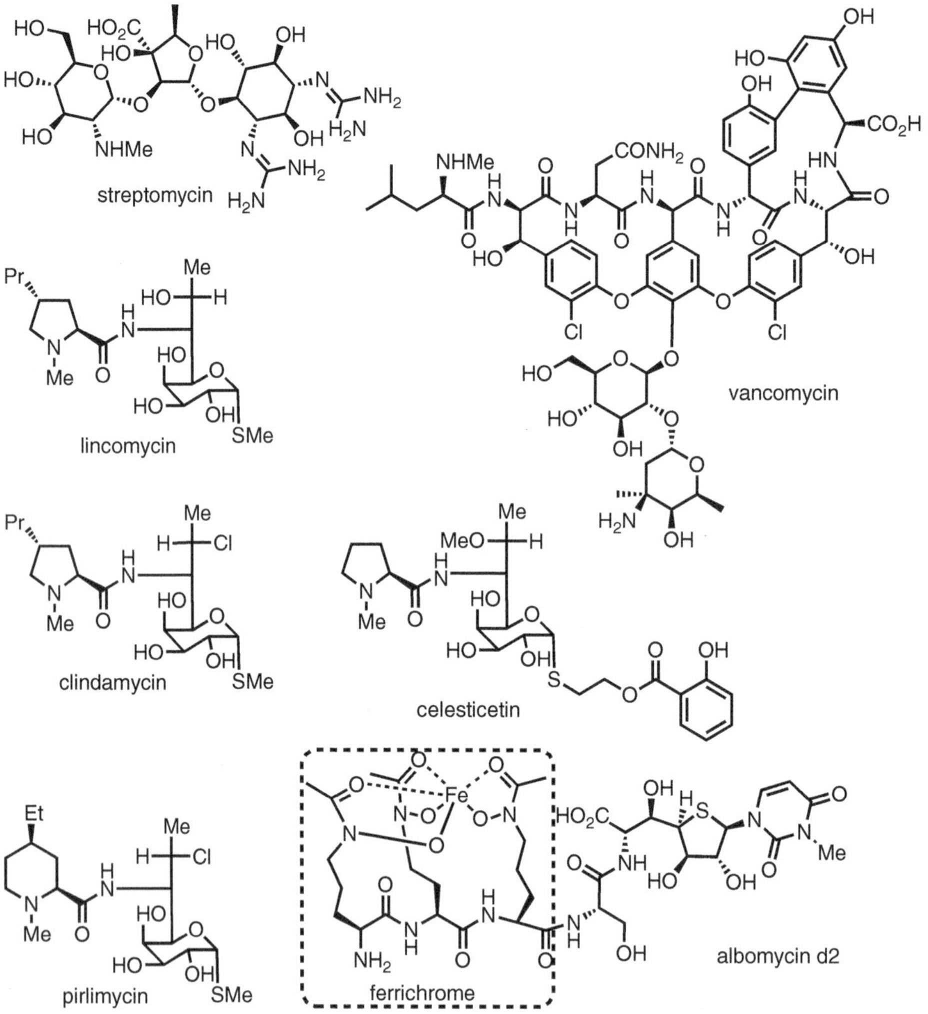

Figure 4.29 *Antibiotics with modified carbohydrates*

Vancomycin is a highly complex structure, which was found in 1956 from a soil sample in Borneo by a team of scientists at the American pharmaceutical company Eli Lilly. Vancomycin is one of the few antibiotics that are effective against methicillin-resistant *Staphylococcus aureus* (MRSA), and has therefore gained a reputation as the antibiotic of last resort against bacteria. However, some bacteria of *Staphylococcus aureus* and *Enterococcus faecalis* have developed vancomycin-resistant strains.

Lincomycin was originally isolated from a soil sample from Lincoln, Nebraska, containing *Streptomyces lincolnensis*. Besides containing a rare amino sugar, it also contains a rare amino acid, 4-propyl proline. Lincomycin has been chemically modified to clindamycin, which has largely superseded lincomycin. Pirlimycin is a structurally related lincosamide antibiotic mainly used for the treatment of mastitis in cattle.

Indiscriminate use of antibiotics has led to the emergence of multiple drug-resistant strains of the most common pathogens. One way to look for an alternative strategy to antibiotics is to utilise sideromycins' mode of action. Sideromycins consist of an antibiotic and an iron-complexing siderophore. They are recognised and taken up by the bacteria through a siderophore-dependent transport system. The antibiotic is thus concentrated inside the microbial cells, hopefully leading to lethal concentrations. This feature makes the sideromycin antibiotic albomycin distinct among antibiotics, which are taken up by diffusion [50].

4.7 Cyclitols

The cyclitols are a diverse group of cyclic polyhydroxylated compounds, which have a cyclohexane skeleton. Although strictly not carbohydrates, inositols are sugars, and biogenetically they are derived from glucose by an enzymatic reaction that performs a reaction mechanistically close to Ferrier cyclisation (Scheme 4.21) [51]. The ring-opened form of D-glucose-6-phosphate is oxidised to a ketoaldehyde, which undergoes a formal intramolecular aldol reaction followed by reduction to *myo*-inositol-3-phosphate.

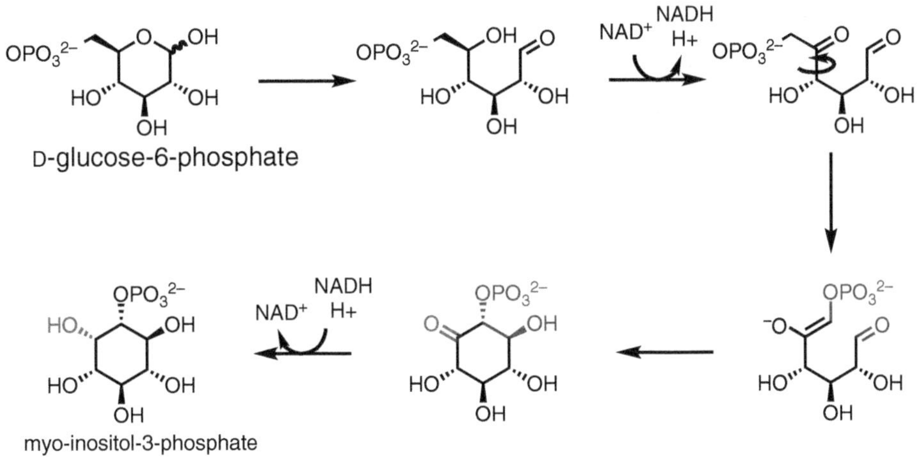

Scheme 4.21 *Biosynthesis of inositol derivatives*

The biological activities of inositols are as varied as their structures, and many of them occur in phosphorylated form. Phosphorylated inositols have been shown to act as second messengers in many intracellular signal transduction processes by mediating the release of calcium from nonmitochondrial stores. Aminodeoxyinositols and -conduritols occur in the aminocyclitol antibiotics, and a number of conduritol derivatives have important physiological actions, such as glycosidase inhibition, antifeedant, antibiotic, tumorstatic, and growth-regulating activities. The synthetic activity is further boosted by the fact that various hydroxylated cyclohexene derivatives act as glycosidase inhibitors [52].

The carbacyclic 1,2,3,4,5,6-cyclohexanehexols are known as inositols (*Gr. inos* = fiber, muscle). There are altogether nine stereoisomers for inositol, of which only two are optically active (Figure 4.30). *myo*-Inositol is the most common

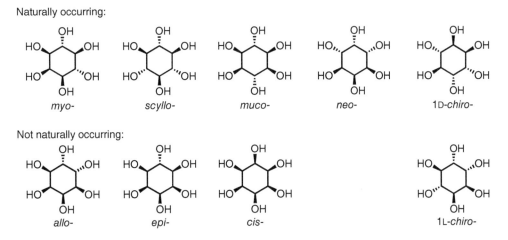

Figure 4.30 *Inositols*

inositol, and it is present in practically all animal and plant species. In animals and microorganisms, *myo*-inositol is usually present in phospholipids, whereas in plants it usually occurs as its hexaphosphate, phytic acid. Of the other inositols, only D-chiro- (D-pinitol), L-chiro (L-quebrachitol), and scylloinositol occur naturally.

Conduritols are the corresponding cyclohexenetetraols, of which only two occur in nature (Figure 4.31). Some conduritol derivatives show hypoglycemic, antifeedant, antibiotic, antileukemic, and growth-regulating activity.

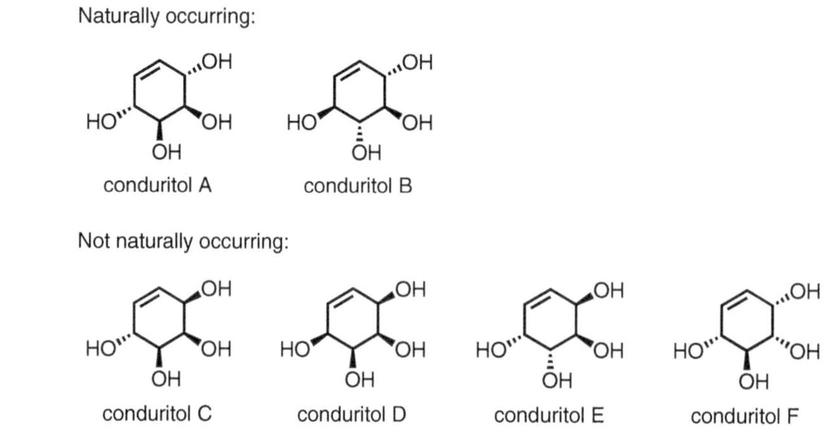

Figure 4.31 *Conduritols*

The synthesis of cyclitols has seen a renaissance, mainly thanks to the rapid development of novel strategies for the construction of polyhydroxylated systems in a stereochemically homogeneous manner [53]. An interesting development is the utilisation of microbial oxidation of benzene and its derivatives. Oxidation of benzene with *Pseudomonas putida* 39-D produces the corresponding *cis*-cyclohexanedienediol [54].

Steven V. Ley (1945–, Cambridge University) has applied this microbial oxidation to the synthesis of pinitol, a feeding stimulant for the larvae of the yellow butterfly *Eurema hecabe mandarina* and an inhibitor of larval growth of *Heliothis zea* on soybeans (Scheme 4.22) [55]. The oxidation product was protected as the dibenzoate and epoxidised with mCPBA to give a mixture of the desired *trans* epoxide (73%) and the undesired *cis* epoxide (17%), which resulted from the attack of the peracid presumably directed by the benzoate group. The major product was then subjected to epoxide opening to give the expected product with methoxide attack at the distal end of the epoxide. Final *cis* hydroxylation of the double bond (OsO$_4$, NMMO) gave the triols in a 5:1 ratio with the desired isomer predominating (through attack of the oxidating reagent from the less hindered α-face).

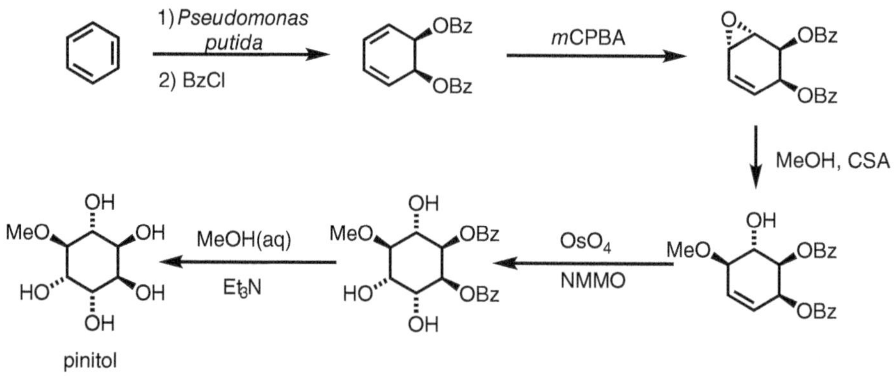

Scheme 4.22 *Synthesis of racemic pinitol*

The microbial oxidation is both stereospecific and enantioselective; if the benzene ring carries a substituent (typically a halogen), the product will be a single enantiomer (such as the bromine-containing compound in Scheme 4.23). These diol derivatives have been further derivatised using standard methodology (protection of the diols followed by sequential epoxidation/epoxide opening and *cis*-hydroxylation) to give (−)-pinitol. Reversal of the order of the oxidation processes gives access to the enantiomeric non-natural (+)-pinitol [56].

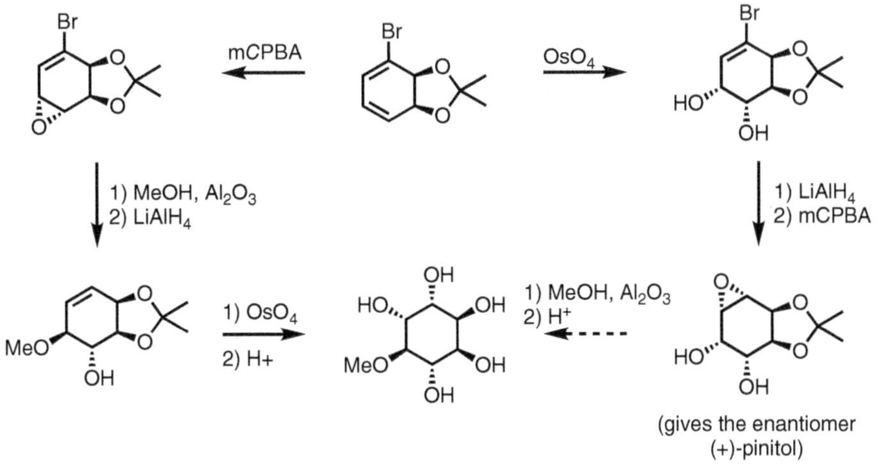

Scheme 4.23   *Synthesis of (−)-pinitol*

A similar strategy was also used in the synthesis of D- and L-*myo*-inositol 1,4,5-triphosphates, as well as the corresponding deoxy, fluoro, and methyl derivatives (Scheme 4.24) [57]. The addition of the nucleophile was completely stereoselective (from the less hindered α-face), but the regioselectivity was less impressive for the small nucleophiles (hydride and fluoride, ca 4:1 to 6:1 ratio, respectively).

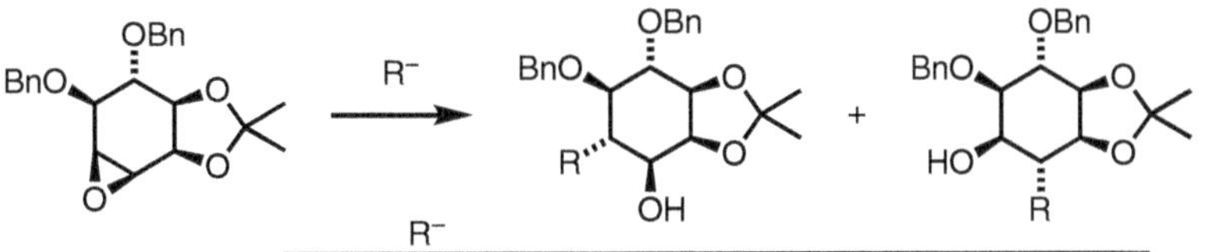

R⁻		
H− (LiAlH₄)	76%	12%
F− (TASF)	59%	15%
Me− (Me₂Cu(CN)Li₂)	73%	3%

Scheme 4.24   *Regioselectivity in epoxide opening*

Starting from chlorobenzene, one can devise a rapid entry to conduritols (Scheme 4.25) [58]. Oxidation gave the *cis*-diol [59], which was regio- and stereoselectively epoxidised with mCPBA (Henbest oxidation, the allylic alcohol directs the oxidation through coordination). Epoxide opening was finally achieved again with high regioselectivity due to polarity control [60].

Employment of the Ferrier rearrangement [61] gives access to the cyclitols from carbohydrates (Scheme 4.26) [62]. The enol acetate was derived from glucose, and mercuric acetate induced the Ferrier cyclisation to give the corresponding ketone. Stereoselective reduction of the inosose intermediate was achieved using Saksena–Evans reduction (intramolecular hydride delivery [63] with sodium triacetoxyborohydride) to give the desired differentially protected hexol, which was converted to the final product.

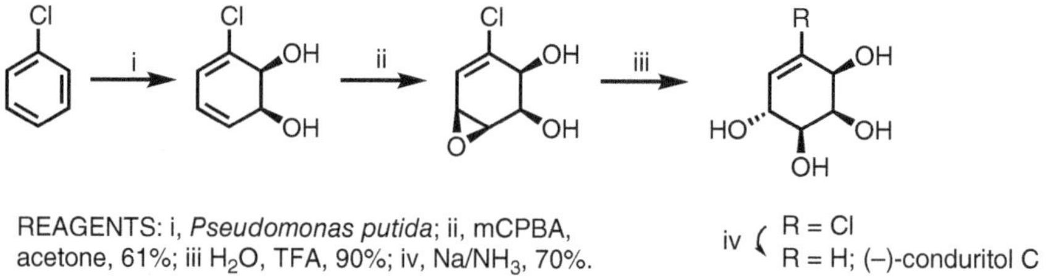

Cl
i
OH
OH
ii
iii
R
HO
iv
REAGENTS: i, *Pseudomonas putida*; ii, mCPBA, acetone, 61%; iii H2O, TFA, 90%; iv, Na/NH3, 70%.
R = Cl
R = H; (−)-conduritol C

Scheme 4.25 *Synthesis of (−)-conduritol*

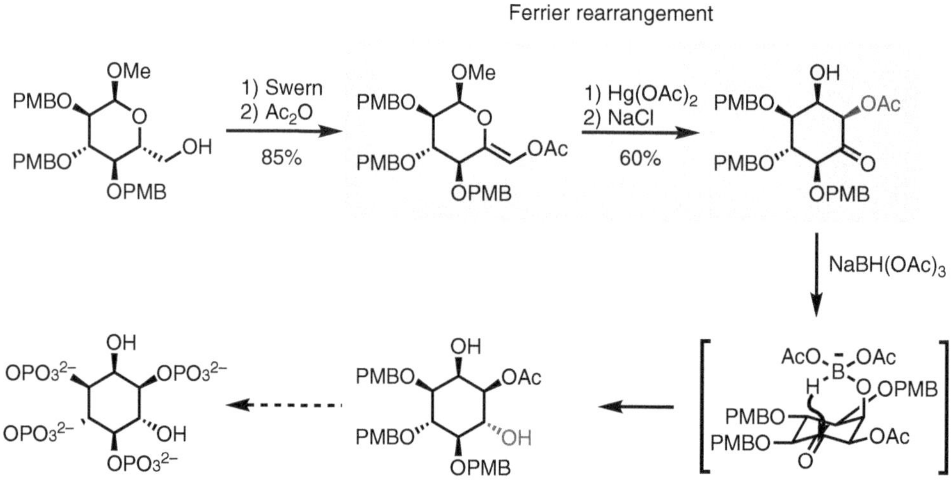

Ferrier rearrangement
OMe
PMBO
O
OH
OPMB
1) Swern
2) Ac2O
85%
OAc
1) Hg(OAc)2
2) NaCl
60%
NaBH(OAc)3
AcO
B
H
PMBO
OPO3 2−

Scheme 4.26 *Synthesis of cyclitols from carbohydrates*

The stereocontrolled reduction occurs through the complex formed from the triacetoxyborohydride and the (necessary) free alcohol β to the ketone. The complex delivers the hydride intramolecularly from the same face as the axial alcohol resides to give the desired equatorial alcohol contaminated with less than 10% of the axial alcohol isomer [64].

4.8 Iminosugars

Iminosugars are related to sugars in the sense that the ring oxygen atom is replaced with a nitrogen atom (Figure 4.32) [65]. Compared to their oxygen containing relatives, these compounds are unstable because of the lower electronegativity of the nitrogen atom, which donates electron density to the anomeric C–O bond, making the bond easier to cleave. One way to make the iminosugars more stable is to remove the anomeric oxygen atom altogether to form a deoxyiminosugar. Iminosugars or iminoglycosides are structural mimics of sugars, and therefore they have found applications as glycosidase inhibitors.

Based on the diversity of sugar-related chemistry, it seems only natural that academic interest in replacing the ring oxygen in monosaccharides by other heteroatoms had already arisen in the 1960s. The earliest attempts were made independently in 1961 by Schwarz [66], Owen [67], and Whistler [68], who succeeded in replacing the ring oxygen of D-xylopyranose with sulfur (Figure 4.33). One year later, Paulsen failed in the synthesis of unprotected ring nitrogen-containing pyranoses: the initially formed iminosugars readily lost three molecules of water in acidic conditions forming the corresponding aromatic pyridines [69]. However, the formation of *N*-acetyl-iminopyranoses were successfully demonstrated the same year independently by Paulsen [70], Jones [71], and Hanessian [72].

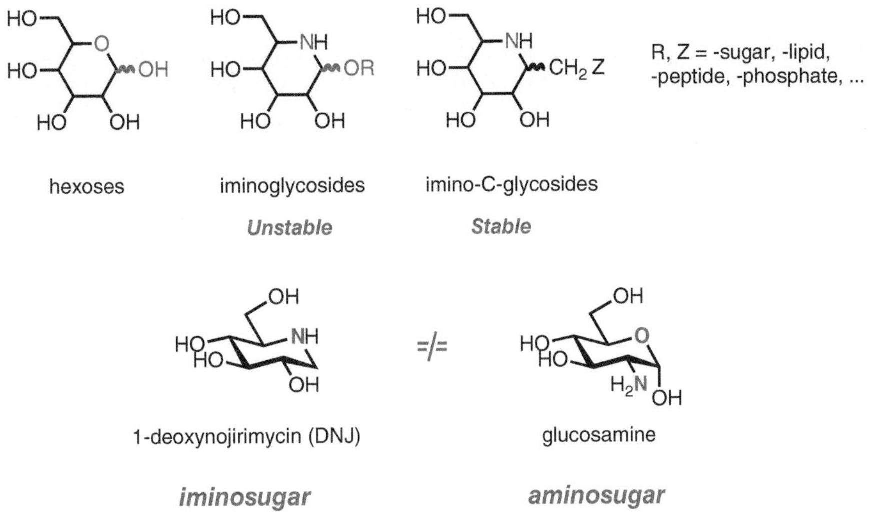

Figure 4.32 *Sugars, iminoglycosides, and deoxyiminosugars*

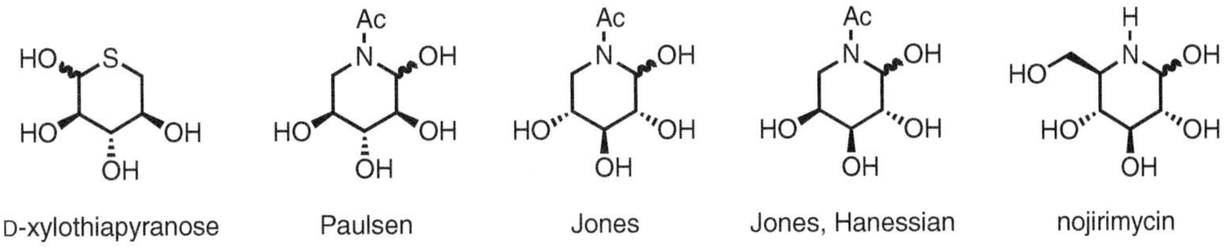

Figure 4.33 *First hetero- and iminosugars*

Iminosugar-containing herbal extracts have been used in traditional Chinese phytomedicine. Moreover, the first industrially produced medicine, Haarlem oil, contained an extract from the leaves of the white mulberry tree, known to be an excellent source of iminosugars. Haarlem oil was recommended for the treatment of diabetes and for whitening one's skin in the seventeenth century [73].

Nojirimycin was isolated from *Streptomyces roseochromogenes* species only after the first syntheses were reported [74] and the structure ascertained in 1966 [75]. The first 1-deoxyanalog of nojirimycin, 1-deoxynojimycin (DNJ), was also synthesised the same year (Figure 4.34) [76]. Nojirimycin itself was first synthesised by Inouye in 1968, along with 1-deoxynojirimycin [77].

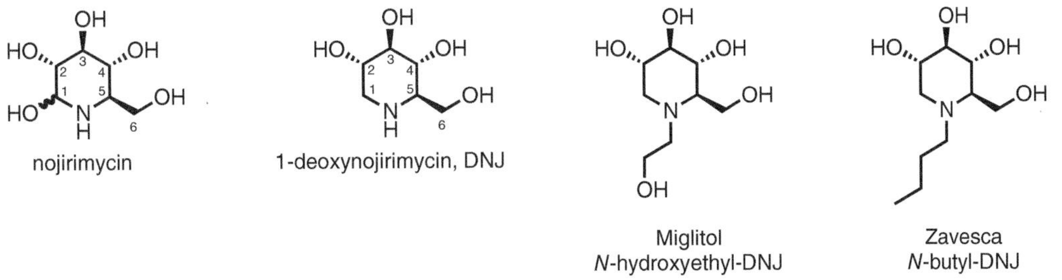

Figure 4.34 *Nojirimycin and 1-deoxynojirimycin*

In 1976, chemists at Bayer discovered DNJ as an α-glucosidase inhibitor [64]. The same year, DNJ was also isolated from the mulberry tree [78]. Bayer's interest in 1-deoxynojirimycin analogs finally lead to the commercialisation of Miglitol as a treatment for non-insulin-dependent diabetes in Europe and the US.

Glycosidases are a vast group of enzymes that catalyse the cleavage of glycosidic bonds in oligosaccharides and other carbohydrate-containing biopolymers. They take part in the digestion of carbohydrates, but they also play various more specialised and sophisticated roles, including the lysosomal catabolism of glycoconjugates within cells. Glycosidases are also involved in the biosynthesis and catabolism of glycoproteins and glycolipids making them essential for all life.

N-butyl-DNJ inhibits the glycosyltransferase involved in the biosynthesis of glycosylceramides [79], and thus prevents the accumulation of glucosylceramide in Gaucher's disease, a lysosomal storage disorder, for the treatment of which it is used under the trademark Zavesca [80].

After the isolation of nojirimycin and 1-deoxynojirimycin from natural sources, two stereoisomers of nojirimycin, mannojirimycin [81] and galactonojirimycin [82], were isolated in 1984 and 1988, respectively. In addition, three more 1-deoxynojirimycins have been isolated: 1-deoxymannojirimycin in 1979, and 1-deoxyaltronojirimycin [83] and 1-deoxygulonojirimycin in 2001. The rest of the 1-deoxynojirimycin stereoisomers and many derivatives thereof have been synthesised and tested for various biochemical indications [64], [72], [84]. A significant part of these are based on converting suitable sugars to desired iminosugars.

Here we will highlight two approaches, one based on conjugate addition of chiral lithio amide, and the second approach using L-serine as the starting material and employing internal asymmetric induction. The first example is from the group of *Stephen G. Davies* (1950–) who has extensively studied the conjugate addition approach to both amino sugars and iminosugars, as well as many other types of natural products [85]. For the synthesis of (–)-1-deoxymannojirimycin (Scheme 4.27) [86], the conjugate addition of the chiral lithium (*R*)-*N*-benzyl-*N*-(*R*-methylbenzyl)amide onto the enoate followed by oxidation of the resultant enolate with (–)-camphorsulfonyloxaziridine [(–)-CSO], gave β-amino α-hydroxy ester in 80% yield and > 99:1 dr. This was converted in a few steps to the benzyl-protected aldehyde, which underwent a moderately Felkin–Anh-selective reaction with vinyl magnesium bromide to give the precursor for the cyclisation to the iminosugar piperidine ring. The cyclisation was affected by treatment with I_2 and $NaHCO_3$, followed by immediate *O*-desilylation with HF·pyridine. Methanolysis of the carbonate functionality and subsequent global hydrogenolytic deprotection in the presence of Pearlman's catalyst [Pd(OH)$_2$/C] gave (–)-1-deoxymannojirimycin in 87% yield and > 99:1 dr.

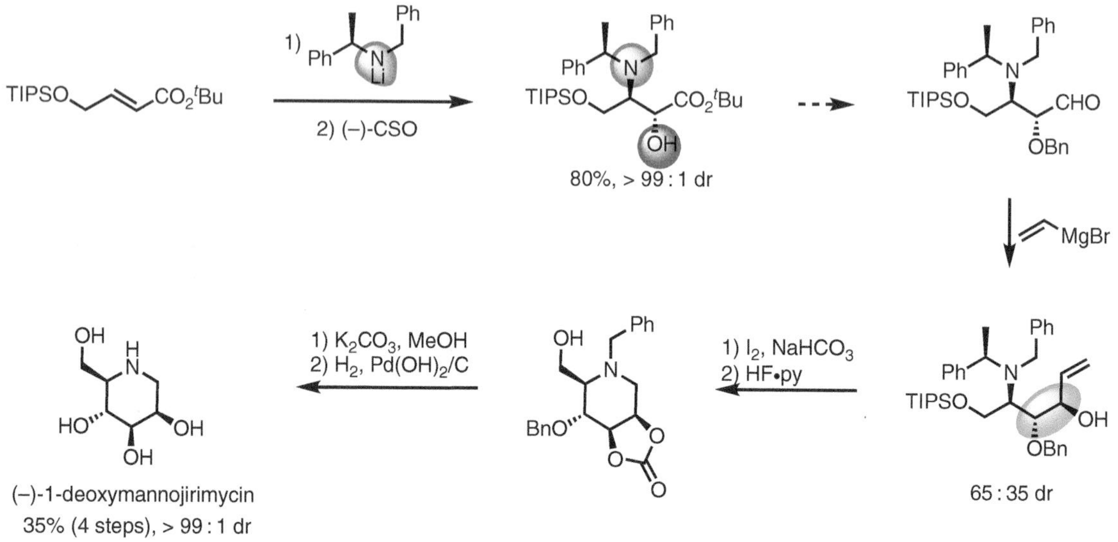

Scheme 4.27   Conjugate addition of chiral lithio amide to DNJ

The second example is the result from the realisation that all eight stereoisomers (Figure 4.35) can be derived from L-serine through highly stereocontrolled reactions. Initial addition of the nucleophile to Garner's aldehyde generates two series of stereoisomers: those where the relative stereochemistry of the hydroxyl group coming from the addition of the nucleophile to the aldehyde function is either *anti* or *syn* relative to the amine containing carbon.

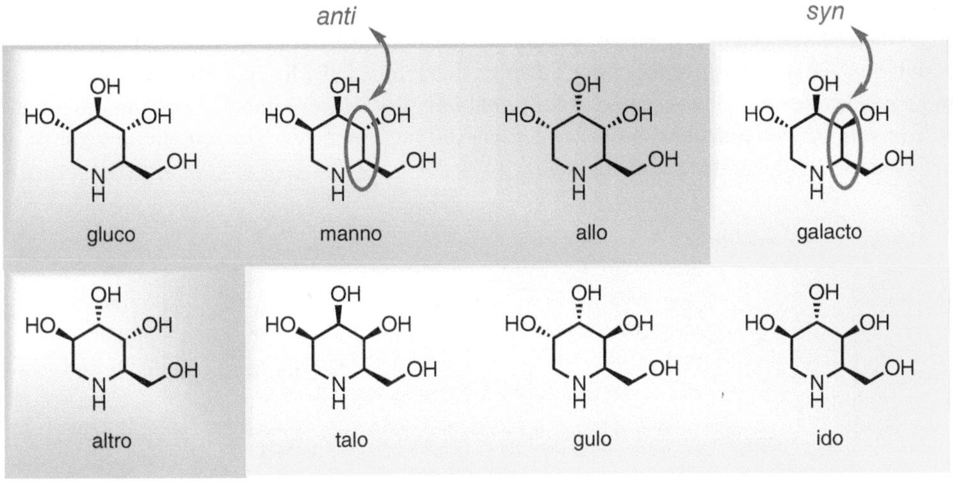

Figure 4.35 *Eight diastereomers of the deoxynojirimycin type*

Addition of a lithioacetylide on Garner's aldehyde (GA) is highly diastereoselective giving the *anti*-product with > 15:1 diastereoselectivity (Scheme 4.28) [87]. Standard operations allowed rapid access to the unsaturated aminoalcohol, a key intermediate for the synthesis of the deoxynojirimycin isomers [88]. The *syn* isomer of the addition was simply obtained by transmetallating the lithio acetylide with zinc bromide before addition of Garner's aldehyde. The *syn*-product was obtained in 82% yield and no *anti*-product was detected. The remainder of the synthesis to provide this *syn*-tetrahydropyridine derivative followed the route depicted in Scheme 4.28.

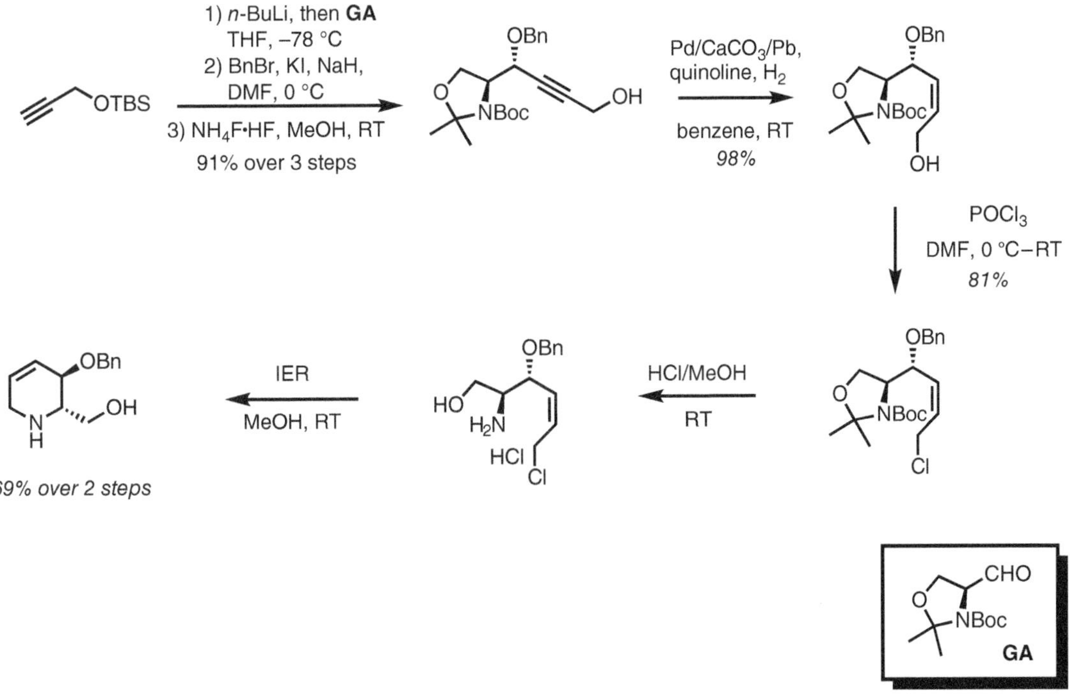

Scheme 4.28 *Internal asymmetric induction with L-serine as the starting material*

Starting from Garner's aldehyde (**GA**), one can thus make either one of the two iastereomeric intermediate aminoal-cohols, which can then be subjected to standard operations: *syn*-dihydroxylation of the allylic alcohol either following the Kishi selectivity, or the *anti*-Kishi-selectivity (Scheme 4.29). *anti*-Dihydroxylation reaction, for instance through epoxidation and epoxide opening, allows access to the remaining four diastereomers [89]. Thus, all eight stereoisomers of the deoxynojirimycins are obtainable from the two advanced intermediates. The remaining eight isomers would be obtainable from D-serine following the same chemistry.

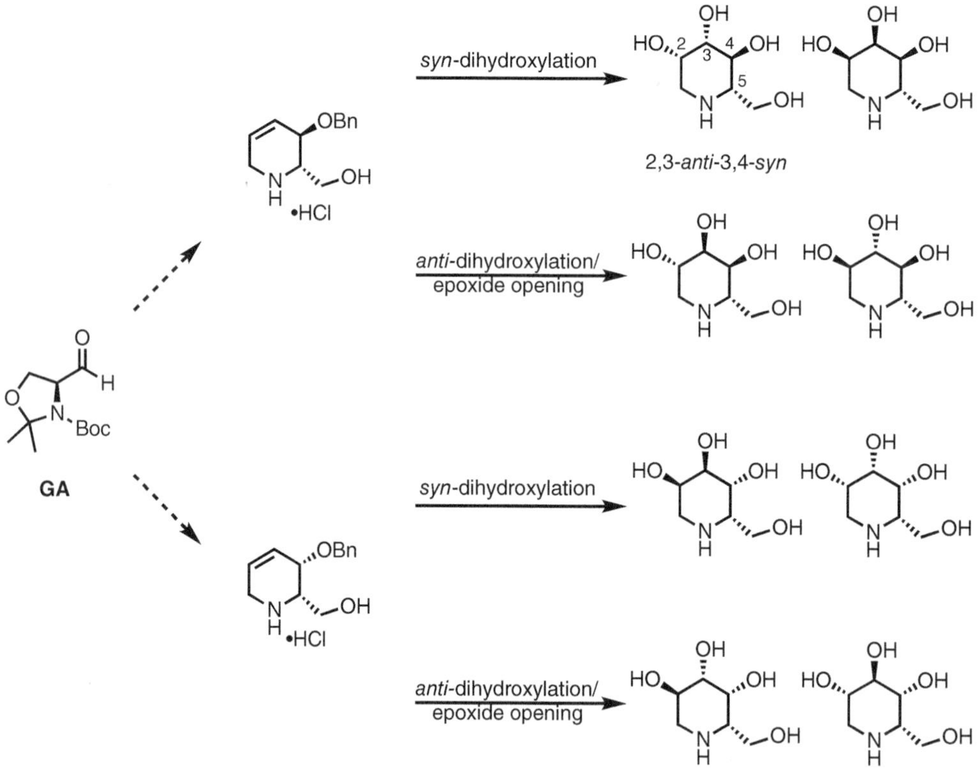

Scheme 4.29   Diastereomeric alcohols to eight diastereomers of DNJs

References

1. Payen, A. *Comptes rendus* **1838**, *7*, 1052–1056.
2. Altona, C., Romers, C. *Acta Crystallogr.* **1963**, *16*, 1225–1232.
3. David, S., Eisenstein, O., Hehre, W.J., Salem, L., Hoffmann, R. *J. Am. Chem. Soc.* **1973**, *95*, 3806–3807.
4. Goekjian, P.G., Wu, T.-C., Kishi, Y. *J. Org. Chem.* **1991**, *56*, 6412–6422.
5. a) Fischer, E. *Ber. Dtsch. Chem. Ges.* **1891**, *24*, 1836–1845; b) Fischer, E. *Ber. Dtsch. Chem. Ges.* **1891**, *24*, 2683–2687; c) Lichtenthaler, F.W. *Angew. Chem. Int. Ed. Engl.* **1992**, *31*, 1541–1556.
6. Bijvoet, J.M., Peerdeman, A.F., van Bommel, A.J. *Nature* **1951**, *168*, 271–272.
7. Rosanoff, M.A. *J. Am. Chem. Soc.* **1906**, *28*, 114–121.
8. Braconnot, H. *Ann. Chim. Phys.* **1811**, *79*, 265–304.
9. Maehr, H., Schaffner, C.P. *J. Am. Chem. Soc.* **1970**, *92*, 1697–1700.
10. a) Parker, K.A., Babine, R.E. *J. Am. Chem. Soc.* **1982**, *104*, 7330–7331; b) Corrected structure for sibiromycin: Leber, J.D., Hoover, J.R.E., Holden, K.G., Johnson, R.K., Hecht, S.M. *J. Am. Chem. Soc.* **1988**, *110*, 2992–2993.
11. Pawlak, J., Sowinski, P., Borowski, E., Gariboldi, P. *J. Antibiot.* **1995**, *48*, 1034–1038.
12. Koskinen, A.M.P., Otsomaa, L.A. *Fortschritte d. Chem. Org. Naturst.* **1998**, *74*, 193–263.

13. Omura, S., Nakagawa, A., Otani, M., Haa, T. *J. Am. Chem. Soc.* **1969**, *91*, 3401–3404.
14. Omura, S., Sadakane, N., Tanaka, Y., Matsubara, H. *J. Antibiot.* **1983**, *36*, 927–930.
15. Sun, C., Wei, J., Huang, J., Jin, W., Wang, Y. *Actinomycetologica* **1999**, *13*, 120–125.
16. a) Stevens, C.L., Gutowski, G.E., Byant, C.P., Glinski, R.P., Edwards, O.E., Sharma, G.M. *Tetrahedron Lett.* **1969**, *10*, 1181–1184; b) Kirst, H.A., Mynderse, J.S., Martin, J.W., Baker, P.J., Paschal, J.W., Rios Steiner, J.L., et al. *J. Antibiot.* **1996**, *49*, 162–167.
17. Hochlowski, J.E., Mullally, M.M., Brill, G.M., Whittern, D.N., Buko, A.M., Hill, P., et al. *J. Antibiot.* **1991**, *44*, 1318–1330.
18. Grauppner, P.R., Martynow, J., Anzeveno, P.B. *J. Org. Chem.* **2005**, *70*, 2154–2160.
19. Paululat, T., Kulik, A., Hausmann, H., Karagouni, A.D., Zinecker, H., Imhoff, J.F., et al. *Eur. J. Org. Chem.* **2010**, 2344–2350.
20. Koskinen, A.M.P., Otsomaa, L.A. *Tetrahedron* **1997**, *53*, 6473–6484.
21. Leblanc, Y., Fitzsimmons, B.J., Springer, J.P., Rokach, J. *J. Am. Chem. Soc.* **1989**, *111*, 2995–3000.
22. a) Friesen, R.W., Danishefsky, S.J. *J. Am. Chem. Soc.* **1989**, *111*, 6656–6660; b) Halcomb, R.L., Danishefsky S.J. *J. Am. Chem. Soc.* **1989**, *111*, 6661–6666.
23. Griffith, D.A., Danishefsky, S.J. *J. Am. Chem. Soc.* **1990**, *112*, 5811–5819.
24. Edmonds, M.K. *Chem. New Zealand* **2009**, *73*, 140–142.
25. a) Magano, J. *Tetrahedron* **2011**, *67*, 7875–7899; b) Magano, J. *Chem. Rev.* **2009**, *109*, 4398–4438.
26. Karpf, M., Trussardi, R. *Angew. Chem., Int. Ed.* **2009**, *48*, 5760–5762.
27. Yeung, Y.-Y., Hong, S., Corey, E.J. *J. Am. Chem. Soc.* **2006**, *128*, 6310–6311.
28. a) Werner, L., Machara, A., Hudlicky, T. *Adv. Synth. Catal.* **2010**, *352*, 195–200; b) Sullivan, B., Carrera, I., Drouin, M., Hudlicky, T. *Angew. Chem., Int. Ed.* **2009**, *48*, 4293–4295.
29. Ishikawa, H., Suzuki, T., Orita, H., Uchimaru, T., Hayashi, Y. *Chem. Eur. J.* **2010**, *16*, 12616–12626.
30. Trost, B.M., Zhang, T. *Chem. Eur. J.* **2011**, *17*, 3630–3643.
31. Trajkovic, M., Ferjancik, Z., Saicik, R.N. *Org. Biomol. Chem.* **2011**, *9*, 6927–6929.
32. Woodward, R.B Sondheimer, F., Taub, D., Heusler, K., McLamore, W.H. *J. Am. Chem. Soc.* **1952**, *74*, 4223–4251.
33. Hayashi, Y. *J. Org. Chem.* **2021**, *86*, 1–23.
34. Schmidt, R.R. *Angew. Chem., Int. Ed. Engl.* **1986**, *25*, 212–235.
35. Mukaiyama, T., Murai, Y., Shoda, S. *Chem. Lett.* **1981**, 431–432.
36. Nicolaou, K.C., Dolle, R.E., Papahatjis, D.P., Randall, J.L. *J. Am. Chem. Soc.* **1984**, *106*, 4189–4192.
37. Nicolaou, K.C., Groneberg, R.D., Miyazaki, T., Stylianides, N.A., Schulze, T.J., Stahl, W. *J. Am. Chem. Soc.* **1990**, *112*, 8193–8195.
38. a) Matsumoto, T., Maeta, H., Suzuki, K., Tsuchihashi, G. *Tetrahedron Lett.* **1988**, *29*, 3567–3570; b) Suzuki, K., Maeta, H., Matsumoto, T., Tsuchihashi, G. *Tetrahedron Lett.* **1988**, *29*, 3571–3574; c) Review on glycosylationa: Zhu, X., Schmidt R.R. *Angew. Chem. Int. Ed.* **2009**, *48*, 1900–1934.
39. Nicolaou, K.C., Bockovich, N.J., Carcanague D.R. *J. Am. Chem. Soc.* **1993**, *115*, 8843–8844.
40. Paulsen, H. *Angew. Chem. Int. Ed. Engl.* **1982**, *21*, 155–173.
41. a) Mootoo, D.R., Konradsson, P., Udodong, U., Fraser-Reid, B. *J. Am. Chem. Soc.* **1988**, *110*, 5583–5584; b) Fraser-Reid, B., Udodong, U.E., Wu, Z., Ottosson, H., Merritt, J.R., Rao, C.S., et al. *Synlett* **1992**, 927–942; c) Fraser-Reid, B., Merritt, J.R., Handlon, A.L., Andrews, C.W. *Pure Appl. Chem.* **1993**, *65*, 779–786.
42. a) Douglas, N.L., Ley, S.V., Lücking, U., Warriner, S.L. *J. Chem. Soc. Perkin Trans. 1* **1998**, 51 –65; b) Koeller, K.M., Wong, C.-H. *Chem. Rev.* **2000**, *100*, 4465–4493; c Jensen, H.J., Pedersen, C.M., Bols, M. *Chem. Eur. J.* **2007**, *13*, 7576–7582.
43. Plante, O.J., Palmacci, E.R., Seeberger, P.H. *Science* **2001**, *291*, 1523–1527.
44. Kunz, H. *Angew. Chem., Int. Ed. Engl.* **1987**, *26*, 294–308.
45. a) Garg, H.G., Jeanloz, R.W. *Adv. Carbohydr. Chem. Biochem.* **1985**, *43*, 135–201; b) Taylor, C.M. *Tetrahedron* **1998**, *54*, 11317–11362.
46. Presper, K.A., Heath, E.C. In *The Enzymology of Post-translational Modofication of Proteins* (Freedman, R.B., Hawkins, H.C., eds.) Academic Press: New York, **1985**, 53–93.
47. Macfarlane, G. *Alexander Fleming The Man and the Myth*. Oxford University Press, **1985**, 185–186.
48. a) Quan, S., Larson, A., Meyer, K. *Soc. Exp. Biol. Med.* **1948**, *66*, 528–532; b) Byrne, W.R., Welkos, S.L., Pitt, M.L., Davis, K.J., Brueckner, R.P., Ezzell, J.W., et al. *Antimicrob. Agents Chemother.* **1998**, *42*, 675–681.
49. Boulanger, L.L., Ettestad, P., Fogarty, J.D., Dennis, D.T., Romig, D., Mertz, G. *Clin. Infect. Dis.* **2004**, *38*, 663–669.
50. Pramanik, A., Stroeher, U.H., Krejci, J., Standish, A.J., Bohn, E., Paton, J.C., et al. *Int. J. Med. Microbiol.* **2007**, *297*, 459–469.
51. Barnett, J.E.G., Rasheed, A., Corina, D.L. *Biochem. J.* **1973**, *131*, 21–30.

52. a) Barton, D.H.R., Dalko, P., Gero, S.D. *Tetrahedron Lett.* **1991**, *32*, 2471–2474; b) McIntosh, M.C., Weinreb, S.M. *J. Org. Chem.* **1991**, *56*, 5010–5012.
53. Duchek, J., Adams, D.R., Hudlicky, T. *Chem. Rev.* **2011**, *111*, 4223–4258.
54. a) Gibson, D.T., Hensley, M., Yoshika, H., Mabry, R. *J. Biochemistry* **1970**, *9*, 1626–1630; b) Gibson, D.T., Mahaderan, V., Davey, J.F. *J. Bacteriol.* **1974**, *119*, 930–936.
55. Ley, S.V., Sternfeld, F. *Tetrahedron* **1989**, *45*, 3463–3476.
56. Hudlicky, T., Price, J.D., Rulin, F., Tsunoda, T. *J. Am. Chem. Soc.* **1990**, *112*, 9439–9440.
57. Ley, S.V., Parra, M., Redgrave, A.J., Sternfeld, F., Vidal, A. *Tetrahedron Lett.* **1989**, *30*, 3557–3560.
58. Carless, H.A. *J. Chem. Soc., Chem. Commun.* **1992**, 234–235.
59. Boyd, D.R., Dorrity, M.R.J., Hand, M.V., Malone, J.F., Sharma, N.D., Dalton, H., et al. *J. Am. Chem. Soc.* **1991**, *113*, 666–667.
60. Ho, T.-L. *Polarity Control for Synthesis*, John Wiley & Sons: New York, **1991**.
61. Ferrier, R.J. *J. Chem. Soc. Perkin Trans. I* **1979**, 1455–1458.
62. Estevez, V.A., Prestwich, G.D. *J. Am. Chem. Soc.* **1991**, *113*, 9885–9887.
63. a) Saksena, A.K., Mangiaracina, P. *Tetrahedron Lett.* **1983**, *24*, 273–276; b) Turnbull, M.D., Hatter, G., Ledgerwood, D.E. *Tetrahedron Lett.* **1984**, *25*, 5449–5452; c) Evans, D.A., Chapman, K.T., Carreira, E.M. *J. Am. Chem. Soc.* **1988**, *110*, 3560–3578.
64. Bender, S.L., Budhu, R.J. *J. Am. Chem. Soc.* **1991**, *113*, 9883–9885.
65. *Iminosugars as Glycosidase Inhibitors Nojirimycin and Beyond*. A.E. Stütz (ed.), Wiley-VCH, Weinheim, **1999**, 397 pp.
66. Schwarz, J.C.P., Yule, K.C. *Proc. Chem. Soc* **1961**, *417*.
67. Adley, T.J., Owen, L.N. *Proc. Chem. Soc* **1961**, *418*.
68. Ingles, D.L., Whistler, R.L. *J. Org. Chem.* **1962**, *27*, 3896–3898.
69. Paulsen, H. *Angew. Chemie Int. Ed. English* **1962**, *1*, 454–454.
70. Paulsen, H. *Angew. Chemie Int. Ed. English* **1962**, *1* 597–597.
71. a) Jones, J.K.N., Turner, J.C., *J. Chem. Soc.* **1962**, 4699–4703; b) Jones, J.K.N., Szarek, W.A. *Can. J. Chem.* **1963**, *41*, 636–640.
72. a) Hanessian, S., Haskell, T.H. *J. Org. Chem.* **1963**, *28*, 2604–2610; b) Hanessian, S. *Chem. Commun.* **1966**, 796–798.
73. Compain, P., Martin, O.R. (eds.), In *Iminosugars: From Synthesis to Therapeutic Applications*; John Wiley & Sons Ltd: London, **2007**; pp 1–6.
74. Nishikawa, T., Ishida, N. *J. Antibiot.* **1965**, *18*, 132–133.
75. Inouye, S., Tsuruoka, T., Niida, T. *J. Antibiot.* **1966**, *19*, 288–292.
76. Paulsen, H. *Angew. Chemie Int. Ed. English* **1966**, *5*, 495–510.
77. Inouye, S., Tsuruoka, T., Ito, T., Niida, T. *Tetrahedron* **1968**, *24*, 2125–2144.
78. Yagi, M., Kouno, T., Aoyagi, Y., Murai, H. *Nippon Nogei Kagaku Kaishi* **1976**, *50*, 571–572.
79. Platt, F.M., Neises, G.R., Dwek, R.A., Butters, T.D. *J. Biol. Chem.* **1994**, *269*, 8362–8365.
80. Cox, T., Lachmann, R., Hollak, C., Aerts, J., van Weely, S., Hrebícek, M., et al. *Lancet* **2000**, *355*, 1481–1485.
81. Niwa, T., Tsuruoka, T., Goi, H., Kodama, Y., Itoh, J., Inouye, S., et al. *J. Antibiot.* **1984**, *37*, 1579–1586.
82. Miyake, Y., Ebata, M. *Agric. Biol. Chem.* **1988**, *52*, 661–666.
83. Asano, N., Yasuda, K., Kizu, H., Kato, A., Fan, J.-Q., Nash, R.J., et al. *Eur. J. Biochem.* **2001**, *268*, 35–41.
84. a) Pearson, M.S.M. Mathé-Allainmat, M., Fargeas, V., Lebreton, J. *Eur. J. Org. Chem.* **2005**, 2159–2191; b) Afarinkia, K., Bahar, A. *Tetrahedron: Asymmetry* **2005**, *16*, 1239–1287; c) Asano, N. *Curr. Top. Med. Chem.* **2003**, *3*, 471–484; d) Cipolla, L., Nicotra, B.L.F. *Curr. Top. Med. Chem.* **2003**, *3*, 485–511.
85. a) Davies, S.G., Smith, A.D., Price, P.D. *Tetrahedron: Asymmetry* **2005**, *16*, 2833–2891; b) Davies, S.G., Fletcher, A.M., Roberts, P.M., Thomson, J.E. *Tetrahedron: Asymmetry* **2012**, *23*, 1111–1153.
86. Davies, S.G., Figuccia, A.L.A., Fletcher, A.M., Roberts, P.M., Thomson, J.E. *Org. Lett.* **2013**, *15*, 2042–2045.
87. Karjalainen, O.K., Koskinen, A.M.P. *Org. Biomol. Chem.* **2011**, *9*, 1231–1236.
88. Karjalainen, O.K., Koskinen, A.M.P. *Tetrahedron*, **2014**, *70*, 2444–2448.
89. Koskinen, A.M.P. *Heterocycles* **2021**, *103*, 609–623.

5

Amino Acids, Peptides, and Proteins

Ich habe nicht mehr Glück als Sie, aber ich probiere mehr als Sie!

Adolf von Baeyer

Peptides and proteins play a central role in the function of cells and organs. Large proteins can support the structure of the cell. Proteins catalysing chemical transformations are called enzymes. Some proteins participate in the transfer of information and signals between cells; hence, they are called receptors. Enzymes and receptors are often extremely sensitive in recognising specific molecules, and by way of their intrinsic asymmetric structure, they exhibit high discrimination towards enantiomers. This phenomenon is known as *substrate selectivity*. Smaller peptides can function as chemical messenger molecules in vital signal transduction processes, e.g. in the central nervous system (neurotransmitters) and hormonal activity (peptide hormones). Whether one considers a large enzyme complex composed of several peptide chains, which catalyses chemical reactions, or a small neurotransmitter peptide, they share the common feature of being constructed of amino acids.

In this chapter we will take a look at the amino acids and polypeptides, especially in terms of structure and synthesis. Finally, we shall briefly review enzymes, their properties and function, and the factors controlling them.

5.1 Amino Acids

Natural peptides and proteins in eukaryotes are built from twenty-two so-called natural or, more appropriately, DNA-encoded or proteinogenic amino acids (Figure 5.1). Twenty of them occur universally in living species, and these are coded by three-codon sequences in DNA (Figure 5.2). Selenocysteine and pyrrolysine (see Scheme 5.14) are coded for in certain species at the mRNA level indirectly by overriding the natural stop codons (TGA for selenocysteine and TAG for pyrrolysine). The amino acids are, with the exception of glycine, optically active, in other words their α-carbon is asymmetrically substituted. With the exception of certain microbial products, all natural amino acids belong to the same stereochemical series (L), where the absolute stereochemistry of the chiral carbon is S. Cysteine makes an exception, where the stereochemical designation is R – this, however, is due to the nature of the systematic nomenclature. Two amino acids, threonine and isoleucine, have a second chiral center. Again, one diastereomer of each of these is predominant in nature.

An important feature for the chemistry of amino acids is that they contain both a basic (NH_2) and an acidic (COOH) group. Because of the difference in the pK_a values of the carboxylic acid and amino groups, each amino acid has a characteristic pH, where the amino acid occurs in doubly ionised form, as a zwitterion. This pH, the *isoelectric point*,

Asymmetric Synthesis of Natural Products, Third Edition. Ari M.P. Koskinen.
© 2023 John Wiley & Sons Ltd. Published 2023 by John Wiley & Sons Ltd.

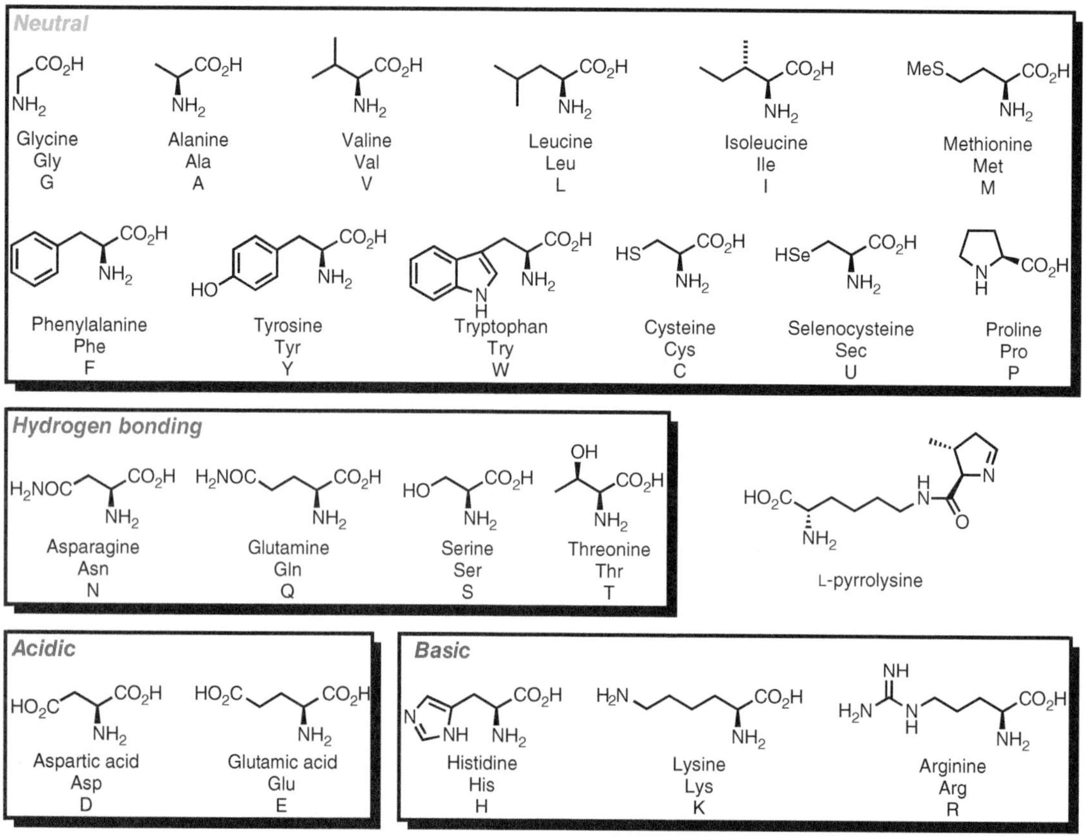

Figure 5.1 *The twenty-two proteinogenic amino acids. The L-series is shown. Below each structure is the name, and the three-letter and one-letter codes for each amino acid*

Second base

		T	C	A	G	
		Phe	Ser	Tyr	Cys	T
	T	Phe	Ser	Tyr	Cys	C
		Leu	Ser	Stop (ochre)	Stop (opal)	A
		Leu	Ser	Stop (amber)	Trp	G
		Leu	Pro	His	Arg	T
	C	Leu	Pro	His	Arg	C
First base		Leu	Pro	Gln	Arg	A
		Leu	Pro	Gln	Arg	G
		Ile	Thr	Asn	Ser	T
	A	Ile	Thr	Asn	Ser	C
		Ile	Thr	Lys	Arg	A
		Met	Thr	Lys	Arg	G
		Val	Ala	Asp	Gly	T
	G	Val	Ala	Asp	Gly	C
		Val	Ala	Glu	Gly	A
		Val	Ala	Glu	Gly	G

Third base

Figure 5.2 *The genetic code (DNA level)*

is the arithmetic mean of the two pK_a values, and it is also the pH where the solubility of the amino acid is lowest. In Table 5.1, the amino acids are classified according to their chemical nature as acidic, neutral, and basic amino acids. As one can observe, the isoelectric point of the neutral amino acids is slightly on the acidic side. The *essential amino acids* (marked with a plus sign) cannot be synthesised by the human body, but have to be obtained with nutrients in order to prevent negative nitrogen balance. Also included are the three- and one-letter codes commonly used for the amino acids.

Table 5.1 *Natural proteinogenic amino acids*

Amino acid	3 lett	1 lett	MW[a]	E.coli[b]	pI[c]	Essential
Acidic						
Aspartic acid	Asp	D	114	9.9	2.8	
Glutamic acid	Glu	E	128	10.8	3.2	
Neutral						
Cysteine	Cys	C	103	1.8	5.1	
Asparagine	Asn	N	114	–	5.4	
Phenylalanine	Phe	F	147	3.3	5.5	+
Threonine	Thr	T	101	4.6	5.6	+
Serine	Ser	S	87	6.0	5.7	
Glutamine	Gln	Q	128	–	5.7	
Methionine	Met	M	131	3.8	5.7	+
Tyrosine	Tyr	Y	163	2.2	5.7	
Tryptophan	Trp	W	186	1.0	5.9	+
Glycine	Gly	G	57	5.7	6.0	
Alanine	Ala	A	71	13.0	6.0	
Valine	Val	V	99	6.0	6.0	+
Leucine	Leu	L	113	7.8	6.0	+
Isoleucine	Ile	I	113	4.4	6.0	+
Proline	Pro	P	97	4.6	6.3	
Selenocysteine	Sec	O	168	–	6.7	(+)
Basic						
Histidine	His	H	137	0.7	7.5	+
Lysine	Lys	K	129	7.0	9.6	+
Arginine	Arg	R	157	5.3	11.2	+

[a]MW = molecular weight;
[b]Occurrence in *Escherichia coli* as percentage;
[c]isoelectric point.

Although the proteinogenic amino acids occur in the L-form, D-amino acids frequently occur in bacterial cell walls, especially in the rigid peptidoglycan net surrounding the cytoplasmic membrane, which is responsible for the shape and strength of the bacterial cell. Besides the D-enantiomers of the proteinogenic amino acids, a large number of other amino acids of varied structures are also known. Several hundred of them occur naturally, but their functions are widely unknown [1]. Structurally and biogenetically the simplest ones are just modifications of the genetically coded ones; e.g. 4-hydroxyproline and 4-hydroxylysine, constituents of the collagen tissue, are produced post-translationally by oxygenation of the intact peptide. The newly introduced hydroxyl groups participate in the formation of the cross links between the peptide chains, giving collagen its characteristic strength. *N*-Methylated amino acids are also found in many peptides and proteins. At least in some cases, methylation is used to impart the peptide certain structural characteristics. More complicated rare amino acids include the cyclopropane-containing amino acids (Figure 5.3). 1-Aminocyclopropane carboxylic acid (ACC) is the biogenetic precursor of the plant growth factor ethylene. Carnosa-dine, from the red alga *Grateloupia carnosa*, is an example of 2,3-methanoamino acids which are conformationally restricted amino acid analogues (see Section 5.4). Coronatine is a phytotoxin produced by the bacterium *Pseudomonas syringae*. It is a functional mimic of the active form of the plant hormone (+)-7-*iso*-jasmonoyl-L-isoleucine. The lack of coronatine makes *P. syringae* less virulent under a diurnal light and dark cycle. It is known to induce hypertrophy and chlorosis, inhibit root elongation, and stimulate ethylene production. The structure of coronatine consists of two distinct structural components that function as biosynthetic intermediates: the polyketide coronafacic acid and the aminocyclopropane carboxylic acid coronamic acid. α-Methylenecyclopropylglycine and hypoglycine A occur in the unripe fruit of *Blighia sapida* (ackee) and lychee (*Litchi chinensis*) seeds. These compounds are so powerful hypoglycemic agents that ingestion of the unripe fruit can be lethal.

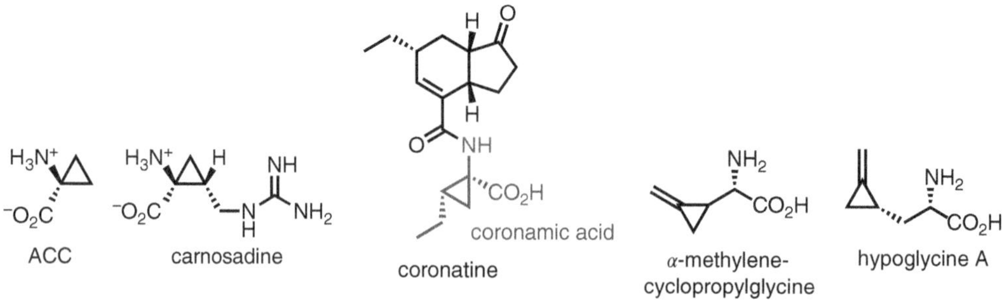

Figure 5.3   *Cyclopropane-containing amino acids*

Several pyrrolidine-containing amino acids of the kainic acid type are also known (Figure 5.4). These amino acids exhibit remarkably potent neuro-excitatory activity via activation of ionotropic glutamate receptors in the brain. Ionotropic glutamate receptors are involved in various neurophysiological processes, including memory and pain transmission. Kainic acid was first isolated in 1953 from the alga *Digenea simplex*, (*kainin-sou* in Japanese) and it is being used extensively as a research tool in neurobiology to impart selective lesions in the brain, and to explore the physiological pharmacology of excitatory transmission. Domoic acid is a related neurotoxin responsible for amnesic shellfish poisoning caused by blooms of the red algae *Chondria armata* (*doumoi* in Japanese). In 1983, Shirahama and Matsumoto isolated acromelic acids A and B from *Clitocybe acromelalga* (*dokusasako* in Japanese). Acromelic acid A is almost 10 times more potent than domoic acid and 100 times more potent than kainic acid. These compounds, therefore, have potential as important biological research tools. The isoxazole-containing amino acid ibotenic acid is, like kainic and domoic acids, an agonist of the glutamate receptor. Ibotenic acid is the toxic compound in the mushroom *Amanita muscaria* (fly agaric), known for its hallucinogenic effects which were used by shamans to incur a state of euphoria and clairvoyance. It derives its name from the first isolation in 1960s from *Amanita ibotengutake* (ringed-bulb amanita in Japanese) in Japan.

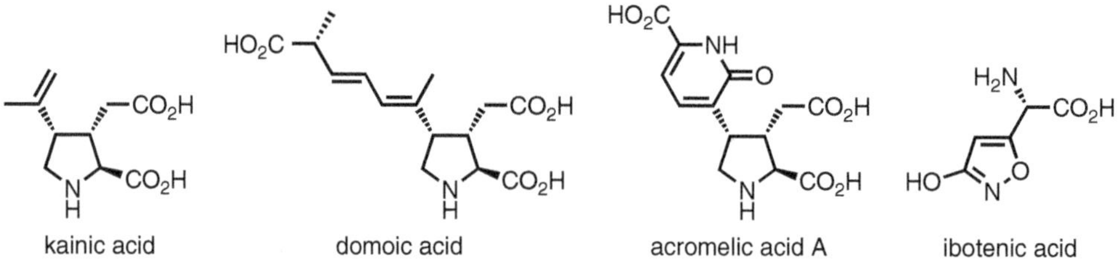

Figure 5.4   *Pyrrolidine-containing and related amino acids*

Unsaturated amino acids, like MeBmt, a constituent of the immunosuppressant antibiotic cyclosporin A, and ADDA, (2*S*,3*S*,8*S*,9*S*)-3-amino-9-methoxy-2,6,8-trimethyl-10-phenyldeca-4*E*,6*E*-dienoic acid, a part of the cyanobacterial hepatotoxins microcystin-LR and nodularin, are examples of more complex amino acids (Figure 5.5).

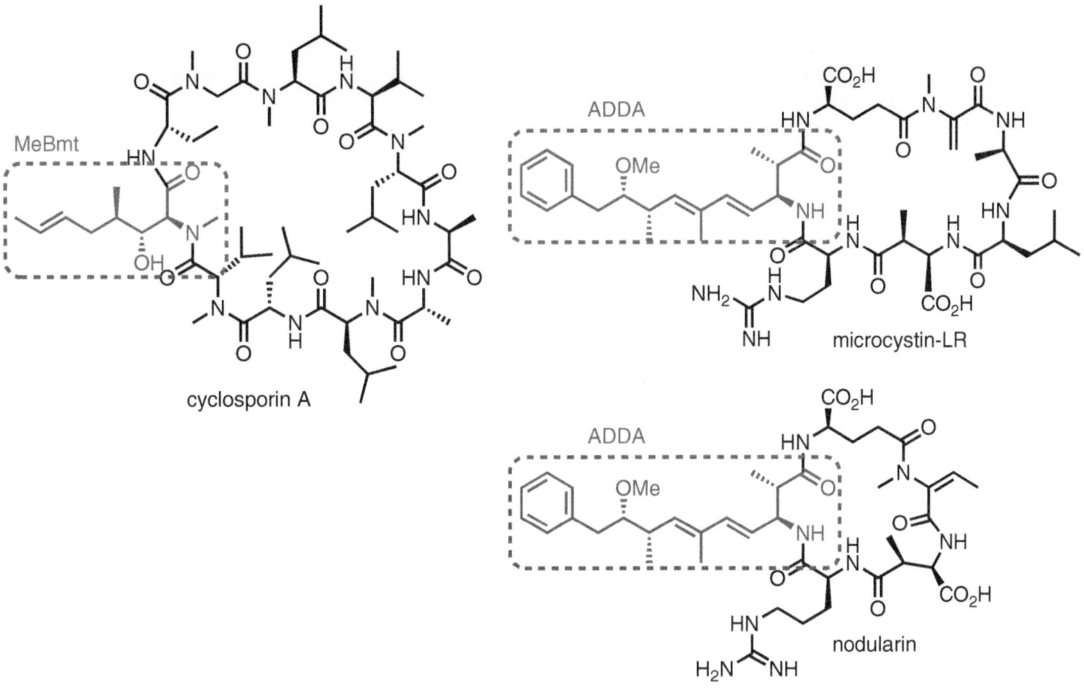

Figure 5.5 *Unsaturated complex amino acids*

5.1.1 Biosynthesis of Amino Acids

Amino acids are mainly biosynthesised from relatively simple precursors from the Calvin cycle (3-phosphoglycerate), glycolysis (3-phosphoglycerate and pyruvate), citric acid cycle (α-ketoglutarate and oxaloacetate), or a combination of glycolysis and Calvin cycle (Table 5.2). Histidine is the odd one standing out; it is biosynthesised from ribose-5-phosphate, which is the product and an intermediate in the pentosephosphate pathway. Ribose-5-phosphate can also be generated from ribulose-5-phosphate (Calvin cycle intermediate) during rapid cell growth when nucleotide synthesis is accelerated.

The amino group is introduced to the structures by converting the keto groups of α-ketocarboxylic acids to an amino group by a transamination reaction (Scheme 5.1). In transamination, an aminotransferase reacts glutamic acid with pyridoxal phosphate (vitamin B_6) to form a Schiff base, and converts them through a sequence of isomerisations and eventual hydrolysis to α-ketoglutarate and pyridoxamine, a storage form of nitrogen in the syntheses of amino acids.

5.1.1.1 3-Phosphoglycerate family

3-Phosphoglycerate (from glycolysis, Scheme 4.1) undergoes oxidation by phosphoglycerate dehydrogenase to produce 3-phosphohydroxypyruvate (Scheme 5.2). Transamination gives 3-phosphoserine, which is hydrolysed by the corresponding phosphatase to serine. Serine hydroxymethyltransferase removes the hydroxymethyl group to tetrahydrofolate and water in a pyridoxal phosphate-mediated reaction with simultaneous production of glycine. For the synthesis of cysteine, serine is condensed with homocysteine to give cystathione which is the substrate for a lyase to produce cysteine and α-ketobutyrate.

Table 5.2 *Metabolic origins of amino acids*

Precursor origin	Metabolic precursor	Amino acid
Calvin cycle (Scheme 1.1)	3-phosphoglycerate	serine glycine cysteine
Glycolysis (Scheme 4.1.1)	pyruvate	alanine valine leucine isoleucine
Citric acid cycle (Scheme 4.1.2)	α-ketoglutarate	glutamic acid glutamine proline arginine
Citric acid cycle (Scheme 4.1.2)	oxaloacetate	aspartic acid asparagine methionine threonine lysine
Glycolysis (Scheme 4.1.1)	phosphoenolpyruvate	phenylalanine
Calvin cycle (Scheme 1.1)	erythrose-4-phosphate	tyrosine tryptophan
Pentose phosphate pathway	ribose-5-phosphate	histidine

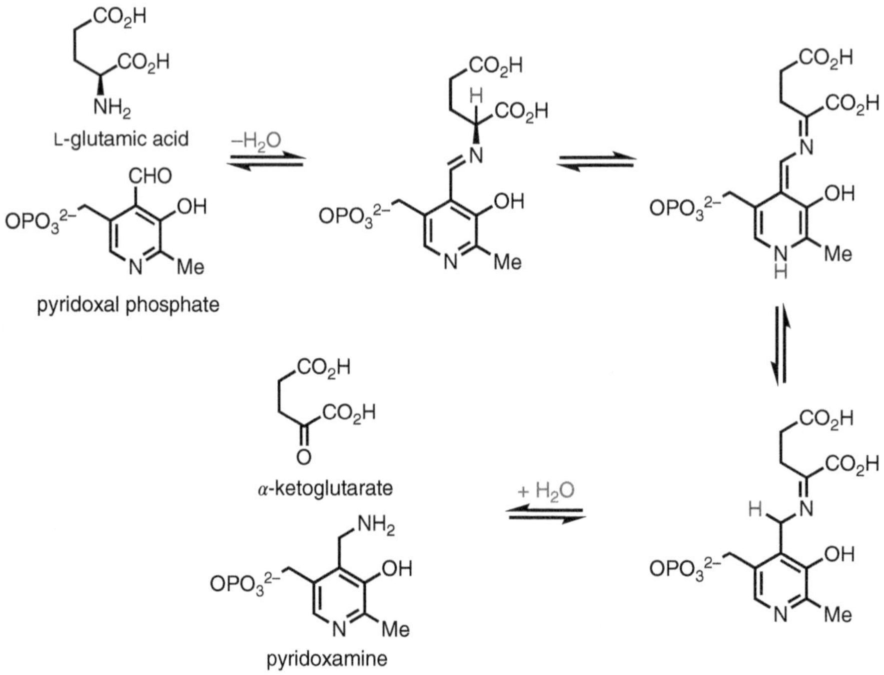

Scheme 5.1   *Transamination*

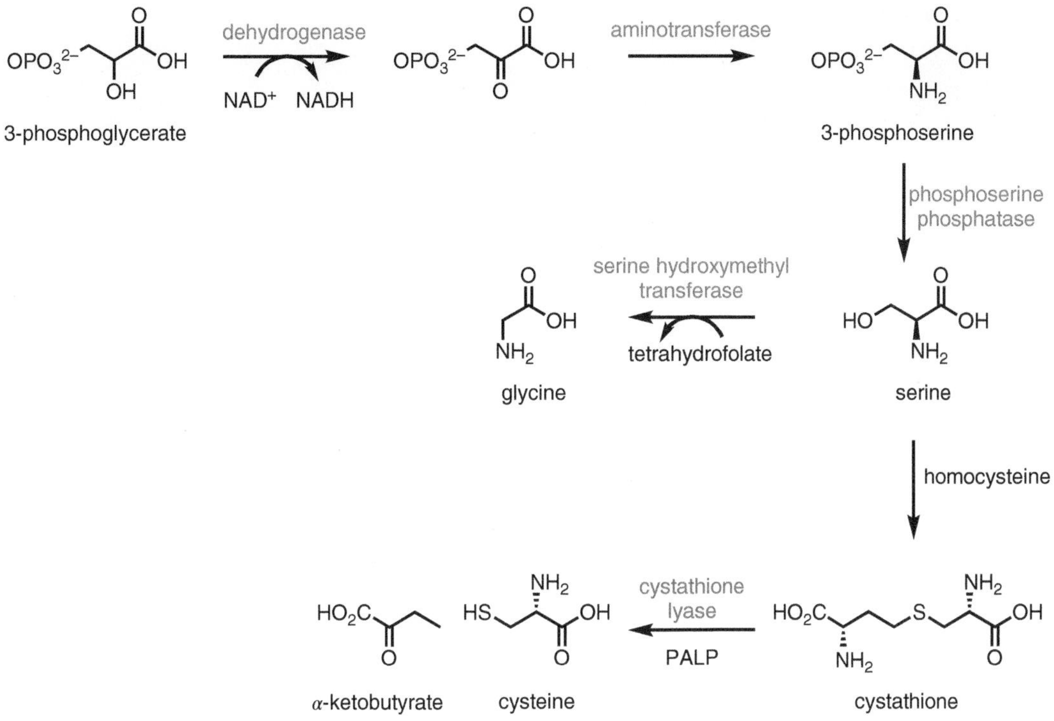

Scheme 5.2   *Biosynthesis of serine, glycine and cysteine*

5.1.1.2   *Pyruvate family*

The key product from glycolysis is pyruvic acid, an α-keto carboxylic acid. The α-keto carboxylic acids provide a direct entry to α-amino acids through a biological reductive amination reaction (Scheme 5.3). Pyruvic acid first condenses with pyridoxamine to form a Schiff base. A series of proton shifts leads to a new Schiff base with the proper stereochemistry of L-alanine. Final hydrolysis liberates L-alanine. Similar transamination reactions are involved in the biosynthesis of the simple aliphatic amino acids valine, leucine, and isoleucine.

5.1.1.3   *From citric acid cycle: α-ketoglutarate and oxaloacetate families*

In the citric acid cycle, pyruvate is converted to two further α-keto acids, α-keto glutarate and oxaloacetate, which undergo transamination to give glutamate and aspartate, respectively. Glutamic acid is the precursor for glutamine, arginine, and proline, whereas aspartic acid is similarly converted to asparagine, methionine, threonine, and lysine (Scheme 5.4).

5.1.1.4   *Aromatic amino acids*

A large number of aromatic natural products are derived from aromatic amino acids phenylalanine and tyrosine. These are formed from a common intermediate, shikimic acid (Figure 5.6), which was originally isolated in 1870 by the Dutch phytochemist *Johan Fredrik Eykman* (1851–1915) from the highly toxic Japanese star anise *Illicium religiosum*, (*shikimi* in Japanese) [2]. Shikimic acid functions in many plants as the starting material for aromatic amino acids (phenylalanine, tyrosine, and tryptophan), and is also the origin of a wide variety of other aromatic products [3].

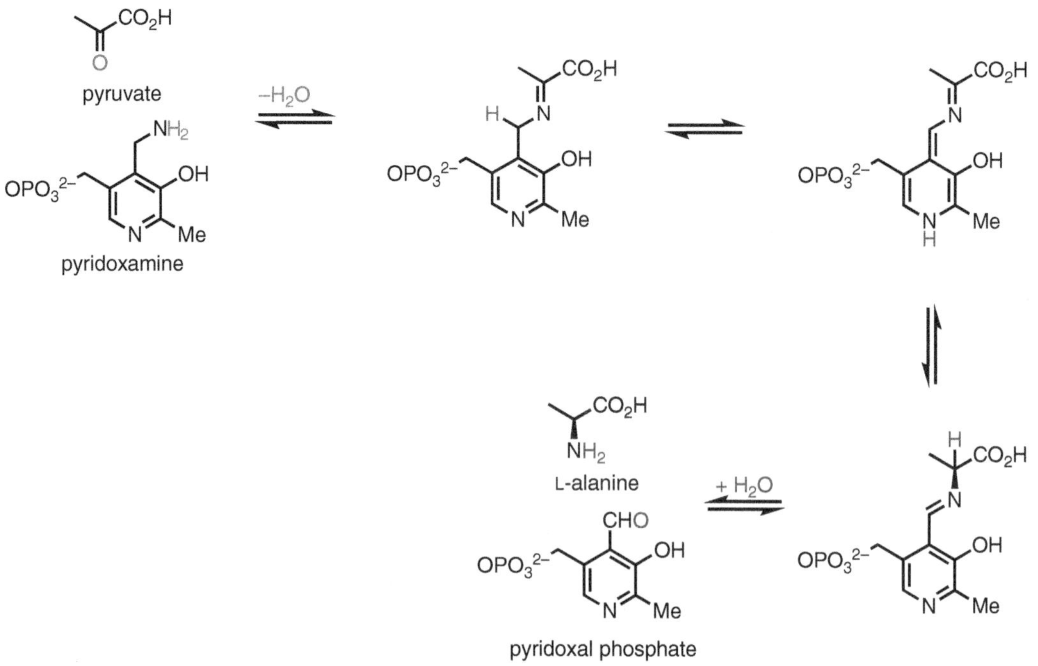

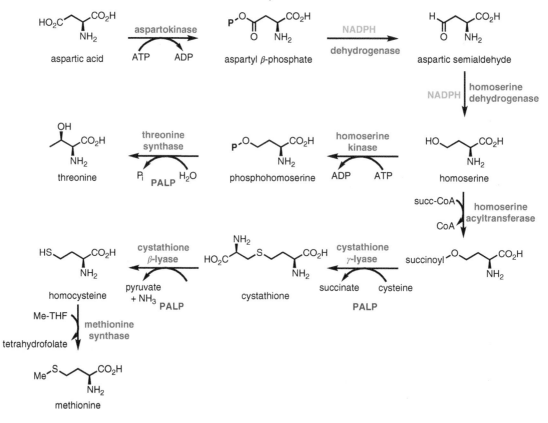

Scheme 5.3   *Biosynthesis of alanine*

Scheme 5.4   *Biosynthesis of aliphatic amino acids*

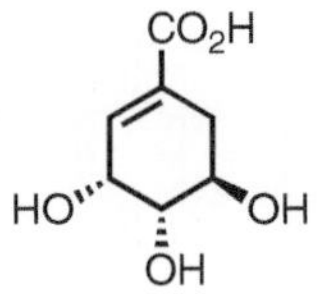

Figure 5.6   *Shikimic acid*

This process is centered on the formation of shikimic acid from pyruvate and D-erythrose (Scheme 5.5). Shikimic acid also serves as a launchpad to a wide range of natural products on their own, phenylpropanoids, which will be discussed in detail in Chapter 7.

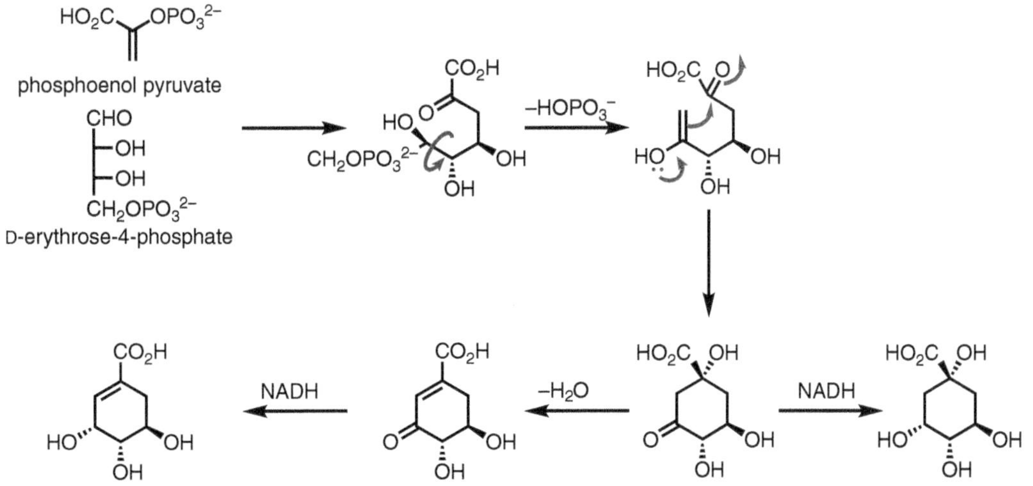

Scheme 5.5   *Biosynthesis of shikimic acid*

Aldol reaction between pyruvate (in the form of phosphoenol pyruvate, PEP) and D-erythrose-4-phosphate leads to the formation of a seven-carbon sugar acid, 3-deoxy-D-arabino-heptulosonic acid-7-phosphate (DAHP). This spontaneously loses phosphate, and the enol form of the resulting diketo acid cyclises to 3-dehydroquinic acid. NADH-mediated reduction gives quinic acid, whereas stereospecific dehydration (with the loss of the hydrogen *syn* to the hydroxy group) gives 3-dehydroshikimic acid and thereby shikimic acid after reduction (NADH).

Several quinones serve as lipid-soluble electron carriers in the membrane-bound electron transport chains of both prokaryotes and eukaryotes. Ubiquinones (coenzyme Q) and plastoquinones are coenzymes involved in one-electron transport processes in living systems (Figure 5.7).

In the biosynthesis, quinic acid is oxidised to *p*-benzoquinone, which was originally obtained by the Russian chemist *Alexander Woskresensky* (1809–1880), in 1838. The quinone structure is then coupled with a number of prenyl units (typically $n = 6$ to 10). Vitamins K (phyloquinones and menaquinoes) are naphthoquinone derivatives. Quinones occur widely in a number of natural products, including pigments and antibiotics (e.g. rifamycin, see macrolide antibiotics, Section 8.3.2).

The biosynthesis of the aromatic amino acids (phenylalanine and tyrosine) proceeds via shikimic acid through chorismic acid and prephenic acid (Scheme 5.6). Shikimic acid is monophosphorylated at the pseudoaxial 3-hydroxyl group, and the 5-hydroxyl group reacts with phosphoenol pyruvate to give chorismic acid after elimination (in a Grob sense) of phosphoric acid. This elimination is completely stereospecific in that only the pro-*R* hydrogen at C6 participates in the elimination. Chorismic acid contains an allyl enol ether moiety required for the ensuing Claisen rearrangement which in turn produces prephenic acid, the ubiquitous intermediate in the biosynthesis of the aromatic amino acids phenylalanine and tyrosine.

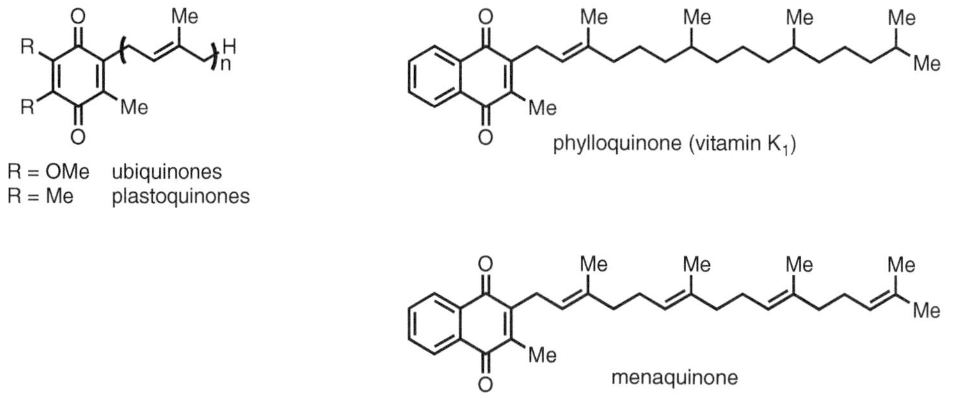

Figure 5.7   *Quinone structures*

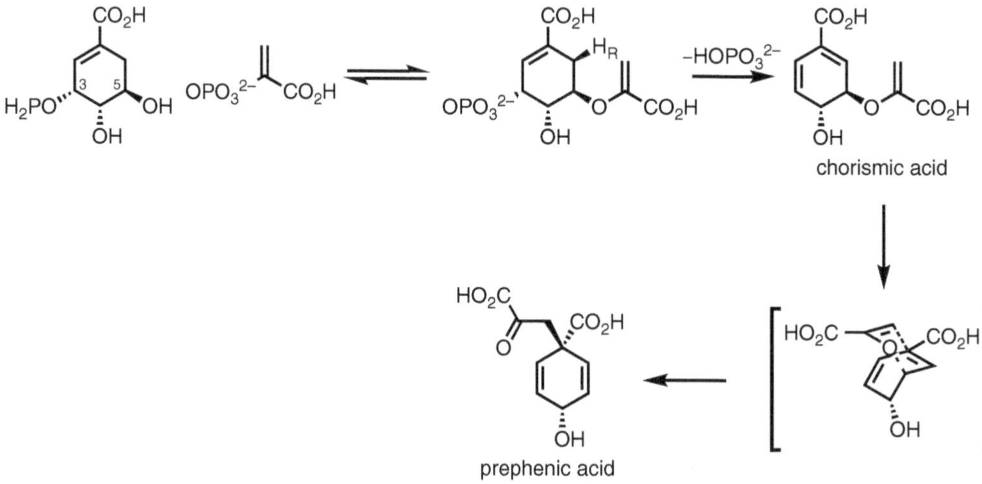

Scheme 5.6   *Biosynthesis of prephenic acid*

Conversion of prephenic acid to the target amino acids requires only the adjustment of the oxidation level and introduction of the amino function. Interestingly, this is done via elimination of the chiral information in prephenic acid as follows (Scheme 5.7). Simultaneous loss of carbon dioxide and water from prephenic acid gives phenylpyruvic acid, which is transaminated enzymatically (pyridoxamine) to phenylalanine. Alternatively, the carboxyl group can be lost oxidatively, and the *p*-hydroxyphenylpyruvic acid gives tyrosine upon transamination.

The biosynthesis of tryptophan (and other indole derivatives) also starts with chorismate. Initial amination and loss of pyruvate leads to anthranilic acid (Scheme 5.8).

The biosynthesis of tryptophan continues by condensation of anthranilic acid with phosphoribosyl pyrophosphate to give an amino sugar intermediate (Scheme 5.9). Opening of the cyclic hemiaminal followed by Amadori rearrangement, which transforms the α-hydroxyimine into an α-amino ketone, gives an intermediate suitable for the cyclisation of the indole ring to give indole-3-glyceryl phosphate. Its reaction with the enamino acid derived from serine gives tryptophan after elimination of glyceraldehyde. Remarkable in this synthesis is the fact that *none* of the original chiral centers of ribose or serine are conserved! The chiral center of tryptophan (corresponding to that of serine) is formed by a stereoselective alkylation of the indole portion.

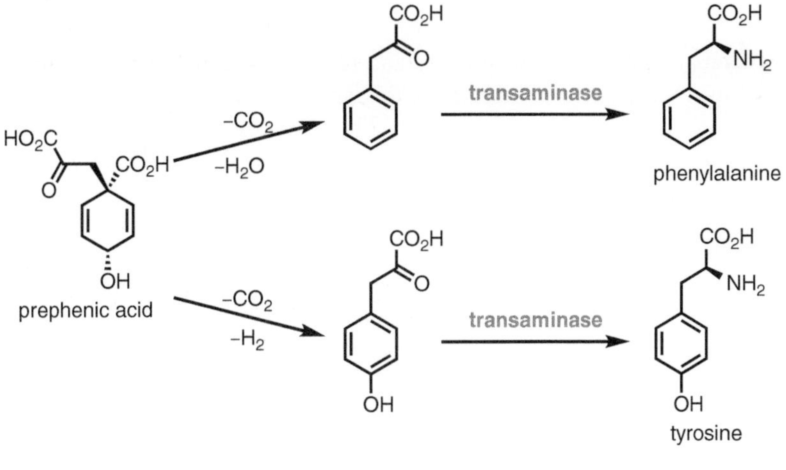

Scheme 5.7 *Biosynthesis of aromatic amino acids*

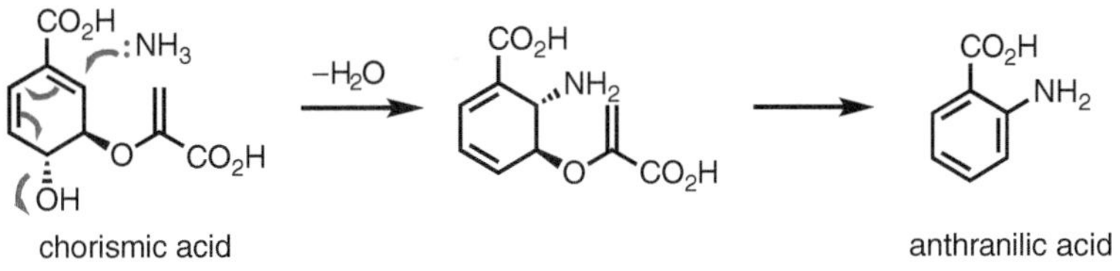

Scheme 5.8 *Biosynthesis of anthranilic acid*

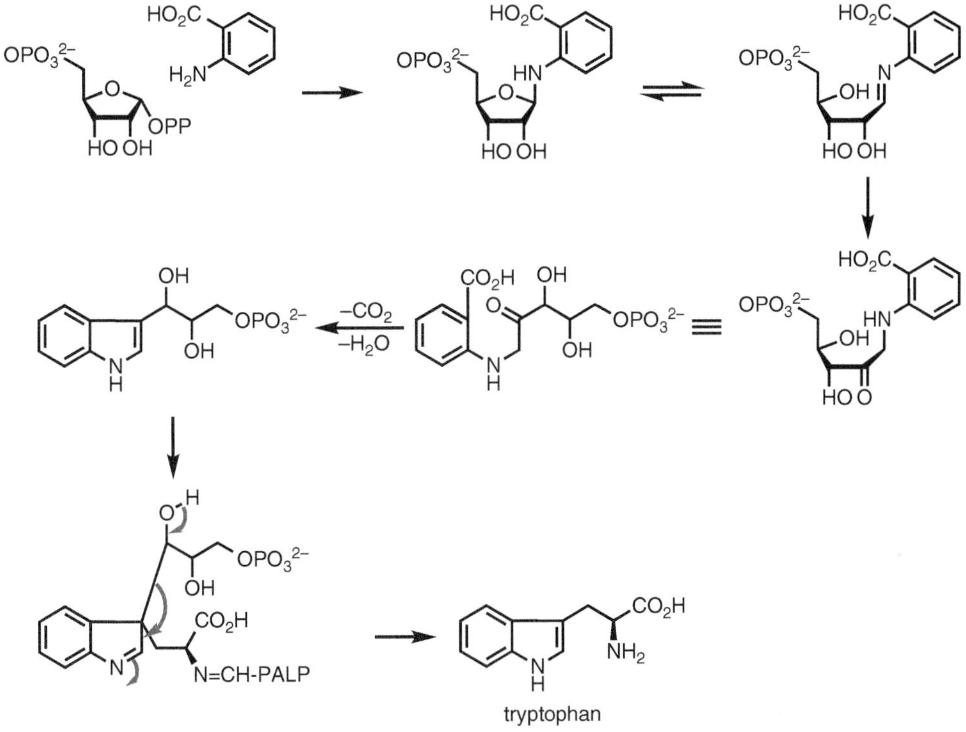

Scheme 5.9 *Biosynthesis of tryptophan*

5.1.1.5   *Histidine*

Histidine is an essential component of protein structure and function, and the biosynthetic pathway for histidine is found only in plants and microorganisms. In mammals, this amino acid serves additional roles as the precursor to histamine, a potent neurotransmitter and immunoregulator. The biosynthesis of histidine deviates considerably from the biosynthesis pathway of all other amino acids (Scheme 5.10) [4]. The imidazole ring has its origins in ATP, which is first ribosylated with phosphoribosyl pyrophosphate to give phosphoribosyl-ATP (PRATP). Hydrolysis of the triphosphate and hydrolysis of the purine ring of adenosine gives the ring-opened intermediate prophosphoribosyl formimino-5-aminoimidazole-4-carboxamide ribonucleotide (5'-ProFAR) which undergoes an enzymatic Amadori rearrangement to phosphoribosyl formimino-5-aminoimidazole-4-carboxamide ribonucleotide (PRFAR). The imidazole ring is closed in the next step in an aminolysis reaction where glutamine serves as the source of ammonia to give imidazoleglycerol phosphate. The other half of PRFAR is converted to 5-aminoimidazole-4-carboxamide ribonucleotide (AICAR), which is returned to de novo ATP synthesis. Loss of water from imidazoleglycerol phosphate gives imidazoleacetol phosphate, which upon transamination gives histidinol phosphate. Dephosphorylation and final NAD-mediated dehydrogenation complete the biosynthesis.

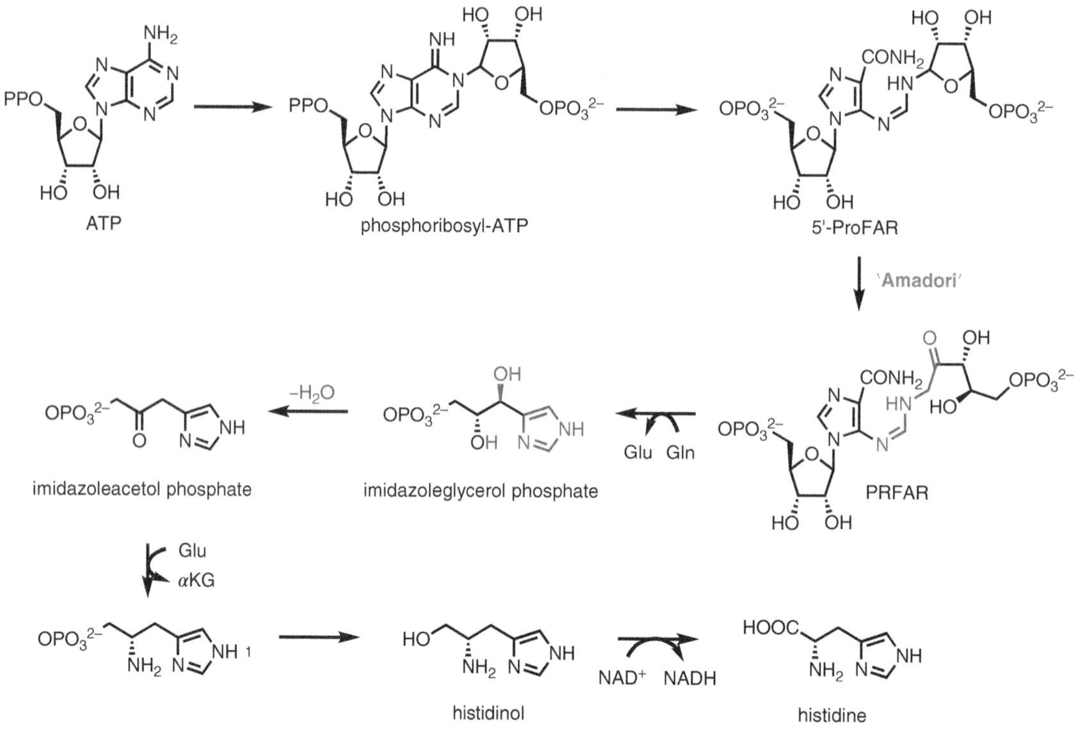

Scheme 5.10   *Histidine biosynthesis*

5.1.1.6   *Rare amino acids: Aminocyclopropanecarboxylic acid and pyrrolysine*

Ethylene is a plant hormone that initiates fruit ripening and regulates many aspects of plant growth and development. In the conversion of methionine to ethylene, the C1 of methionine is converted to CO_2, C2 to formic acid, and C3,4 to ethylene. The sulfur atom, however, is retained in the tissue. Studying the ethylene formation in detail, *Shang Fa Yang* (1932–2007) at UC Davis discovered that aminocyclopropanecarboxylic acid (ACC) is an intermediate in the formation of ethylene (Scheme 5.11, Yang cycle) [5]. The key reaction here is the PALP-mediated cyclisation of *S*-adenosylmethionine to ACC and 5'-methylthioadenosine (Scheme 5.12).

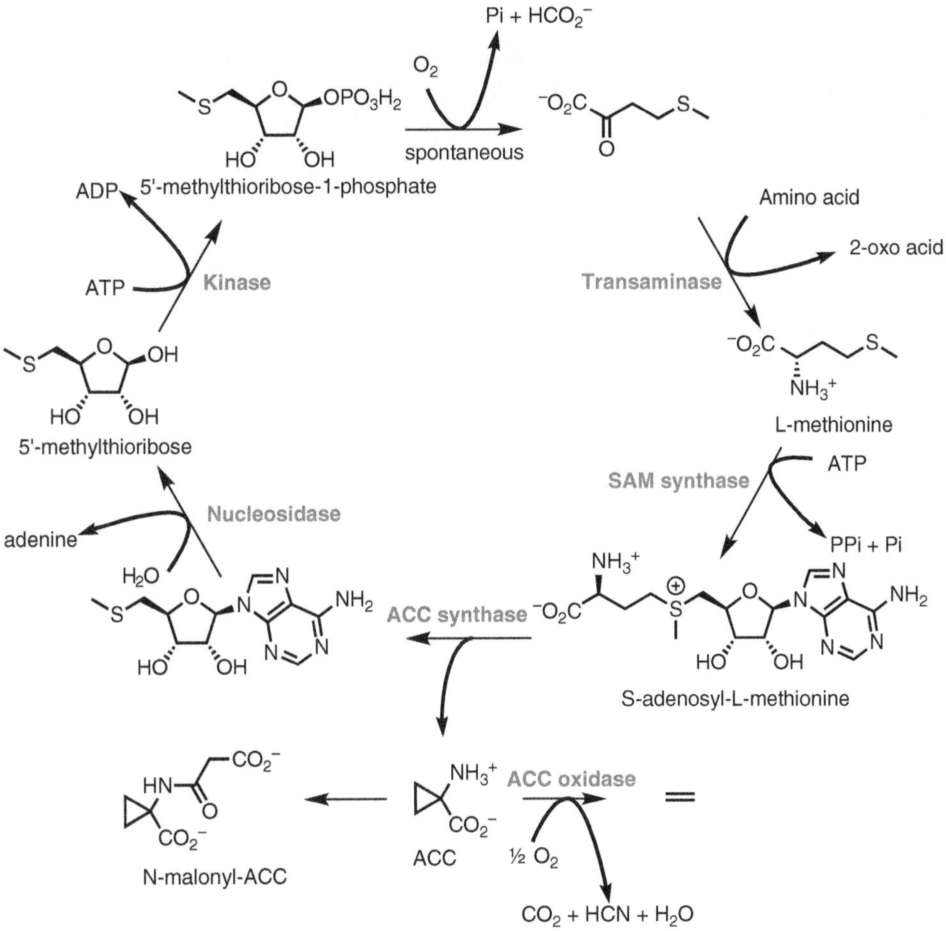

Scheme 5.11 *Yang cycle*

The biosynthesis of the plant phytotoxin coronatine exemplifies an interesting novel class of O_2- and *R*-ketoglutarate-requiring mononuclear nonheme iron halogenases which can halogenate unactivated sp^3 carbon centers. This enzyme produces the γ-Cl-L-allo-Ile moiety which cyclises to 1-amino-1-carboxy-2-ethylcyclopropane (coronamic acid, Scheme 5.13). Coronatine, a bacterial mimic of the plant hormone jasmonic acid, is an amino acid-polyketide hybrid [6].

The final example of the biosynthesis of a rare amino acid is that of pyrrolysine, the twenty-second genetically encoded amino acid. This sounds strange, but some archaebacteria actually use the amber codon UAG to code for this particular amino acid to be incorporated into proteins [7]. The biosynthesis of this amino acid has recently been elucidated and is shown in Scheme 5.14 [8]. Two molecules of ornithine are used to make pyrrolysine: one is used as such; and the other one is used for the rearrangement of lysine to (3*R*)-3-methyl-D-ornithine via a lysine mutase reaction.

5.1.2   Asymmetric Synthesis of Amino Acids

α-Alkylation of amino acids restricts the conformational mobility of the amino acid side chains, thereby enhancing the probability to induce defined 3D structures to peptides and proteins. The synthesis of α-alkylated amino acids is therefore important in their chiral form, and numerous routes have been designed towards this end [9]. We will briefly review some of these methods.

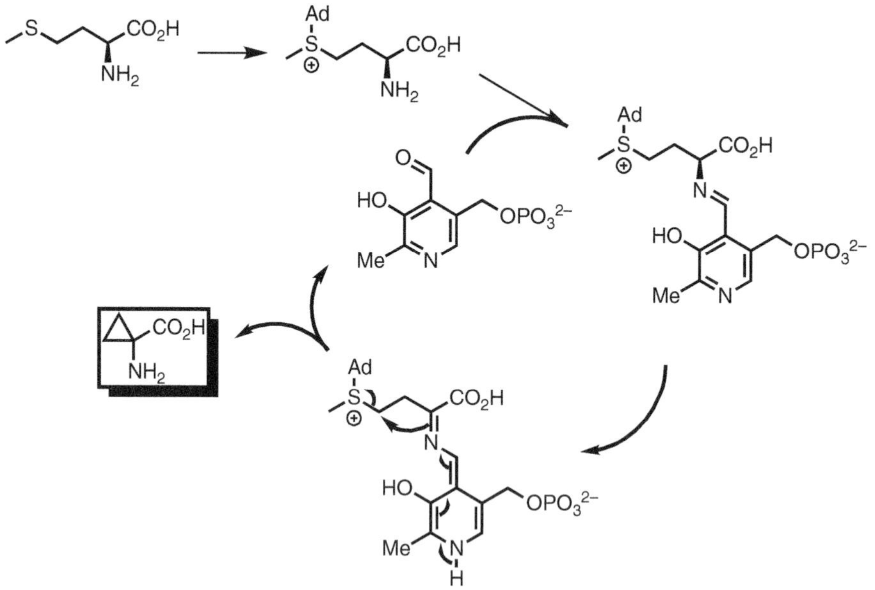

Scheme 5.12   *Synthesis of aminocyclopropane carboxylic acid*

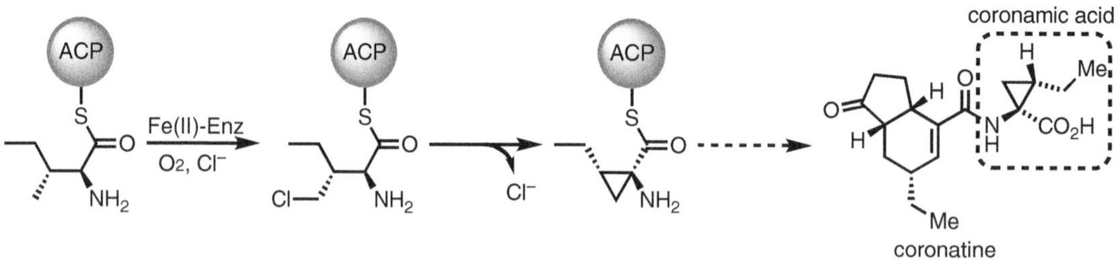

Scheme 5.13   *Biosynthesis of coronamic acid*

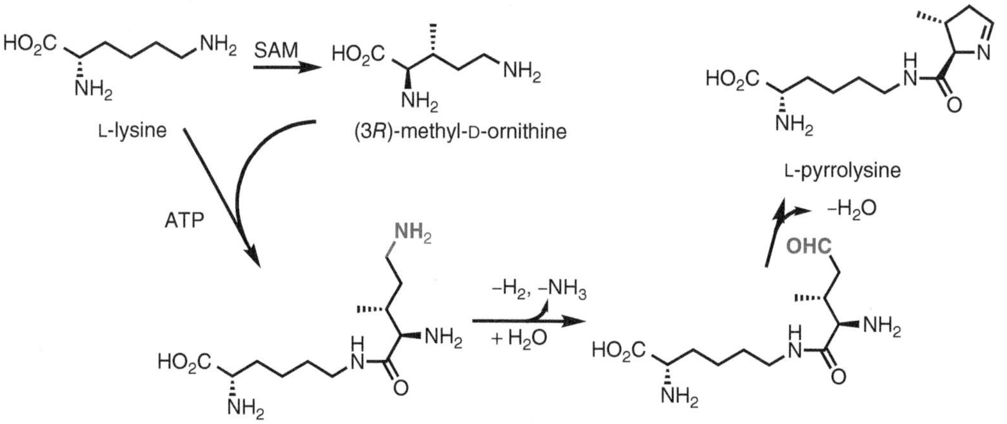

Scheme 5.14   *Biosynthesis of pyrrolysine*

5.1.2.1 α-Alkylation

Perhaps the most commonly used alkylation route is that developed by *Ulrich Schöllkopf* (1928–1998) which relies on the use of bislactim ethers (Scheme 5.15) [10]. The requisite reagents are simply derived from diketopiperazines through *O*-alkylation (e.g. with trimethyl oxonium tetrafluoroborate). Treatment of the bislactim ether with a base yields an anion which is selectively alkylated on one side. The other face of the lactim ring is blocked effectively by the alkyl substituent on the remaining chiral center. Hydrolysis of the product yields two amino acids of which the α-alkylated amino acid now formally has the original α-hydrogen replaced with the alkyl group.

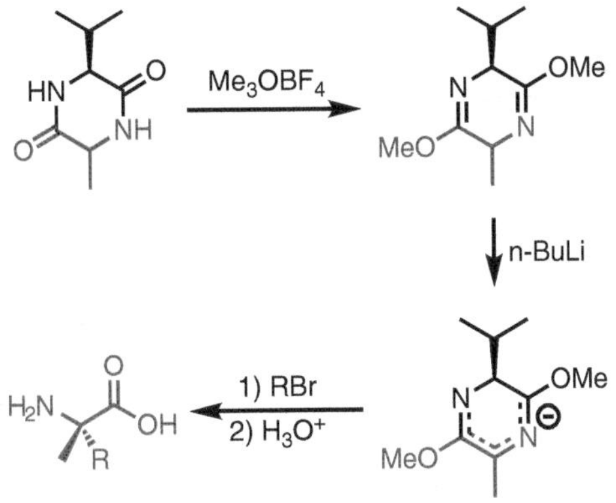

Scheme 5.15   *Schöllkopf alkylation*

The bislactim ethers also react with α,β-unsaturated carbonyl compounds in a Michael sense to give, after unraveling the amino acid moiety, β-substituted glutamic acid derivatives with high diastereoselectivity (> 150:1, Scheme 5.16) [11]. An improved and environmentally safer method of producing the Schöllkopf bislactim ethers has been reported [12].

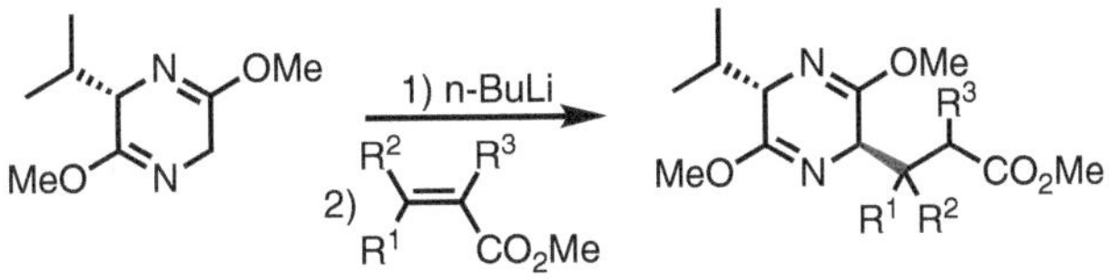

Scheme 5.16   *Conjugate addition with bislactim ethers*

Seebach developed an ingenious method utilising self-regeneration of stereocenters in enolate alkylation in the context of α-alkyl amino acid synthesis (Scheme 5.17) [13]. In a typical application of this method, proline is first converted to an oxazolidinone with pivalaldehyde. A new stereocenter is created highly selectively in such a manner that the bulky *tert*-butyl group is placed *exo* (convex face) with respect to the oxazabicyclo[3.3.0]octane ring system. Enolization of the lactone converts the original proline stereocenter to a planar sp^2 center. The bulky *tert*-butyl group directs the alkylation to occur from the convex face of the bicyclic system. The alkylated product can be hydrolysed to the free amino acid by acidolysis. Thus, the overall effect is the replacement of the α-hydrogen with an alkyl group with apparent retention of configuration.

This method has gained considerable applications [14], and we only show a few examples. It has been used in a synthesis of more elaborate conformationally constrained amino acid analogues for structural and antibody recognition studies (Scheme 5.18) [15]. The allyl proline derivative was coupled with an amino acid, and processing of the olefin gave the spirocycle. Lemieux–Johnson oxidation (OsO$_4$, NaIO$_4$) with reductive work-up gave the alcohol which was cyclised under Mitsunobu conditions. The spirocyclic amino acid analogue functions as a rigid mimic of a β-turn, as shown by solution-phase NMR studies as well as molecular modeling.

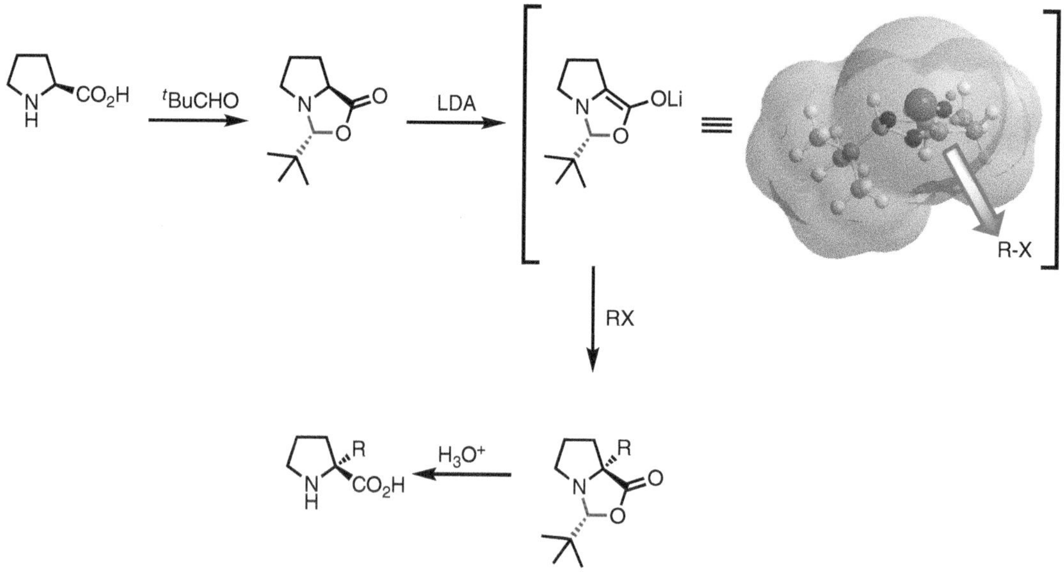

Scheme 5.17   *Seebach self-regeneration of stereocenters*

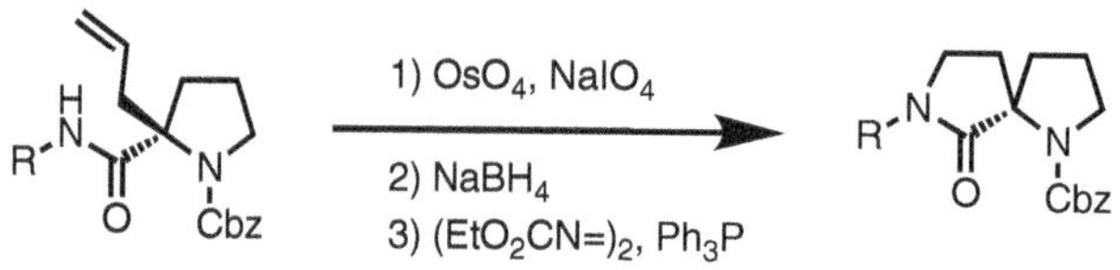

Scheme 5.18   *Spirocyclic amino acid analogue*

Robert M. Williams (1953–2020) has applied this alkylation in the enantioselective synthesis of *Penicillium brevicompactum* toxin (−)-brevianamide (Scheme 5.19) [16]. Allylation of the proline derivative was followed by the formation of the diketopiperazine moiety of the target. Ozonolytic cleavage of the allyl chain was followed by chain extension and eventual cyclisation to furnish the target compound.

The method has precedent in the synthesis of α-hydroxy acids, where the formation of a bicyclic intermediate is prohibited (Scheme 5.20) [17]. Alkylation of the dioxolanone furnishes the α-alkyl α-hydroxy acid with retention, as in the case of amino acids, but this time the reasons for the stereoselectivity are reversed: the *tert*-butyl group directs the alkylation on the opposite face of the five-membered ring.

Interestingly, the stereochemical outcome is sensitive to the substituents in the temporary stereocenter (Scheme 5.21) [18]. When proline anilide is converted to the aminal with isobutyraldehyde, benzaldehyde, or 1-naphthaldehyde, and these imidazolinones are alkylated (LDA, then benzyl bromide), the stereochemistry of the product depends on the aminal substituent: with the aliphatic aldehyde-derived aminal, the expected *syn* product is obtained; whereas 1-naphthaldehyde leads predominantly to the *anti* product. With benzaldehyde, a nearly 1:1 mixture of *anti* and *syn* products is obtained. When the proline amide was changed to butylamide, alkylations with several alkyl halides consistently gave > 91:9 diastereoselectivities.

An extension of the methodology allows for the synthesis of both α- and β-amino acid derivatives starting with chiral glycine and β-alanine enolates (Scheme 5.22) [19]. Again, the stereochemical information resides in the temporary aminal center. The chiral 1-benzoyl-2(*S*)-*tert*-butyl-3-methylperhydropyrimidin-4-one can be prepared from asparagine in a few steps [20]. Both the glycine and β-alanine enolates alkylate on the face opposite to the existing *tert*-butyl group, in contrast to the bicyclic proline enolate case discussed previously.

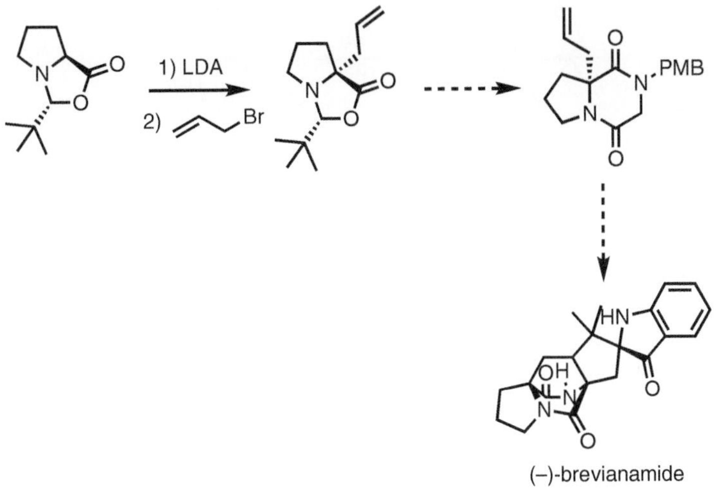

Scheme 5.19   *Synthesis of brevianamide*

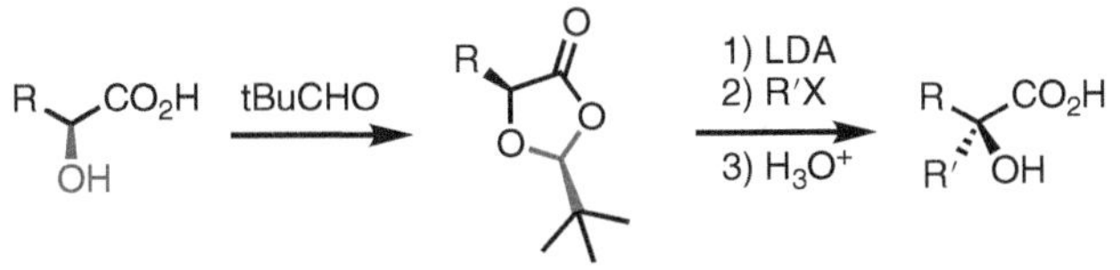

Scheme 5.20   *Self-regeneration of stereocenters in the synthesis of hydroxy acids*

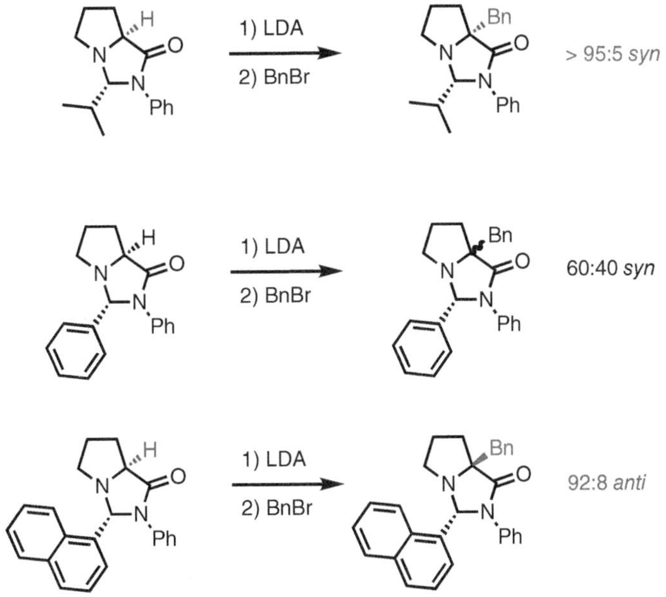

Scheme 5.21   *Complementary stereochemical outcome*

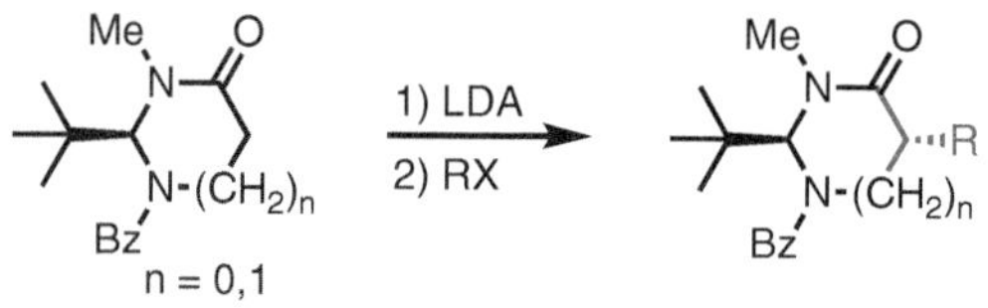

Scheme 5.22   *β-Amino acids*

Amino acid-derived aryl ureas undergo a highly stereoselective rearrangement upon treatment with base to generate α-arylated quaternary amino acids (Scheme 5.23) [21]. A number of amino acids were converted with pivalaldehyde and triphosgene to the *trans-N*-chloroformylimidazolidines, which reacted with a number of *N*-methylanilines to give the imidazolidinoureas shown. When these were treated with 1.5 equivalent of KHMDS, the arylated products were formed in high yields (81–98%) and high diastereoseletivities (> 93:7 dr). The complementary *cis-tert*-butyl imidazolidones were obtained from the same amino acids using a modified route, and they gave rise to the enantiomeric quaternary α-aryl amino acids with similar yields and stereoselectivities.

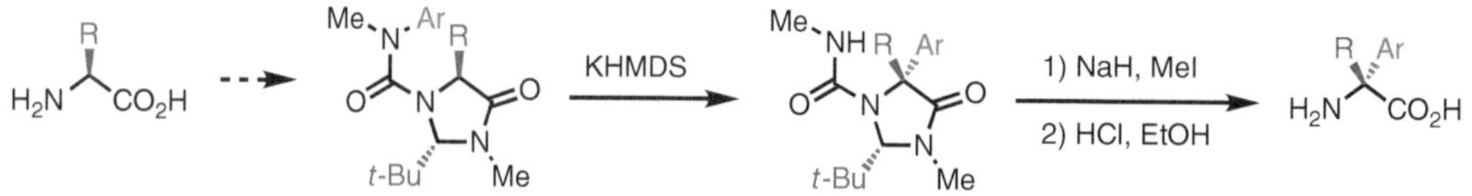

Scheme 5.23   *Asymmetric α-arylation of amino acids*

Another method for introducing alkyl substituents at the α-position of an existing amino acid relies on the use of azetidinyl derivatives (Scheme 5.24) [22]. Deprotonation with LDA leads to the ester enolate which can be alkylated with high selectivity (typically > 93% *de*) with a variety of electrophiles. The β-lactam moiety is destroyed by hydrogenolysis followed by acidolysis to liberate the α-alkyl amino acid derivative. The overall applicability of this sequence is somewhat overshadowed by the facts that a rather elaborate route is needed to construct the starting material, and that after the requisite transformations, the chiral auxiliary is completely lost.

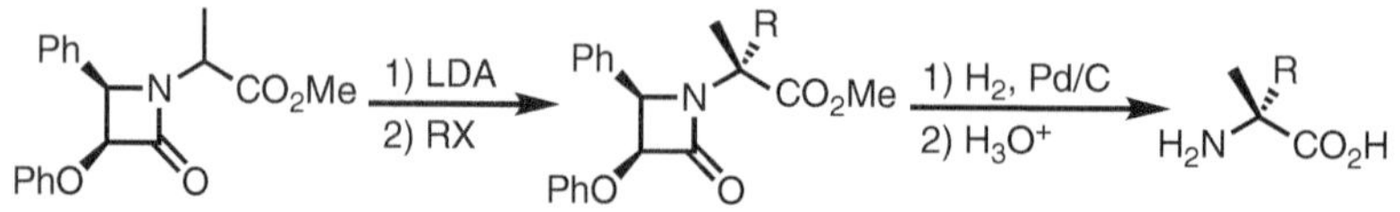

Scheme 5.24   *β-Lactams in the synthesis of amino acids*

For the synthesis of simple, non-natural amino acids bearing only a single alkyl substituent at the α-carbon, Williams has developed an interesting route relying on the use of 2,3-diphenyl morpholinones as starting materials (Scheme 5.25) [23]. The chiral morpholinone is commercially available in both enantiomeric forms. Treatment with NBS gives the bromo compound with high stereoselectivity. This compound can be alkylated with retention of stereochemistry with suitable organometallic nucleophiles. Besides organozinc compounds (or zinc catalysis), allyl silanes and silyl enol ethers may also be used. The chiral auxiliary is again cleaved destructively by hydrogenolysis of the benzylic bonds (catalytic hydrogenolysis or by dissolving metal reduction), or oxidatively (NaIO$_4$) after deprotection of the nitrogen and hydrolysis of the lactone (TMS-I followed by aqueous acid). The overall enantioselectivities (diastereoselectivities at the stage of alkylation) are high but not of exceptional levels. Allyl silane gave a modest 44% *de*, and the TMS enol ether of acetophenone gave a 66% *de* in the alkylation.

A more versatile route giving higher enantioselectivities is based on the alkylation of the morpholinone enolate with an electrophile (Scheme 5.26) [24]. Both mono- and disubstituted amino acids can be synthesised in high yields. In the first alkylation it is critical to use either lithium or sodium hexamethyldisilazane (HMDS), as stronger bases (including

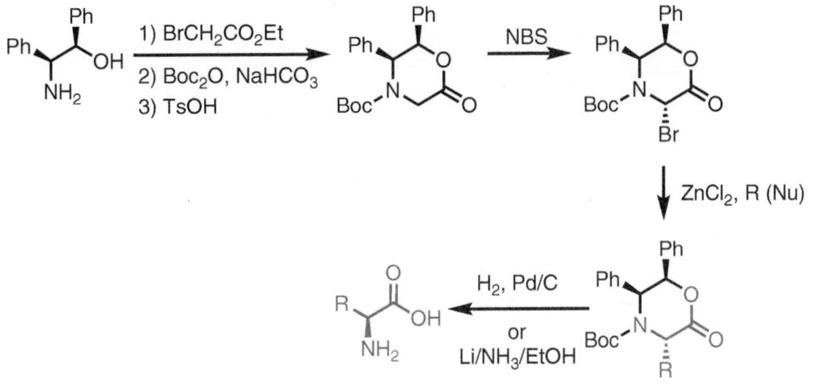

Scheme 5.25   *Morpholinones in the synthesis of amino acids*

potassium HMDS) cause decomposition of the oxazinone. The second alkylation, correspondingly, requires a stronger base. Activated electrophiles are also a necessary condition for both alkylations.

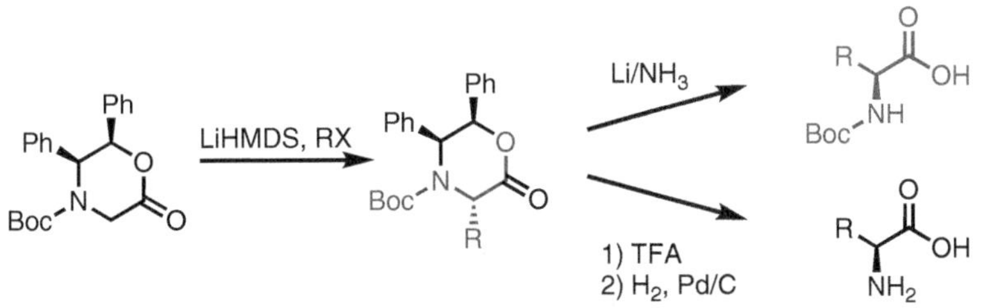

Scheme 5.26   *Direct alkylation of morpholinones*

The Evans chiral oxazolidinone auxiliaries are also well suited for use as chiral glycine enolate synthons. The utility of these chiral auxiliaries has been studied during the synthesis of MeBmt, the rare amino acid from cyclosporine A. As the chiral glycine equivalent, the authors used the isothiocyanate derived from the chloroacetyloxazolidinone (Scheme 5.27) [25]. The enolate of the acyloxazolidinone must be formed with stannous triflate (Sn(OTf)$_2$) since the corresponding lithium and boron enolates gave disappointingly low diastereoselectivities. Stannous triflate secured high selectivities (91:9 to 99:1) for the formation of *syn* aldol products. Epimerization is further suppressed by the formation of a cyclised adduct, preventing retro-aldol reaction as was seen in the case of Seebach's MeBmt synthesis [26]. It is also of interest to note that a chiral center α to the aldehyde function did not alter the *syn/anti* selectivity of the reaction to any notable degree. The reaction can thus be regarded as reagent controlled. When the reaction was performed using 2R-2-methyl-5-hexenal, the intermediate aldol product could be transformed into MeBmt in three steps.

The titanium–carbohydrate complex-mediated addition of a glycine enolate to aldehydes gives an efficient and economical access to *syn-β*-hydroxy-α-amino acids (Scheme 5.28) [27]. The high *syn*-selectivity and highly preferred addition of the enolate from the *Re*-face of the aldehydes (typical diastereoselectivities > 96% *de*, enantioselectivities > 96% *ee*) accompanied by the easy availability of the chiral alcohol ligand and the recoverability of the reagents makes this method very attractive in many applications.

The gold(I)-catalysed aldol type reaction of aldehydes and α-isocyanoacetate esters in the presence of chiral ferrocenylamine ligands possessing both central and planar chirality gives rise to optically active oxazolines in high enantiomeric purity (Scheme 5.29) [28]. The mechanism of the reaction has been studied and a transition state model to account for the observed stereoselectivity has been advanced [29].

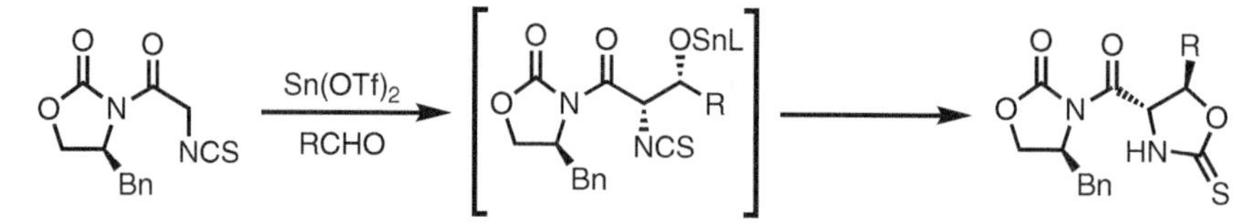

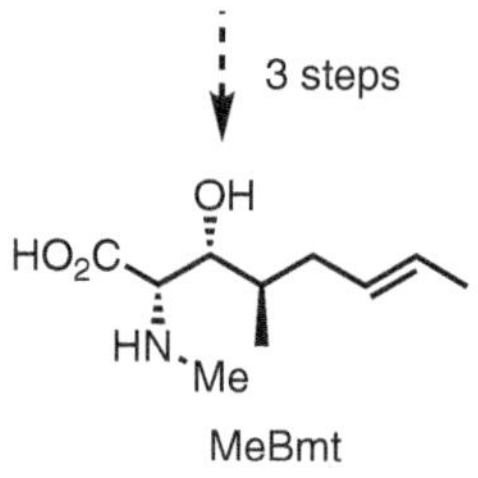

RCHO	syn:anti	yield, %
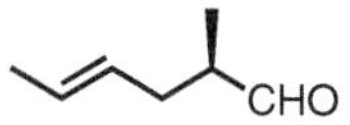 (2-methyl-4-hexenal) CHO	94:6	73
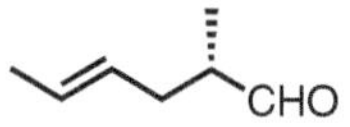 (2-methyl-4-hexenal) CHO	97:3	71
 (4-hexenal) CHO	93:7	81
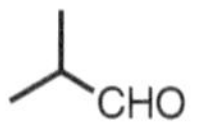 (isobutyraldehyde) CHO	99:1	92
MeCHO	91:9	75
PhCHO	99:1	91

Scheme 5.27   *Chiral oxazolidinones*

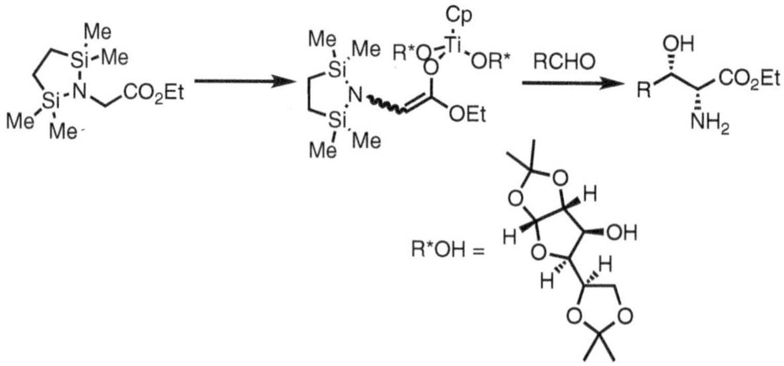

Scheme 5.28   *Titanium aldol of glycine esters*

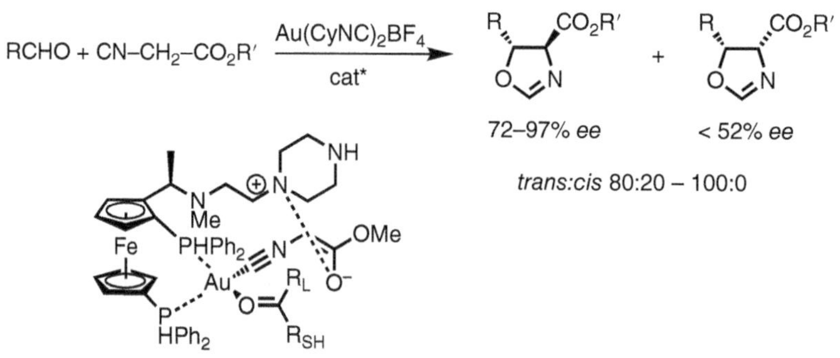

$$RCHO + CN{-}CH_2{-}CO_2R' \xrightarrow[\text{cat*}]{Au(CyNC)_2BF_4}$$

72–97% *ee*   < 52% *ee*

trans:cis 80:20 – 100:0

Scheme 5.29   *Au-catalysed aldol*

A highly stereoselective Strecker-type reaction was devised by Ellman et al. (Scheme 5.30) [30]. Addition of 5-methylfuryllithium to sulfinyl ketimines in the presence of $AlMe_3$ affords the sulfinamides in 75–97% yields and with diastereoselectivities ranging from 75:25 to 99:1. Subsequent oxidative degradation of the furyl group to carboxylate with $RuCl_3$/$NaIO_4$ affords *tert*-butanesulfonyl (Bus)-protected α,α-disubstituted amino acids in 62–69% yields. The attractiveness of this reaction lies in the conversion of the addition product to the (sulfonyl-protected) amino acid: the carboxylic group is formed by oxidation of the furyl moiety with sodium iodate under ruthenium catalysis, thus avoiding the harsh conditions required for the hydrolysis of the aminonitriles obtained in conventional Strecker reactions using cyanide as the nucleophile.

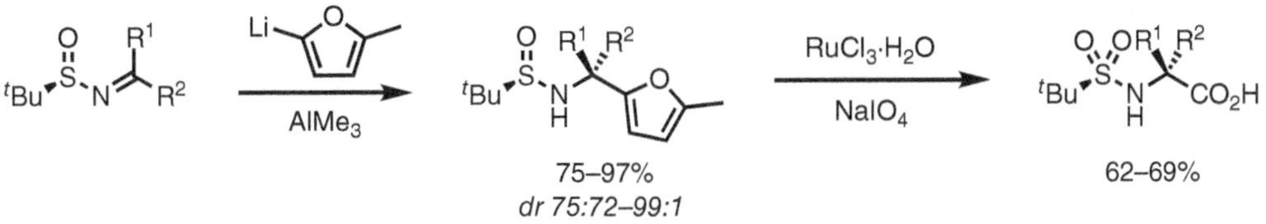

75–97%
dr 75:72–99:1

Scheme 5.30 *Diastereoselective Strecker-type alkylation*

5.1.2.2 *Amination of ester enolates*

Direct amination of ester enolates can be used for the synthesis of α-amino acids. These methods rely on the use of a suitable chiral auxiliary on the acid equivalent, e.g. the Evans chiral acyl oxazolidinones can be used efficiently in this conversion. The oxazolidinyl enolates can be aminated directly with sulphonyl azide, or their corresponding α-bromo derivatives can be treated with azide, followed by reduction of the azido function to the amino group.

Chiral arylglycines are constituents of the glycopeptide antibiotics vancomycin and ristocetin. The arylglycine moiety is very prone to racemisation due to the enhanced acidity of the α-protons. Thus, their synthesis provides a challenge to any synthetic route designed for these antibiotics. The Evans oxazolidinones have been used with success in the synthesis of the parent antibiotics as well as some analogues (Scheme 5.31) [31]. Azidation (KHMDS, Trisyl–N_3)[32] of the acyloxazolidinone yields a sulphonyltriazene intermediate which can be decomposed with potassium acetate to give the α-azido compound in high yield and high diastereoselectivity (88:12 $\rightarrow$ 95:5). Catalytic reduction of the azido group in the presence of di-*tert*-butyl pyrocarbonate gives the *N*-Boc-amino derivative.

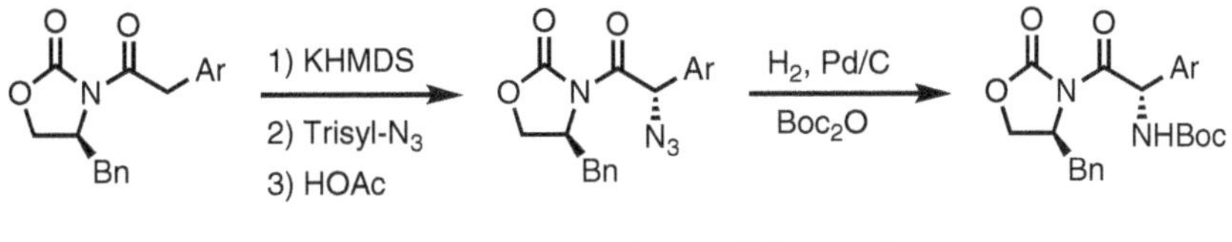

Scheme 5.31 *Amination of oxazolidinone derivatives*

The azidation has also been applied to the synthesis of diphthamide; the most complex posttranslationally modified amino acid known to date (Figure 5.8) [33]. Diphthamide is the target amino acid for ADP-ribosylation of protein synthesis elongation factor EF-2 triggered by diphtheria toxin. The inhibition of protein synthesis is the explanation at molecular level for the cytotoxicity of diphtheria toxin.

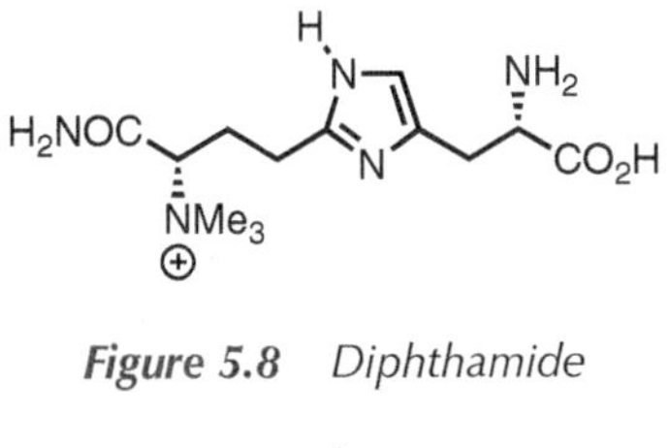

Figure 5.8 *Diphthamide*

Few practical sources of an electrophilic amino group exist. Chloronitroso alkanes provide a feasible reagent for this conversion, and 1-chloro-1-nitrosocyclohexane has been used for the synthesis of α-amino carbonyl compounds (Scheme 5.32) [34]. The enolate derived from an acylated Oppolzer camphor sultam exhibited excellent facial selectivity (> 99% *de*). The product hydroxylamine can be reduced to the amine by zinc.

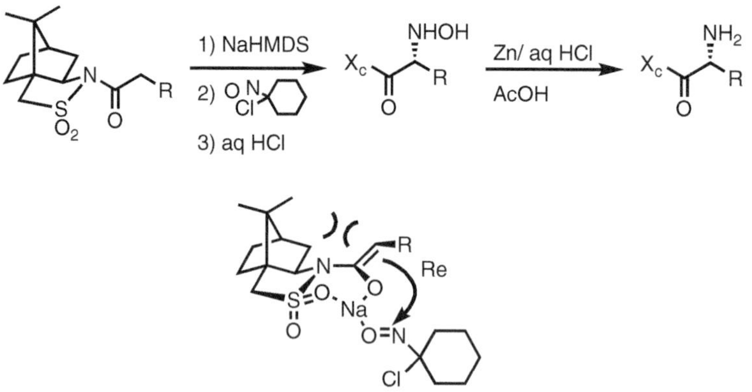

Scheme 5.32   Oppolzer camphor sultams in amino acid synthesis

The extremely high facial selectivity is explained to arise through a chelation-controlled process where the Z-enolate is held in the conformation shown. The formation of the *E*-enolate is suppressed by unfavorable steric interactions with the camphor-3-methylene moiety. The lower *Re* face is now open for the approach of the electrophile giving the observed high face selectivity.

Utilising a similar strategy, Oppolzer also developed a chiral aminating reagent based on the α-chloro-α-nitroso reagents capable of aminating prochiral carbonyl compounds with high enantiofacial differentiation (Scheme 5.33) [35]. The product α-amino ketones can be reduced in high yield to the corresponding *anti*-β-aminols, which are valuable synthetic intermediates.

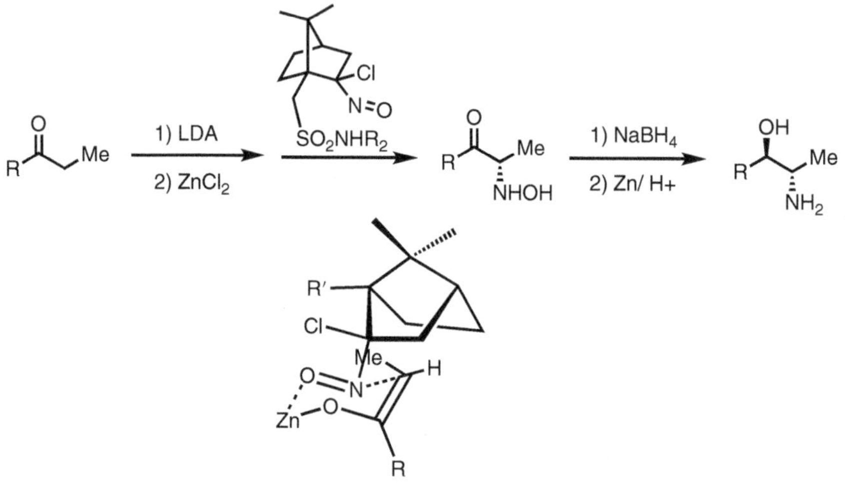

Scheme 5.33   Direct amination with chiral nitroso reagent

Transmetallation of the enolate to the Zn enolate is necessary for the successful outcome of the reaction, as the lithium enolates give eroded enantioselectivities. The stereochemical outcome has been rationalised as occurring through a Zn-chelated transition state of the *Z*-enolate. The corresponding *E*-enolates derived from cyclic ketones or 2,6-dimethylphenyl propionate react sluggishly to give a complex mixture of products [36].

5.1.2.3 Catalytic hydrogenation

In fact, the first asymmetric synthesis of amino acids was reported already in 1956 and was based on catalytic hydrogenation (Scheme 5.34) [37]. *Shīro Akabori* (1900–1992) was able to synthesise optically active amines and amino acids using an asymmetric catalyst obtained by reduction with a protein–palladium complex prepared by adsorption of palladium chloride on silk fibroin fibre. Optically active glutamic acid, phenylalanine and diphenylethylenediamine were formed by hydrogenating their precursors, diethyl α-acetoximino-glutarate, ethyl α-acetoximinophenylpropionate, 4-benzilidene-2-methyloxazol-5-one and α-benzildioxime, respectively, in the presence of this catalyst. The observed enntioselectivities were low by today's standards.

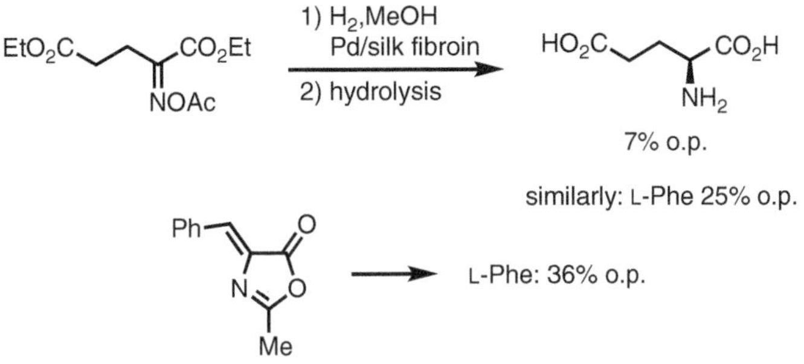

Scheme 5.34 *First asymmetric synthesis of amino acids by catalytic hydrogenation*

Catalytic asymmetric hydrogenation of α-aminoacrylic acid derivatives is an industrially important process. In this process, a soluble Rh catalyst is used, which is ligated with chiral diphosphine ligands. One of the earliest efficient ligands, DIOP, was developed by French *Henri B. Kagan* (1930–), a pioneer in asymmetric catalysis (Scheme 5.35) [38]. This catalyst system gave already good levels of asymmetric induction, but further developments of the catalyst were needed before a commercial process could be achieved. The so-called Monsanto process relies on the diphosphine DIPAMP, chiral at phosphorus, and this is used for the synthesis of L-DOPA [39].

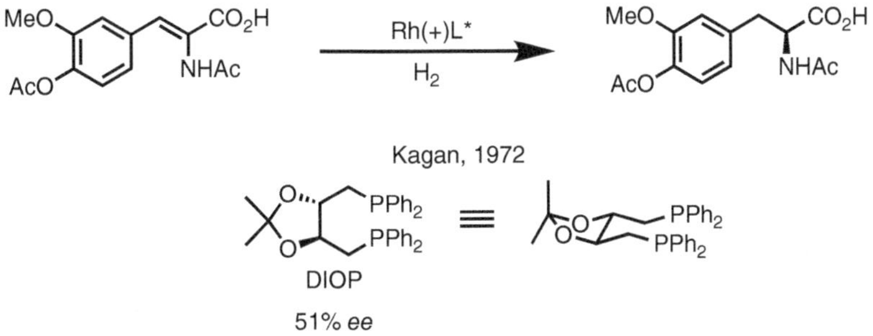

Scheme 5.35 *Monsanto process*

Numerous other ligands have been developed with the aim of achieving a more general catalytic system for the asymmetric reduction of double bonds (Figure 5.9). The mechanistic details of the reaction are sufficiently well understood so that this can help the development work [40]. It is known that the enamide forms two complexes with the Rh-DIPAMP catalyst, of which the minor one reacts with hydrogen much faster than the major one.

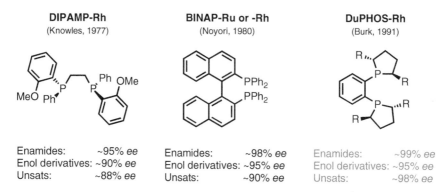

DIPAMP-Rh
(Knowles, 1977)

BINAP-Ru or -Rh
(Noyori, 1980)

DuPHOS-Rh
(Burk, 1991)

Enamides: ~95% *ee*	Enamides: ~98% *ee*	Enamides: ~99% *ee*
Enol derivatives: ~90% *ee*	Enol derivatives: ~95% *ee*	Enol derivatives: ~95% *ee*
Unsats: ~88% *ee*	Unsats: ~90% *ee*	Unsats: ~98% *ee*

Figure 5.9   *Ligands for Rh- and Ru-catalysts*

Dynamic kinetic resolution has also been applied in the synthesis of amino acids. The α-center of α-acetamido-β-ketobutyrate is susceptible to epimerisation (Scheme 5.36). The use of ruthenium-BINAP catalyst in dichloromethane allows nearly complete enantio- and diastereoselection to give *N*-acetyl threonine in nearly quantitative yield (see Section 2.7) [41].

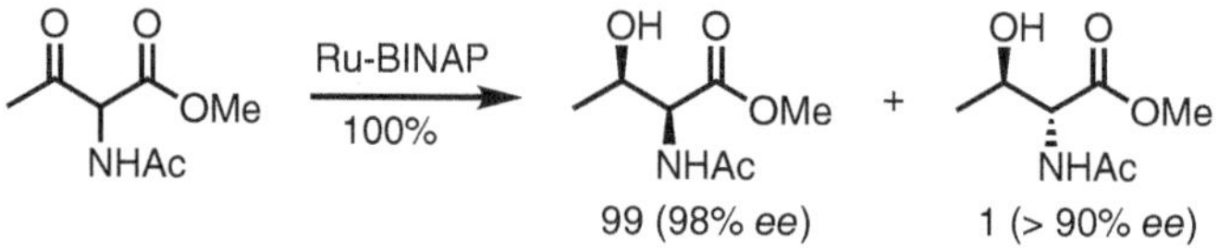

Scheme 5.36   *Noyori dynamic kinetic resolution*

Enantioselective chiral quaternary ammonium carboxylate-mediated phase-transfer transamination of α-ketoesters to the corresponding α-amino acid derivatives proceeds smoothly and with high enantioselectivity in the absence of any external base additives (Scheme 5.37) [42]. *p*-Nitrobenzylamine proved best as the amine donor. The catalytic reaction was used in the formal synthesis of (+)-monomorine, a trail pheromone of the tropical Pharaoh ant *Monomorium pharaonic* L.

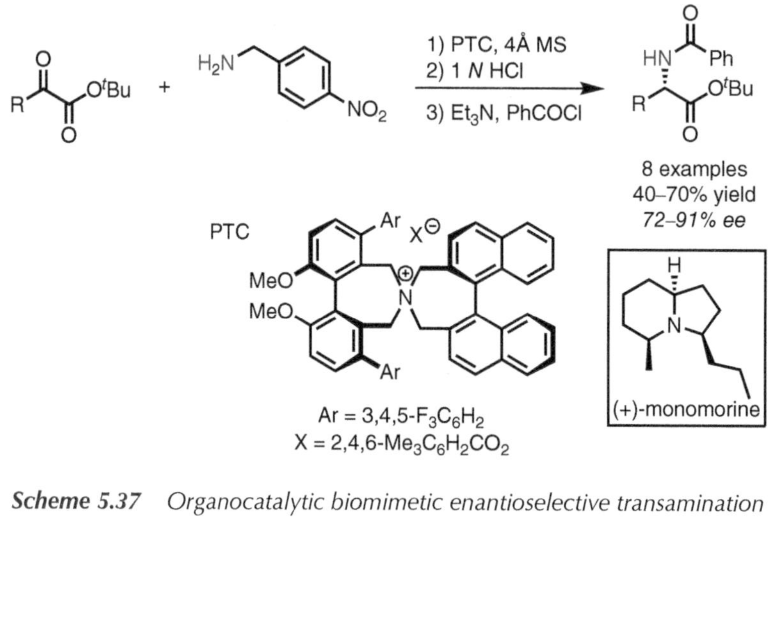

Scheme 5.37   *Organocatalytic biomimetic enantioselective transamination*

5.1.2.4   *Miscellaneous*

Corey developed a general catalytic method for the highly enantioselective synthesis of amino acids utilising the CBS reduction. Trichloromethyl ketones are reduced with high enantioselectivity with catecholborane and the (*S*)-oxazaborolidine catalyst to the (*R*) secondary alcohols (Scheme 5.38) [43]. Treatment of these with a basic solution of NaN$_3$ gives the α-azido carboxylic acids with clean inversion of configuration at the α-center. The azido group is finally converted to an amino group in a standard manner. The method has been applied to the synthesis of widely varying groups R (including *tert*-butyl), and the observed enantioselectivities are high (> 92% *ee*). The overall conversion from the carbonyl compound is typically 70–80% [44].

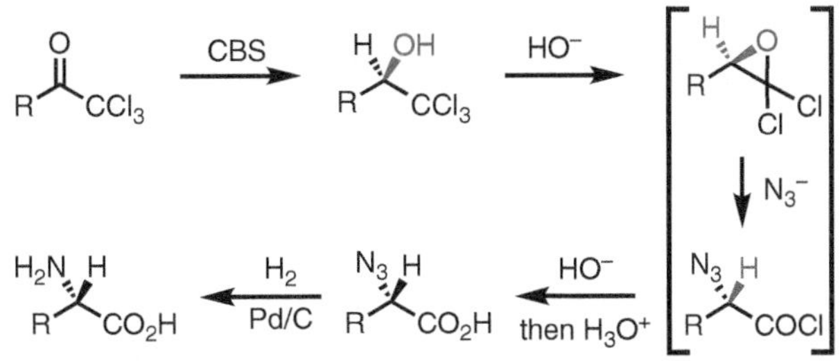

Scheme 5.38   *Corey reduction/epoxide formation route*

Serine-derived cyclic sulphamidates and sulphamidites function as alanyl β-cation equivalents (Scheme 5.39) [45]. The sulphamidates are synthesised from serine in five steps in about 50% yield and undergo highly regioselective nucleophilic ring-opening at the β-carbon with a variety of soft nucleophiles. It is interesting to note the similarity of this strategy to the one nature adopts in the biosynthesis of tryptophan: in both cases, the β-carbon of serine is activated towards a reaction with a soft nucleophile.

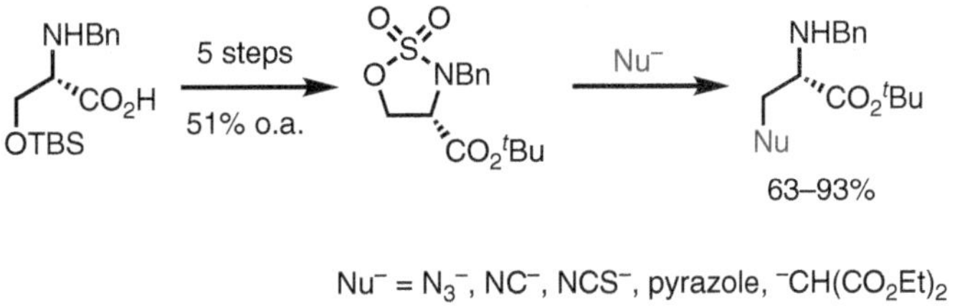

Nu$^-$ = N$_3^-$, NC$^-$, NCS$^-$, pyrazole, $^-$CH(CO$_2$Et)$_2$

Scheme 5.39   *Alkylation of serine derivatives*

β-Amino acids can be synthesised efficiently from achiral imines utilising the chiral boron ester enolate developed by Corey (Scheme 5.40) [46]. *S-tert*-Butyl thiopropionate, on reaction with the chiral diazaborolidine, gives rise to the *E*-enolate which undergoes rapid addition to imines to give the *anti* product. The diazaborolidine catalyst secures high levels of enantioselectivity (> 90% *ee*).

The high levels of enantioselectivity and diastereoselectivity are explained by steric factors: the thermodynamically less favorable *Z*-aldimine complexes preferentially with the boron enolate and the addition proceeds via a Zimmermann–Traxler-type six-membered transition state.

5.1.2.5   *Natural products*

Amino acids are an important class of chiral pool compounds, and their use in natural product syntheses have been reviewed [47]. Here, we will take a look at the enantioselective (or diastereoselective) syntheses of two rare amino acid natural products, kainic acid and acromelic acid, mainly to demonstrate how the synthesis strategies have evolved in a rather short period of time. Kainic acid syntheses have been reviewed [48].

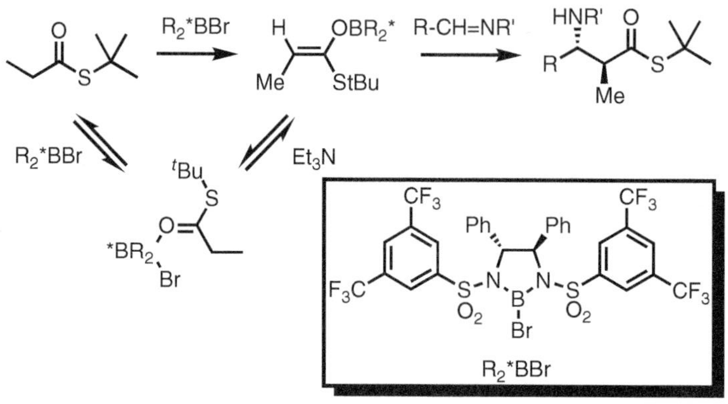

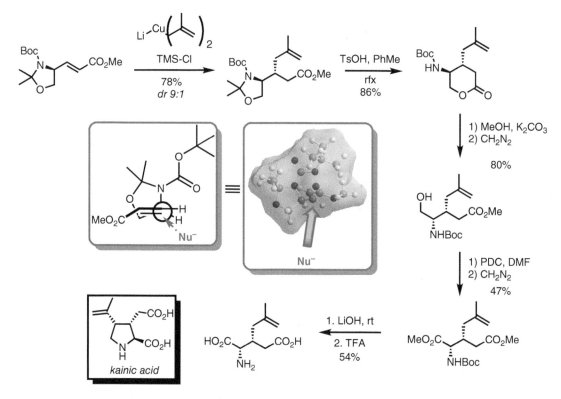

Scheme 5.40   *Corey imine addition*

The first example, from 1991, illustrates the diastereoselective conjugate addition of nucleophiles to the D-serine-derived enoate (Scheme 5.41) [49]. Propyl, benzyl, allyl, and methallyl lithium dialkylcuprates, in the presence of TMSCl, gave the conjugate addition products in *syn:anti* ratios ranging from 4:1 to 19:1. The diastereoselectivity can be explained with the help of the Felkin–Anh model: the axis of the carbon–nitrogen bond of the enoate is perpendicular to that of the carbon–carbon double bond, and the nucleophile approaches the electrophile from the *Si* face. As shown in the illustration with a Connolly molecular surface, the BOC group efficiently shields

Scheme 5.41   *Synthesis of secokainic acid by conjugate addition*

the *Re* face from the approach of the nucleophile. Finally, acid-catalysed cleavage of the aminal led to cyclisation to the corresponding lactones, where only the major isomer cyclised. The synthesis of *seco*-kainic acid was completed through a sequence of standard operations.

A radically different approach was devised in 2012, where again the serine-derived enone was first subjected to a Johnson–Corey–Chaykovsky cyclopropanation to give a nearly one-to-one diastereomeric mixture of cyclopropanoketones (Scheme 5.42) [50]. When this mixture was subjected to single-electron reduction with SmI$_2$, a [3 + 2] cycloaddition ensued to give the bicyclic product with high diastereoselectivity with respect to the benzoyl-bearing center. Moreover, complete *trans* relationship was observed between the C2/C3 stereocenters. Base treatment (DBU) moved the double bond to conjugation with exclusive formation of the *cis*-fused bicyclic system. With all the stereocenters now set, the remainder of the synthesis was straightforward. Ozonolysis, followed by oxidative work-up and esterification, gave the methyl ketone which was methylenated with the Tebbe reagent. Final oxidation and protecting-group removals gave kainic acid in a synthesis which consisted of 15 steps with an overall yield of 24%.

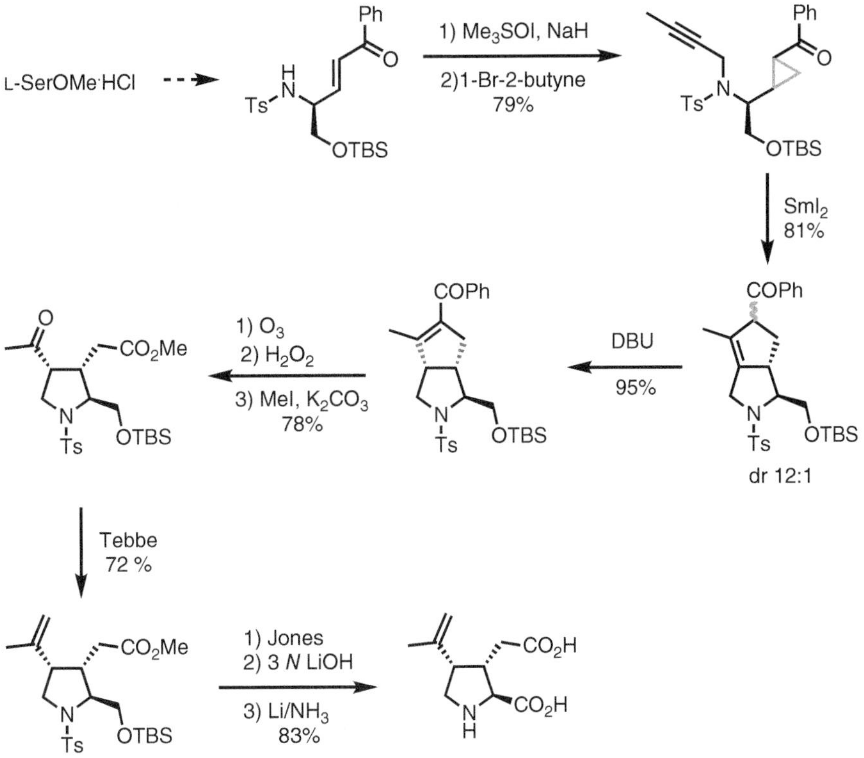

Scheme 5.42 Radical cyclisation in the synthesis of kainic acid

The final example illustrates the synthesis of acromelic acid A (Scheme 5.43) [51]. The starting nitroalkene was obtained regioselectively from 2,6-dichloropyridine in four steps. Stereoselective construction of the vicinal stereocenters at the C-3,4 positions was achieved by a Ni-catalysed asymmetric conjugate addition of α-ketoesters to the nitroalkenes. Construction of the pyrrolidine ring was realised in a single operation via a sequence consisting of reduction of the nitro group, intramolecular condensation with the ketone, and reduction of the resulting ketimine. Final steps involved standard operations to remove the protecting groups. Thus, a practical total synthesis of acromelic acid A was accomplished in 13 steps and 36% total yield. A modified sequence gave acromelic acid B in 17 steps and 6.9% total yield.

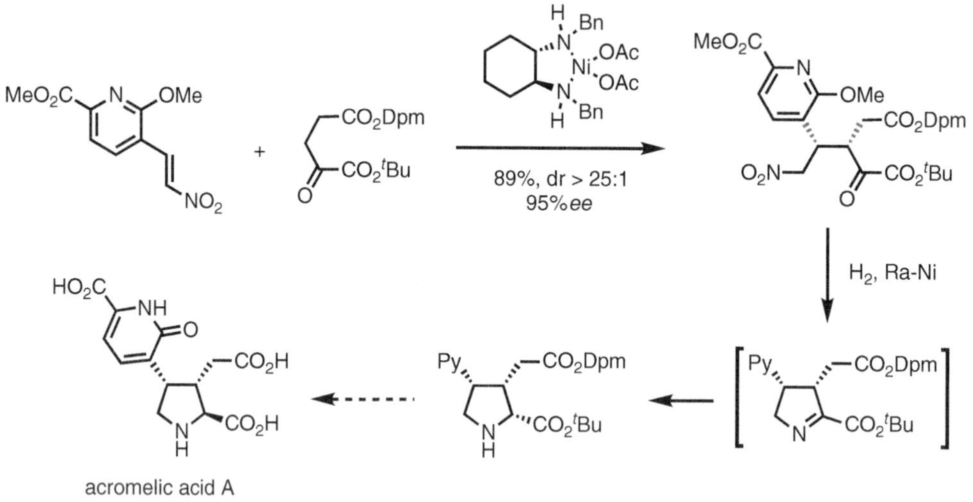

Scheme 5.43   *Ni-catalysed conjugate addition route to acromelic acid A*

5.2   Peptides and Proteins

Condensation of an amine with a carboxylic acid leads to the formation of an amide bond. When this bond exists in a peptide or a protein, it is called a peptide bond. In the amide bond, the carbonyl oxygen has a partial negative charge and the amide nitrogen a partial positive charge: a small dipole is set up (Figure 5.10). Therefore, the colored atoms in the amide bond are in plane. Virtually all peptide bonds occur in the *trans* configuration. Proline makes an exception: rotation around the amide bond is possible (barrier *ca.* 90 kJ/mol). The *cis* and *trans* conformers are only about 4 kJ/mol apart!

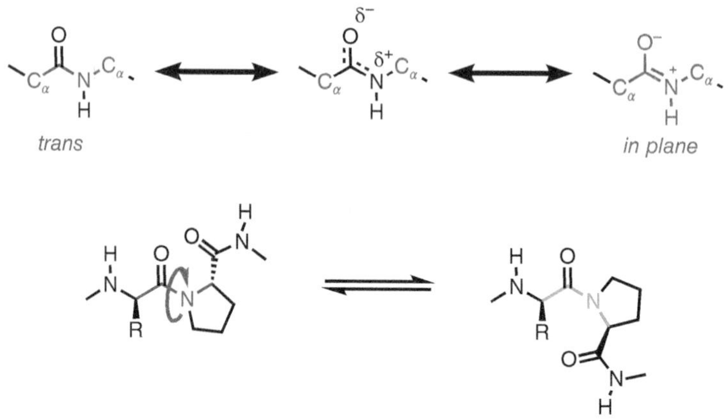

Figure 5.10   *Amide bond*

Cholecystokinin or pancreozymin (CCK or CCK-PZ; from Greek *chole* < bile; *cysto* < sac; *kinin* < move) is a regulatory peptide hormone acting through two distinct G-protein-coupled receptors: the CCKA (CCKA-R) and CCKB (CCKB-R) receptors located predominantly in the gastrointestinal tract and in the central nervous system, respectively. The cholecystokinin octapeptide (CCK 26-33 or CCK-8) and larger forms display high affinity for the two receptors, while shorter C-terminal CCK fragments are more selective for CCKB-R, although with reduced affinity to the gastrointestinal system responsible for stimulating the digestion of fat and protein [52].

CCK is synthesised in the duodenum, the first segment of the small intestine. It induces the release of digestive enzymes and bile from the pancreas and gallbladder, respectively, and also acts as a hunger suppressant, and this property is being utilised for the development of treatment for both anorexia and obesity. As an example of peptides, we shall inspect the strongly blood pressure-raising CCK-8, the *C*-terminal octapeptide of cholecystokinin (Figure 5.11).

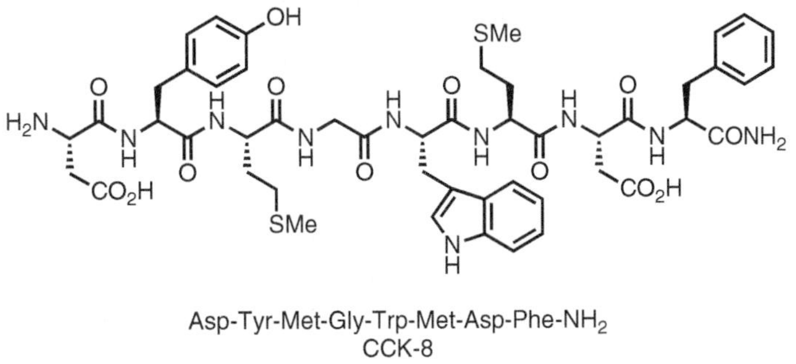

Asp-Tyr-Met-Gly-Trp-Met-Asp-Phe-NH$_2$
CCK-8

Figure 5.11 *CCK-8*

By convention, the structure of a peptide is represented by drawing the amine-containing *N*-terminus to the left, and the carboxylic acid–containing *C*-terminus to the right. In naming, one follows the same order. Thus, CCK-8 can be represented in the three-letter codes as Asp-Tyr-Met-Gly-Trp-Met-Asp-Phe-NH$_2$, or in the one-letter representation as DYMGWMDF-NH$_2$. The larger proteins, and especially the functional proteins and receptors, are often formed from several hundreds of amino acid residues. For these larger proteins, it is difficult to use systematic names, and one usually uses either three- or one-letter codes for the amino acid residues.

Each amino acid has its characteristic chemical and physical properties. When the amino acids combine to form a polypeptide, this also has its own characteristic properties which are partly dependent on the amino acid sequence. This sequence order of amino acids is called the *primary structure*, and it describes the order in which the amino acids are joined together.

The sequence can be determined in a straightforward manner by splitting off one amino acid residue at a time and analysing the amino acid thus obtained. The Edman degradation is a common method for the determination of the sequence starting from the *N*-terminal end (Scheme 5.44). Phenylisocyanate, or more conveniently, phenylisothiocyanate, is reacted with the free *N*-terminal amino group to form a thiourea. Under acidic conditions, the sulfur atom displaces the protonated amino group to give a thiazolidinone, and a peptide with one less amino acid residue. Acid-catalysed opening of the thiazolidinone followed by re-closure to reveal the phenylthiohydantoin concludes the reaction sequence. The degradation techniques are very highly developed, based on classical organic analytical reactions, and the sequence analysis is usually performed with a fully automated instrument [53].

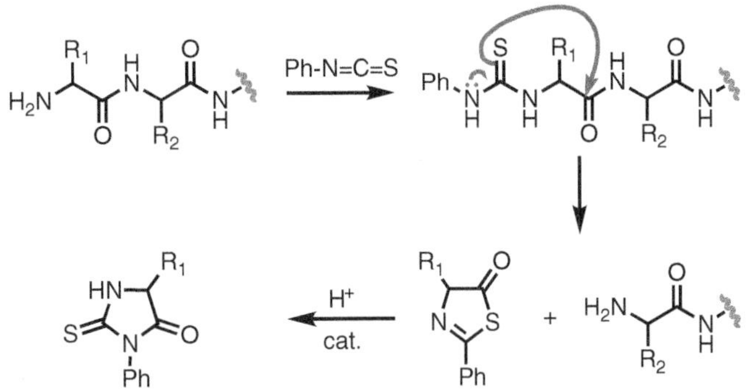

Scheme 5.44 *Edman degradation*

The primary structure of peptides is not the only, if even the most important, structural feature for the function of the peptide. The next structural level is called the *secondary structure*, which refers to the local structures of spans of several amino acids-long sequences. The peptide chain folds into various possible conformations, structural motifs: α-helix, β-sheet, and various forms of turns (Figures 5.12 and 5.13) [54].

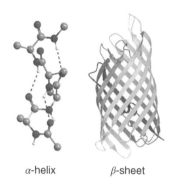

α-helix   β-sheet

Figure 5.12   *α-Helix and β-sheet*

The α-helices are formed through an intramolecular hydrogen bond connecting the amino group of the residue $i + 4$ to the carbonyl group of the residue i through a hydrogen bond. An intramolecular 13-membered hydrogen-bonded ring is formed. In an antiparallel β-sheet, the peptide chains form a 'carpet' held together by interstrand hydrogen bonds. Figure 5.12 (right) shows a 12-stranded β-barrel of an autotransporter protein produced by *Escherichia coli*.

Not so long ago, the regions joining helices and sheets were described simply as random loops. However, they are neither random nor loops. Protein structures have become available both in solid state (crystallography) and solution state by NMR and they are found to contain certain often repeated patterns, including β-turns [55]. In the apparently random regions of the peptide we can distinguish regions where the protein chain folds back on itself, forms a turn.

Two α-helices can be joined to each other by a sequence, which on its own would assume a more or less random form. The associative forces between the two helices can help this sequence to assume a particular type of turn. The turns, and especially the combinations of turns, sheets, and helices, have been shown to be important for many recognition processes in life.

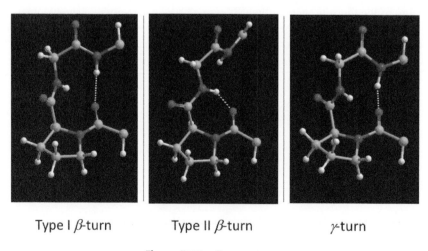

Type I *β*-turn   Type II *β*-turn   *γ*-turn

Figure 5.13   *Reverse turns*

The relative stabilities of secondary structures are highly dependent on both local and global energetic components, such as hydrogen bonding, salt bridges, and van der Waals forces. Although these interactions are typically small, their combined action can lead to remarkably high-energetic contributions. If one conformation is more favorable than the others, one talks about a local or a global energy minimum. This simply means that the conformation is energetically more favorable than the other ones either close to it on the reaction path (local minimum) or it is the absolute energy minimum (global minimum). The global minimum conformation is the one that can be expected to be obtainable on isolation of the protein.

A crystallographic study at high resolution established that the active site of an aspartic proteinase from *Rhizopus chinensis* contains two β-turns: both of the catalytic asparagines are involved in turns [56]. Approximately 60% of the β-turns are located on the surface of the protein. The pronounced occurrence of β-turns on the protein surface makes them viable candidates for molecular recognition, as exemplified by antigenic recognition, cell–cell recognition and protein–DNA recognition [57].

The structures of some structurally important turns are shown in Figure 5.13. Two major categories of turns are defined: β-turn (reverse turn, β-bend, inverse turn, U turn) and γ-turn. The former is distinguished by a hydrogen bond between residue i(C=O) and $i + 3$(N–H) (forming a 10-membered ring), and the latter similarly between i(C=O) and $i + 2$(N–H) (7-membered ring) [58]. The β-turns are further classified into a number of sub-types, depending on the nature of the side chains which affects the actual geometry of the turn.

A further point to be taken into account is the fact that these substitutions usually have a pronounced effect on the conformational freedom of the two key peptide torsion angles, the ϕ and ψ angles, which define the secondary structure of the peptide (Figure 5.14). Roughly speaking, the (ϕ,ψ) angles for the various secondary structures are: right handed α-helix (−60,−60); left-handed α-helix (60,60); parallel β-pleated sheet (−120,120); antiparallel β-pleated sheet (−140,140); extended chain (180,180); collagen helix (−60,140); and 3_{10}-helix (−60,−30). The β-turns are characterised by the angles of two consecutive amino acid residues, for instance, for the type II β-turn, the first ones are around (−60,120) and those of the second around (80,0). The angle ω is 0 for *cis*-peptide bonds, and 180 for *trans* peptide bonds. The direction of the side chain is described with the angle χ which can vary to a much larger degree than the ϕ, ψ, or ω angles.

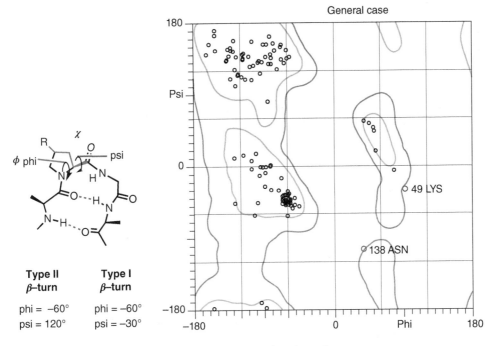

Figure 5.14 *Ramachandran plot*

The ϕ,ψ-angle dependency of the peptide chain conformation is usually presented graphically in the form of a Ramachandran plot. The panel on the right in Figure 5.14 displays a Ramachandran plot where the individual dots correspond to individual amino acids. The example protein is a staphylococcal nuclease (Protein Data Bank code 1SYB). The structure is clearly rich in type II β-turns as well as α-helical regions.

As one continues with the combination of secondary structural motifs, the final result is the *tertiary structure* of the protein. This overall-folded form of the peptide chain is thus the structure that best describes the protein in its natural environment. This structure is often so stable that it can be isolated and even crystallised in its original form.

If the protein is formed from more than one peptide chain, one talks about a *quaternary structure* for the intact protein complex. This describes the overall structure of a complex formed from two or more peptide chains which are joined together by non-covalent bonds to form the protein. Hemoglobin is an example of such a multi-chain protein complex. It is formed from four chains ($\alpha_2\beta_2$), each in its own tertiary structural form, which are then joined to form a complex which in turn binds four molecules of hemin. It is only this very large complex of eight individual components that can assume the functional form able to transport oxygen in the blood. Even then, only the correctly assembled complex is functional. Even very small changes in the structure of the peptide chain can dramatically alter the function of a complete protein complex. Hemoglobin S is a mutant of hemoglobin, which has suffered a single-point mutation at its sixth amino acid residue in the β-chain (Glu to Val). This causes changes throughout the three-dimensional structure of the hemin complex, with the end result that at low oxygen pressures, the red blood cells change their shape from a biconcave disc into an elongated form. In other words, a change of a single amino acid in the primary structure is reflected in a change of the quaternary structure of the protein complex. These sickle-shaped cells can block the capillaries and thus disturb blood circulation, causing serious tissue damage. Sickle cell anemia, as the disease is known, is a serious hereditary disease, especially in central Africa.

5.3 Enzymes and Receptors

Enzymes are naturally occurring polypeptides or proteins which catalyse reactions. The most important property of enzymes for their function is their outstanding substrate specificity. Out of a set of structurally closely related substrates, a typical enzyme can select a single one whose reaction it will catalyse. This is a key feature in the function of enzymes and also in the direction evolution has taken from primordial times.

Why are enzymes so selective? We have learned that each peptide chain adopts an individual form and shape; a three-dimensional structure. Enzymes (and receptors) recognise their substrates (usually small molecules) in the active center, which can be regarded as a pocket, a glove, or a groove (Figure 5.15). The substrate can bind only to a pocket of the right size and shape. This is, in simplified form, the Emil Fischer lock-and-key model [59], which was the prevailing theory for enzyme–substrate recognition until the 1960s. According to the model modified by *Daniel Edward Koshland*, Jr. (1920–2007), we know that this is only partly true, as the key also forces the lock to change its spatial structure [60].

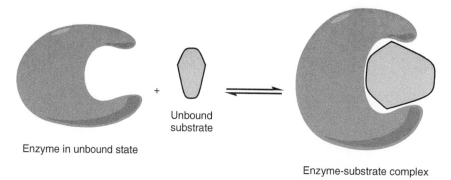

Enzyme in unbound state

Unbound substrate

Enzyme-substrate complex

Figure 5.15   Induced-fit model for enzyme recognition

As the substrate binds to the active site, there is a structural change in the overall complex, and in each binding partner. As a result, a suitable catalytic group of the enzyme is brought close to the reactive site in the substrate and a chemical reaction takes place. After the chemical transformation has occurred, the modified substrate (product) dissociates from the active site since it cannot fit the site as perfectly as the substrate can. The active site is believed to be most closely complementary (bind tightest) to the transition state of the reaction, which equals the action of lowering the free energy of the transition state. According to the Gibbs function, this has a rate-accelerating effect, which in many enzyme-catalysed reactions, has been observed to be on the order of 10^9. The reaction chain can proceed over and over again, in a truly catalytic fashion, giving high turn-over numbers. Since the active site is very selective due to its structural (both steric and electronic) features, enzyme catalysis of organic reactions has been explored quite actively [61].

We have already mentioned another type of proteins, the receptors. These are structurally akin to the enzymes. The most notable difference is that unlike enzymes, receptors do not participate in chemical reactions.

The binding of an active compound to the receptor site causes a structural change not unlike that observed in the case of an enzyme discussed above. The receptor proteins are usually located in the cell membrane, as represented in simplified form in Figure 5.16 for the platelet-derived growth factor (PDGF) receptor [62]. Activation of the receptor leads to a complex cascade of events leading ultimately to intracellular phosphorylation by mitogen-activated protein kinases (MAPK, thick arrow). This phosphorylation is modulated by a number of other signals, including input from the phorbol–ester receptor protein kinase C (PKC) and guanosine triphosphatase activating protein (GAP).

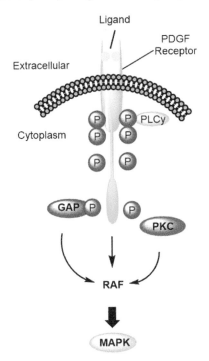

Figure 5.16   *Receptor proteins are usually located in membranes*

The function of receptors is to receive and filter various signals coming from inside the cell or from its surroundings, and then transmit the correct information further to another site, where an enzymatic reaction is usually affected. Only the correct agonist can trigger the transmission of the signal. Receptors, like enzymes, are abundant all over the organism. For instance, the function of the nervous system is based on a very delicate balance of a multitude of receptors.

The control function of receptors is also important in controlling the intracellular syntheses, whether that of peptides, proteins, or secondary metabolites. As a need for a particular chemical transformation is raised within the cell, this information is passed on, through receptors and chemical messengers, to the enzymes, which initiate (or terminate) the synthesis. This system can be used advantageously in medicine, since most maladies are caused by the malfunction of enzymes or receptors. This property has also been harnessed for the synthesis of natural products, employing bacterial cells, plant cell cultures and even isolated enzymes.

5.4   Chemical Modifications of Peptides

After protein synthesis, many natural proteins have not yet achieved their final functional form – they must be subjected to post-translational modifications. A typical example is the existence of 4-hydroxyproline in collagen and plant cell membrane proteins. The hydroxy function is brought into proline, already incorporated in a protein, by an enzymatic reaction, and is a necessary condition for the cross-linking of the peptide chains to form a rigid matrix.

Many hormones and neurotransmitters are relatively small peptides. It is known that these signal transmitters bind to their respective receptors in a highly defined conformation. The pharmaceutical utility of peptides – although usually readily available by isolation, solid-phase synthesis, or recombinant DNA techniques – is limited by several factors. They can easily assume a wide range of conformations, leading to multiple pharmacophore presentation and thereby lack of specificity. Natural-like peptides are also easily hydrolysed by thermal and proteolytic means. These factors lead to low oral activity, low membrane penetration, and often problems associated with potential antigenicity. Modified (non-natural) amino acids have been shown to be useful in various ways for the modification of existing peptide structures, especially in medicinal chemistry. Bioisosteric replacements of the peptide bond with a methylene group, alkene isostere, thioamide, or the so-called retro-inverso peptide (Figure 5.17) can often alleviate the hydrolysis sensitivity.

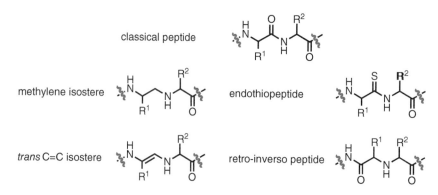

Figure 5.17   *Bioisosteric replacements*

One can also devise peptide mimetics with modified α-centers (Figure 5.18), resulting in reduced sensitivity towards hydrolysis. For instance, steric congestion at the α-center causes neo-pentyl type interactions at the tetrahedral intermediate of hydrolysis, thus lowering the hydrolysis rate. Simultaneously, the chiral center is also blocked against epimerisation, which is often of considerable aid in the synthesis of such modified peptides. The steric factors also impose strong conformational bias, and different kinds of secondary structural features can be dialed in by proper choice of substitution. The narrower conformational space available for the peptide also means that the problem of multiple-pharmacophore presentation is lessened. Because of the stability and the possibility of adjusting the lipophilicity of the drug candidate, oral activity and membrane penetration can usually be improved. The major problem, of course, is that such peptidomimetics are available only through total synthesis.

An example of a conformationally constrained peptide mimic is compound L-364,718 (devazepide, Figure 5.19) which has been designed using computer-aided molecular modeling and design methods. This molecule was designed to mimic the action of the active conformation of CCK-8 (see Figure 5.11).

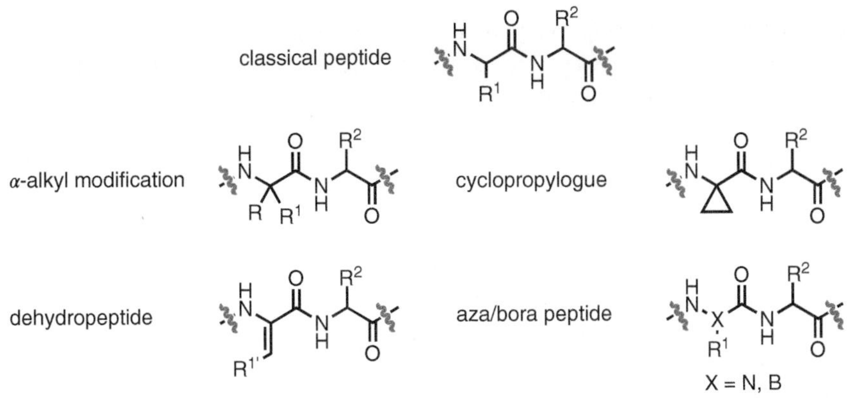

Figure 5.18 *α-Carbon modifications*

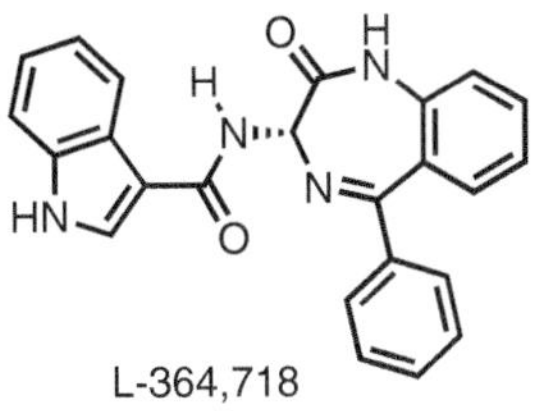

L-364,718

Figure 5.19 *CCK-8 mimic*

Although practically all of the original active peptide structure has been abolished, L-364,718 binds to the same receptor site, thus blocking the action of CCK-8 (in other words it is an antagonist). The compound has an appetite-increasing action, and its use for the treatment of anorexia is being studied.

5.4.1 Amino Alcohols

Hydrolysis of the peptide bond goes through the tetrahedral intermediate, which makes it highly attractive to mimic the tetrahedral intermediate through an sp^3-hybridised carbon atom in place of the amide carbonyl group. Besides the earlier mentioned peptide modifications, amino alcohols provide one alternative to achieve this.

Kidneys participate in the regulation of blood pressure in many ways. One of them is to release a hormone that acts directly on blood vessels. The process starts chemically with the release of a large peptide, angiotensinogen, which has hardly any effect on its own. In response to a suitable signal, another protein released by kidneys, the enzyme renin, cleaves angiotensinogen to a ten-amino acid angiotensin I. This is further cleaved by angiotensin-converting enzyme (ACE) into an eight-amino acid peptide, angiotensin II, which is a strong vasoconstrictor and thus raises blood pressure. Renin and ACE are examples of a wide group of enzymes known as proteases.

With the discovery of the wide-spectrum acid protease inhibitor pepstatin, a naturally occurring hexapeptide which incorporates a γ-amino-β-hydroxy acid moiety, much interest has been generated towards the synthesis of compounds bearing this structural moiety. The natural product blocks the cleavage of pepsin (a member of food-digesting enzymes) at a site similar to that at which renin cleaves angiotensinogen. These studies shed much light on the development of peptidomimetics of this structural class. The most generally accepted synthetic strategy relies on the reaction of an ester enolate with a chiral α-amino aldehyde. The chemistry of α-amino aldehydes has been reviewed [63], and we shall only take a few representative examples of the synthesis of γ-amino-β-hydroxy acids.

The diastereoselectivity of additions onto amino aldehydes is generally very low. The Felkin–Anh transition state model predicts the formation of the *anti* product, whereas the chelation-controlled transition state model favors the formation of the *syn* product (Scheme 5.45).

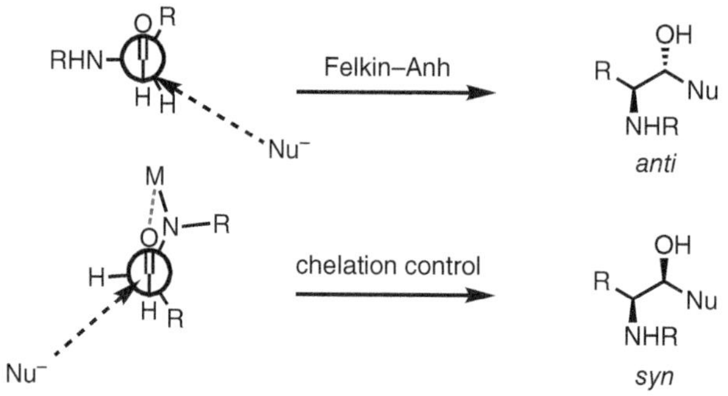

Scheme 5.45 *Diastereoselectivity in addition*

The open transition state usually correlates with the obtained results, but employing methods favoring the chelation-controlled addition, the *syn* products can usually be formed with acceptable diastereoselectivity (*ca.* 4:1, Scheme 5.46) [64].

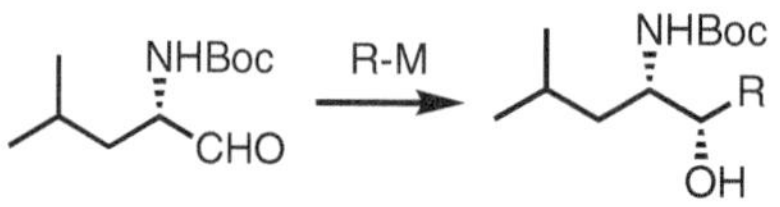

Scheme 5.46 *Diastereoselective addition to amino aldehydes*

The addition of *O*-methyl *O*-trimethylsilyl ketene acetal to carbamate-protected amino aldehydes under Lewis-acidic conditions leads to the formation of the *syn* product in a diastereomer ratio 94:6 (Scheme 5.47) [65]. The high diastereoselectivity was achieved by using titanium tetrachloride as the Lewis acid. Other chelating Lewis acids also gave the *syn* product, albeit in a diminished diastereoselectivity.

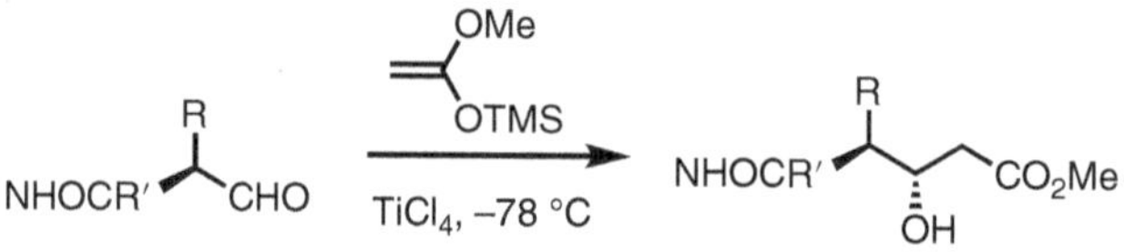

Scheme 5.47 *Mukaiyama aldol*

In the case of amino acids not capable of chelation, the open Felkin–Anh model can predict the outcome of the reaction quite reliably. This was beneficially utilised in the synthesis of dolaproine, a rare amino acid forming a part of the structure of the reportedly antineoplastic pentapeptide dolastatin 10 (Scheme 5.48) [66]. Reaction of *N*-Boc-(*S*)-prolinal with the *Z*-boron enolate of (*S*)-phenyl thiopropionate gave the *syn,anti* product in 64% yield. The *syn,syn* and *anti,syn* isomers were formed in 10 and 1% yields, respectively. The aldol selectivity is thus extremely high (74:1), and the Felkin–Anh model is followed with 64:11 selectivity.

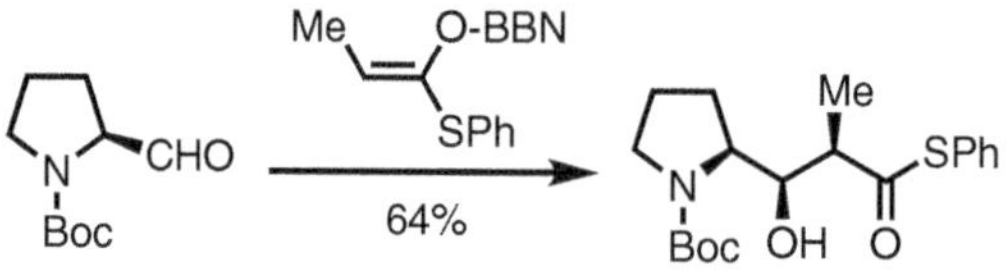

Scheme 5.48 *Dolaproine synthesis*

References

1. Wagner, I., Musso, H. *Angew. Chem., Int. Ed. Engl.* **1983**, *22*, 816–828.
2. Eykman, J.F. *Pharm. J. Trans.* **1870**, *11*, 1046–1050, translated in: *J. Chem. Soc., Abstr.* **1881**, *40*, 918–919.
3. Jiang, S., Singh, G. *Tetrahedron* **1998**, *54*, 4697–4753.
4. Ames, G.F., Roth, J.R. *J. Bacteriol.* **1968**, *96*, 1742–1749.
5. Adams, D.O., Yang, S.F. *Proc. Natl. Acad. Sci. USA* **1979**, *76*, 170–174.
6. Vaillancourt, F.H., Yeh, E., Vosburg, D.A., O'Connor, S.E., Walsh, C.T. *Nature* **2005**, *436*, 1191–1195.
7. a) Srinivasan, G., James, C. M., Krzycki, J.A. *Science* **2002**, *296*, 1459–1462; b) Hao, B., Gong, W., Ferguson, T.K., James, C.M. Krzycki, J.A., Chan, M.K. *Science* **2002**, *296*, 1462–1466; c) Atkins, J.F., Gesteland, R. *Science* **2002**, *296*, 1409–1410.
8. Gaston, M.A., Zhang, L., Green-Church, K.B., Krzycki, J.A. *Nature* **2011**, *471*, 647–651.
9. Vogt, H., Bräse, S. *Org. Biomol. Chem.* **2007**, *5*, 406–430.
10. a) Schöllkopf, U., Hartwig, W., Groth, U. *Angew. Chem.* **1979**, *91*, 863–864; b) Groth, U., Schöllkopf, U. *Synthesis* **1983**, 37–38; c) Schöllkopf, U. *Pure Appl. Chem.* **1983**, *55*, 1799–1806.
11. Pettig, D., Schöllkopf, U. *Synthesis* **1988**, 173–175.
12. Chen, J., Corbin, S.P., Holman, N.J. *Org. Proc. Res. Dev.* **2005**, *9* 185–187.
13. a) Seebach, D., Boes, M., Naef, R., Schweizer, W.B. *J. Am. Chem. Soc.* **1983**, *105*, 5390–5398; b) review: Seebach, D., Sting, A.R., Hoffmann, M. *Angew. Chem., Int. Ed. Engl.* **1996**, *35*, 2708–2748.
14. Calaza, M.I., Cativiela, C. *Eur. J. Org. Chem.* **2008**, 3427–3448.
15. Hinds, M.G., Welsh, J.H., Brennand, D.M., Fisher, J., Blennie, M.J., Richards, N.G.J., et al. *J. Med. Chem.* **1991**, *34*, 1777–1789.
16. Williams, R.M., Glinka, T., Kwast, E., Coffman, H., Stille, J.K. *J. Am. Chem. Soc.* **1990**, *112*, 808–821.
17. Frater, G., Muller, U., Gunther, W. *Tetrahedron Lett.* **1981**, *22*, 4221–4224.
18. Knight, B.J., Stache, E.E., Ferreira, E.M. *Org. Lett.* **2014**, *16*, 432–435.
19. a) Seebach, D., Juaristi, E., Miller, D.D., Schickli, C., Weber, T. *Helv. Chim. Acta* **1987**, *70*, 237–261; b) Juaristi, E., Quintana, D., Lamatsch, B., Seebach, D. *J. Org. Chem.* **1991**, *56*, 2553–2557; c) Seebach, D., Lamatsch, B., Amstutz, R., Beck, A.K., Dobler, M., Egli, M., et al. *Helv. Chim. Acta* **1992**, *75*, 913–934.
20. Juaristi, E., Balderas, M., Ramírez-Quirós, Y. *Tetrahedron: Asymmetry* **1998**, *9* 3881–3888.
21. Leonard, D.J., Ward, J.W., Clayden, J. *Nature* **2018**, *562*, 105–109.
22. Ojima, I., Qiu, X. *J. Am. Chem. Soc.* **1987**, *109*, 6537–6538.
23. Williams, R.M.,; Sinclair, P.J., Zhai, D., Chen, D. *J. Am. Chem. Soc.* **1988**, *110*, 1547–1557.
24. Williams, R.M., Im, M.-N. *J. Am. Chem. Soc.* **1991**, *113*, 9276–9286.
25. Evans, D.A., Weber, A.E. *J. Am. Chem. Soc.* **1986**, *108*, 6757–6761.
26. Blaser, D., Ko, S.Y., Seebach, D. *J. Org. Chem.* **1991**, *56*, 6230–6233.
27. Bold, G., Duthaler, R.O., Riediger, M. *Angew. Chem., Int. Ed. Engl.* **1989**, *28*, 497–498.
28. Ito, Y., Sawamura, M., Hayashi, T. *J. Am. Chem. Soc.* **1986**, *108*, 6405–6406.
29. Togni, A., Pastor, S.D. *J. Org. Chem.* **1990**, *55*, 1649–1664.
30. Borg, G., Chino, M., Ellman, J.A. *Tetrahedron Lett.* **2001**, *42*, 1433–1435.
31. Evans, D.A., Evrard, D.A., Rychnovsky, S.D., Fruh, T., Whittingham, W.G., DeVries, K. M. *Tetrahedron Lett.* **1992**, *33*, 1189–1192.
32. Evans, D.A., Britton, T.C., Ellman, J.A., Dorow, R.L. *J. Am. Chem. Soc.* **1990**, *112*, 4011–4033.
33. Evans, D.A., Lundy, K.M. *J. Am. Chem. Soc.* **1992**, *114*, 1495–1496.
34. Oppolzer, W., Tamura, O. *Tetrahedron Lett.* **1990**, *31*, 991–994.
35. Oppolzer, W., Tamura, O., Sundarababu, G., Signer, M. *J. Am. Chem. Soc.* **1992**, *114*, 5900–5902.
36. a) Heathcock, C.H. In *Asymmetric Synthesis* (Morrison, J.D., ed.) Academic Press: New York, **1984**; vol. *3*, 111–212; b) Masamune, S., Ellinghoe, J.W., Choy, W. *J. Am. Chem. Soc.* **1982**, *104*, 5526–5528.
37. Akabori, S. *Nature* **1956**, *178*, 323–324.
38. Dang, T.P., Kagan, H.B. *J. Chem. Soc., Chem. Commun.* **1971**, *481*.
39. Vineyard, B.D., Knowles, W.S., Sabacky, M.J., Bachman, G.L., Weinkauf, D.J. *J. Am. Chem. Soc.* **1977**, *99*, 5946–5952.
40. a) Ohkuma, T., Kitamura, M., Noyori, R. In *New Frontiers in Asymmetric Catalysis*, Mikami, K., Lautens, M. (eds.), John Wiley & Sons, New Jersey, **2007**, pp. 1–32; b) Genet, J.-P. In *Asymmetric Synthesis*, 2nd ed., Christmann, M., Braese, S. (eds.) Wiley-VCH, Weinheim, **2008**, pp. 282–287.
41. Noyori, R. *Science* **1990**, *248*, 1194–1199.

42. Kang, Q.-K., Selvakumar, S., Maruoka, K. *Org. Lett.* **2019**, *21*, 2294–2297.
43. Corey, E.J., Bakshi, R.K. *Tetrahedron Lett.* **1990**, *31*, 611–614.
44. Corey, E.J., Link, J.O. *J. Am. Chem. Soc.* **1992**, *112*, 1906–1908.
45. Baldwin, J.E., Spivey, A.C., Schofield, C.J. *Tetrahedron: Asymmetry* **1990**, *1*, 881–884.
46. Corey, E.J., Decicco, C.P., Newbold, R.C. *Tetrahedron Lett.* **1991**, *32*, 5287–5290.
47. Paek, S.-M., Jeong, M., Jeyun Jo, J., Heo, Y.M., Han, Y.T., Yun, H. *Molecules* **2016**, *21*, 951.
48. Stathakis, C.I., Yioti, E.G., Gallos, J.-K. *Eur. J. Org. Chem.* **2012**, 4661–4673.
49. Jako, I., Uiber, P., Mann, A., Wermuth, C.-G., Boulanger, T., Norberg, B., et al. *J. Org. Chem.* **1991**, *56*, 5729–5733.
50. Luo, Z., Zhou, B., Li, Y. *Org. Lett.* **2012**, *14*, 2540–2543.
51. Inai, M., Ouchi, H., Asahina, A., Asakawa, T., Hamashima, Y., Kan, T. *Chem. Pharm. Bull.* **2016**, *64*, 723–732.
52. Pellegrini, M., Mierke, D.F. *Biochemistry* **1999**, *38*, 14775–14783.
53. Bodanszky, M. *Peptide Chemistry; A Practical Textbook,* Springer-Verlag: Berlin, **1988**.
54. Pauling, L., Corey, R.B., Branson, H.R. *Proc. Nat. Acad. Sci. USA* **1951**, *37*, 205–211.
55. a) Venkatachalam, C.M. *Biopolymers* **1968**, *6*, 1425–1436; b) Crawford, J.L., Lipscomb, W.N., Schellman, C.G. *Proc. Nat. Acad. Sci.* **1973**, *70*, 538–542; c) Lesczynski, J.F., Rose, G.D. *Science* **1986**, *234*, 849–855; d) Milner-White, E.J. *Biochim. Biophys. Acta* **1987**, *911*, 261–265.
56. Suguna, K., Davies, D.R. *J. Mol. Biol.* **1987**, *196*, 877–900.
57. Ollis, D.L., White, S.W. *Chem. Rev.* **1987**, *87*, 981–996.
58. Venkatachalam, C.M. *Biopolymers* **1968**, *6*, 1425–1436.
59. Fischer, E. *Ber. Dtsch. Chem. Ges.* **1894**, *27*, 2985–2993.
60. Koshland, D.E. *Proc. Natl. Acad. Sci.* **1958**, *44*, 98–104.
61. a) Santaniello, E., Ferraboschi, P., Grisenti, P., Manzocchi, A. *Chem. Rev.* **1992**, *92*, 1071–1140; b) Gupta, M.N., Roy, I. *Eur. J. Biochem.* **2004**, *271*, 2575–2583; c) Azerad, R. *Adv. Org. Synth.* **2005**, *1*, 455–518; d) Kayser, M.M. *Tetrahedron* **2009**, *65*, 947–974.
62. Pelech, S.L., Sanghera, J.S. *Science* **1992**, *257*, 1355–1356.
63. Jurczak, J., Golebiowski, A. *Chem. Rev.* **1989**, *89*, 149–164.
64. Holladay, M.W., Rich, D.H. *Tetrahedron Lett.* **1983**, *24*, 4401–4404.
65. Takemoto, Y., Matsumoto, T., Ito, Y., Terashima, S. *Tetrahedron Lett.* **1990**, *31*, 217–218.
66. Tomioka, K., Kanai, M., Koga, K. *Tetrahedron Lett.* **1991**, *32*, 2395–2398.

6

Nucleosides, Nucleotides, and Nucleic Acids

It has not escaped our notice that the specific pairing we have postulated immediately suggests a possible copying mechanism for the genetic material.

J.D. Watson, F.H.C. Crick, 1953

Nucleic acids deoxyribonucleic acid (DNA) and ribonucleic acid (RNA) play important roles in life. The former is the carrier of the cellular and organism's hereditary information, and the latter function as signal transduction molecules for the synthesis of proteins. The information embedded in DNA is read and transcribed into RNA within the nucleus of the cell, and this information is then passed on to other parts of the cell where the sequence information in the messenger RNA is translated into a protein sequence.

In this chapter we shall not study the genetic functions of DNA or RNA, fascinating as they are in their chemistry, but these are left to the realm of biochemistry and molecular biology. Many biologically active natural compounds and pharmaceuticals derived from them interact directly with DNA or RNA. It is therefore important to familiarise ourselves with the structural features of these biological macromolecules in order to understand their interactions with small molecules at molecular level. We shall briefly inspect the structural features of the heterocyclic nucleobases and their naturally occurring relatives, as well as molecules structurally related to the nucleic acid constituents with medicinal importance.

In eukaryotic cells, the hereditary information is stored in the cell nucleus in the supercoiled duplex form of DNA, which is further arranged in supramolecular structures with histone proteins to form nucleosomes. The human genome consists of approximately 3.1 billion base pairs, which encode approximately 20 000 proteins (Table 6.1). It is interesting to note that the length of the DNA does not correlate with a number of proteins coded. It is also of interest to note that of the coded proteins less than 1000 are presently targets for drugs [1].

6.1 Building Blocks of DNA and RNA

The nucleic acids are formed from three structural building blocks; the nucleobase, the sugar, and phosphoric acid. The base is a heterocyclic compound which belongs either to the purine or the pyrimidine bases (Figure 6.1). The purine bases adenine and guanine are common to both DNA and RNA, but there is variation in the pyrimidine bases: cytosine and thymine are found in DNA; and in RNA thymine is replaced with the demethylated analogue, uracil.

The sugar units in the DNA and RNA form the foundation of the structural and, to some extent, functional differences between the two types of nucleic acid heteropolymers. In DNA, the carbohydrate is 2-deoxyribose, whereas in RNA it is ribose (Figure 6.2). The combination of the sugar unit and the nucleobase gives rise to the *nucleosides* (in RNA) and *deoxynucleosides* (in DNA). Thus, the *nucleotides* or *deoxynucleotides* are the actual building blocks of RNA and DNA, respectively.

Asymmetric Synthesis of Natural Products, Third Edition. Ari M.P. Koskinen.
© 2023 John Wiley & Sons Ltd. Published 2023 by John Wiley & Sons Ltd.

Table 6.1   *Length of genome and number of proteins coded in different eukaryotes.*

Species		M bp	Coded proteins
Human	*Homo sapiens*	3100	20 000
Rapeseed	*Brassica napus*	125	25 500
Roundworm	*Chaenorhabditis elegans*	97	19 000
Fruit fly	*Drosphila melanogaster*	180	13 600
Baker's yeast	*Saccharomyces cerevisiae*	12,1	5800

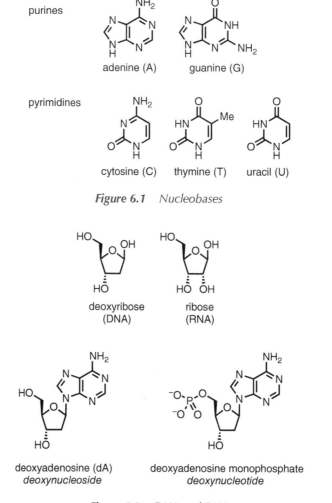

purines

adenine (A)   guanine (G)

pyrimidines

cytosine (C)   thymine (T)   uracil (U)

Figure 6.1   *Nucleobases*

deoxyribose
(DNA)

ribose
(RNA)

deoxyadenosine (dA)
deoxynucleoside

deoxyadenosine monophosphate
deoxynucleotide

Figure 6.2   *DNA and RNA*

The nucleosides are joined together through phosphodiester linkages between the 5'- and 3'-hydroxyl groups of consecutive sugars (Figure 6.3). The 2'-hydroxyl group of the RNA can participate as a neighboring group in the hydrolytic cleavage of the adjacent phosphodiester. This intrinsic instability of the RNA polymer is important in preventing RNA from accumulating in the cell.

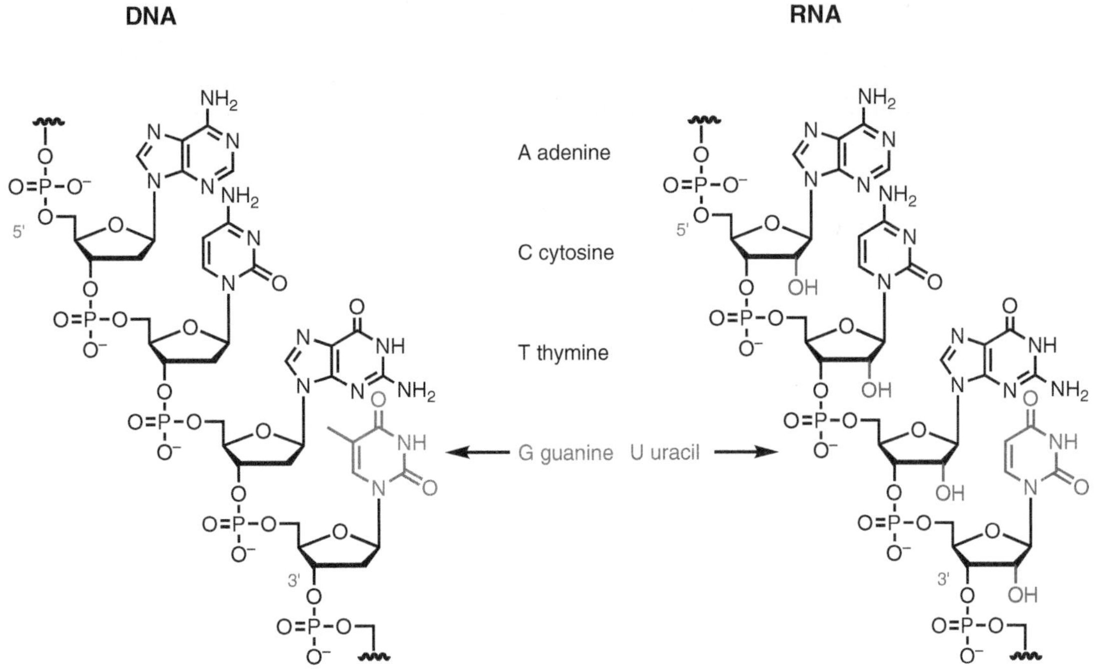

DNA   **RNA**

Figure 6.3 DNA and RNA chains

The nucleic acids differ in their stability, structure, and function. The three-dimensional structures of DNA and RNA are different, and this is the structural reason for their different functions in cells. DNA is essentially the information storage molecule in cells; the genetic information retained from one generation of cells to another is stored in the architecture of DNA. When the synthesis of a particular protein is needed, the information content of the DNA segment that codes the protein is *transcribed* into the RNA language: a messenger RNA (mRNA) is synthesised. The mRNA molecule is equipped with a number of signaling devices which relay the requisite information on the target site of the following cellular events (transport from the nucleus to cytosol, transport to the different organelles in the cytosol, information regarding whether the peptide or protein that is synthesised is to be retained within the cell or to be excreted, etc.). The RNA signal shall have a relatively short lifetime. The presence of the hydroxyl group at C2 in ribose labilises the phosphate ester through neighboring-group-assisted hydrolysis, and thus ensures that the RNA will decompose rapidly enough [2].

For the storage of genetic information, it is obviously important to have a relatively stable structure which is resistant to deleterious random chemical transformations. At the same time, however, the structure must be accessible at the time this information is processed, such as transcription to the RNA language. DNA must survive much longer periods of time under sometimes rather heavy bombardment from its environment. Through evolution, nature has developed a finely tuned system capable of conserving the information (and even repairing some of the damages caused by external agents), at the same time retaining an efficient machinery for unraveling the information rapidly and accurately as it is needed. We shall take a brief look at the structural basis for the interactions and basic recognition events that form the foundation of molecular biology, i.e. the structure of DNA.

6.2 Structural Features of DNA

The protein–DNA interactions ensure that the DNA remains protected and closely packed inside the cell nucleus. Besides proteins, a large number of small molecules also interact directly with the DNA duplex. In the following, we will briefly examine the interactions between small molecules and the Watson–Crick induced B DNA duplex [3].

The shape of the DNA molecule is dictated by the backbone of the sugar chain, i.e. the deoxyribose units connected through phosphodiester linkages. The most stable conformation of the sugar phosphodiester backbone consists of two salient structural properties. First, the phosphodiester units adopt the energetically favorable gauche-gauche conformation, and secondly, the O–C–C–O torsion angle in the deoxyribose unit is approximately 80 degrees. Together these two features impart helicity to the DNA strand, which is further strengthened by the inter-strand hydrogen bonds. With relatively small changes in these three structural parameters, one can arrive at the three basic structural motifs of double-stranded DNA, which are known as the A-, B- and Z-form double helices (Figure 6.4) [4].

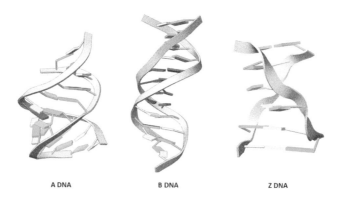

A DNA   B DNA   Z DNA

Figure 6.4   *The structures of A-, B-, and Z-forms of DNA*

The detailed structure of the B-form DNA (Figure 6.5) shows two further structural features which are of importance in recognition events. The clefts between the two carbohydrate backbone strands, the so-called major and minor grooves, provide different environments for the interacting species [5]. The grooves differ at molecular level; whereas the depths of the grooves are both approximately 8.5 Å, the major groove is nearly twice as wide as the minor groove (12 Å vs. 7 Å), thus offering comparatively strong binding to ligands (proteins and large molecules) into the major groove. The major groove also exposes the phosphodiester backbone, and this is where countercations (typically Ca^{2+}) reside. The minor groove is usually untenanted and thus available for occupation by small molecules. As most of the antibiotic and anticancer drugs are small molecules, the minor groove is their main binding site.

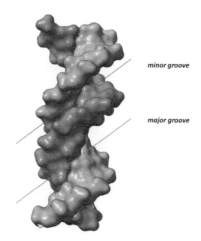

minor groove

major groove

Figure 6.5   *Minor and major grooves of B-DNA duplex*

A closer look at the nucleobases shows that all the bases contain functional groups capable of hydrogen bonding, participating either as hydrogen bond donors or acceptors. Such considerations led *James D. Watson* (1928–) and *Francis Crick* (1916–2004) to propose, in 1953, that DNA can form structures where one strand of the polynucleotide is matched with a complementary strand to form the maximum number of such inter-strand hydrogen bonds, which will impart stability to the complex of two strands relative to isolated strands [6]. According to the Watson–Crick model, the base pairing is highly regular: adenosine (A) only forms stable pairs with thymine (T); and similarly, guanine (G) pairs with cytidine (C). In the purines, the N1 and C6 sides *anti* to the oligosaccharide chain participate in the base pairing with the pyrimidines. The Watson–Crick base pair structures are shown in Figure 6.6 (dR = deoxyribose). In the figure, the hydrogen bond donors are colored blue, and the hydrogen bond acceptors red.

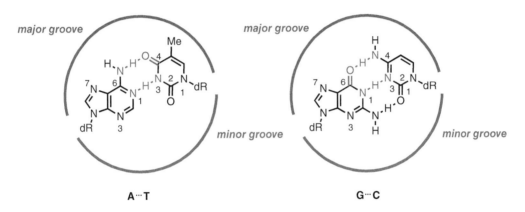

Figure 6.6 *Watson–Crick base pairing<.FC>*

In Section 6.3, we will discuss the DNA-binding natural products. The majority of reactions that involve covalent modification of DNA can be divided into two main categories: (i) those involving reactions of electrophiles with nucleophilic sites on DNA; (ii) and those involving radicals reacting with DNA. From Figure 6.6, you can see that electrophilic modification at sites N7 or N3 on purine residues are dependent on the port of entry of the electrophiles; if the initial approach occurs through the minor groove, the N3 of adenine and N3 and the exocyclic N2 amino group of guanine may be attacked. Major group-binding electrophiles react with the N7 of the purines. Electrophilic modification of the purine residues results in labilisation of the glycosidic bond, eventually leading to the rupture of the base from the sugar backbone. Under alkaline conditions, such abasic sites are rapidly converted to strand breaks in DNA. Typical radical reactions involve abstraction of hydrogen atoms from the deoxyribose backbone of DNA, and this almost always leads to the cleavage of the sugar backbone through a complex reaction cascade.

An alternative hydrogen bonding pattern, the Hoogsteen base pairing, is also possible (Figure 6.7). In the Hoogsteen base pair, the N7 of the purine base acts as a hydrogen bond acceptor and the C6 amino group as a donor, which bind the Watson–Crick (N3–C4) face of the pyrimidine base. This side is *syn* to the oligosaccharide backbone, and leaves the Watson–Crick base pairing still possible, thus allowing the formation of a triple helix by binding either a thymidine to an AT pair or a cytidine to a GC pair. The Watson–Crick base pairings are shown as green rings, the hydrogen bond donors are colored blue, and acceptors are red in the figure. Make note that the formation of the Hoogsteen base pairing occurs from the major groove side of the Watson–Crick base pairing.

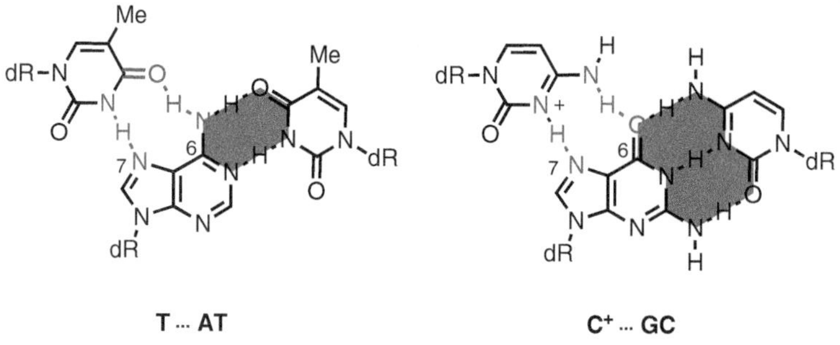

Figure 6.7 *Hoogsteen base pairing*

6.3 DNA-binding Natural Products

Depending on the nature of binding modes, anticancer drugs are broadly divided into two main categories: *non-covalent binders*, which physically interact with DNA either through the minor groove, major groove, or by intercalation between the base pairs, and thereby demonstrate their cytotoxic effects by altering DNA function temporarily; and *covalent binders*, which cause permanent damage to DNA and its functions. The mode of action of individual molecules is not this straightforward but can be a combination of several events. The compound can, for instance, navigate its way to a certain location along the DNA sequence selectively through interactions with the major groove or minor groove. Then, part of the molecule or the whole molecule can slip in between the base pairs to intercalate, and finally react chemically with the bases to cross-link them. We shall now examine each of these categories through some examples.

6.3.1 Minor-groove Binding Agents

The minor-groove binding agents have concave-shaped aromatic framework that fits in the convex minor groove and has functional groups capable of hydrogen bonding. The high negative electrostatic potential of helical B-DNA in the AT-rich regions interacts with the protonated amines of ligands under physiological pH. The complex is further stabilised by strong van der Waals interactions [7].

The typically crescent shapes of these molecules are apparent even in the two-dimensional structures of natural products netropsin, distamycin, and antibiotic CC-1065, which bind selectively to AT-rich sequences (Figure 6.8). Mithramycin (also known as plicamycin), an antibiotic produced by *Streptomyces plicatus*, binds to GC-rich sequences through the minor groove.

The hydrophobic minor groove provides a very snug fit for hydrophobic 'flat' molecules, as exemplified by the X-ray structure of netropsin bound to DNA (Figure 6.9) [8]. The structure of distamycin is composed of three pyrrole rings joined by amide bonds, which provide the hydrogen-bonding partners. The amidino side chain plays an important role in the electrostatic interactions with double-helical DNA. The crescent shape assures the amide group N–Hs facing towards the base of the minor groove. Finally, the positively charged amidine group at the end enables it to interact through electrostatic and/or hydrogen bonding with the phosphate while the remaining part is inserted into the minor groove for interaction with adenine–thymine (AT)-rich sequences.

Enediyne antibiotics are a scarce group of natural products, and only 13 have been structurally characterised over the past 30 years. Figure 6.10 shows a few common examples. With the exquisite mode of action and the extraordinary cytotoxicity, the enediynes have been successfully translated into clinical drugs, including FDA-approved antibody-drug conjugates (ADCs) (e.g. Mylotarg® (gemtuzumab ozogamicin) is a compound where *N*-acetyl-calicheamicin γ_1^I is covalently linked with a hydrolysable linker to a humanised murine antibody that targets cells with the CD33 myeloid antigen. The antibiotic payload is released selectively at the target organ).

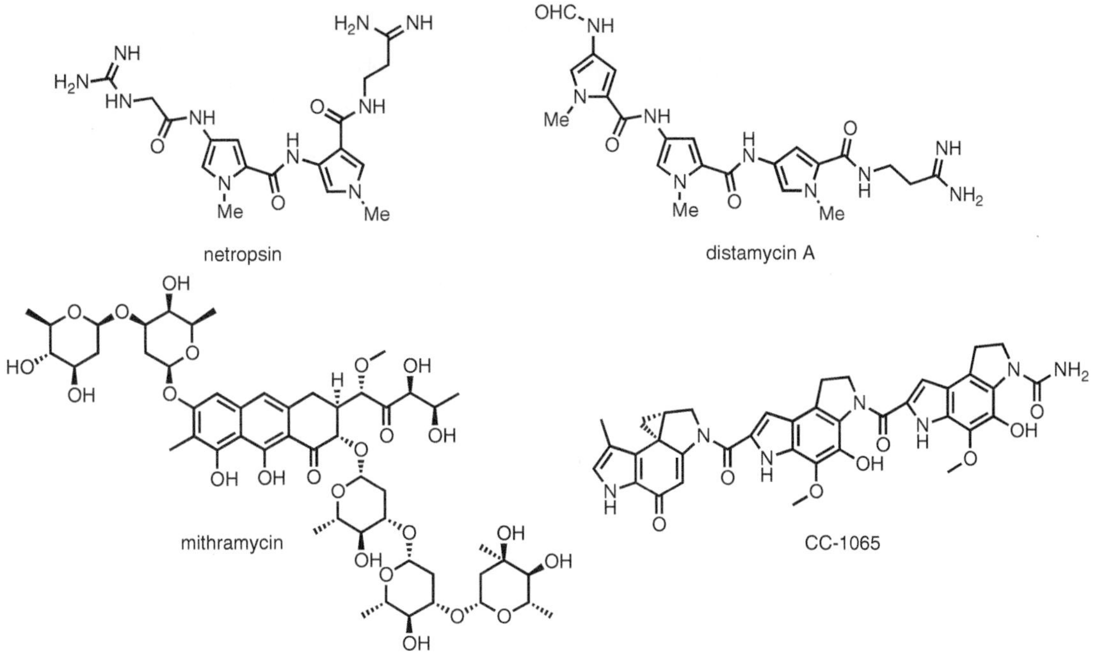

netropsin

distamycin A

mithramycin

CC-1065

Figure 6.8   *Minor-groove binding agents*

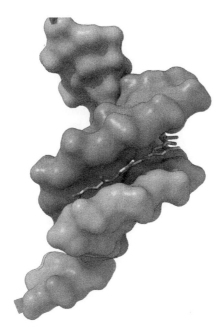

Figure 6.9   *Netropsin bound to the minor groove of the B-form DNA (1Z8V)*

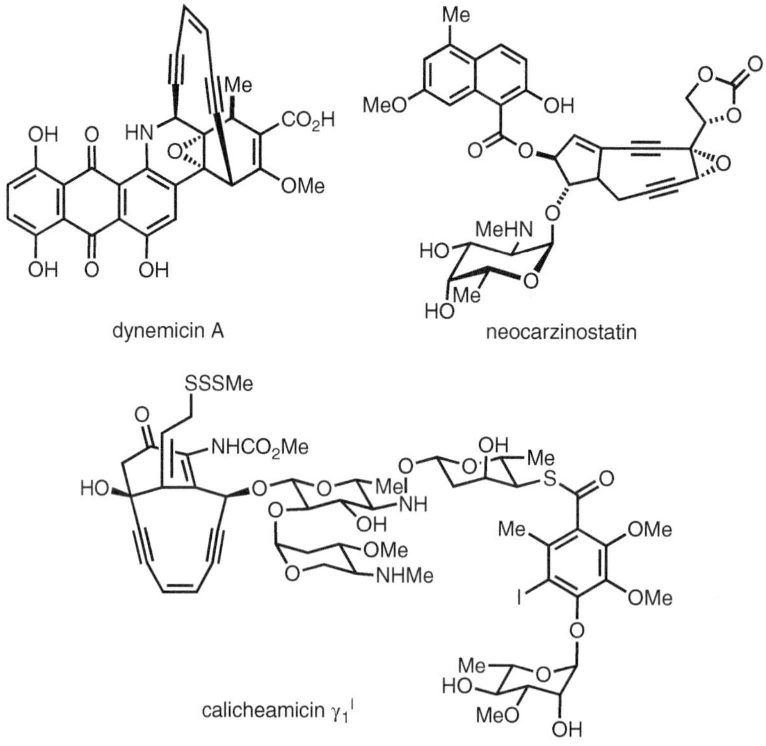

dynemicin A

neocarzinostatin

calicheamicin γ$_1$^{I}

Figure 6.10   *Enediyne antibiotics*

The enediynes share a common mode of action; they undergo an electronic rearrangement to generate a transient benzenoid diradical. This is illustrated in Scheme 6.1 for the neocarzinostatin molecule. The trisulfide is initially attacked by a nucleophile and a thiol is released. This thiol attacks in a conjugate fashion to the enone to ring-close the five-membered ring. Change of hybridisation from sp^2 to sp^3 brings the termini of the enediyne closer, which facilitates the Masamune–Bergman cyclisation to produce a benzenoid diradical. When positioned within the minor groove of DNA, the diradical abstracts hydrogen atoms from the deoxyribose backbone to afford DNA-centered radicals that cause interstrand crosslinks, or react with molecular oxygen leading to DNA double-strand breaks, or both.

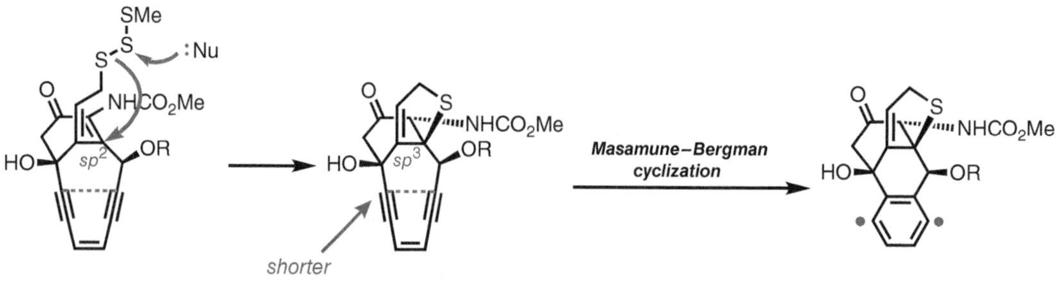

Scheme 6.1   *Masamune-Bergman cyclisation in the neocarzinostatin*

6.3.2 Major-groove Binding Agents

Larger proteins, and occasionally a third DNA strand, can only bind to the more hydrophilic *major groove*, but a limited group of small molecules also interact with the DNA through the major groove [9]. Then naphthoquinone derivatives altromycin B and hedamycin, which belong to the family of pluramycins, react with the DNA alkylating the N7 of guanine residues (Figure 6.11). In a similar vein, azinomycin B from *Streptomyces* strains interacts with duplex DNA and crosslinks guanine N7 amino groups at the epoxide and aziridine functionalities. The covalent attachment of the strands leads to cell cytotoxicity.

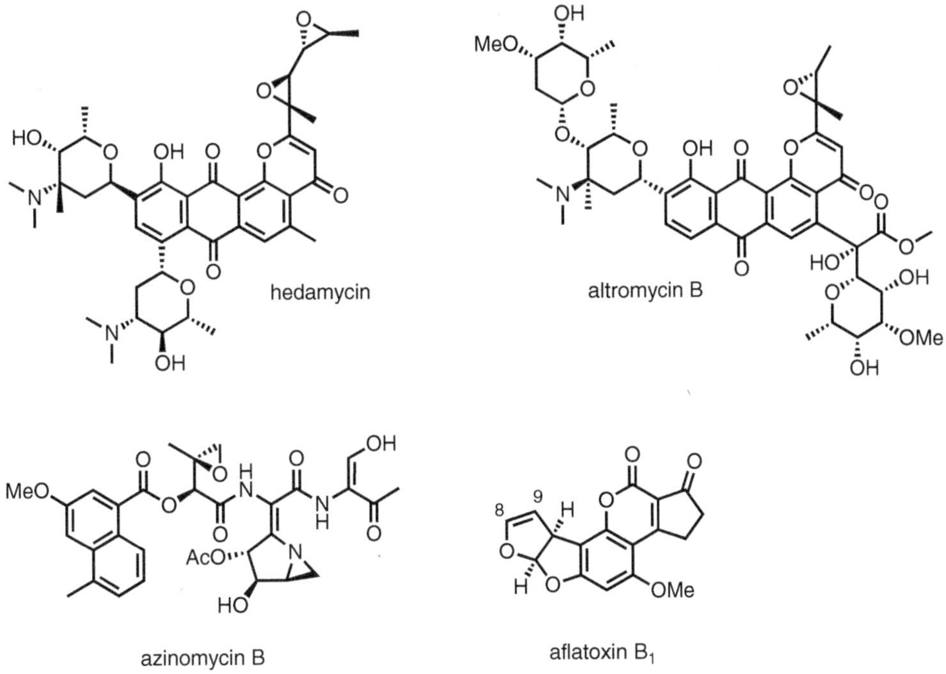

hedamycin

altromycin B

azinomycin B

aflatoxin B$_1$

Figure 6.11   *Major-groove binding agents*

Aflatoxins are poisonous carcinogens and mutagens that are produced by certain fungi (*Aspergillus flavus* and *Aspergillus parasiticus*) which grow in soil, decaying vegetation, hay, and grains. Large doses of aflatoxins lead to acute poisoning (aflatoxicosis) that can be life-threatening, usually through liver damage. Afltoxin B$_1$ is first oxidised to the epoxide at the C8–C9 double bond. This epoxide then reacts with guanine N7. Figure 6.12 shows an X-ray structure of this aflatoxin adduct with DNA polymerase IV [10].

6.3.3 Cross-linking Agents

We have already encountered molecules that will react covalently with the DNA strands to bind the two strands together. The enediynes are brought to the vicinity of the bases through the minor groove, whereas azinomycin B was an example where the natural product first finds its way to the nucleobases through binding to the major groove. Simple alkylating agents like nitrogen mustard mustine, chlorambucil, and melphalan act by this mechanism. However, in this section we will take a look at the interesting mechanism displayed by mitomycins.

Mitomycins are a group of aziridine-containing natural products isolated from *Streptomyces* species. The biosynthesis route resembles the shikimic acid pathway (Scheme 6.2). First, phosphoenolpyruvate (PEP) reacts with erythrose-4-phosphate (E4P) and the product is aminated. Ring closure catalysed by dihydroquinate synthase gives aminodehydroquinate. Double oxidation gives aminodehydroshikimate and aromatisation gives

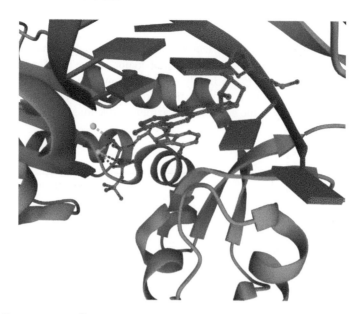

Figure 6.12   *Aflatoxin B₁ metabolite bound to the DNA groove (3PW4)*

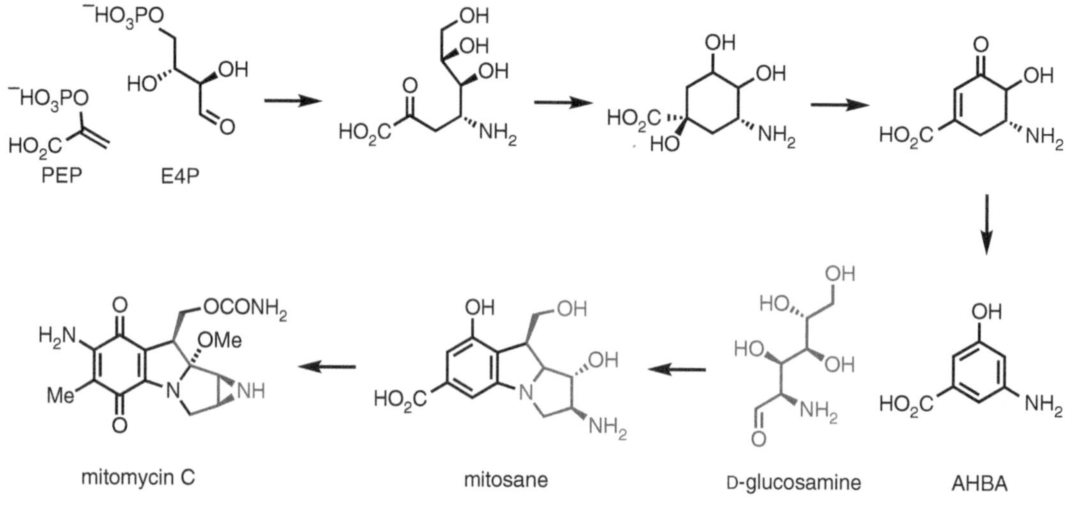

Scheme 6.2   *Biosynthesis of mitomycin C*

3-amino-5-hydroxybenzoic acid (AHBA). Condensation with glucosamine, mediated by yet unknown enzymes, gives the mitosane skeleton which is then elaborated to mitomycin C.

The alkylation of genomic deoxyguanosine residues at the 7-position is considered to be the primary event in the chemically induced carcinogenesis/mutagenesis resulting from the action of agents such as aflatoxins and N-mustards, as well as the mode of action of some DNA-alkylating agents used in cancer chemotherapy. Mitomycins are such DNA-alkylating antitumor antibiotics that bind through the minor groove and alkylate the DNA strands (Scheme 6.3) [11]. Rupture of the deoxyguanosine leads to loss of the nucleotide, thus causing destruction of the secondary structure of DNA at the alkylation site.

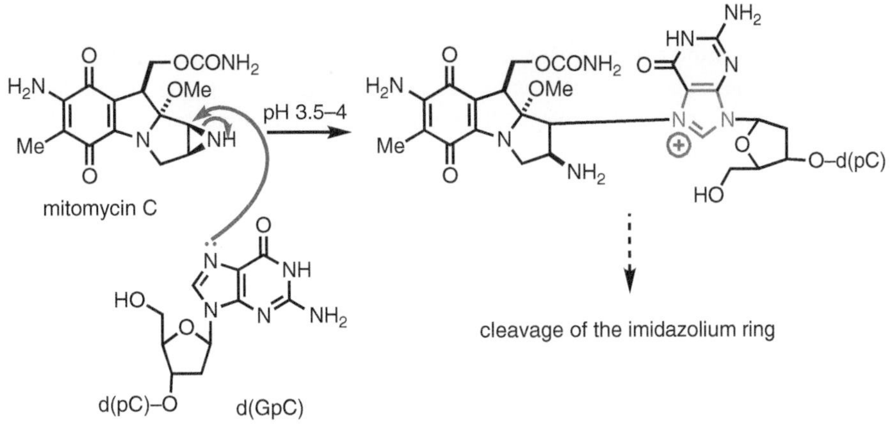

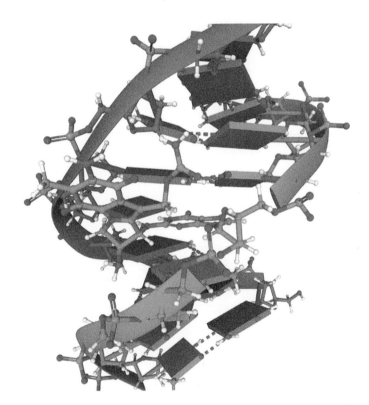

Scheme 6.3 *Mitomycins induce cleavage of DNA*

Figure 6.13 *Mitomycin C bound to DNA (NMR structure 199D)*

The solution structure of the monoalkylated mitomycin C-DNA complex has been determined by extensive NMR-molecular dynamics studies (Figure 6.13) [12]. The covalent linkage from the guanidine N-7 is clearly visible, as is the fact that the carbamoyl group is positioned opposite to the cytidine in the adjacent chain. The mitomycin C-ring remains positioned in the minor groove with the indoloquinone aromatic ring system at an angle of approximately 45° relative to the helix axis and directed towards the 3'-direction on the unmodified strand.

6.3.4 DNA-intercalating Agents

The minor groove also provides the entry site for smaller, typically aromatic molecules which insert between two consecutive 'layers' of the bases. This *intercalation* causes a structural change in the double helix leading to lower stability and higher susceptibility to strand cleavage. Typical intercalating natural products are shown in Figure 6.14. The enediyne antibiotics (e.g. calicheamicin g_1^I, see Section 6.3.1) initially bind into the minor groove, and when the enediyne has undergone the Masamune−Bergman cyclisation, the flat diradical can intercalate between the base pairs and cause rupture of the DNA through radical reactions. Many classical anti-tumor compounds such as the anthracyclinones, e.g. daunomycin and adriamycin (Section 4.6), and alkaloids such as ellipticine, exert their action this way.

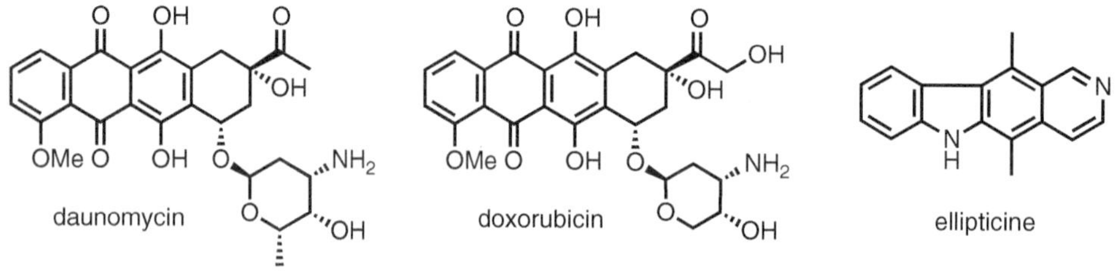

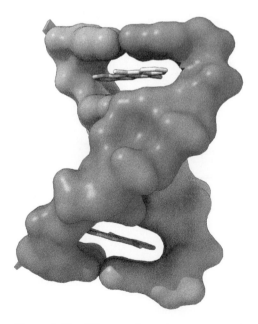

daunomycin   doxorubicin   ellipticine

Figure 6.14   *DNA-intercalating agents*

Figure 6.15 displays the binding of the alkaloid ellipticine to DNA. One prevalent mechanism of action of ellipticine antitumor, mutagenic, and cytotoxic activities has been suggested to be due to the intercalation into DNA and inhibition of DNA topoisomerase II activity.

Figure 6.15   *Ellipticine bound to DNA (1z3f)*

6.4 Nucleosides and Energy

The intracellular energy storage system is provided by adenosine triphosphate, ATP. In this compound, adenosine is esterified with a high-energy triphosphate unit whose hydrolysis liberates a considerable amount of energy (*ca.* 31 kJ/mol). This energy can be used in, e.g. protein synthesis, where the formation of each peptide bond consumes *ca.* 2 kJ/mol.

In biological systems, one often finds formal hydride reductions, which resemble usual sodium borohydride or lithium aluminum hydride reductions. One biological reductant capable of hydride delivery is nicotinamide adenine dinucleotide hydride (NADH), which functions as a coenzyme in such reactions (see Figure 6.16). The oxidised form, nicotinamide adenine dinucleotide (NAD) is a coenzyme found in all living cells. NAD is a dinucleotide because it consists of two nucleotides joined through their phosphate groups. One nucleotide contains an adenine nucleobase and the other nicotinamide. NAD is involved in redox reactions, carrying electrons from one reaction to another. The cofactor is found in two forms in cells; the oxidised form NAD^+ accepts electrons from other molecules and becomes reduced to NADH, which can then be used as a reducing agent to donate electrons. These electron transfer reactions are the main function of NAD.

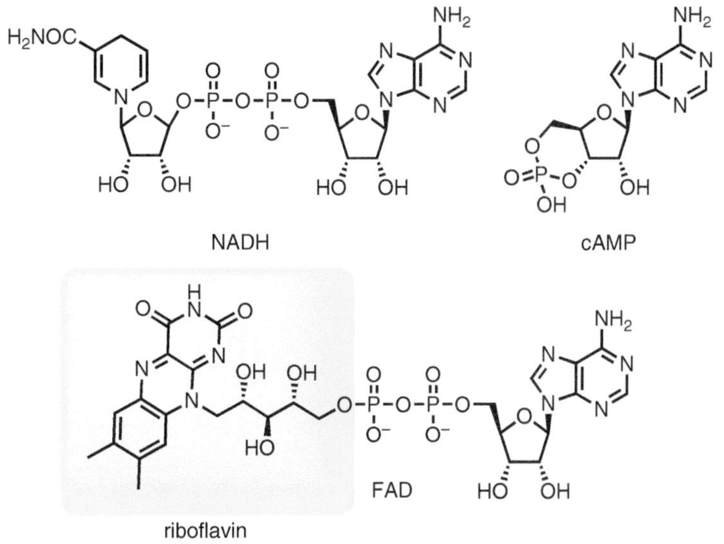

Figure 6.16 *NADH and cAMP*

Cells communicate with each other by passing chemical messenger molecules (first messengers), such as hormones or neurotransmitters. Peptide hormones and neurotransmitters are typically hydrophilic molecules, and thus cannot physically cross the phospholipid bilayer to induce action within the cell directly (steroid hormones can pass through the cell membrane). This functional limitation requires the cell to have signal transduction mechanisms to convert the extracellular first-messenger signal into intracellular second-messengers, which are released into the cytosol when the external factor is bound to a membrane-embedded receptor. Thus, the signal may be propagated into the cell to trigger physiological changes at cellular level such as proliferation, differentiation, migration, survival, apoptosis, and depolarisation.

In chemical signal transduction, many transmitters function by coupling their action with the hydrolysis of cyclic adenosine monophosphate (cAMP). An important and large group of proteins called G-protein-coupled receptors (GPCRs), or seven-transmembrane domain receptors (7-TM), is dependent on either cAMP or phosphatidylinositol.

These receptors are activated by hormones, neurotransmitters, pheromones, odors, and light-sensitive compounds. Nearly one-third of currently used pharmaceuticals target GPCRs.

Flavin adenine dinucleotide (FAD), along with flavin mononucleotide (FMN, riboflavin-5'-phosphate), is an important cofactor in many enzymatic reactions. FAD and FMN both originate from riboflavin (vitamin B_2). Bacteria, fungi, and plants can produce riboflavin, but other eukaryotes, including humans, have lost the ability to make it. Therefore, humans must obtain riboflavin from dietary sources. FAD is a redox-active coenzyme with a more positive reduction potential than NAD^+ and is therefore a very strong oxidising agent. It is involved in several enzymatic reactions including components of the succinate dehydrogenase complex, α-ketoglutarate dehydrogenase, fatty acyl CoA dehydrogenase, and a component of the pyruvate dehydrogenase complex.

FAD can exist in four redox states, which are the flavin-N(5)-oxide, quinone, semiquinone, and hydroquinone. FAD, in its fully oxidised quinone form, accepts two electrons and two protons to become $FADH_2$ (hydroquinone form).

6.5   Nucleoside Antibiotics

There are currently more than 30 nucleoside or nucleotide analogs approved for treating viruses, cancers, and parasites, as well as bacterial and fungal infections. The development of nucleoside analogue antivirals has been recently reviewed [13]. Some of the first nucleosides found to have medicinal properties were arabinose (Ara) derivatives. The first Ara nucleotides were natural products isolated from the Caribbean sponge *Tethya crypta*. These analogues, spongothymidine and spongouridine, never became useful drugs. However, they inspired the synthesis of several analogues, including AraC or Cytarabine, which is included in the Model List of Essential Medicines[14] by the World Health Organization. Cytarabine has pronounced activity against many cancers including non-Hodgkin's lymphoma, and most importantly against several leukemias such as myeloid leukemia, acute lymphatic leukemia, and chronic myelogenous leukemia (Figure 6.17). The uridine derivative doxifluridine has been brought to market in China and South Korea, and it shows excellent efficiency against forms of breast, stomach, and intestinal cancers. The first AIDS drug, 2'-deoxy-3'-azidothymidine (AZT, zidovudine), is also a nucleoside analogue. However, AZT suffered from various side effects and problems with toxicity and, most importantly, the development of viral resistance.

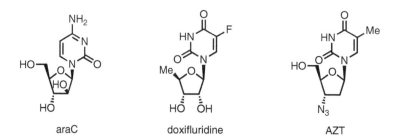

Figure 6.17   *Pyrimidine nucleoside antibiotics*

Puromycin is an adenosine derivative which inhibits the translation of RNA in protein synthesis. The compound is an effective antibiotic against trypanosomiasis and amaebosis, which are diseases typical of tropical regions (trypanosomiasis accounts for roughly 25% of deaths in these regions). The compound, however, is very toxic, and therefore cannot be widely used in preventive medicine.

Nucleobases can also combine with several rare sugars, amino sugars, and amino acids to form complex nucleoside antibiotics [15]. Polyoxins and nikkomycins (Figure 6.18) are peptidyl nucleosides, where the sugar terminus contains an amino acid structure which is further linked to another amino acid. Polyoxins are agricultural fungicides, and they were the first nucleoside antibiotics found to inhibit fungal cell wall chitin biosynthesis. Their structures mimic UDP-N-acetylglucosamine, a substrate for chitin synthetase. Nikkomycins were isolated from culture filtrates of *Streptomyces tandae*. Acaricidal and insecticidal activities have been reported for these compounds. Capuramycin is a unique uracil nucleoside having an unsaturated uronic acid as well as a hydroxyacid in the sugar terminus. It showed only weak antibacterial activities [16]. Biosyntheses of these classes of nucleoside antibiotics have been reviewed [17].

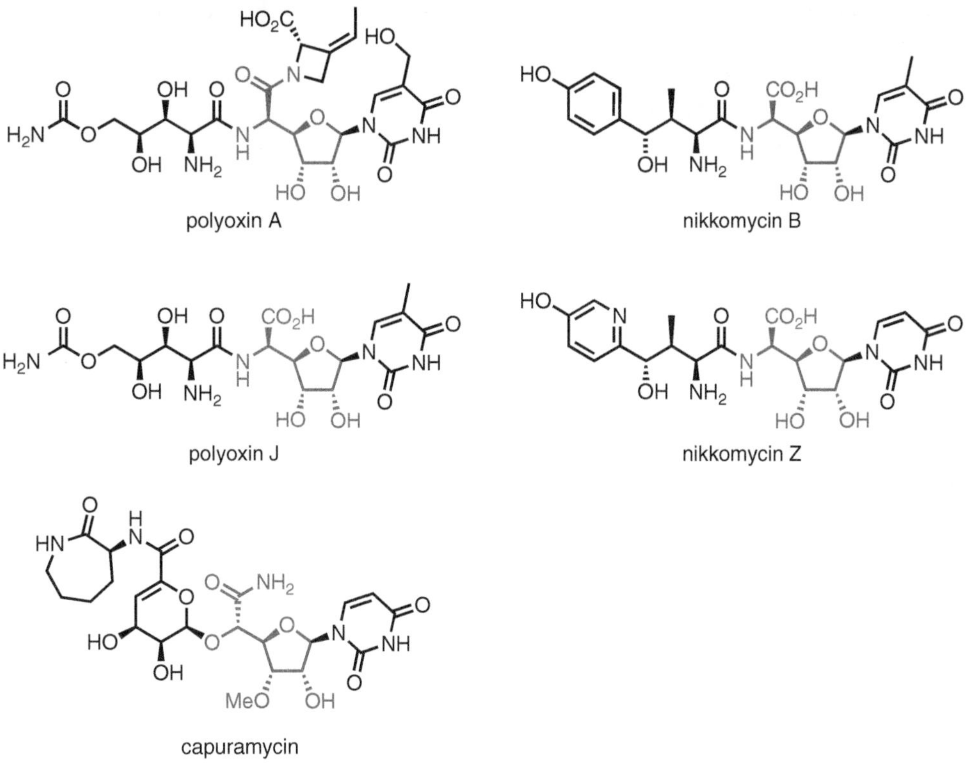

Figure 6.18   *Pyrimidine peptidyl nucleoside antibiotics*

Recently, the discovery of natural nucleoside antibiotics has been accelerated because of the need to combat the increase in global antibiotic resistance. Several nucleoside antibiotic biosynthetic gene clusters (BGCs) of actinobacterial origin have been reported thanks to the development of genome sequencing technology. This has resulted in significant advancements for yield enhancement of nucleoside antibiotic production, as well as the production of designer nucleoside antibiotics with improved or alternative biological properties. Pyrimidine nucleoside antibiotics hikizimycin, tunicamycin, gougerotin, and mureidomycin A (Figure 6.19) are examples of compounds containing both rare amino sugars and amino acids [18].

Biological methylations are involved in several biosynthetic processes. The methyl group is derived from *S*-adenosyl methionine[19] (SAM, for structure, see Scheme 5.8, Yang cycle), and a number of methyl transferases catalyse these reactions. Consequently, analogues of SAM could have potential as methyltransferase inhibitors. Sinefungin, an antifungal nucleoside isolated from strains of *Streptomyces griseolus*, shows antifungal and antimalarial activities. It is a structural analogue of SAM. Herbicidin, isolated from the actinomycete *Streptomyces saganonensis*, is a nucleoside antibiotic showing herbicidal activity. The 11-carbon uronic acid structure of herbicidin is biosynthesised by the condensation of adenosine and glucose. These purine nucleoside antibiotics are shown in Figure 6.20 (for biosynthesis, see ref. 17c).

The nucleobase portion can also be modified. Tubercidin and toyocamycin are examples where a nitrogen atom has been deleted from the purine base. In formycin, one nitrogen is shifted to obtain a C-glycoside. In coformycin and deoxycoformyin (pentostatin), the adenine ring has expanded to a diazepine ring, gaining the extra carbon atom from the C1 of D-ribose. The latter compounds are tight-binding naturally occurring inhibitors of adenosine deaminase, a

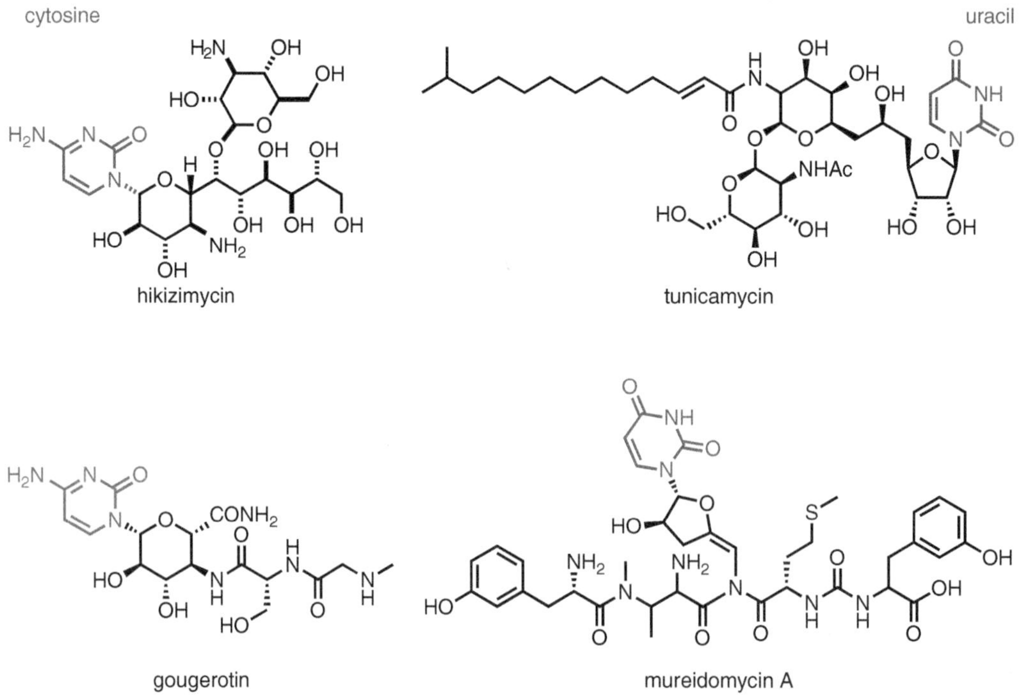

Figure 6.19 *Pyrimidine nucleoside antibiotics*

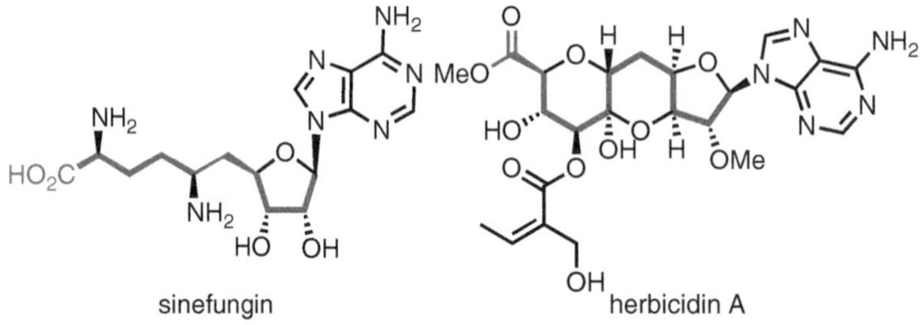

Figure 6.20 *Purine nucleoside antibiotics with a unique uronic acid structure*

ubiquitous mammalian enzyme with a central role in the purine salvage pathway. Pentostatin is used to treat hairy cell leukemia. It has been used to treat steroid-refractory acute and chronic graft-versus-host disease, and it is also used in chronic lymphocytic leukemia (CLL) patients who have relapsed.

In 1966, an unusual nucleoside with a cyclopentyl ring in place of the furanose sugar moiety was synthesised (Figure 6.21) [20]. This analogue, later named aristeromycin, was subsequently isolated from *Streptomyces citrocolour* and found to have potent antiviral properties against viruses such as measles, parainfluenza, vaccinia virus, and vesicular stomatitis. Similarly, a closely related structural analogue was first isolated from *Ampullariella regularis* in 1981, and later synthesised and named neplanocin A. Similar to aristeromycin, neplanocin A also has a cyclopentyl ring in place of the furanose sugar; however, in neplanocin A the cyclopentyl ring possesses a double bond between C4' and C6'. Both neplanocin A and aristeromycin were highly active; however, they were also quite toxic, as their triphosphate forms closely mimicked ATP, thus leading to misincorporation and disruption of many important biological systems, and thus their development was discontinued.

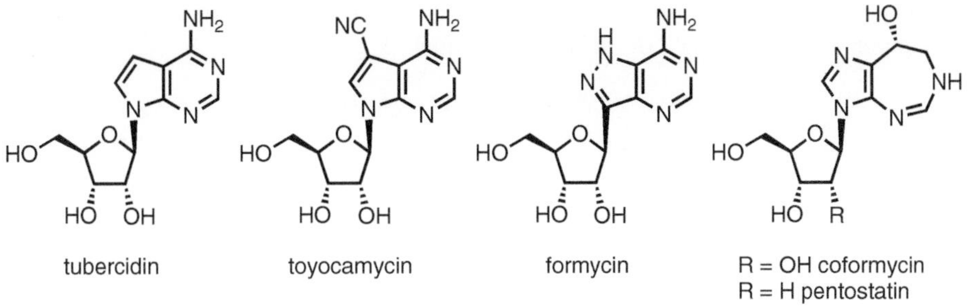

Figure 6.21   *Purine-modified nucleoside antibiotics*

Carbocyclic nucleoside analogues are some of the most potent inhibitors against *S*-adenosyl-L-homocysteine hydrolase (SAHase) – a key enzyme to many biological methylations. (−)-Carbovir proved to be a potent anti-HIV therapeutic, but its cytotoxicity, low solubility, and poor oral bioavailability limited its use in clinical applications. In order to alleviate some of these issues, further studies led to the development of abacavir, a carbocyclic analogue with a unique modification on the exocyclic amine group at the 6-position of the nucleobase. The cyclopropylamino group is eventually hydrolysed to give the bioactive (−)-carbovir 5'-triphosphate, thus making abacavir a prodrug of the active form of (−)-carbovir (Figure 6.22).

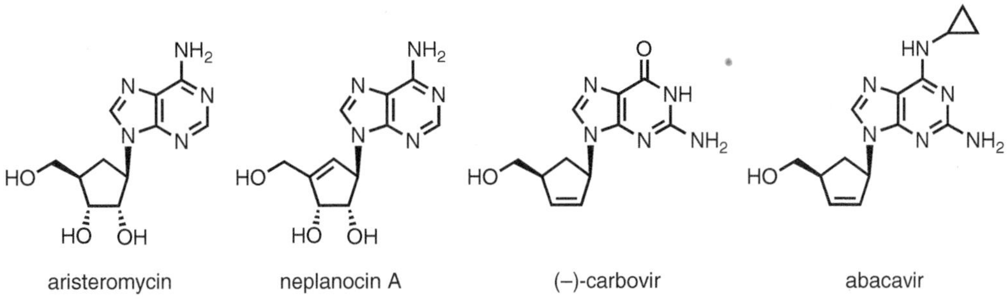

Figure 6.22   *Purine carbanucleoside antibiotics*

6.5.1   Synthesis of Nucleoside Antibiotics

Because of the acknowledged importance of nucleoside antibiotics and especially their synthetic modifications, it is clear that the literature contains a large number of enantioselective syntheses of these compounds. In the following sections, we will highlight some syntheses of these compounds to showcase different approaches of asymmetric synthesis that were covered in Chapter 3. Many efficient syntheses are omitted, and especially the ones starting from chiral-pool starting materials will be left out.

6.5.1.1   *Hikizimycin*

Inspection of the intrinsic symmetry of the target compounds can often lead to extremely useful insights in the synthesis design. Realisation of the latent symmetry[21] in the molecules and application of two-directional synthesis strategies[22] can provide creative new ways of constructing even highly complex molecules. A pioneering example is discussed in Chapter 7.3.2 in Still's synthesis of the ansa bridge of rifamycin S [23]. Here, we follow the hikizimycin synthesis by Schreiber (Scheme 6.4)[24] The original source of chiral information comes from L-diisopropyl tartrate, which was elongated through olefination of the dialdehyde to the bisenoate. Highly diastereoselective Upjohn dihydroxylation following the Kishi rule[25] gave the tetraol with excellent selectivity. After terminus differentiation, the dihydroxylation was applied for a second time, and again the diastereoselectivity was superb. The final conversion to hikizimycin served to confirm the structure of the natural product.

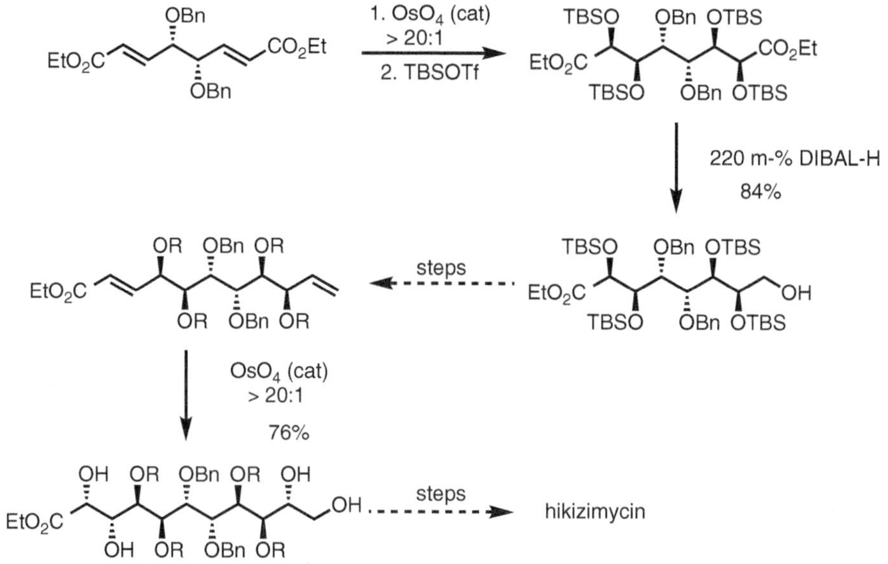

Scheme 6.4 Schreiber two-directional synthesis of hikizimycin

6.5.1.2 Sinefungin

The synthesis of sinefungin gained added interest, since the compound itself was too toxic in clinical studies. We will briefly inspect four different strategies to introduce the correct stereochemistry at C6, the isolated stereocenter, whose absolute stereochemistry required synthesis to be proven.

Three synthetic approaches were developed already in the early and mid-1980s, however, without much attention paid to the stereochemistry at this crucial C6 center. The first synthesis to pay attention to this problem was by *Henry Rapoport* (1918–2002) in 1990 (Scheme 6.5) [26]. The synthesis was based on utilisation of commercial chiral-pool starting materials, L-ornithine and D-ribose, as well as adenine. The L-ornithine-derived nitroalkane reacted with the D-ribose-derived aldehyde in a potassium–fluoride-catalysed nitro aldol reaction to give an inseparable mixture of

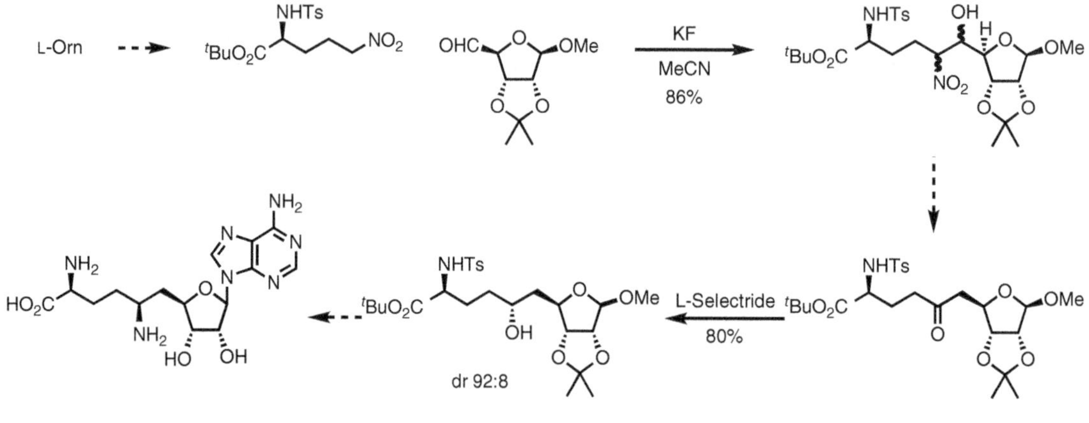

Scheme 6.5 Rapoport synthesis of sinefungin

diastereomers. These were converted to the key intermediate ketone, whose reduction with L-selectride gave the desired alcohol diastereomer with high selectivity. This selectivity was surprising and serendipitous, as different protecting groups on the nitrogen gave poorer results (e.g. Boc: 64:36) and changing the borane to sodium borohydride gave a 55:45 mixture. The tosylate derived from the alcohol underwent a smooth inversion with azide to give the proper stereochemistry at the then-unconfirmed chiral center. Standard protocols to introduce the adenine and remove the protecting groups finally gave sinefungin.

Radical reactions are seldom used in asymmetric syntheses, but an elegant example is provided by the Barton synthesis of sinefungin [27]. Barton's pyridinethione-mediated decarboxylative radical generation from carboxylic acids was utilised twice in the following synthesis (Scheme 6.6). Thus, irradiation of the *N*-hydroxythiopyridone derivative of the adenosine acid gave the radical, which underwent an addition reaction with the acrylamide derived from aspartic acid. Reductive desulfurization led to a 1:1 mixture of epimers at C6, which could be isolated and converted individually through Curtius rearrangements to the sinefungin skeleton. Removal of protecting groups gave sinefungin.

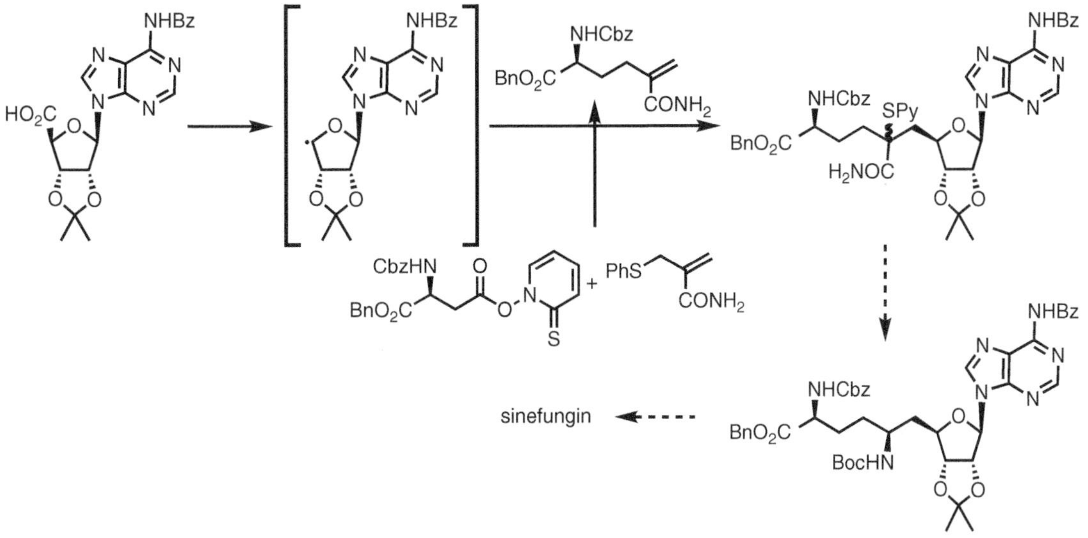

Scheme 6.6 *Barton synthesis of sinefungin*

The example of sinefungin synthesis by Ghosh illustrates the use of combined tactics in the construction of the stereocenters (Scheme 6.7) [28]. Ribose was used as the obvious core for the chiral part of sinefungin. Oxidation of the primary alcohol to an aldehyde was followed by a Horner–Wadsworth–Emmons reaction. Reduction of the double bond, ester hydrolysis, and conversion to a chiral oxazolidinone derivative set the stage for a highly diastereoselective allylation (a single diastereomer was observed). After replacement of the chiral auxiliary with a Cbz group through a Curtius rearrangement, followed by further *N*-benzyl protection, the terminal alkene was converted to a 5:1 *Z:E* mixture of *N*-acetyl enamides. [Rh(COD)(R,R-DIPAMP)]$^+$BF$_4$$^-$-catalysed asymmetric hydrogenation[29] of the *E/Z*-mixture underwent a clean catalyst-controlled hydrogenation to a single *S*-enantiomer at C9 in 95% yield and 98% *de* [30]. Final conversion of this intermediate to sinefungin utilised the powerful Vorbruggen reaction to achieve the adenylation with high β-selectivity [31].

The Shair synthesis of 9'-epi-sinefungin, a new analogue of sinefungin, was based on a catalytic asymmetric Overman rearrangement of trichloro acetimidate to set the C6 stereochemistry (Scheme 6.8) [32]. D-Ribose acetonide was first triflated and then alkynylated with a propargylic nucleophile. Cleavage of the TBS-protection, hydroalumination of the alkyne to the *E*-allylic alcohol, and onversion to the trichloroacetimidate set the stage for the Overman rearrangement. Treatment of the *E*-allylic trichloroacetimidate in CH$_2$Cl$_2$ with Overman's [R–COP–Cl][33] effected the desired [3,3]-sigmatropic rearrangement, delivering allylic trichloroacetamide in 85% yield with a 14:1 dr. Wacker–Tsuji oxidation with *tert*-butyl nitrite as an organic redox cocatalyst then gave the terminal aldehyde.

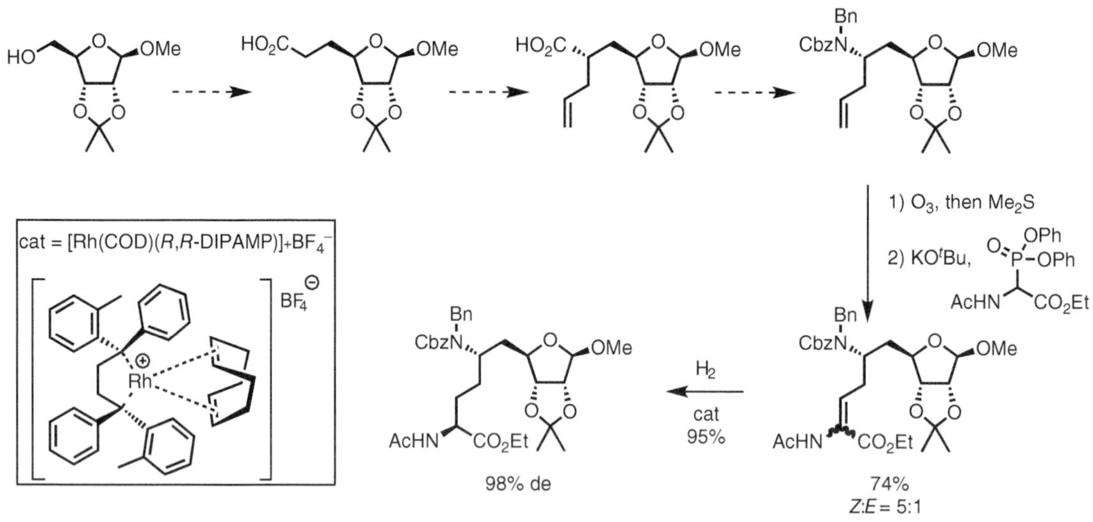

Scheme 6.7 *Ghosh synthesis of sinefungin*

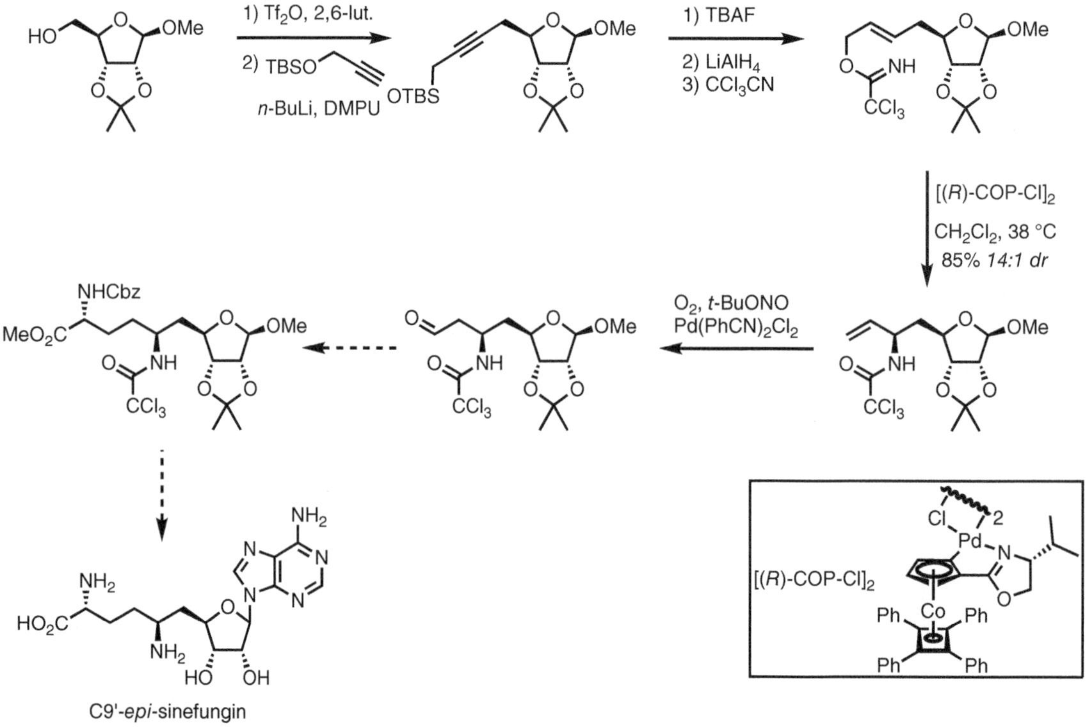

Scheme 6.8 *Shair synthesis of C9′-epi-sinefungin*

A Horner–Wadsworth–Emmons olefination with Cbz-α-phosphonoglycine trimethyl ester yielded the enamide, and the C9'–amino stereocenter was set using a rhodium-catalysed asymmetric hydrogenation with (S,S)–Ph–BPE–Rh, a Burk-type 1,2-bis(phospholano)ethane ligand,[34] providing R-carbamate as a single diastereomer in 70% yield. Finally, Vorbrüggen glycosylation followed by protecting-group removals provided 9'-*epi*-sinefungin.

6.5.1.3 Pentostatin

In this section, we will compare two syntheses of the tetrahydroimidazodiazepinol nucleoside antibiotic pentostatin. The first, by Rapoport, is based on the use of chiral starting materials, thus relying on internal asymmetric induction [35]. Using L-vinylglycine, readily available enantiopure in quantity from L-methionine, was used as the starting material. Diastereoselective epoxidation gave the desired *syn*-epoxide in 70% isolated yield along with 17% of the *anti*-epoxide. After exchange of the nitrogen protecting group, the epoxide was opened with azide. After conversion of the ester functionality to a nitrile, ring closure to the amino imidazole was achieved by reaction with an imidate ester. Final ring closure to the diazepine was achieved by first converting the amino group to an imidate and then reducing the azide with propanedithiol which also affected the ring closure. The pentostatin/coformycin aglycone was thus obtained in *e.r.* > 99:1 (Scheme 6.9).

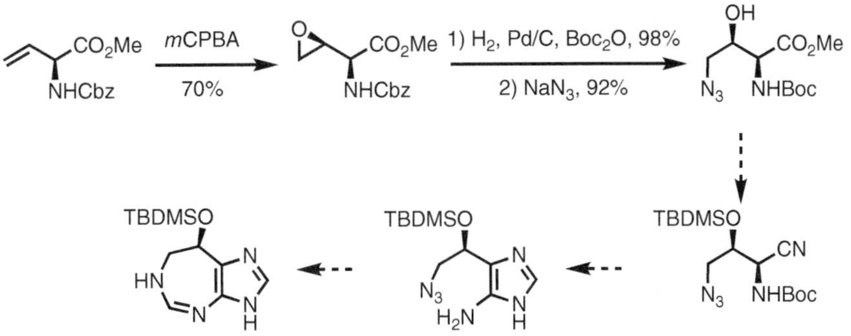

Scheme 6.9 *Rapoport synthesis of the pentostatin aglycone*

The second example is one of external asymmetric induction, and it is the industrial production method by Synbias Pharma (Scheme 6.10) [36]. The synthesis starts with enzymatic exchange of the nucleobase in deoxyuridine for 6,7-dihydroimidazo-[4,5-*d*]-[1,3]diazepin-8(3*H*)-one catalysed by commercial nucleoside deoxyribosyltransferase in 78% yield. The crucial stereochemistry-defining step is the reduction of the ketone to an alcohol, which was achieved with the Noyori RuCl(p-cymene)[(R,R)-Ts-DPEN] catalyst in 83% yield and > 99:1 enantioselectivity.

6.5.1.4 Caprazamycin

Tuberculosis (TB) remains a serious infectious disease, especially with the emergence of extensively multidrug-resistant TB (XDR-TB) for which there is a paucity of effective medications. In 2003, a group of liponucleoside antibiotics called caprazamycins were reported as promising anti-TB natural products. Caprazol, with its diazepinone ring, forms the core structure of caprazamycins, and is itself also produced during the fermentation process for caprazamycin production. The first asymmetric synthesis of caprazol was reported by Matsuda and Ishikawa in 2007 (Scheme 6.11) [37]. The synthesis started with a protected uridine derivative which was converted in a few steps to the α,β-unsaturated ester. The key stereocenters of the amino alcohol were introduced using the Sharpless asymmetric aminohydroxylation reaction, which gave a diastereomeric mixture of the desired product and its diastereomer (not shown) in an isolated ratio of 52:9.

In the Shibasaki synthesis of caprazol (Scheme 6.12), the same stereocenters were introduced using a diastereoselective isocyanoacetate aldol reaction [38]. It is noteworthy that the uridine NH did not require protection. Thus, treatment

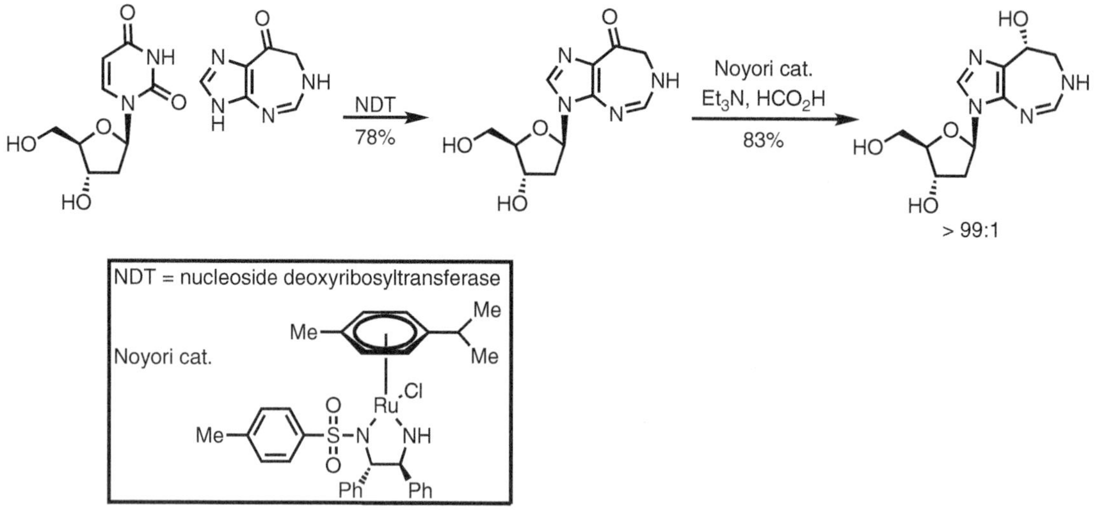

Scheme 6.10 *Synbias Pharma synthesis of the pentostatin aglycone<.FC>*

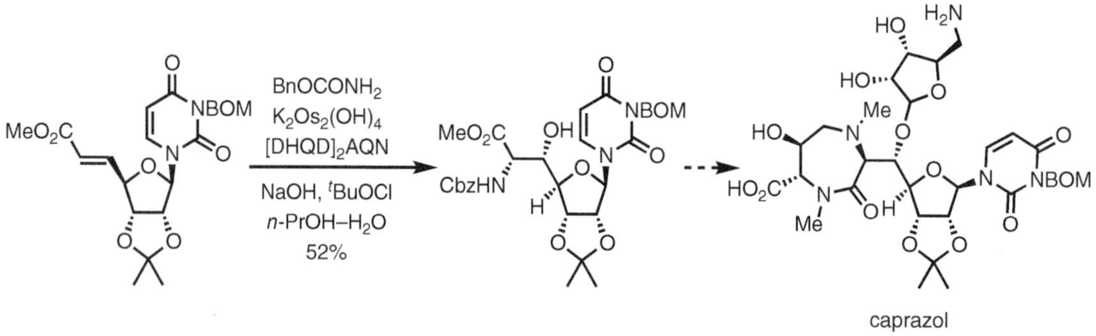

Scheme 6.11 *Asymmetric aminohydroxylation route to caprazol*

of the uridine aldehyde with 5 mol-% CuCl and 10 mol-% of Ph₃P allowed for the aldol process to proceed cleanly, with high diastereoselectivity. The oxazoline ring was cleaved and the amino group protected to give an intermediate resembling that of the Ishikawa synthesis.

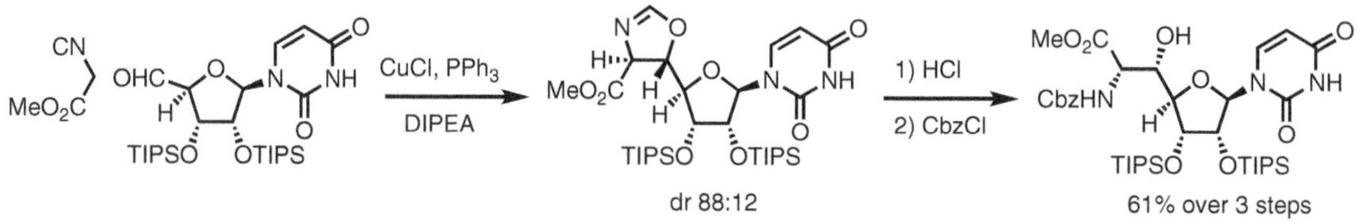

Scheme 6.12 *Catalytic asymmetric isocyanato aldol route to caprazol*

6.5.1.5 *Carbanucleosides*

The first enantioselective synthesis of (−)-aristeromycin and (−)-neplanocin A was reported by Ohno and was based on an enzymatic approach [39]. Here, we highlight a synthesis by Koizumi, which is based on an asymmetric Diels–Alder cycloaddition to set the stereochemistries properly (Scheme 6.13) [40]. The Diels–Alder cycloaddition of $(S)_S$-3-(2-pyridylsulfinylpropenoate with cyclopentadiene in the presence of a Lewis acid gave the cycloadduct in 96% yield almost as a single diastereomer. The alkene was dihydroxylated, protected as the acetonide, converted to the sulfone, and eliminated to give the unsaturated ester. Ozonolysis of the double bond with reductive work up gave the triol, which was transformed oxidatively to the Ohno lactone.

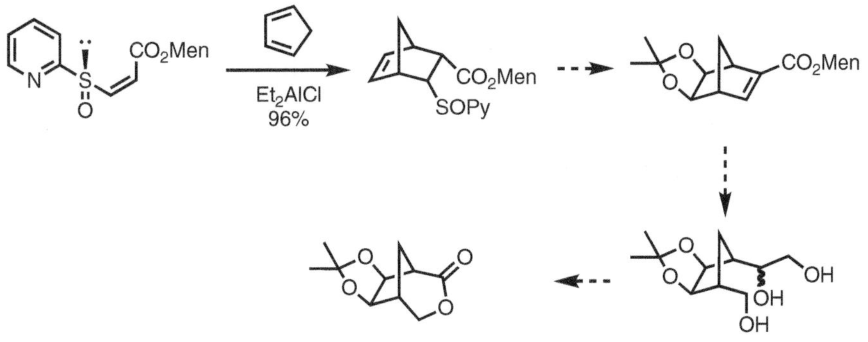

Scheme 6.13   *Asymmetric Diels–Alder route to (−)-aristeromycin and (−)-neplanocin A*

An early synthesis of (−)-carbovir relied on the enantioselective deprotonation of *trans*-4-*t*-butyldimethylsiloxy methyl-1,2-epoxycyclopentane (Scheme 6.14) [41]. This was achieved with a chiral lithium amide, lithium (S)-2-(pyrrolidin-1-ylmethyl)pyrrolidide, to afford $(1S,4S)$-*trans*-4-*t*-butyldimethylsiloxymethyl-2-cyclopenten-1-ol in 83% *ee*. Adenosylation was effected with chloroadenosine under Mitsunobu conditions to adjust the stereochemistry properly to *cis*. Cleavage of the TBS protection and basic hydrolysis of the chloroimine completed the synthesis.

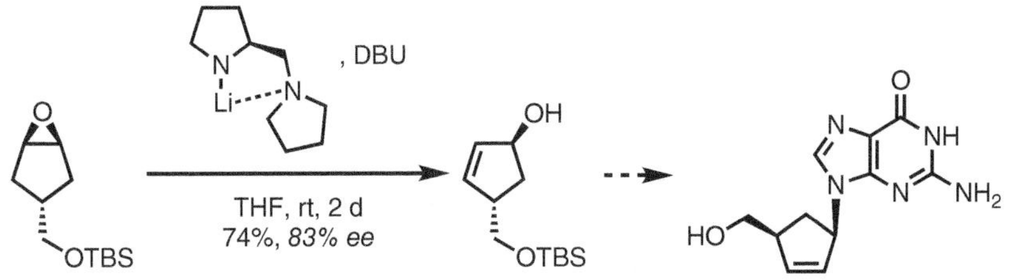

Scheme 6.14   *Synthesis of (−)-carbovir*

A general strategy for the synthesis of carbanucleosides has been developed by Trost, based on the powerful asymmetric allylic alkylation reaction (Scheme 6.15) [42]. In the first-generation synthesis, the nucleobase (Nu) was first reacted with cyclopentadiene monoepoxide. A second Pd(0)-catalysed alkylation with the anion of phenylsul-fonyl(nitro)methane introduced a one-carbon side chain which could be converted into the hydroxymethyl group. Although rather short, this synthesis gave racemic carbanucleosides. Development of a series of chiral modular ligands allowed for the development of an asymmetric second-generation synthesis, but the installation of the hydroxymethyl side chain and the nucleobase required several steps. Eventually, initial highly enentioselective introduction of the nuclobase on the bisbenzoate using the rigid bidentate ligand gave the desired carbanucleosides in high *ee*'s. This strategy was used in the syntheses of (−)-aristeromycin, (−)-neplanocin A, and (−)-carbovir [43].

Crimmins developed a divergent synthesis of carbovir and abacavir which combined three key transformations: the relative and absolute configuration of the pseudo sugar was secured by an asymmetric aldol reaction; the five-membered ring was formed through ring-closing metathesis; and finally, the aromatic base was attached through a Trost-type palladium-mediated substitution reaction (Scheme 6.16). Thus, the titanium tetrachloride and (−)-sparteine aldol reaction of the thiopyrrolidone and crotonaldehyde gave the desired *syn*-aldol adduct in 82% yield and > 99% *de*. Reductive cleavage of the chiral auxiliary followed by acetylation of the two alcohols provided the precursor for the ring-closing

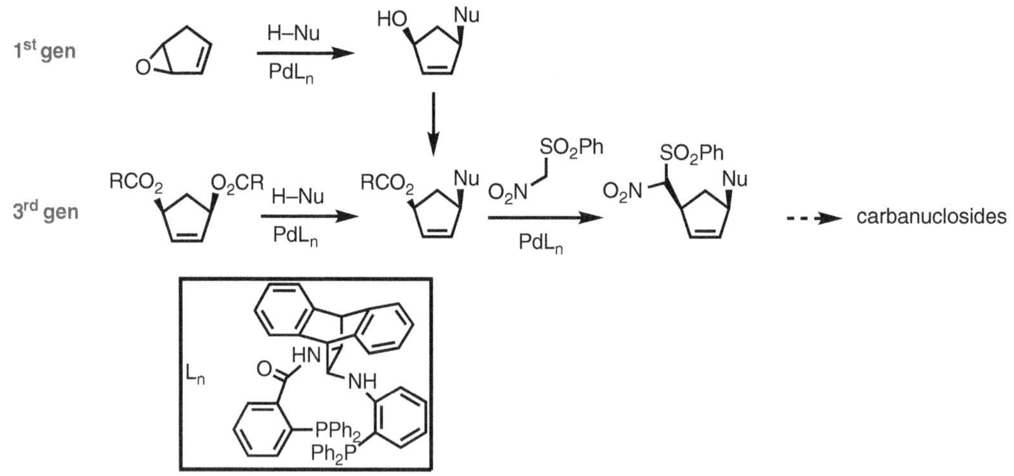

Scheme 6.15 *Trost asymmetric allylic alkylation in the synthesis of carbanucleosides*

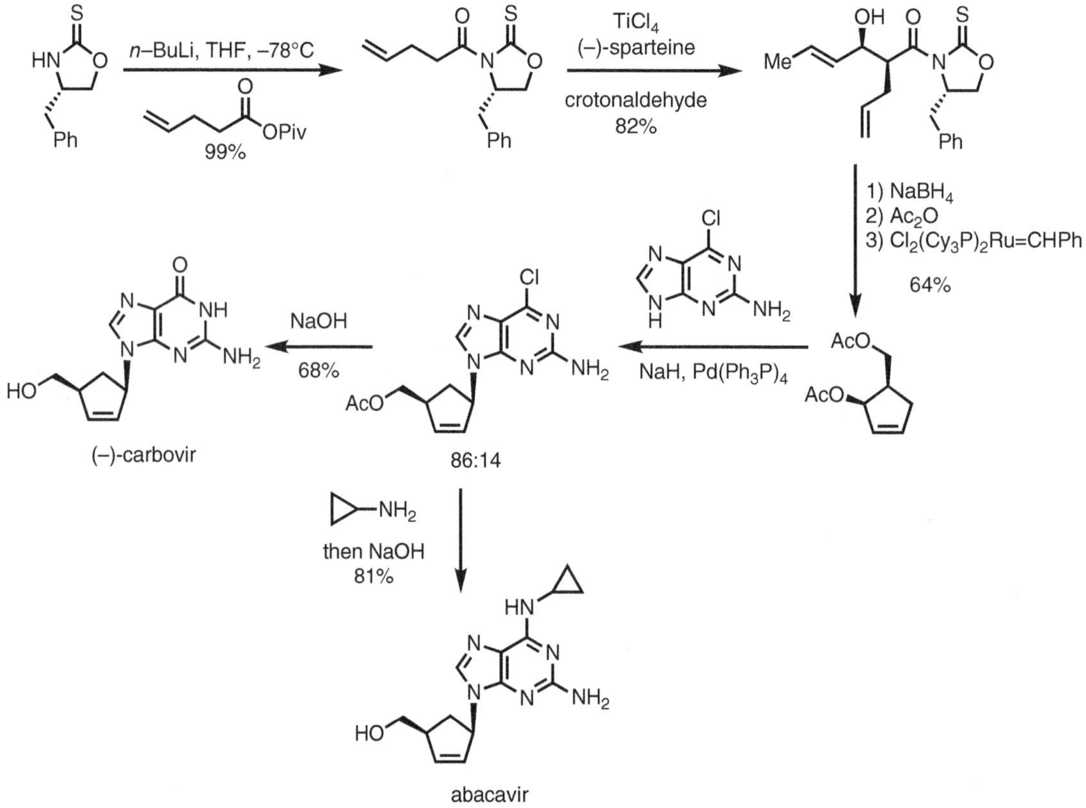

Scheme 6.16 *Crimmins synthesis of (−)-carbovir and abacavir*

metathesis. Finally, adenosylation with 2-amino-6-chloropurine was achieved using the Trost protocol ($Pd(PPh_3)_4$, NaH) to give a 86:14 mixture of the N9 and N7 isomers, the desired product being obtained in 65% isolated yield. Basic hydrolysis provided carbovir directly, and treating the chloropurine derivative first with cyclopropyl amine followed by base hydrolysis provided abacavir [44].

References

1. Santos, R., Ursu, O., Gaulton, A., Bento, A.P., Donadi, R.S., Bologa, C.G., et al. *Nature Rev. Drug Disc.* **2017**, *16*, 19–34.
2. Sanger, W. *Principles of Nucleic Acid Structure*, Springer; Berlin–Heidelberg, **1983**.
3. Boer, D.R.; Canals, A.; Coll, M. *Dalton Trans.* **2009**, 399–414.
4. Structures from RSCB Protein Data Bank (www.pdb.org) Berman, H.M., Westbrook, J., Feng, Z., Gilliland, G., Bhat, T.N., Weissig, H., et al. *Nucleic Acids Research* **2000**, *28*, 235–242; PDB ID's for a) A-DNA: 1ANA Conner, B.N., Yoon, C., Dickerson, J.L., Dickerson, R.E. *J. Mol. Biol.* **1984**, *174*, 663–695; b) B-DNA: 1BNA Drew, H.R., Wing, R.M., Takano, T., Broka, C., Tanaka, S., Itakura, K., et al. *Proc. Natl. Acad. Sci. USA* **1981**, *78*, 2179–2183; c) 436D Tereshko, V., Minasov, G., Egli, M. *J. Am. Chem. Soc.* **1999**, *121*, 470–471; d) Z-DNA: 3P4J Brzezinski, K., Brzuszkiewicz, A., Dauter, M., Kubicki, M., Jaskolski, M., Dauter, Z. *Nucleic Acids Res.* **2011**, *39*, 6238–6248.
5. Wing, R., Drew, H., Takano, T., Broka, C., Tanaka, S., Itakura, K., et al. *Nature* **1980**, *287*, 755–758.
6. Watson J.D., Crick F.H.C. *Nature* **1953**, *171*, 737–738.
7. a) Khan, G.S., Shah, A., Zia-ur-Rehman, Barker, D. *J. Photochem. Photobiol. B: Biology* **2012**, *115*, 105–118; b) Mišković, K., Bujak, M., Baus Lončar, M., Glavaš-Obrovac, L. *Arh. Hig Rada Toksikol.* **2013**, *64*, 593–602.
8. Van Hecke, K., Nam, P.C., Nguyen, M.T., Van Meervelt, L. *FEBS J.* **2005**, *272*, 3531–3541.
9. Hamilton, P.L., Arya, D.P. *Nat. Prod. Rep.* **2012**, *29*, 134–143.
10. Banerjee, S., Brown, K.L., Egli, M., Stone, M.P. *J. Am. Chem. Soc.* **2011**, *133*, 12556–12568.
11. Tomasz, M., Lipman, R., Verdine, G.L., Nakanishi, K. *J. Am. Chem. Soc.* **1985**, *107*, 6120–6121.
12. Sastry, M., Fiala, R., Lipman, R., Tomasz, M., Patel, D.J. *J. Mol. Biol.* **1995**, *247*, 338–359.
13. a) Seley-Radtke, K.L., Yates, M.K. *Antiviral Research* **2018**, *154*, 66–86; b) Seley-Radtke, K.L., Yates, M.K. *Antiviral Research* **2019**, *162*, 5–21.
14. WHO electronic Essential Medicines List (eEML), World Health Organization, **2020**. https://list.essentialmeds.org/ (beta version 1.0). Licence: CC BY 3.0 IGO.
15. Knapp. S. *Chem. Rev.* **1995**, *95*, 1859–1876.
16. Isono, K. *J. Antibiot.* **1988**, *41*, 1711–1739.
17. a) Winn, M., Goss, R.J.M., Kimura, K., Bugg, T.D.H. *Nat. Prod. Rep.* **2010**, *27*, 279–304; b) Niu, G., Tan, H. *Trends in Microbiol.* **2015**, *23*, 110–119; c) Shiraishi, T., Kuzuyama, T. *J. Antibiot.* **2019**, *72*, 913–923.
18. Gong, R., Yu, L., Qin, Y., Price, N.P.J., He, X., Deng, Z., et al. *Biotechnol. Adv.* **2021**, *46*, 107673.
19. Cantoni, G.L. *J. Am. Chem. Soc.* **1952**, *74*, 2942–2943.
20. Shealy, Y.F., Clayton, J.D. *J. Am. Chem. Soc.* **1966**, *88*, 3885–3887.
21. a) Hudlicky, T., Rulin, F., Tsunoda, T., Price, J.D. *J. Am. Chem. Soc.* **1990**, *112*, 9439–9440; b) Hudlicky, T. *Pure Appl. Chem.* **1992**, *64*, 1109–1113; c) Hudlicky, T. *Chem. Rev.* **1996**, *96*, 3–30; d) Hudlicky, T., Reed, J.W. *The Way of Synthesis*, Wiley-VCH, Weinheim, **2007**, pp. 146–154.
22. Poss, C.S., Schreiber, S.L. *Accts. Chem. Res.* **1994**, *27*, 9–17.
23. Still, W.C., Barrish, J.C. *J. Am. Chem. Soc.* **1983**, 2487–2489.
24. Ikemoto, N., Schreiber, S.L. *J. Am. Chem. Soc.* **1990**, *112*, 9657–9659.
25. Cha, J.K., Christ, W.J., Kishi, Y. *Tetrahedron* **1984**, *40*, 2247–2255.
26. Maguire, M P., Feldman, P.L., Rapoport, H. *J. Org. Chem.* **1990**, *55*, 948–955.
27. Barton, D.H.R., Géro, S.D., Quiclet-Sire, B., Samadi, M. *J. Chem. Soc. Perkin Trans. I* **1991**, 981–985.
28. Ghosh, A.K., Liu, W. *J. Org. Chem.* **1996**, *61*, 6175–6182.
29. (a) Knowles, W.S., Sabacky, M.J., Vineyard, B.D., Weinkauff, D.J. *J. Am. Chem. Soc.* **1975**, *97*, 2567–2568; (b) Vineyard, B.D., Knowles, W.S., Sabacky, M.J., Bachman, G.L., Weinkauff, D.J. *J. Am. Chem. Soc.* **1977**, *99*, 5946–5952.
30. Scott, J.W., Keith, D.D., Nix, G. Jr., Parrish, D.R., Remington, S., Roth, G.P., et al. *J. Org. Chem.* **1981**, *46*, 5086–5093.
31. Vorbruggen, H., Krolikiewicz, K., Bennua, B. *Chem. Ber.* **1981**, *114*, 1234–1255; b) Vorbruggen, H., Hofle, G. *Chem. Ber.* **1981**, *114*, 1256–1268.
32. Decultot, L., Policarpo, R.L., Wright, B.A., Huang, D., Shair, M.D. *Org. Lett.* **2020**, *22*, 5594–5599.

33. Cannon, J.S., Overman, L.E. *Acc. Chem. Res.* **2016**, *49*, 2220–2231.
34. Burk, M.J. *Acc. Chem. Res.* **2000**, *33*, 363–372.
35. Truong, T.V., Rapoport H. *J. Org. Chem.* **1993**, *58*, 6090–6096.
36. Zabudkin, O., Schickaneder, C., Matviienko, I., Sypchenko, V. EP 3115461 A1, 7.7.**2015**.
37. Hirano, S., Ichikawa, S., Matsuda, A. *J. Org. Chem.* **2007**, *72*, 9936–9946.
38. Gopinath, P.; Wang, L., Abe, H., Ravi, G., Masuda, T., Watanabe, T., et al. *Org. Lett.* **2014**, *16*, 3364–3367.
39. Arita, M., Adachi, K., Ito, Y., Sawai, H., Ohno, M. *J. Am. Chem. Soc.* **1983**, *105*, 4049–4055.
40. Arai, Y., Hayashi, Y., Yamamoto, M., Takayama, H., Koizumi, T. *Chem. Lett.* **1987**, 185–186.
41. Asami, M., Takahashi, J., Inoue, S. *Tetrahedron: Asymmetry* **1994**, *9*, 1649–1652.
42. Trost, B.M., Kuo, G.H., Bennche, T. *J. Am. Chem. Soc.* **1988**, *110*, 621–622.
43. Trost, B.M., Madsen, R., Guile, S.D., Brown, B. *J. Am. Chm. Soc.* **2000**, *122*, 59447–5956.
44. Crimmins, M.T., King, B.W., Zuercher, W.J., Choy, A.L. *J. Org. Chem.* **2000**, *65*, 8499–8509.

7

Phenylpropanoids

Lignin is mainly formed by oxidative radical polymerization of phenylpropionyl units, especially those derived from coniferyl alcohol, the alcohol corresponding to ferulic acid.

Holger Erdtman, 1933

In Chapter 5 we discussed the biosynthesis of aromatic amino acids phenylalanine and tyrosine from carbohydrates through the shikimic acid pathway. Phenylalanine plays a central role in the biosynthesis of several important aromatic compounds; the phenylpropanoids. There is a somewhat overlapping definition for polyphenols, sometimes misleadingly called 'vegetable tannins.' In this text we adopt the definition by Quideau [1], according to which *"The term 'polyphenol' should be used to define plant secondary metabolites derived exclusively from the shikimate-derived phenylpropanoid and/or the polyketide pathway(s), featuring more than one phenolic ring and being devoid of any nitrogen-based functional group in their most basic structural expression."*

In connection with phenylpropanoid compounds or natural phenolic compounds, you will often encounter the word tannin. This word derives from the process of converting animal skins into leather, in which process Oak (genus *Quercus*) galls (oak apples) were crushed and the extract (tannic acid) was used to treat the skin. Etymologically, the word derives from the ancient Celtic word *tann* for Oak. Sufficiently large molecules (500–3000 Da) with sufficiently many phenolic groups are capable of forming crosslinked structures with collagen molecules through hydrogen-bonding networks. Based on chemical structures, tannins are classified in two main groups: hydrolysable tannind and condensed tannins. Hydrolysable tannins are further divided in two groups; gallotannins and ellagitannins.

A number of phenolic compounds are difficult to classify (Figure 7.1). The axially chiral gossypol can be extracted from the seeds of the cotton plant (*Gossypium hirsutum*). It suppresses sperm production in men; in 1929,

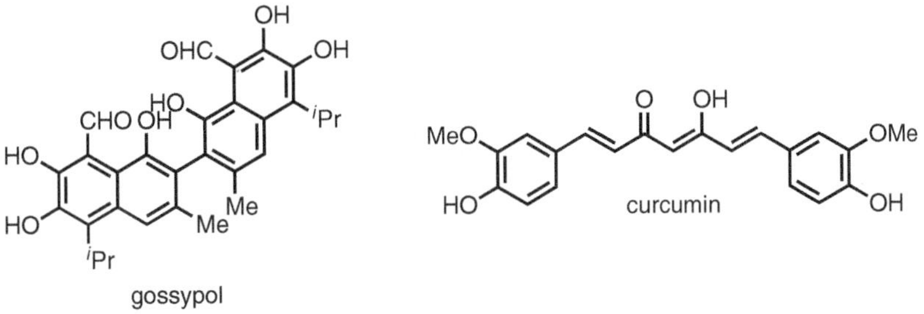

Figure 7.1 *Phenolic compounds of mixed biogenetic origins*

Asymmetric Synthesis of Natural Products, Third Edition. Ari M.P. Koskinen.
© 2023 John Wiley & Sons Ltd. Published 2023 by John Wiley & Sons Ltd.

in Jiangxi, China, consumption of cottonseed oil was correlated with low fertility in men, and therefore it was considered as a possible male contraceptive pill. Serious side effects (fatigue and even paralysis due to the hypokalemic effects of gossypol) led WHO to recommend discontinuation of the tests, in 1998 [2]. Gossypol is biosynthesised from farnesyl pyrophosphate through cadinene (see Chapter 9, terpenes). Curcumin is biosynthesised through the phenylpropanoid pathway. It is one of the earliest isolated natural products [3]. Curcumin is the yellow pigment in turmeric (*Curcuma longa*), a member of the ginger family (*Gingiberaceae*).

Phenylpropanoids can be divided in four larger groups (Figure 7.2); phenolic acids, flavonoids, stilbenes, and lignans. Of these groups, flavonoids, and to some extent lignans, are the only ones that contain chiral molecules. In this chapter we will briefly explore the structures of these compounds and discuss typical asymmetric synthesis of a few selected examples.

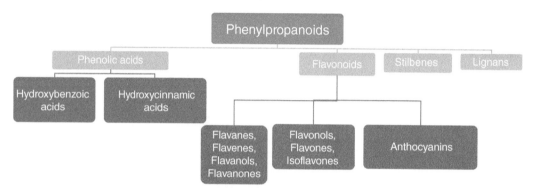

Figure 7.2 *Classification of polypropionates (polyphenols)*

7.1 Phenylpropanoid Biosynthesis

Scheme 7.1 displays the outline of the biosynthetic pathways leading to the different phenylpropanoids [4]. The starting point is shikimic acid whose conversion to phenylalanine was discussed in Chapter 5. Deamination is achieved by phenylalanine ammonia lyase (PAL) in a reaction resembling a biomimetic Hofmann degradation. Salicylic and benzoic acids are formed from cinnamic acid through oxidative processes, and are commonly found in many berries. These compounds are bacteriostatic (inhibit the growth of bacteria), and thus, e.g. lingonberry, cranberry, and cloudberry can be preserved without added preservatives.

Towards the more complex phenylpropanoids, *trans*-cinnamic acid is hydroxylated by cinnamate-4-hydroxylase (C4H) to coumaric acid. Coumaric acid is a precursor of coumarins and vanillin. The action of malonyl coenzyme A three times on coumaric acid gives naringenin chalcone and therefrom naringenin. Isoflavone synthase (IFS), a cytochrome P450 enzyme, oxidises naringenin to genistein. Naringenin provides the entry to flavonoids, flavanols, anthocyanines, and condensed tannins, whereas genistein leads to isoflavones and isoflavanones.

The oxidation of cinnamic acid occurs first at the *p*-position. The mechanism of the aromatic hydroxylation proceeds via disrupture of the aromaticity and epoxidation of the aromatic ring, followed by a 1,2-shift of a hydrogen atom in the aromatic ring. This shift, being of unprecedented type when it was discovered, is known after its site of discovery as the NIH shift (Scheme 7.2) [5]. Further oxidation of coumaric acid at the *meta*-position gives caffeic acid.

7.2 Phenolic Acids

The C_6C_1 benzoic acids, of which salicylic acid (SA) is a familiar example, are included in discussions of phenylpropanoid natural products because of their presumed biosynthetic origin via side-chain shortening of hydroxycinnamic acids. However, this may not be the only route for their synthesis in plants.

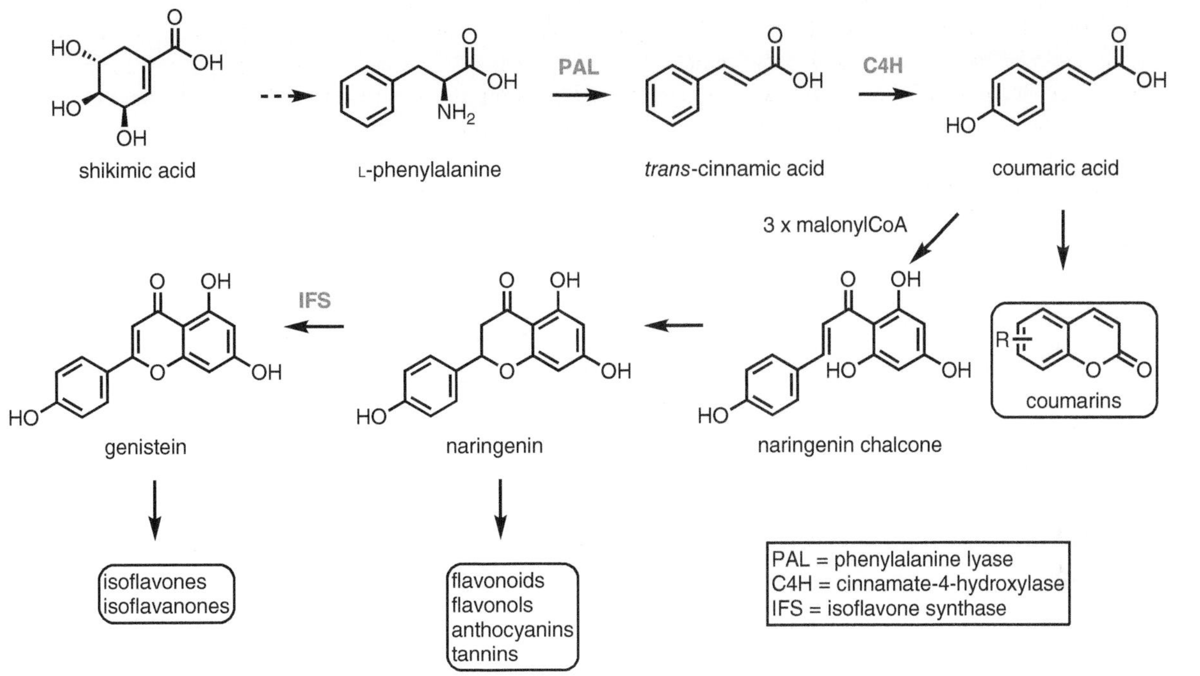

Scheme 7.1   *Phenylpropanoid biosynthesis*

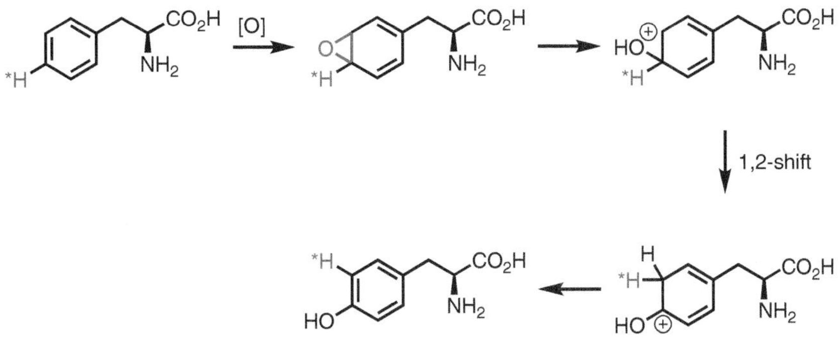

Scheme 7.2   *NIH shift*

Some examples of plant phenolic acids are shown in Figure 7.3. Protocatechuic acid (3,4-dihydroxybenzoic acid) is a major metabolite of antioxidant polyphenols found in green tea. Gallic acid (3,4,5-trihydroxybenzoic acid) is present in Oak galls, from where it derives its name. Ellagic acid is basically a dimerised gallic acid. Together these two occur in gallotannins and ellagitannins, which are peresters of glucose and either gallic or ellagic acid.

7.2.1   Hydroxycinnamic Acids

Caffeic and ferulic acids occur in the tannic acids of coffee as their quinic acid esters (Figure 7.4). 3-Caffeoylquinic acid (chlorogenic acid or caffeotannic acid) was isolated in crystalline form from coffee as early as 1846 by *Anselme Payen* [6]. The composition and relative abundances of the glycoconjugates vary depending on the type of coffee bean.

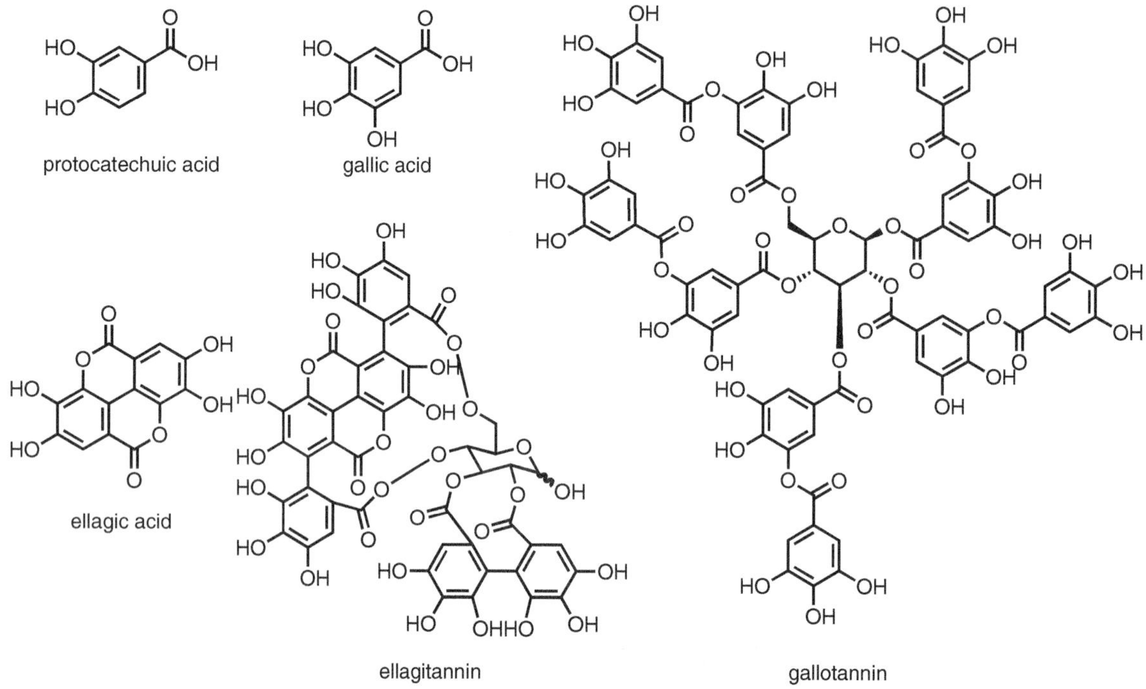

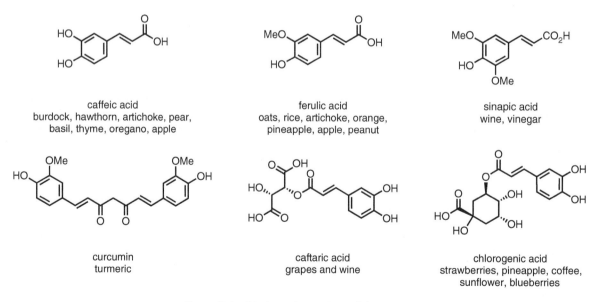

Figure 7.3   *Phenolic acids*

caffeic acid
burdock, hawthorn, artichoke, pear,
basil, thyme, oregano, apple

ferulic acid
oats, rice, artichoke, orange,
pineapple, apple, peanut

sinapic acid
wine, vinegar

curcumin
turmeric

caftaric acid
grapes and wine

chlorogenic acid
strawberries, pineapple, coffee,
sunflower, blueberries

Figure 7.4   *Hydroxycinnamic acid derivatives*

7.2.2 Coumarins

Coumarins and quinones are derived from coumaric acid. Most natural coumarins contain a hydroxyl group at C7. The *o*-hydroxylation is formulated to occur by way of a spirolactone followed by a rearrangement step (Scheme 7.3). This has been shown by tracer experiments utilising ^{18}O-labeled *p*-hydroxycoumarin acid.

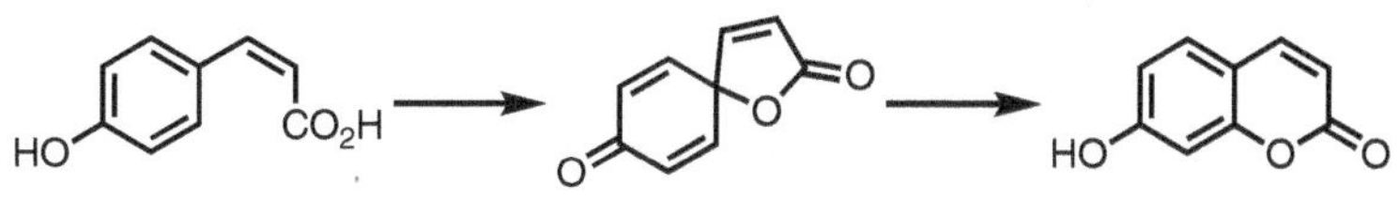

Scheme 7.3 *Biosynthesis of umbelliferone*

Psoralens are linear furocoumarins derived from umbelliferone through alkylation (prenylation) with dimethylallyl pyrophosphate, followed by epoxidation, cyclisation, and cleavage of the side chain (Scheme 7.4). Psoralens are widely distributed in *Umbeliferae* and *Rutaceae* plants, and the most common examples are psoralen and bergapten. Bergamot oil, obtained from the peel of *Citrus aurantium*, contains up to 5% bergapten. Sun tan products often contain bergamot oil, since the psoralen chromophore absorbs in the near-UV region and stimulates the formation of melanin pigments.

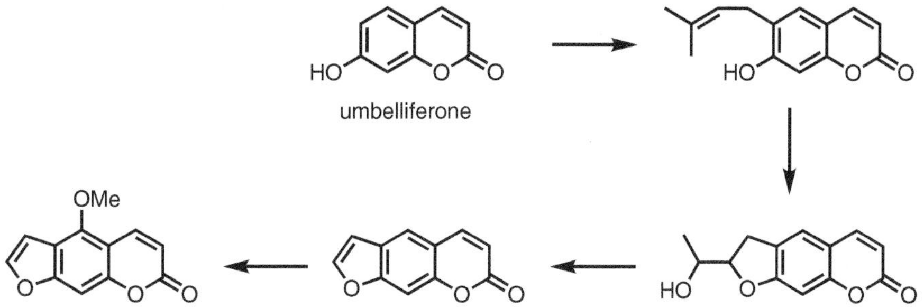

Scheme 7.4 *Biosynthesis of psoralens*

Psoralen is used for the treatment of psoriasis, mainly in the form known as PUVA-treatment (psoralen-UV). The psoralen molecule intercalates with DNA and then binds with the two DNA strands upon UV irradiation, which triggers a [2 + 2] photochemical cycloaddition, resulting in the formation of covalent bonds with both of the DNA strands [7]. Thus, replication of the DNA is prohibited.

7.3 Flavonoids

Flavanoids are a subclass of polypropionates with more than 8000 structures isolated to date. They derive their name from the fact that they often are yellow pigments (latin *flavus*, yellow). They are structurally divided into several subtypes, which in the following discussion will be grouped together depending on their oxygenation levels.

7.3.1 Flavanes, Flavenes, Flavanols, and Flavanones

Flavene, flavane, flavanol, and flavanone skeleta (Figure 7.5) comprise a large class of biologically active natural products, such as tephrowatsin E, hilgartene, catechin, and taxifoline (dihydroquercetin). Catalytic asymmetric synthesis of 2-aryl chromenes including 2*H*-flavenes have been reviewed [8].

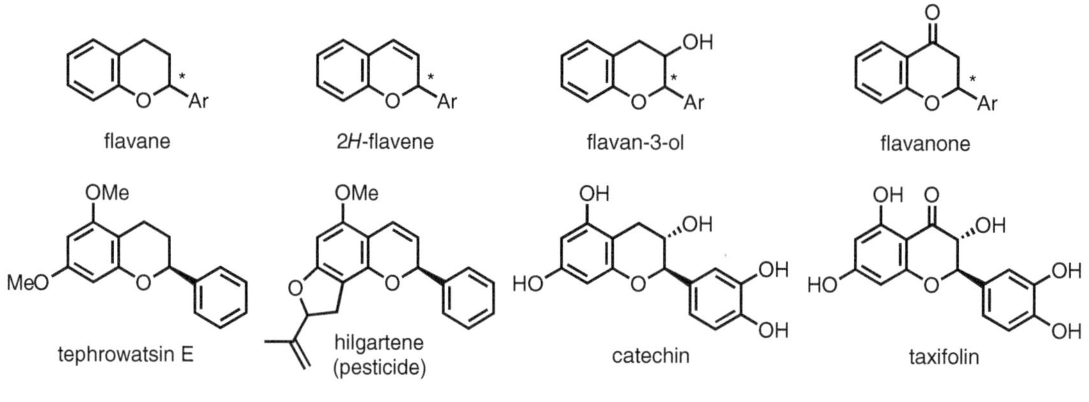

Figure 7.5 *Typical flavanes, flavenes, flavanols and flavanones.*

The first asymmetric organocatalytic synthesis of 2*H*-flavenes was achieved through a domino reaction between a salicylaldehyde derivative and a cinnamaldehyde derivative (Scheme 7.5) [9]. The reaction involves an initial oxa-Michael addition followed by an intramolecular aldol condensation. Catalysis by the Jørgensen prolinol organocatalyst provided the 2-aryl-2*H*-chromene-3-carbaldehydes in promising enantioselectivities and acceptable yields.

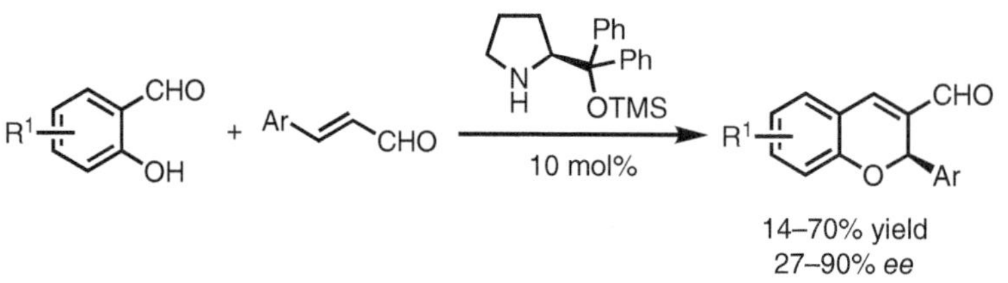

Scheme 7.5 *Organocatalytic synthesis of 2-aryl-2H-chromene-3-carbaldehydes*

Chiral organic contact ion-pair catalysis has been applied in the synthesis of 2*H*-flavenes in excellent enatiopurity (up to 96% *ee*) (Scheme 7.6) [10]. The authors propose a mechanism for the asymmetric induction: in the intramolecular allylic alkylation of allylic alcohols, the highly acidic chiral Brønstedt acid catalyst *N*-triflylphosphoramide protonates the allylic alcohol, which subsequently dehydrates to yield a carbocation which is associated with the phosphoramide anion in a chiral contact ion pair.

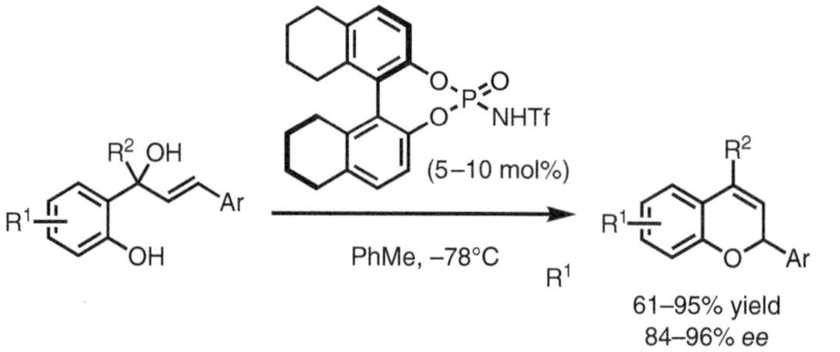

Scheme 7.6 *Chiral phosphoramidite catalysed allylic substitution*

Highly enantioselective synthesis of 2*H*-flavenes can be achieved via a palladium-catalysed asymmetric redox-relay Heck reaction (Scheme 7.7) [11]. Pd(MeCN)$_2$Cl$_2$ with a PyrOx ligand was used as the catalyst for the reaction of

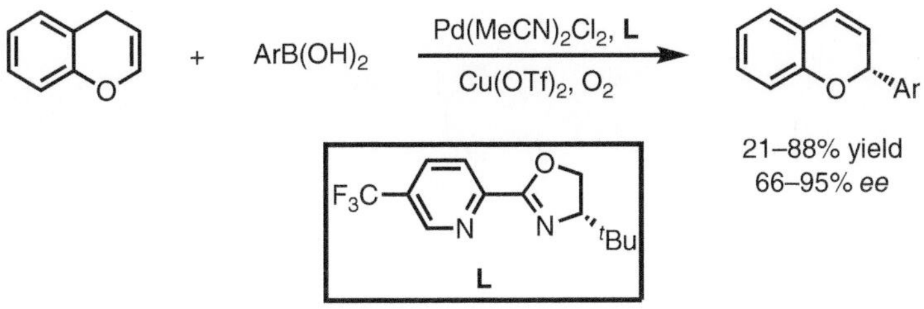

Scheme 7.7 *Redox-relay Heck approach to flavenes*

4*H*-chromene with various arylboronic acids, after optimisation, and a series of chiral 2*H*-flavenes were obtained in good to high enantioselectivity.

Flavan-3-ols (2-phenylchroman-3-ol) and their derivatives are an important class of flavonoids isolated from several plants including *Camellia sinensis* (green tea). Freshly harvested tea leaves are a rich source of flavonoids, especially flavan-3-ols, known collectively as catechins, including (Figure 7.6); (+)-catechin, (+)-gallocatechin, (−)-epicatechin, and (−)-epigallocatechin-3-gallate (EGCG), with the latter one being a major constituent of green tea. The content of about 20% flavan-3-ols (by dry weight) in tea is enormous, classifying tea as an extraordinary plant species with respect to flavonoids. During the manufacture of black tea and oolong tea EGCG is oxidised to form orange-red pigments; theaflavins, a mixture of theaflavin itself and its gallate esters.

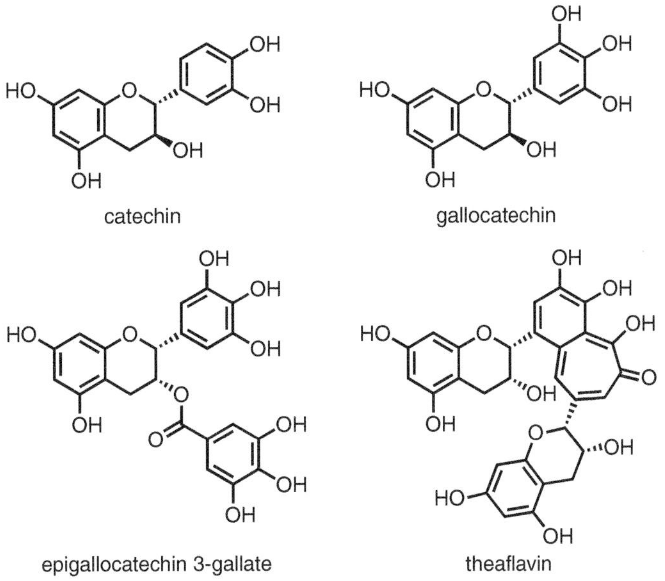

Figure 7.6 *Flavan-3-ols in teas*

The biosynthesis of flavonoids in tea has been elucidated including the enzymes involved [12]. As an example of the biosynthesis of flavanols, we take that of catechin (Scheme 7.8). Naringenin (see Scheme 7.1) is first hydroxylated in the aromatic ring to give eriodictyol. A second hydroxylation by flavanone-3-hydroxylase gives dihydroquercetin. Two consecutive NADPH-dependent reductive steps lead to leucocyanidine and eventually catechin.

Enantioselective synthesis of flavan-3-ols have been reviewed, and synthetic protocols typically involve Sharpless asymmetric epoxidation (SAE) and dihydroxylation (SAD), as well as Shi asymmetric epoxidation strategies [13].

An illustrative example of the SAD strategy is given in Scheme 7.9 [14]. The cinnamyl alcohol derivative was subjected to Sharpless asymmetric dihydroxylation with ADMIX-α, and the triol was cyclised to the epoxy alcohol

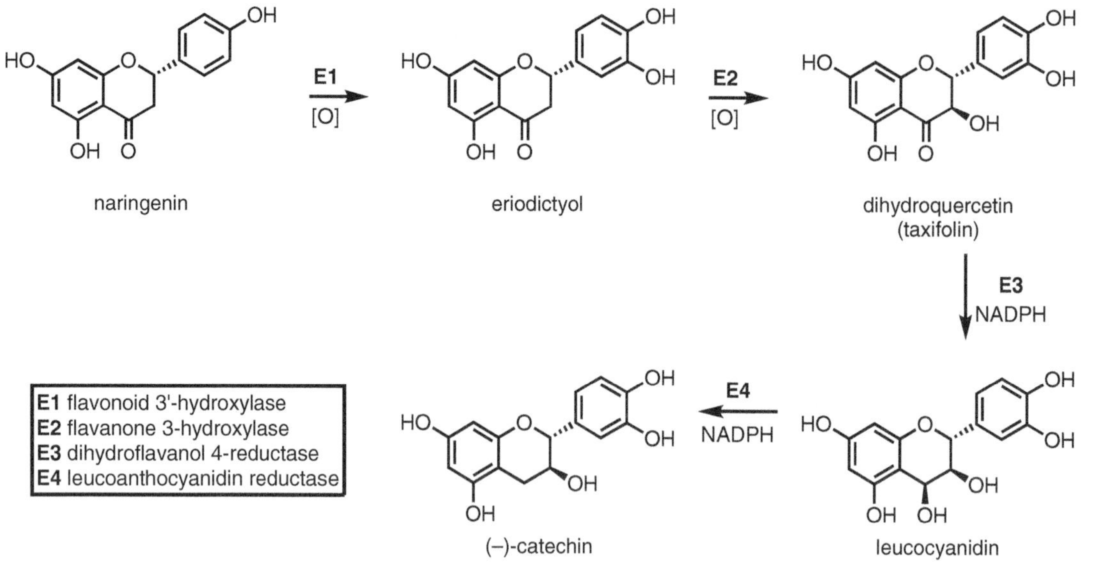

Scheme 7.8   Biosynthesis of catechin

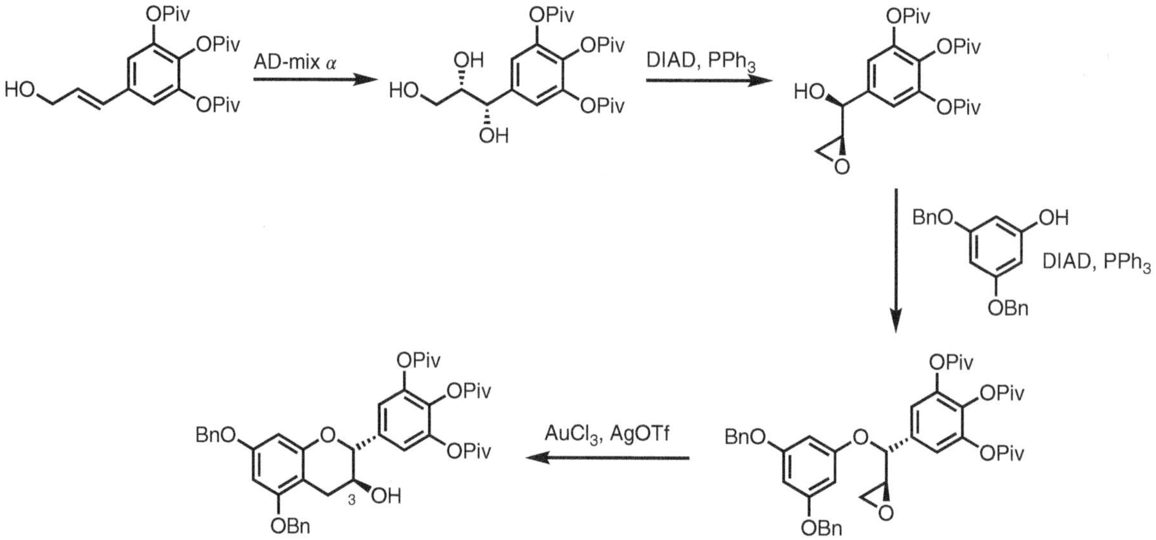

Scheme 7.9   Asymmetric dihydroxylation in the synthesis of epigallocatechin

using the Mitsunobu reaction. Mitsunobu esterification with dibenzylphloroglucinol gave the epoxy ether, which was then cyclised to the flavan-3-ol using gold catalysis to give the gallocatechin stereochemistry. The 3-position could be inverted by oxidation/reduction protocol to reach the epigallo series.

Tanaka et al. used a strategy where the chiral information was derived from epichlorohydrin (Scheme 7.10) [15]. Sequential opening of the epichlorohydrin with two different aryl Grignard reagents followed by esterification give the intermediate shown, which upon oxidative cyclisation with DDQ (2,3-dichloro-5,6-dicyano-*p*-benzoquinone) provided the epigallocatechin product. The stereochemistry at the 3-position is dependent on which epichlorohydrin one chooses as the starting material.

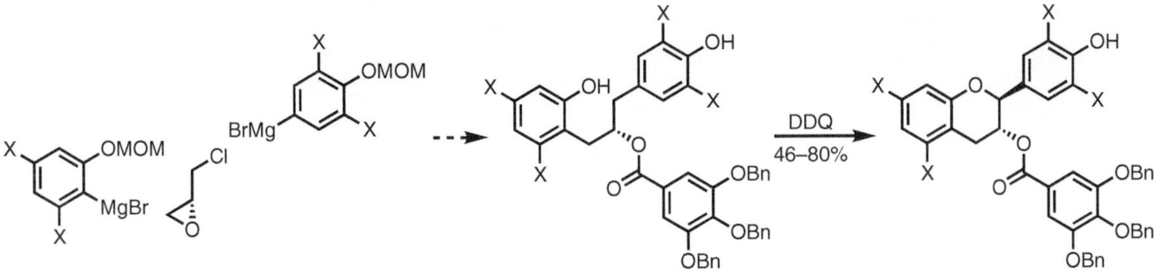

Scheme 7.10 *Tanaka oxidative cyclisation*

Flavanones (Figure 7.7) are a relatively small group of flavonoids, widely occurring in plant families, especially in Compositae, Leguminosae, and Rutaceae. Naringenin, eriodictyol, and hesperetin are most readily associated with citrus fruits, for example naringenin in grapefruit, which has been widely studied for its health benefits. Enantioselective methods for the synthesis of flavanones and chromanones have been reviewed [16].

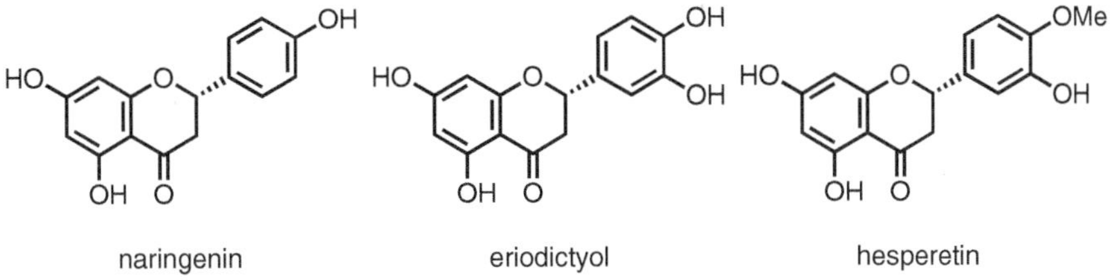

Figure 7.7 *Typical flavanones*

Nearly enantiopure prenylated flavanones were obtained through kinetic resolution of the racemic β-chiral ketones with a chiral rhodium(III) complex in a mixture of formic acid and triethylamine in high yield (Scheme 7.11). The S-isomer was hydrolysed enzymatically by a *Pseudomonas* lipase to (S)-8-prenylnaringenin in 99% *ee*. The R-isomer was oxidised to the ketone, and protecting groups were cleaved by transesterification with methanol, catalysed by the Verkade base, to give (R)-8-prenylnaringenin in 95% *ee* [17].

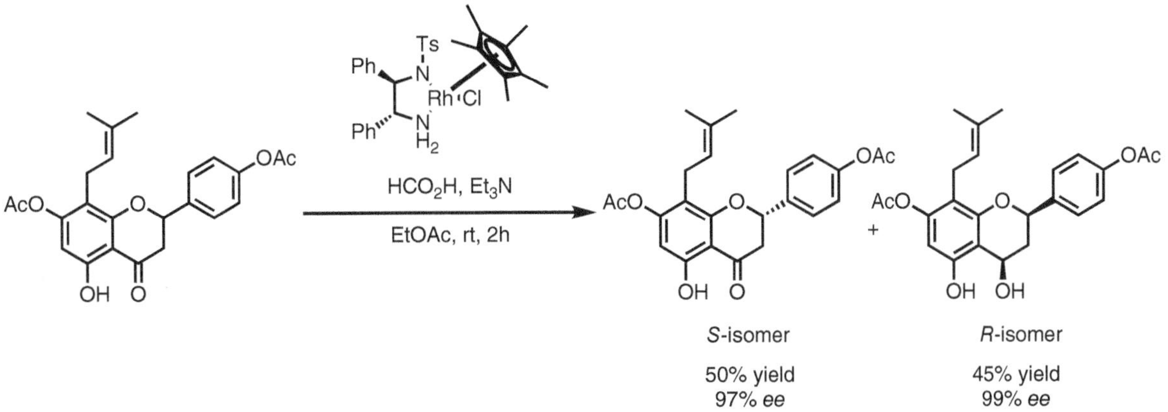

Scheme 7.11 *Synthesis of enantiopure flavanones by catalytic asymmetric transfer hydrogenation*

The extracts of the fruit of milk thistle (*Silybum marianum*) consist of a complex mixture of flavonolignans and flavonoids including isosilybin B. The extracts have been used for centuries for the treatment of liver disorders and as a hepatoprotectant. The first asymmetric total synthesis of (−)-isosilybin A was reported in 2015 (Scheme 7.12), and the stereochemical information was established by the cinchona alkaloid urea catalyst to give the cyclisation product in an *er* 85:15. Diastereoselective enolate oxygenation (LiHMDS, TMSCl followed by DMDO) and cleavage of the protecting groups completed the first asymmetric synthesis of (−)isosilybin A [18].

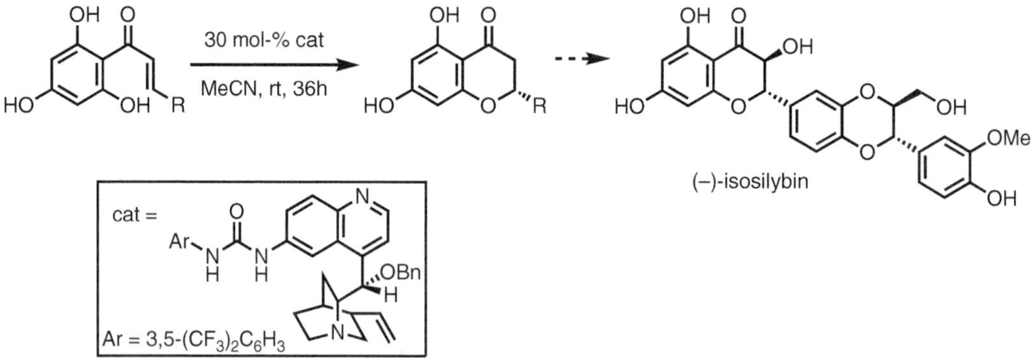

Scheme 7.12   *Organocatalytic conjugate addition route to (−)-isosilybin*

7.3.2   Flavones, Isoflavones, and Flavonols

The usually highly colored flavones form the largest group of oxygen-heterocyclic compounds found in plants (more than 5000 congeners have been identified) [19]. The flavonoids show a mixed biogenesis from shikimates and polyketides, and their structures vary very widely. Their biological role is to interfere with insect pollinating or feeding on plants. Some flavonoids also have a characteristically bitter taste, which makes them repel caterpillars. Anthocyanidins are derived from flavonoids, and they usually have a strong red, violet, or blue color. The magnificent colors seen in autumn leaves are due to flavonoids, certain terpenes (yellow), and anthocyanidins.

Apigenin (Figure 7.8) is particularly abundant in parsley, celery, celeriac, and chamomile tea. In the flowers of chamomile plants, it constitutes 68% of total flavonoids. Luteolin is the principal yellow dye compound obtained from mignonettes, particularly the dyer's wood (*Reseda luteola*), which has been used as a source of the dye since at least the first millennium B.C. Luteolin was first isolated in pure form, and named in 1829 by the French chemist Michel Eugène Chevreul [20].

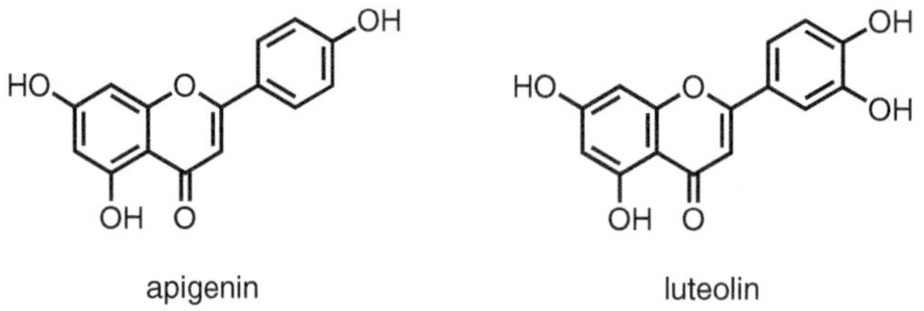

apigenin   luteolin

Figure 7.8   *Flavones*

Isoflavones are isomeric to the flavones in the sense that the 2-aryl group is shifted to the 3-position. They are produced almost exclusively by the members of the bean family, *Fabaceae (Leguminosae)*. Daidzein, genistein, and glycitein (Figure 7.9) are typical members of this class of natural products. They exhibit phytoestrogenic properties, in other words, they are plant derived xenoestrogens.

Flavonols (Figure 7.10) are the most oxidised flavonoids, and they also often provide deep colors in the plant they occur. They have a hydroxy group at the 3-position of flavones. Kaempferol is abundant in capers, saffron, kale, beans,

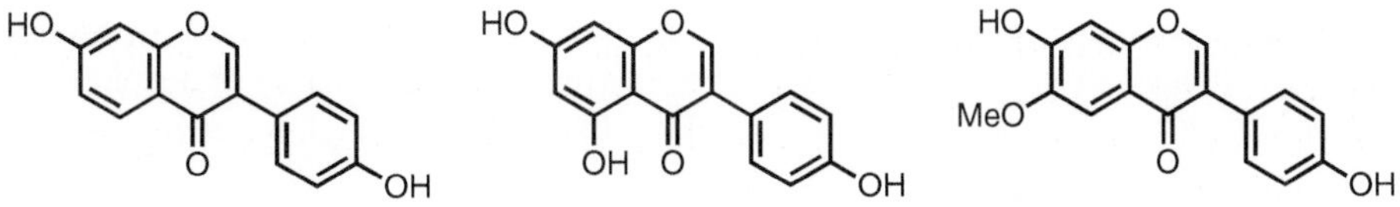

Figure 7.9 *Isoflavones*

tea, spinach, and broccoli. Quercetin can be found in kale and red onions, and myricetin is a constituent of tomatoes, oranges, and red wine.

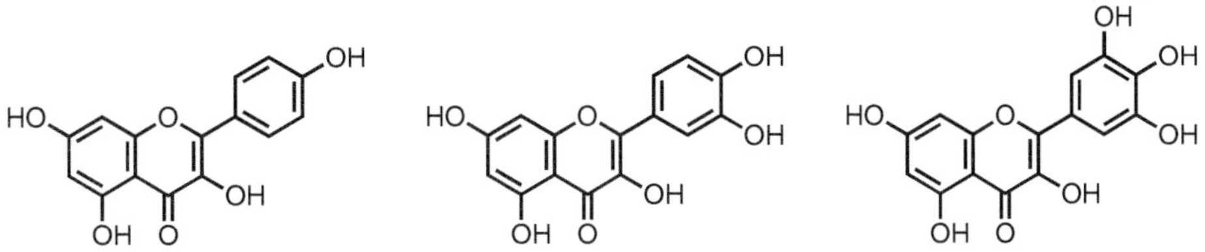

Figure 7.10 *Flavonols*

7.3.3 Anthocyanidins

Anthocyanidins (Table 7.1) are water soluble, primary plant pigments which very often give fruits and berries their distinctive colors. The colors are typically pH dependent and thus many of the anthocyanidins are used as natural pH indicators. Several are used as foodstuff coloring agents, and they also have antioxidant properties. Anthocyanidins are made even more water soluble by glycosylation, in which case the sugar-conjugate is called anthocyanin.

Table 7.1 *Anthocyanidins*

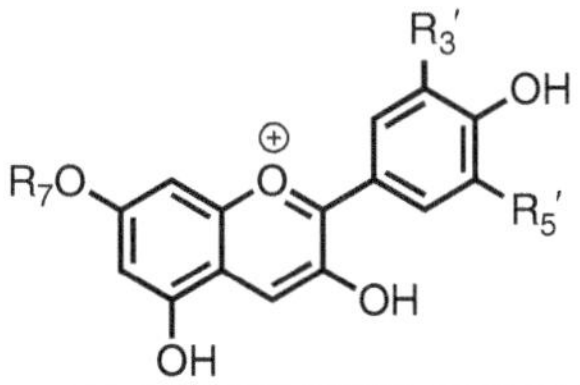

Anthocyanidin	R_3'	R_5'	R_7	Typical Occurrence
Pelargonidin	H	H	H	Yellow; many red berries
Cyanidin	OH	H	H	Many red berries
Delphinidin	OH	OH	H	Blue; cabernet sauvignon
Petunidin	OH	OMe	H	Many red berries
Malvidin	OMe	OMe	H	*Vitis vinifera* (red wine color)
Peonidin	OMe	H	H	Peony
Rosinidin	OMe	H	Me	*Catharantus roseus*

The color exhibited by anthocyanins was first explained in 1939 by *Linus Pauling* (1901–1994), who proposed that the resonant structure of the flavylium ion caused the intensity of their color. For example, the color changes of cyanidin are shown in Scheme 7.13. The flavylium cation is colored red, but it only occurs at highly acidic pH. When the pH is raised, the cation hydrates to form the colorless carbinol pseudobase. Close to neutral pH the pseudobase loses water to form the conjugated quinonoidal base, which is violet. At slightly basic conditions one of the phenols is deprotonated to form the blue anionic form of the quinonoid base. When the pH is further elevated, water adds to the quinonoid form and the heterocycle opens to form the yellow-colored chalcone [21].

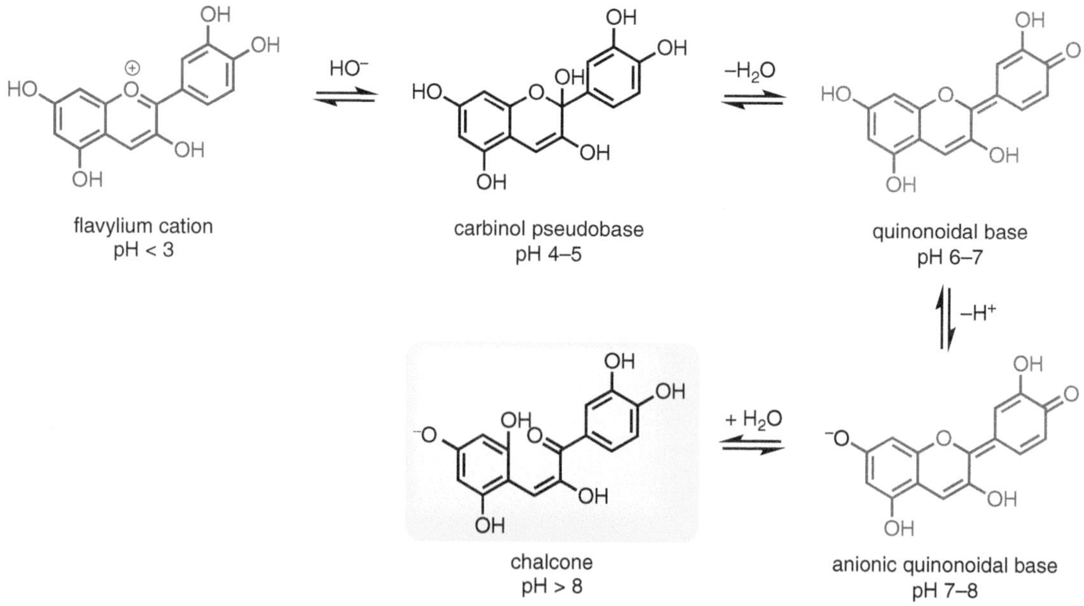

Scheme 7.13   *Color changes of cyanidin as a function of pH*

7.4   Stilbenes

Stilbenes are a relatively small group of phenylpropanoids. Many stilbenes are synthetised in plants as a response to stress caused by fungal infections, ozone, and physical damage, such as pinosylvin (Figure 7.11). *trans*-Resveratrol (3,5,4-trihydroxy-*trans*-stilbene) occurs widely in the skin of grapes, blueberries, raspberries, mulberries, and peanut. Piceatannol and its *O*-glucoside astringin occur in the roots of the Norway spruce (*Picea abies*). It is also a metabolite of *trans*-resveratrol [22].

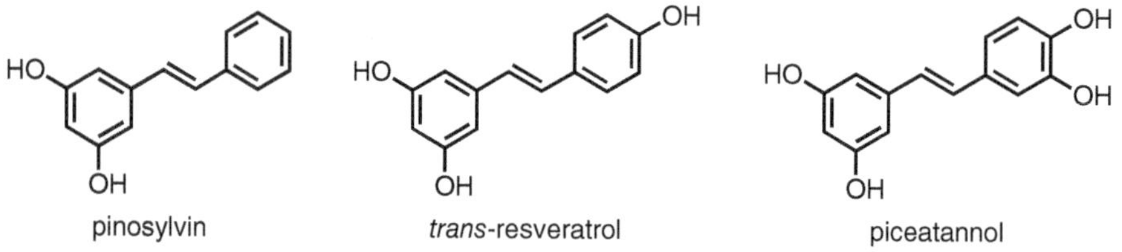

Figure 7.11   *Stilbenes*

7.5   Lignin

Compounds structurally related to shikimates have an important role in the formation of lignin, the binding material of wood. Lignin is a complex plant-derived biopolymer that accounts for *ca.* 20% of all terrestrial biomass, 30% of all non-fossil carbon, and up to 40% of the energy content of the lignocellulosic material. The molecular structure is largely composed of phenylpropanoid residues (*p*-hydroxyphenyl, guaiacyl, and syringyl) derived from three hydroxycinnamyl alcohol-based monomer;: *p*-coumaryl, coniferyl, and sinapyl alcohols. These molecular units are combined in different ways, but the most common linkage is an ether linkage between the *β*-carbon of one phenylpropanoid and the 4-phenolic group of another one. This *β*-O-4 linkage represents *ca.* 45–50% of the linkages in softwood and 60% in hardwood (Figure 7.12).

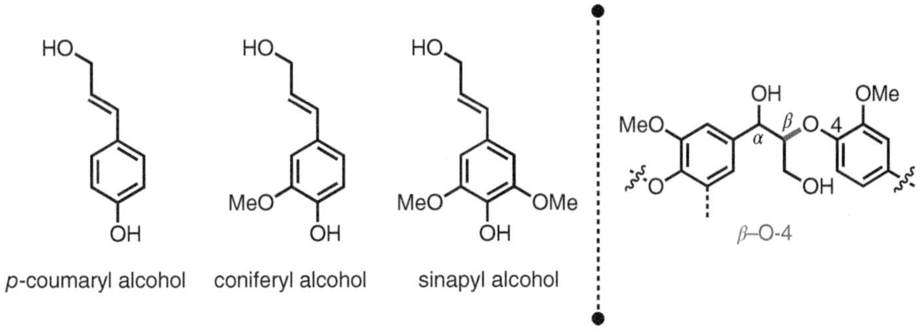

Figure 7.12 *Lignin basic structural units*

The composition of hardwood and softwood lignins varies in the proportions of the sinapyl, coniferyl, and *p*-coumaryl alcohols (Figures 7.13 and 7.14). While about 90% of soft-wood lignin is constructed from coniferyl alcohol, nearly equal proportions of sinapyl and coniferyl alcohols are found in hardwood lignin, although sinapyl units usually dominate (accounting for 45–75% proportion) depending on the hardwood species.

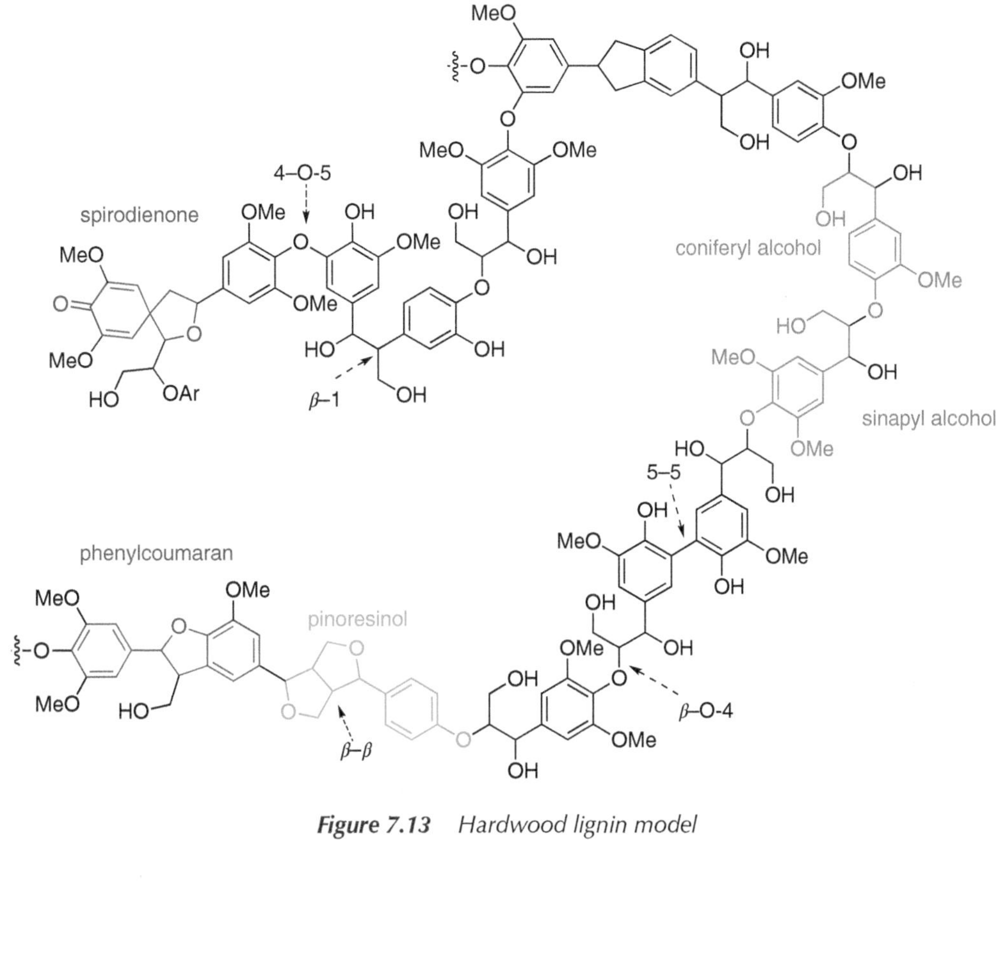

Figure 7.13 *Hardwood lignin model*

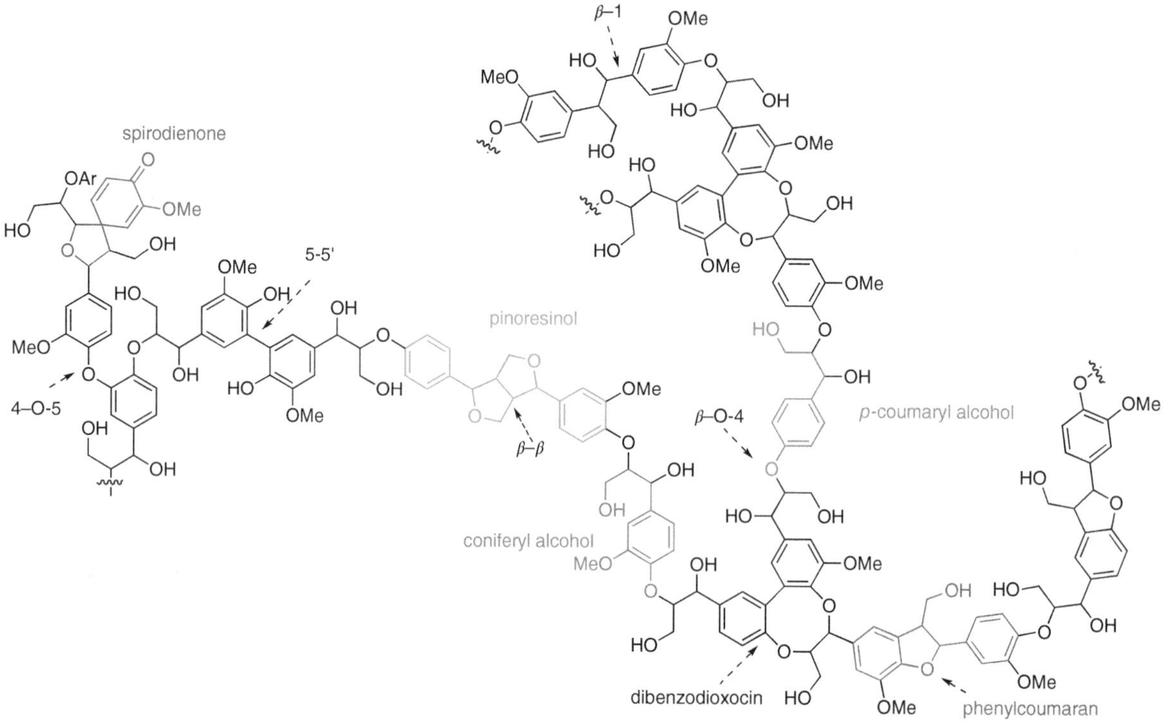

Figure 7.14   *Soft-wood lignin model*

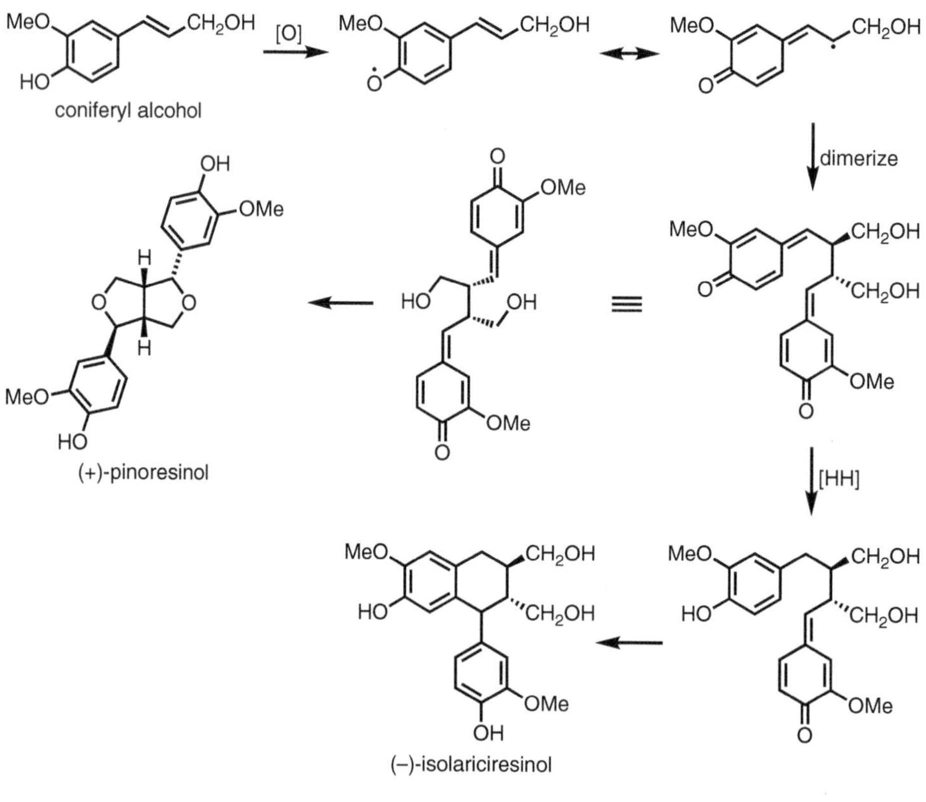

Scheme 7.14   *Biosynthesis of lignans from coniferyl alcohol*

The Swedish chemist *Holger Erdtman* (1902–1989) suggested, in 1933, that lignin is mainly formed by oxidative radical polymerisation of phenylpropionyl units, especially those derived from coniferyl alcohol, the alcohol corresponding to ferulic acid [23]. Because of the several possible precursor alcohols, and the possibility for various ways of radical combinations, lignin has a highly heterogeneous structure (Scheme 7.14).

References

1. Quideau, S., Deffieux, D., Douat-Casassus, C., Pouységu, L. *Angew. Chem. Int. Ed.* **2011**, *50*, 586–621.
2. Coutinho, E.M. *Contraception* **2002**, *65*, 259–263.
3. Vogel, H., Pelletier, J. *Journal de Pharmacie* **1815**, *1*, 289–303.
4. Dixon, R.A., Achnine, L., Kota, P., Liu, C.-J., Srinisava Reddy, M.S., Wang, L. *Mol. Plant Physiol.* **2002**, *3*, 371–390.
5. a) Guroff, G., Reifsnyder, A., Daly, J.W. *Biochem. Biophys. Res. Commun.* **1966**, *24*, 720–724; b) Guroff, G., Daly, J.W., Jerina, D.M., Renson, J., Witkop, B., Udenfriend, S. *Science* **1967**, *157*, 1524–1530.
6. Payen, A. *Compt. Rend. Acad. Sci.* **1846**, *24*, 724–737.
7. Kanne, D., Straub, K., Hearst, J.E., Rapoport, H. *J. Am. Chem. Soc.* **1982**, *104*, 6754–6764.
8. Yang, Q., Rui Guo, R., Wang, J. *Asian J. Org. Chem.* **2019**, *8*, 1742–1765.
9. Govender, T., Hojabri, L., Moghaddam, F.M., Arvidsson, P.I. *Tetrahedron: Asymmetry* **2006**, *17*, 1763–1767.
10. Rueping, M., Uria, U., Lin, M.-Y., Atodiresei, I. *J. Am. Chem. Soc.* **2011**, *133*, 3732–3735.
11. Jiang, Z.-Z., Gao, A., Li, H., Chen, D., Ding, C.-H., Xu, B., et al. *Chem. Asian J.* **2017**, *12*, 3119–3122.
12. Punyasiri, P.A.N., Abeysinghe, I.S.B., Kumar, V., Treutter, D., Duy, D., Gosch, C., et al. *Arch. Biochem. Biophys.* **2004**, *431*, 22–30.
13. Yang, Z., Xiao, F., Zhang, Y., Wu, Z., Zheng, X. *Nat. Prod. Res.* **2019**, *33*, 2995–3010.
14. Lin, G., Chang, L., Liu, Y., Xiang, Z., Chen, J., Yang, Z. *Chem. Asian J.* **2013**, *8*, 700–704.
15. Shiraishi, N., Kumazoe, M., Fuse, S., Tachibana, H., Tanaka, H. *Chem. Eur. J.* **2016**, *22*, 13050–13053.
16. Nibbs, A.E., Scheidt, K.A. *Eur. J. Org. Chem.* **2012**, 449–462.
17. Lemke, M.-K., Schwab, P., Fischer, P., Tischer, S., Witt, M., Noehringer, L., et al. *Angew. Chem. Int. Ed.* **2013**, *52*, 11651–11655.
18. McDonald, B.R., Nibbs, A.E., Scheidt, K.A. *Org. Lett.* **2015**, *17*, 98–101.
19. Oyama, K., Yoshida, K., Kondo, T. *Curr. Org. Chem.* **2011**, 2567–2607.
20. Chevreul, M.E. "30e Leçon, Chapitre XI. De la Gaude" in *Leçons de Chimie Appliquée à la Teinture*, Paris, France: Pichon et Didier, **1829**, pp. 143–148.
21. a) Castañeda-Ovando, A., Pacheco-Hernández, L., Páez-Hernández, E., Rodríguez, J.A., Galán-Vidal, C.A. *Food Chem.* **2009**, *113*, 859–871; b) Roy, S., Rhim, J.-W. *Crit. Rev. Food Sci. Nutr.*, DOI: 10.1080/10408398.2020.1776211.
22. Chong, J., Poutaraud, A., Hugueney, P. *Plant Sci.* **2009**, *177*, 143–155.
23. Erdtman, H. *Biochem. Z.* **1933**, *258*, 172–180.

8

Polyketides

The synthesis of erythromycin A is probably the most extensive single project in the history of synthetic organic chemistry. … the complexity of the molecule's structure, the plethora of stereocenters and functional groups, and the magic of the medium ring has fascinated … large research groups worldwide for more than a decade.

Johann Mulzer, 1991

Lengthening the carbon chains of biological molecules frequently occurs by adding a two-carbon acetate unit into the molecule. Those natural products whose biosynthesis involves few other transformations, and which are distinguishably derived from the two-carbon fragments by straightforward chain extensions and reductions/oxidations are typically grouped into polyketides. The name implies that the acetate units can be replaced with longer carbon chains, such as propionate. However, it is the (substituted) two-carbon units that form the main chain of the natural products, as shown below for the structure of erythronolide B, a biogenetic precursor of all the erythromycins (Figure 8.1). The bold lines indicate the propionate units which come intact from propionyl-CoA.

Polyketides are classified based on their structures and biochemical origins according to the classification shown in Figure 8.2. The major classes are fatty acids, polypropionates, and aromatic polyketides. We shall follow this classification in our journey.

Polyketides are formally synthesised by aldol reactions, and therefore it comes as no surprise that this class of natural products has made aldol reaction the oldest systematically studied reaction [1]. The need for stereoselective and specific generation of 1,2-dioxygenated systems requires thoroughly developed 1,2-oxygenation methods (epoxidation or dihydroxylation); 1,3-dioxygenated systems similarly require the methodology for the aldol and related reactions. These reactions have provided the cornerstones in the era of racemic synthesis, and they also play a significant role in the era of enantioselective synthesis.

8.1   Biosynthesis

The elongation of carbon chains usually occurs through the addition of two carbon atoms at a time. In Chapter 1, we saw the biosynthesis of acetic acid from carbon dioxide and water. In the chain-elongation process, the activation of acetic acid both as the nucleophile and electrophile is achieved by forming thioester bonds. The activation is performed in a dual form by the reaction of the acetate with coenzyme A (CoA, Figure 8.3) to form a thioester. The protons at the α carbon of the thioester are more acidic than the corresponding protons in oxyesters, thus making the enolisation more

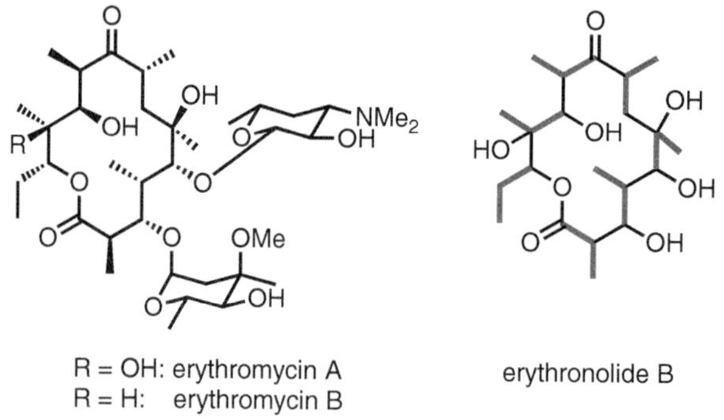

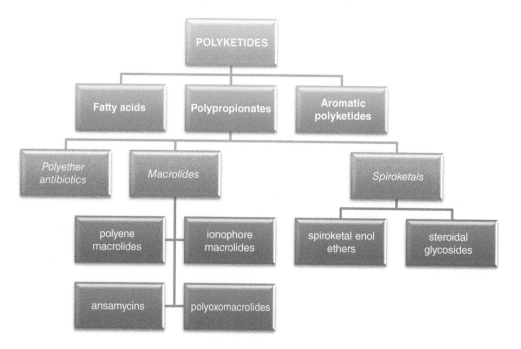

R = OH: erythromycin A
R = H: erythromycin B

erythronolide B

Figure 8.1   *Polyketide units in erythromycins*

Figure 8.2   *Classification of polyketides*

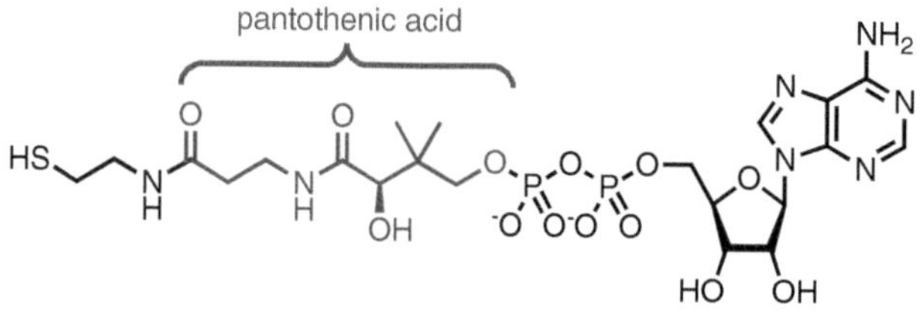

Figure 8.3   *Structure of coenzyme A*

facile. CoA is also relatively stable but still easily cleaved under physiological conditions. Substituents are acceptable, and this facilitates the biosynthesis of, e.g. polypropionates.

The coenzyme also functions to activate the acyl group towards nucleophilic attack: just like in syntheses we try to achieve this by means of a variety of activating groups. Since acetate (or thioacetate) α protons still have relatively low acidities, further activation is often needed. This can be brought about by further carboxylation to give a malonate unit (Scheme 8.1). Biotin participates in this process by first being carboxylated with carbon dioxide in an ATP-dependent process. The carboxylate group is then transferred to an enolised form of acetyl-CoA to form malonyl-CoA, a much more powerful nucleophile than acetyl-CoA.

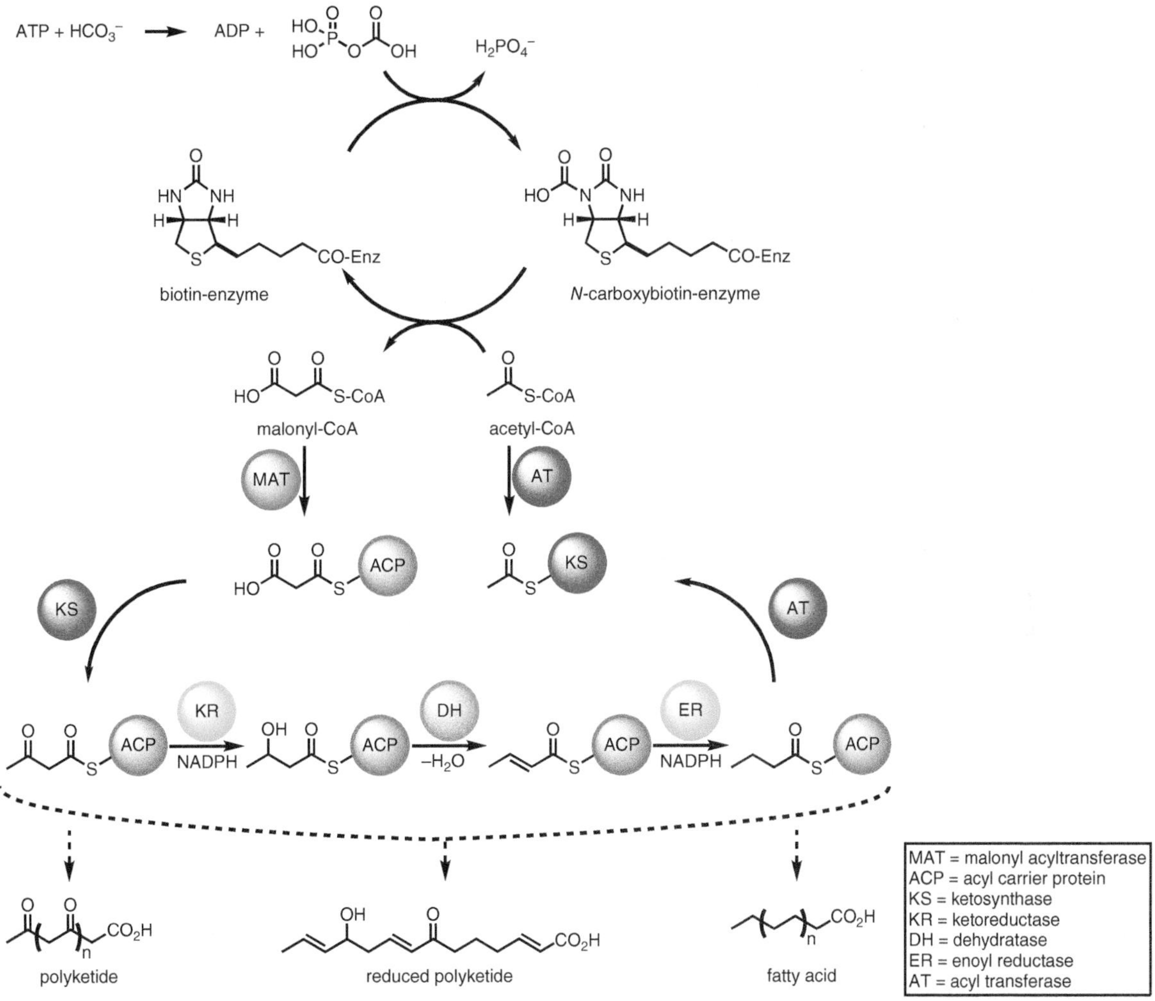

Scheme 8.1   Generalised biosynthesis of polyketides and fatty acids

The polyketide and fatty acid biosynthesis is carried out by multidomain enzymes, the enzyme fatty acid synthase (FAS) and polyketide synthase (PKS), respectively, containing all the requisite catalytic activities (Scheme 8.1) [2]. The growing chain is attached to the active site of ketosynthase (KS) by an acyltransferase (AT) protein, and the malonate is transferred (through the action of malonyl acyltransferase, MAT) to an acyl carrier protein (ACP). The chain lengthening then proceeds by the malonyl-ACP attacking ac(et)yl-KS in a Claisen ester condensation (with simultaneous loss of CO_2), to give an ACP-bound β-ketoester. This can then be processed through two NADPH-dependent reduction reactions (ketoreductase, KR, and enoylreductase, ER) with intermediate loss of water with the action of dehydratase (DH), to a new chain-lengthened acyl-ACP. The latter can re-enter the cycle to form the next two-carbon elongation. Of course, the intermediate reduction and elimination steps do not need to occur every time. This makes it possible to generate skipped diketones, which easily undergo intramolecular condensations to form aromatic polyke-tides (see Scheme 7.14 for biosynthesis of aflatoxins). Only partial application of the reduction/elimination processes leads to reduced polyketides. When the acyl groups transferred are not acetyl groups but propionyl groups (analogously through methylmalonyl-CoA etc.), a very large group of polypropionates is accessed.

8.2   Fatty Acids

Michel Eugène Chevreul (1786–1889), a student and successor of *Louis Nicolas Vauquelin* (1763–1829) at the Muséum National d'Histoire Naturelle (National Museum of Natural History) in the Jardin des Plantes, Paris, is considered the pioneer of the study of animal fats. He isolated cholesterol from gall stones in 1815. In 1823, he published his major work '*Recherches sur les corps gras d'origine animale,*' where he described the true nature of soap. He also elucidated the composition of stearin, the white substance found in the solid parts of most animal and vegetable fats; and olein, the liquid part of any fat; and isolated stearic and oleic acids, the names of which he invented. This work led to important improvements in the processes of candle-manufacture.

Fatty acids, fatty alcohols. and phospholipids belong to polyketides. Typical fatty acids contain a long alkyl chain with an even number of carbon atoms in the main chain. The chain is seldom branched, but unsaturation is often encountered (Figure 8.4). For instance, the 18-carbon straight-chain acids palmitic, oleic, linoleic, and linolenic acids are all known and occur for instance in soybean lecithin. The longer polyunsaturated C_{20} acids form the basis of the arachidonic acid cascade, which will be discussed in connection with prostaglandins, thromboxanes, and leukotrienes. The unsaturation is typically near the center of the chain (Δ^9 is the common site, regardless of the chain length; for instance, compare palmitoleic and oleic acids), and the double bond geometry is *cis*. Polyunsaturated fatty acids (PUFAs) typically contain skipped polyene units, where the two olefinic bonds are separated with a methylene unit (e.g. linoleic and linolenic acids). The unsaturation is brought to the chain only after construction of the carbon chain by fatty acid desaturase enzymes, which are oxygen-dependent non-heme iron enzymes [3].

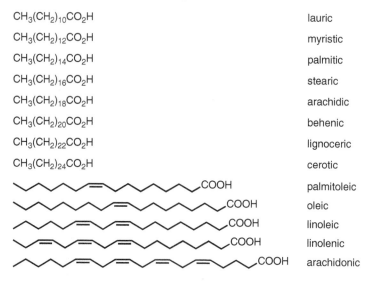

$CH_3(CH_2)_{10}CO_2H$	lauric
$CH_3(CH_2)_{12}CO_2H$	myristic
$CH_3(CH_2)_{14}CO_2H$	palmitic
$CH_3(CH_2)_{16}CO_2H$	stearic
$CH_3(CH_2)_{18}CO_2H$	arachidic
$CH_3(CH_2)_{20}CO_2H$	behenic
$CH_3(CH_2)_{22}CO_2H$	lignoceric
$CH_3(CH_2)_{24}CO_2H$	cerotic

palmitoleic
oleic
linoleic
linolenic
arachidonic

Figure 8.4   *Fatty acids*

Fatty acids usually occur in cell membranes, bound to glycerol phosphate through an ester bond. The four main classes of phospholipids are shown in Figure 8.5. Phospholipids function as the storage sites for fatty acids, which are released

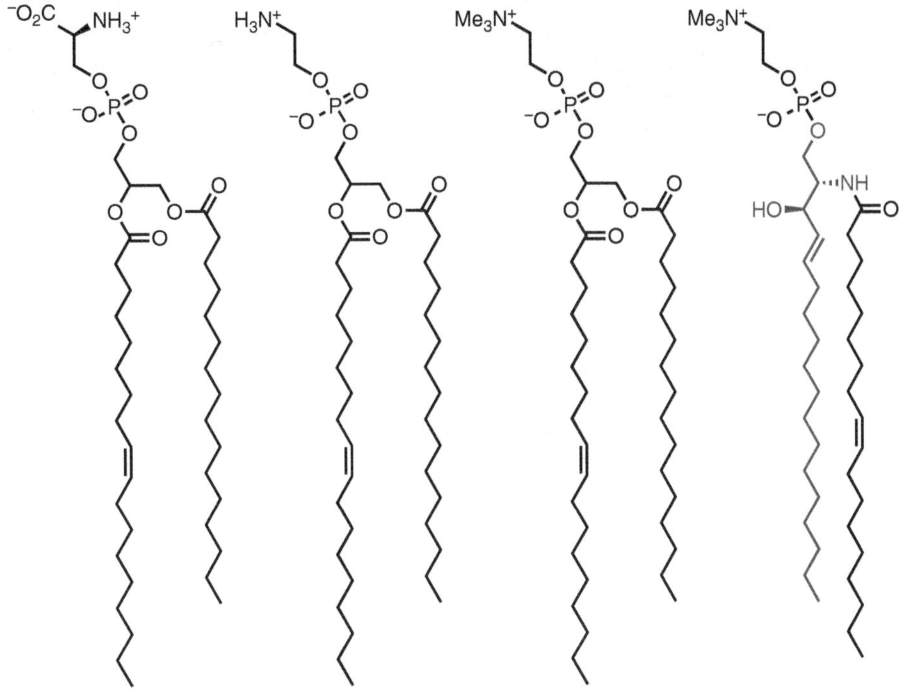

Figure 8.5   The four main phospholipids

through the action of phospholipases as needed. During nerve excitation, phosphatidylcholine and sphingomyelin can liberate choline, which on acetylation gives acetylcholine, an important neurotransmitter.

Terpenoids will be discussed separately in Chapter 9, but it is wise to make a note here of the fact that cholesterol binds to the phospholipid bilayer membrane and thus rigidifies its structure. The dimensions and polarity characteristics of cholesterol are very similar to those of the lipid bilayer (see Figure 8.6): the hydrophilic part is attached to a hydrophobic part of approximately 20 Å length and 7 Å girth. The rigid cholesterol molecule can easily slip into the lipid bilayer and, being itself a rigic molecule, it thus rigidifies the membrane.

The fatty acids in phospholipids can be exchanged rather easily through a transesterification process. This facilitates the intermediary storage function of the membranes. Arachidonic acid, a polyunsaturated fatty acid containing 20 carbon atoms, is particularly important in this respect. After its release from the phospholipid, arachidonic acid participates in controlling several biological phenomena through a series of chemical reactions.

8.2.1   Prostaglandins, Thromboxanes, Leukotrienes

In 1935, the Swedish physiologist *Ulf von Euler* (1905–1983) and the Irish-English physiologist *Maurice Walter Goldblatt* (1895–1967) independently isolated a fraction from human seminal fluid which had both blood pressure-lowering and smooth muscle-contracting properties [4]. Since the compound was also isolated from prostate gland extracts, von Euler called these substance prostaglandins. A large number of these have been isolated, and their physiological roles are varied ranging from blood pressure regulation (PGI_2 or epoprostenol), control of relaxation and contraction of the smooth muscles of uterus (PGE_2, or dinoprostone, and $PGF_{2\alpha}$, or dinoprost, which are used to induce labor or as abortifacients), as well as many other functions under clinical investigations. Many drugs

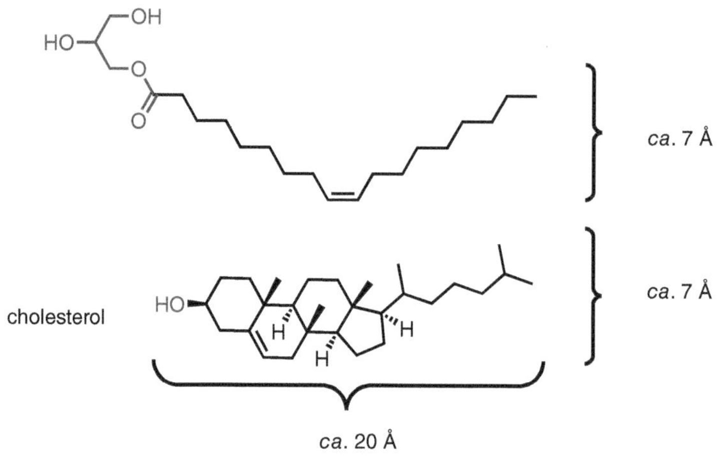

Figure 8.6   *Cholesterol and lipid bilayer*

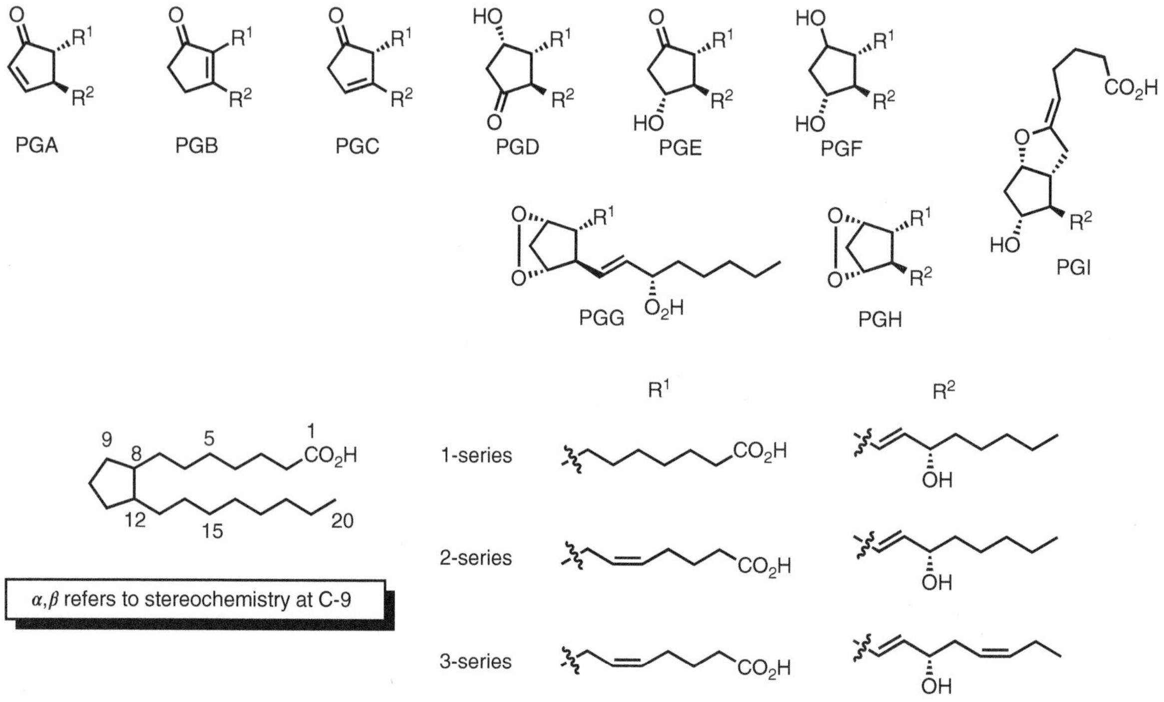

Figure 8.7   *Structures of prostaglandins*

have been developed from the prostanoid leads, e.g. a synthetic $PGF_{2\alpha}$ analogue, latanoprost, is used for the treatment of glaucoma. A shortcut key to naming the various prostaglandins is shown in Figure 8.7.

Two decades later, the Swedish group led by *Sune K. Bergström* (1916–2004) and *Bengt I. Samuelsson* (1934–) identified several compounds from this prostaglandin fraction, and they showed that the compounds were derivatives of arachidonic acid [5]. The total syntheses of these compounds were achieved in rapid pace, mainly through the efforts of E.J. Corey at Harvard University, and the biosynthetic connections between arachidonic acid, prostaglandins, leukotrienes (slow-reacting substances), and thromboxanes has been elucidated [6].

Upon excitation of the respective cellular receptors, arachidonic acid is released from its phospholipid conjugate through the action of phospholipase A, which leads to the initiation of the arachidonic acid cascade. Depending on the stimulus, cell type, and desired function, arachidonic acid is ultimately converted into prostaglandins, leukotrienes, or thromboxanes (Scheme 8.2). 5-Lipoxygenase (5-LO) oxidises arachidonic acid to 5-hydroperoxyeicosatetraenoic acid (5-HPETE) which is dehydrated to leukotriene A_4 (LTA_4). Coupling with glutathione-S-transferase (GST) gives the peptidoleukotriene LTC_4. An alternative oxidation of arachidonic acid by cyclooxygenases (COX-1 and COX-2) gives prostaglandin endoperoxide PGH_2 which is the precursor for both prostaglandins and thromboxanes.

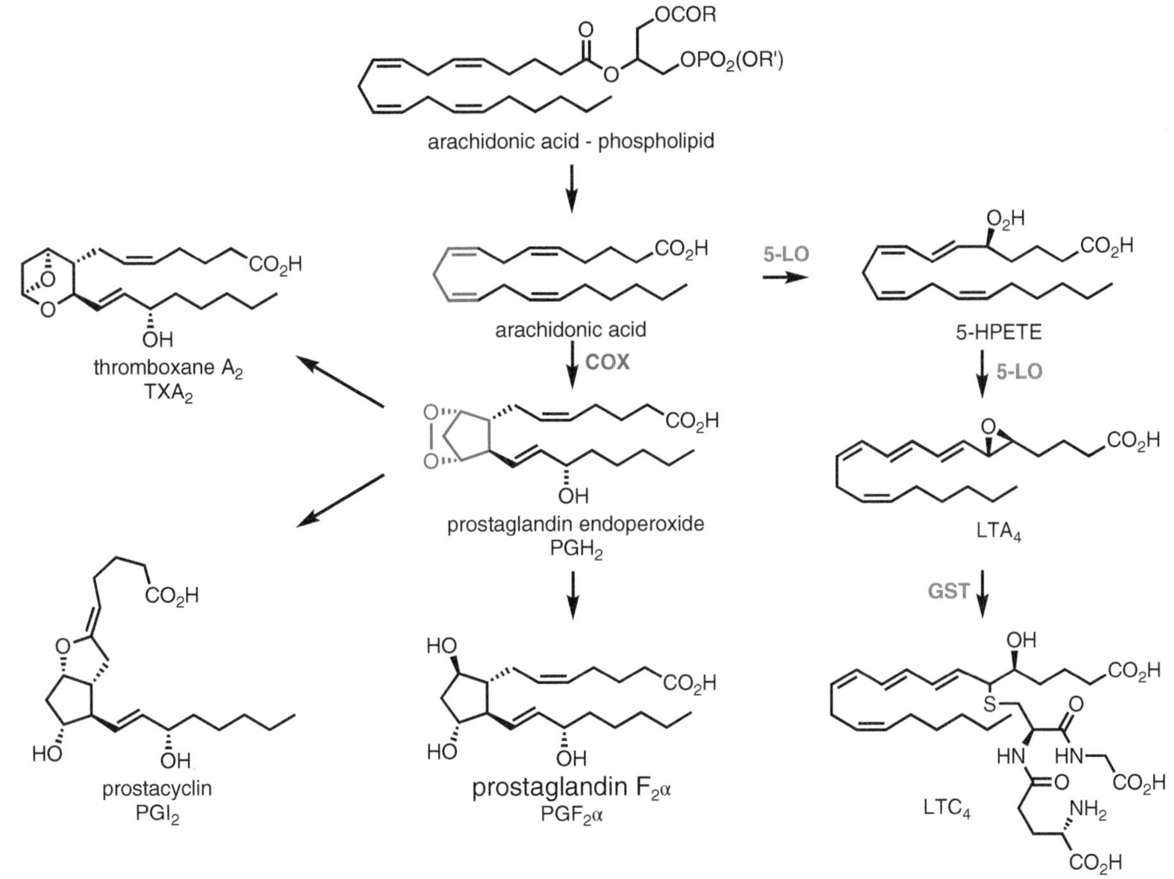

Scheme 8.2 Arachidonic acid cascade

Thromboxane A_2 (TXA_2) is a vasoconstrictor and is involved in blood platelet aggregation and coagulation. Leukotrienes were initially discovered in leukocytes (hence their name). Particularly leukotriene D_4 triggers contractions in the smooth muscles lining the trachea; its overproduction is a major cause of inflammation in asthma. Many anti-inflammatory agents, including cortisone derivatives, inhibit the release of arachidonic acid from phospholipids. A large class of non-steroidal anti-inflammatory agents (NSAID's) inhibit cyclooxygenase, and thus also the synthesis of prostaglandins. Typical examples are salicylates (acetylsalicylic acid), propionic acid derivatives (ibuprofen, ketoprofen, and naproxen), acetic acid derivatives (indomethacin and diclofenac), oxicams (piroxicam), and fenams (tolfenamic acid).

The broad diversity of biological activities of prostaglandins or their synthetic analogs, prostanoids, have spurred wide interest in the synthesis of these compounds. For medicinal chemistry, the challenges are given by the fact that prostanoids are chemically rather instable, they are metabolised rapidly, and they are often associated with numerous side effects. The different synthetic aspects have been widely reviewed [7]. In the synthesis of prostaglandins, the construction of the cyclopentane skeleton posed initially formidable challenges. Several solutions were developed, culminating eventually to two highly efficient, industrially applicable enantioselective approaches. The first one, by Corey, relies on a Lewis acid-catalysed Diels–Alder reaction (Scheme 8.3). The initial version employed an aluminum reagent, where the Diels–Alder adduct was obtained in 96% *ee*. During later operations, the enantiopurity could be improved by crystallisation of a suitable intermediate. Later improvements have found the (*R*)-tryptophan-derived oxazaborolidine to be a superior catalyst, providing the cycloadduct in *ca.* 200:1 enantioselectivity [8]. Note that the formyl C–H ⋯ O bond plays a key role in organising the aldehyde in a specific way. The indole ring of the tryptophan moiety effectively shields one face of the bromoacrolein, and the cyclopentadiene approaches with the sterically less encumbered face forward. The side chains at C8 and C12 could then be elaborated using several strategies to give the target prostaglandins or their derivatives.

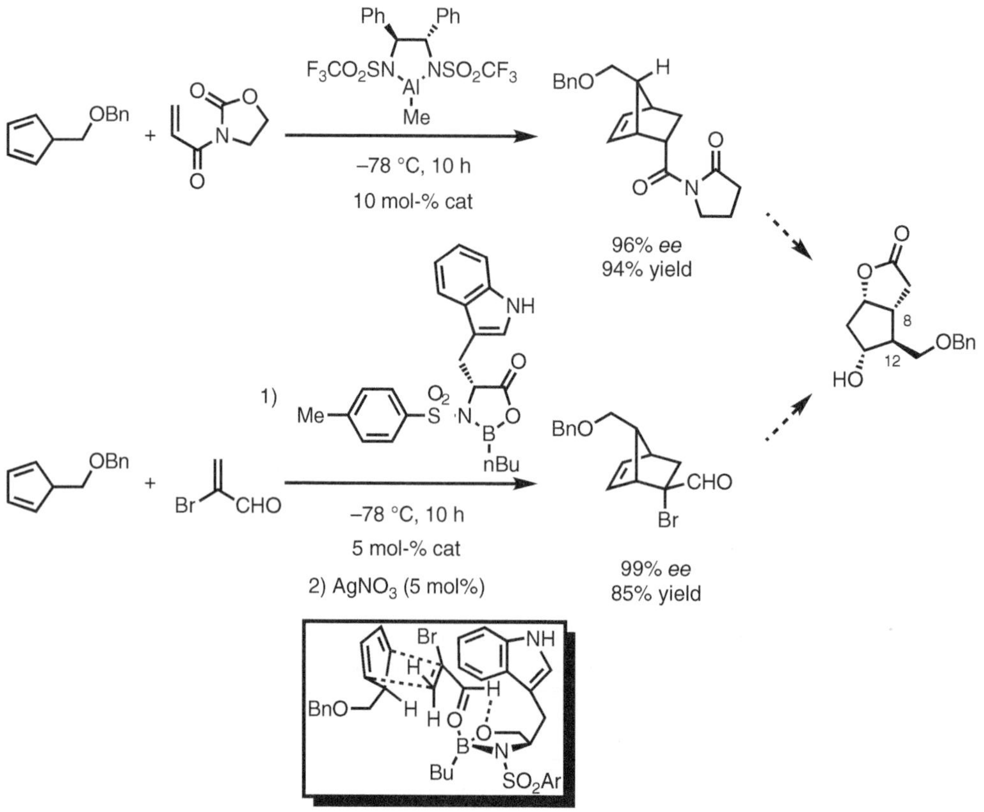

Scheme 8.3   *Catalytic asymmetric Diels–Alder in the synthesis of the Corey lactone*

Noyori devised another synthetic protocol for the construction of the prostaglandin skeleton based on an efficient *three-component synthesis* of a chiral cyclopentenone derivative with a vinyl metal nucleophile, followed by trapping of the intermediate enolate with a suitable electrophile (Scheme 8.4) [9]. The nucleophilic vinyl lithium compound was prepared in 98% *ee* through BINAL-H reduction, as discussed in Section 3.2.1.1. The chiral cyclopentenone can be made either by enzymatic kinetic resolution of a precursor acetate or kinetic resolution of the racemic allylic alcohol with Ru-BINAP hydrogenation [10].

The second challenge that prompted much experimentation before acceptable enantioselective syntheses were achieved was the construction of the correct absolute stereochemistry at C15 (Figure 8.8) [11].

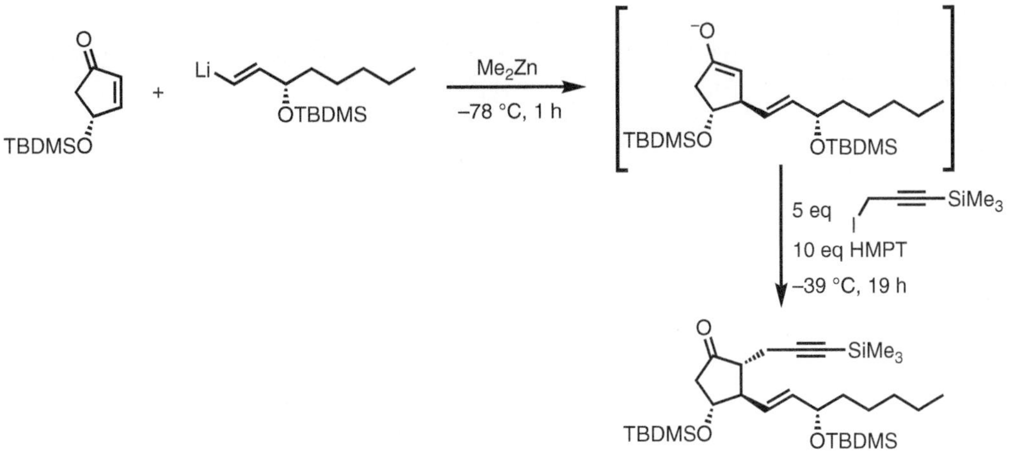

Scheme 8.4 *Noyori three-component prostaglandin synthesis*

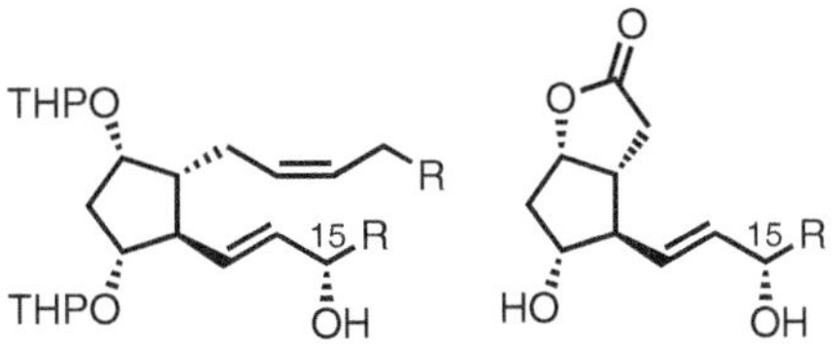

Figure 8.8 *The crucial C15 stereocenter*

BINAL-H reduction of either the bicyclic lactone intermediate (with either THP or acetyl protection on the C11 hydroxyl group) or of the monocyclic prostaglandin $F_{2\alpha}$ intermediate gave practically exclusive selectivity for the production of the desired 15-*S* isomer (Scheme 8.5) [12]. Besides the Noyori BINAL-H reduction mentioned previously, several direct selective reduction protocols for the intact side chain have been developed, and some of the most efficient ones involve bulky lithium borohydride reducing agents in the presence of a suitable controller group on the C11 hydroxyl group [13]. With the advent of the oxazaborolidine reduction, catalytic reduction with the D-proline-derived reagent gave a satisfactory diastereoselectivity. Note the use of the bulky *p*-phenylbenzoyl (PBz) protecting group of the C11 hydroxyl to enhance the steric effect of the existing chiral center.

The importance of these studies in the prostaglandin area has been recognised in the Nobel Prize in Physiology or Medicine in 1970, awarded jointly to Sir Bernard Katz, Ulf von Euler, and Julius Axelrod *"for their discoveries concerning the humoral transmitters in the nerve terminals and the mechanism for their storage, release and inactivation,"* and the Nobel Prize in Physiology or Medicine 1982, awarded jointly to Bergström, Samuelsson, and John R. Vane *"for their discoveries concerning prostaglandins and related biologically active substances."* Furthermore, both *E.J. Corey* (1990) and *Ryoji Noyori* (2001) were awarded Nobel Prizes in Chemistry.

The intriguing and complex structures of prostaglandins have evoked intense research activity into their synthesis, comparable to that generated from *β*-lactam antibiotics and steroids. Prostaglandins are involved in several biological processes including inflammation, pain response, and fever. These compounds have gained widespread applications as pharmaceuticals for the treatment of diseases like pulmonary arterial hypertension and glaucoma. Currently more than 20 pharmaceuticals derived from prostaglandins are on the market, including the antiglaucoma compounds bimatoprost and latanoprost, as well as the stomach pain relieving lubiprostone (Figure 8.9). Annual sales of these compounds reach several hundred million US dollars, and they are produced in quantities up to 40 kgs per year (bimatoprost).

The synthetic strategies based on Corey lactone typically involve 20 or more chemical steps, and therefore are not practical for large-scale synthesis in an economic fashion. Recently, several ingenious approaches have been developed based on modern synthetic protocols; we will take a look at two of these strategies. The first one was developed by

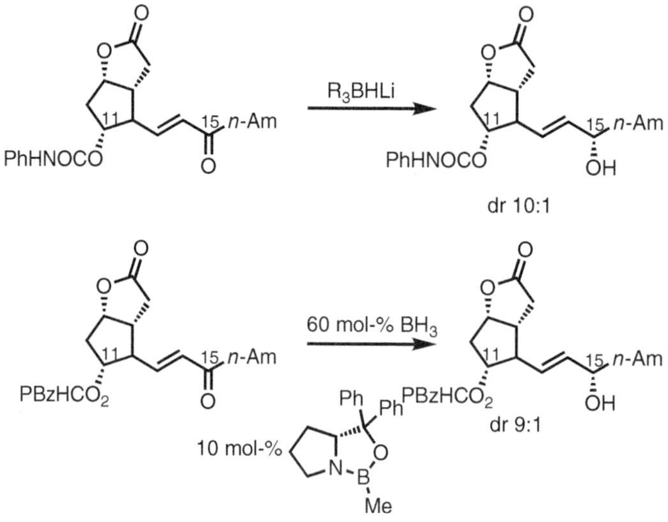

Scheme 8.5   *Setting the stereochemistry at C15*

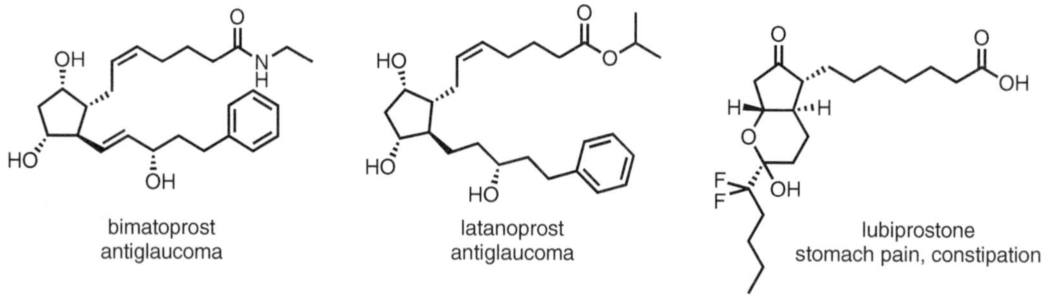

Figure 8.9   *Active pharmaceutical ingredients derived from prostaglandins*

Aggarwal and relies on a cascade aldol condensation to construct the bicyclic core in a highly enantioselective fashion (Scheme 8.6) [14]. Under optimised conditions[14b] the two-stage organocatalytic aldol reaction of succinaldehyde gave the bicyclic enal in 29% isolated yield and 98% *ee*. Conjugate addition of the C12 side chain was performed after protecting the hemiacetal. Installation of the C8 sidechain was achieved using standard Wittig chemistry. Overall, prostaglandin $F_{2\alpha}$ was synthesised in only seven steps.

   Another short and practical synthesis has been reported by Chen and Zhang (Scheme 8.7) [15]. The crucial chiral center was introduced by asymmetric hydrogenation of an enone using an iridium catalyst with a tridentate chiral ampfox-ligand. After conversion of the Weinreb amide to an ynone, the key enyne cycloisomerisation provided the core five-membered ring. Sequential hydrosilylation followed by hydride reduction and hydrolysis of the acetal provided the key intermediate for the construction of prostaglandin $F_{2\alpha}$. Thus, cross metathesis using the Hoveyda–Grubbs second generation catalyst installed the C12 sidechain, and the C8 sidechain was constructed using standard Wittig protocol. The authors also report the synthesis of six other prostaglandin analogs in six to eight steps in overall yields ranging from 5 to 20%.

   A slight modification of the prostaglandin synthesis allowed Aggarwal also to synthesise thromboxane B_2 using the organocatalytic aldol reaction (Scheme 8.8) [16]. The aldol product from Scheme 8.6 was converted to a *p*-methoxy benzyl acetal in 81% yield. This time, conjugate addition of the side chain was followed by ozonolysis to give the ketone. Baeyer–Villiger oxidation provided the thromboxane six-membered ring in 63% overall yield over three steps. The remaining steps included reduction of the lactone to the lactol followed by protecting-group manipulations and sidechain installation. The 12-step synthesis provides the product in 5% overall yield.

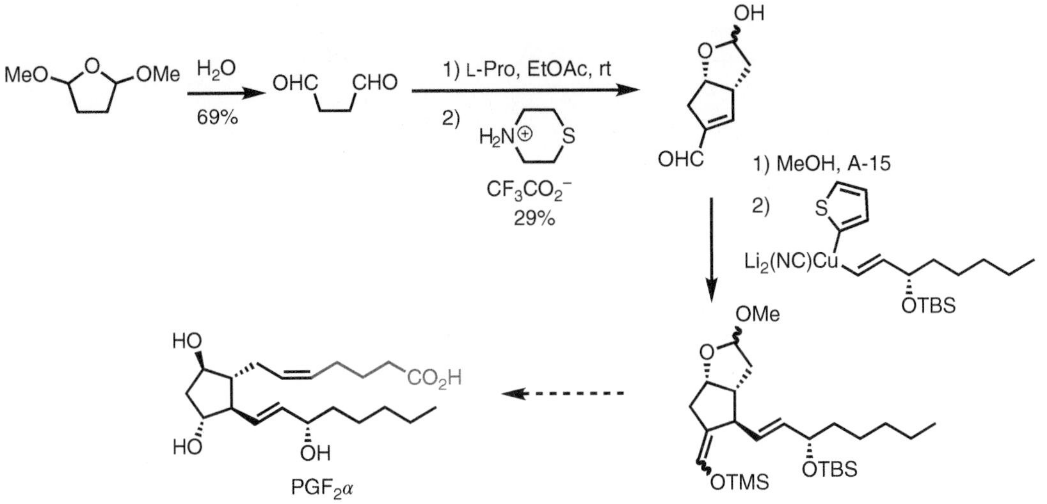

Scheme 8.6 *Aggarwal organocatalytic synthesis of prostaglandin F$_{2\alpha}$*

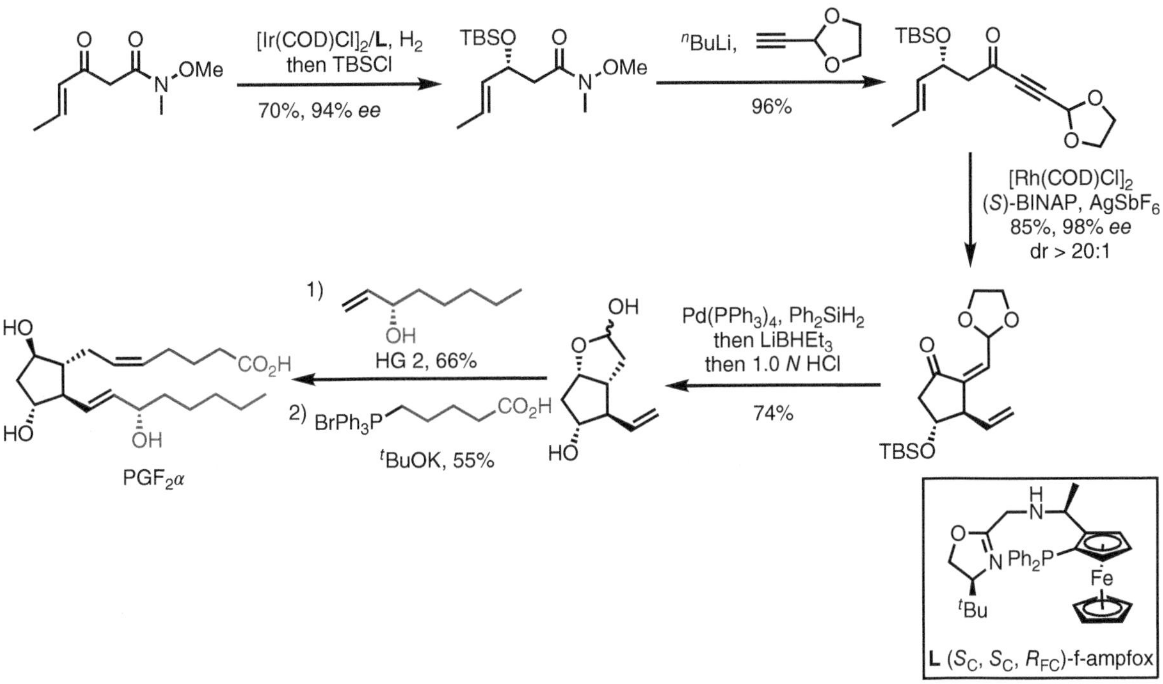

Scheme 8.7 *Zhang cycloisomerisation route to the synthesis of prostaglandin F$_{2\alpha}$*

8.2.2 Sphingolipids

The German-born chemist and clinician *Johann Ludwig Wilhelm Thudichum* (1829–1901) isolated an enigmatic new waxy lipid from human brain in 1884 [17]. He gave the name "sphingosin" to the backbone of these sphingolipids after the Greek monster Sphinx. Sphingosine provides another extremely abundant structural class of polyketides which are prevalent in cell membranes. It has been estimated that up to 70% of the dry weight of the myelin sheath (the protective layer surrounding nerve cells) is composed of sphingosine-containing components.

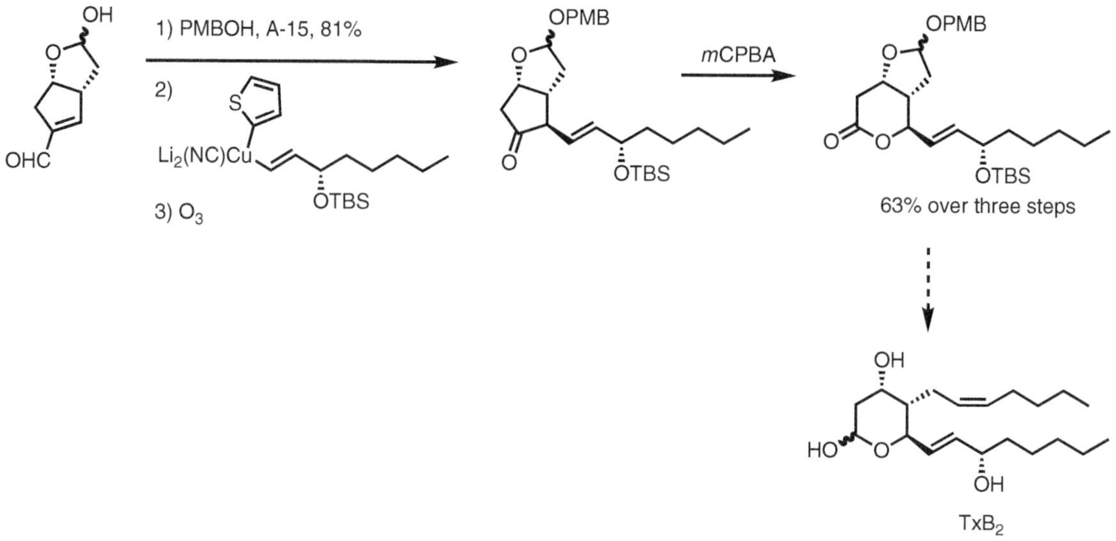

Scheme 8.8   *Aggarwal synthesis of thromoboxane B$_2$*

Sphingolipids are ubiquitous membrane components of essentially all eukaryotic cells, and are abundantly located in all plasma membranes as well as in some intracellular organelles (endoplasmic reticulum (ER), Golgi complex, and mitochondria) [18]. Structurally, sphingolipids are formed from three units; the basic amino alcohol sphingosine, a polar head group, and a fatty acid. The structural unit common to almost all sphingolipids in eukaryotic cells is an amino alcohol D-erythro-sphingosine [(2S,3R)-2-amino-3-hydroxy-4-(E)-octadecenol)] (Figure 8.10). The saturated phytosphingosine is produced by yeast (*Saccharomyces cerevisiae*).

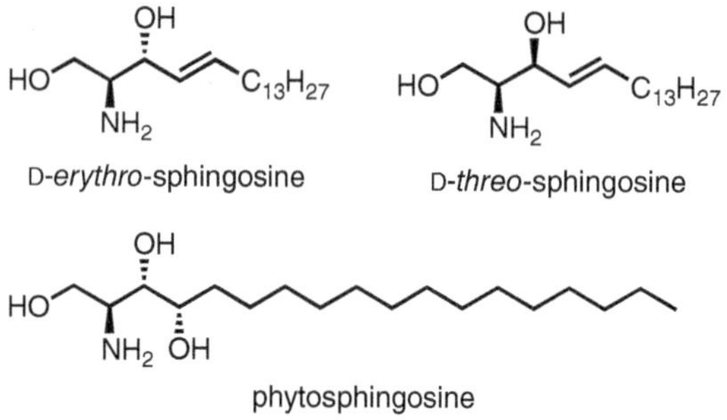

Figure 8.10   *Sphingosine derivatives*

In addition to D-erythro-sphingosine, there are nearly 100 other sphingoid base structures found elsewhere in nature [19]. Phytosphingosines constitute the major base component of higher plants, protozoa, yeast, and fungi, and have also been found in human kidney cerebrosides and in some cancer cell types. The free sphingoid bases occur only in small quantities but are connected through an amide bond to long-chain fatty acids to form ceramide (Cer). Attachment of different polar head groups forms sphingomyelin (Figure 7.5) and glucosyl- and galactosylceramides. More complex glycosphingolipids can contain several dozens of sugar residues. These include gangliosides and the blood group antigens (see Section 4.4) [20].

Sphingosine biosynthesis *de novo* (Scheme 8.9) starts with L-serine, which is decarboxylated with the pyridoxal phosphate-dependent reaction path. Acylation with palmitoyl-CoA occurs with regeneration of the stereocenters to give a ketone intermediate. After reduction of the ketone and attachment of the acyl side chain on the amino group to

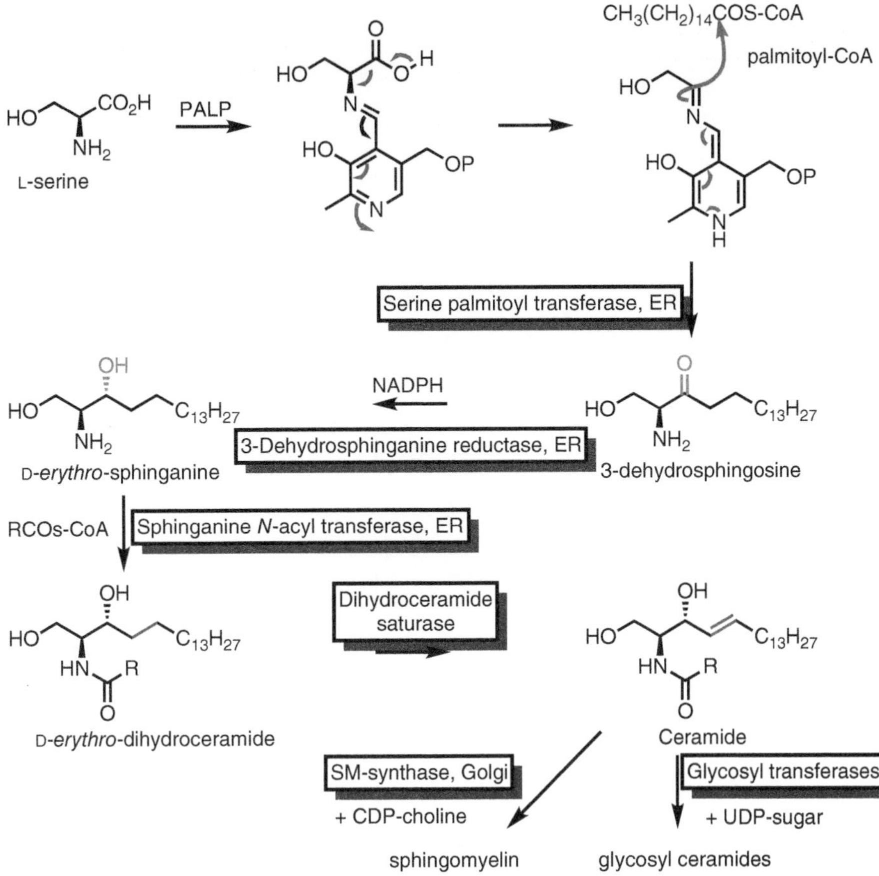

Scheme 8.9 *Biosynthesis of ceramide from L-serine*

form a dihydroceramide, the C4–C5 unsaturation is brought about by a desaturase enzyme. The ceramide thus formed is transferred from the endoplasmic reticulum to the Golgi apparatus, where it is further processed to sphingomyelin or glycosyl ceramides.

Sphingomyelin is the storage molecule for sphingosine, and its breakdown follows the path shown in Scheme 8.10. Sphingomyelinase cleaves off phosphatidylcholine, and the ensuing ceramide is hydrolysed by ceramidase to give free sphingosine. Followed by sphingosine kinase-mediated phosphorylation of the terminal hydroxyl group to form sphingosine-1-phosphate [21], itself an important messenger molecule in cells, phosphatidylethanolamine is cleaved, and the hexadecenal that is formed is processed in fatty-acid metabolic pathways. This latter cleavage process is formally the reverse of the PALP-mediated condensation brought about by serine palmitoyl transferase (Scheme 8.9).

Besides classical sphingolipids, nature abounds with several C_{18} amino alcohols (Figure 8.11). Several of these compounds have been isolated from marine sources, and little of their biological roles is still known. Myriocin is an example of a broad class of sphingolipids containing a quaternary stereocenter at C2 [22]. Myriocin itself was observed to be immunosuppressive, and it has served as the lead molecule in the development of fingolimod, a 3-billion-dollar-selling compound (2020) for the treatment of multiple sclerosis [23]. Jaspine was the first anhydrosphingosine isolated (in 2000) and has been investigated for its inhibitory effects on cell proliferation. Amaminol in turn is a compound isolated only once, from an unidentified tunicate, which was isolated in quantities too small to allow complete pharmacological profiling. We will briefly look at the syntheses of the latter two compounds to illustrate new aspects of enantioselective synthesis.

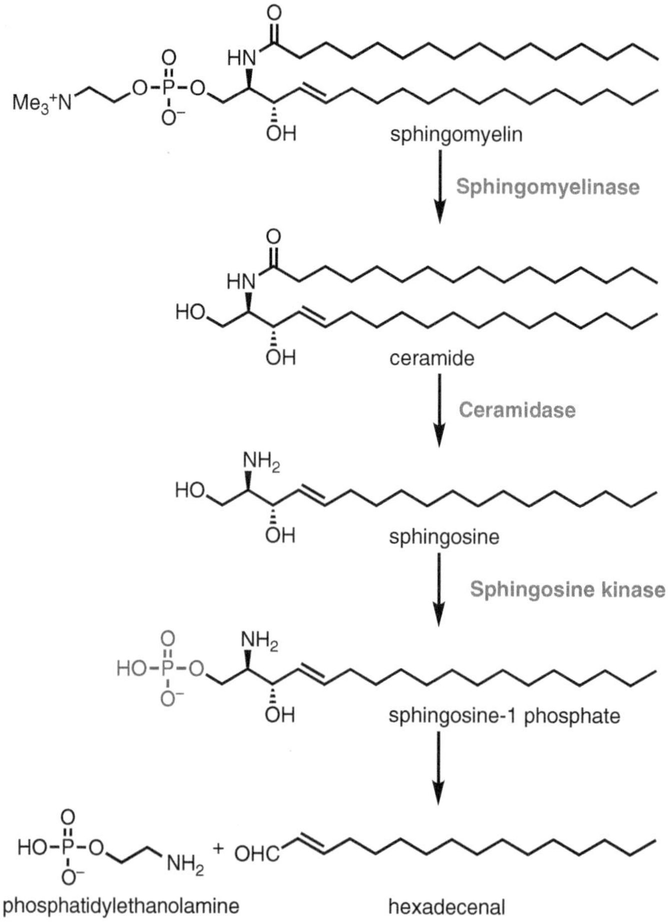

Scheme 8.10   *Degradation of sphingomyelin*

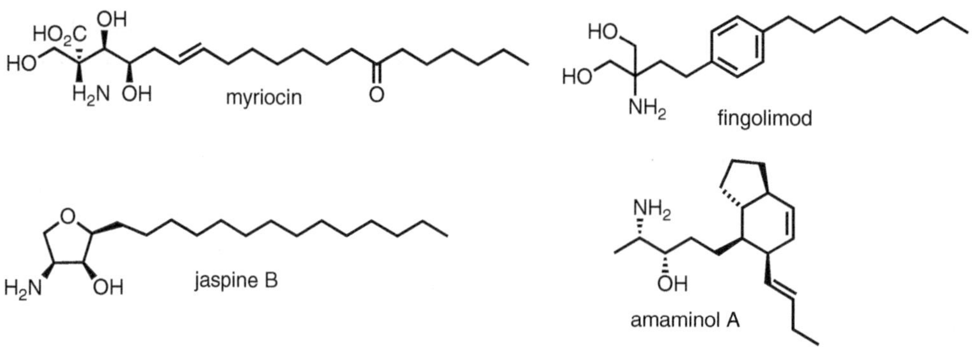

Figure 8.11   *Structures of naturally occurring C_{18} amino alcohols*

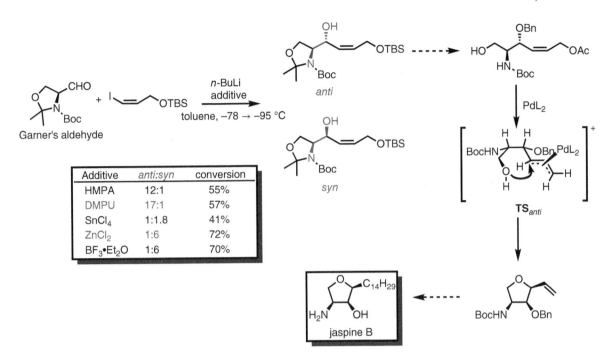

Scheme 8.11 Synthesis of jaspine B

Garner's aldehyde (Scheme 8.11) is a versatile starting material for several amino alcohols. The main worry with amino aldehydes is that they are very prone to epimerisation, and this vulnerability casts serious doubts on their use in larger-scale production [24]. The synthesis of jaspine B illustrates how this vulnerability can be avoided by careful consideration of reaction parameters [25]. Diastereoselectivity of the addition of the vinyl lithium compound onto Garner's aldehyde is readily modulated by additives. Thus, DMPU (a deaggregating ligand for Li) and $ZnBr_2$ (transmetallation to a less basic vinyl zinc species) give opposite diastereomers as major products. Each could be separately converted to the hydroxy allyl acetates (one diastereomer shown), which undergo highly diastereoselective palladium-mediated cyclisations. Final elaboration of the side chain (cross metathesis), reduction, and deprotection lead to jaspine B. From the divergence point at the vinyl lithium addition, one can arrive at all four diastereomers of the natural product, employing chiral information derived only from natural L-serine.

Amaminols A and B are cytotoxic bicyclic aminoalcohols isolated in 1999 from an unidentified tunicate of the family *Polyclinidae*, with an IC_{50} value of 2.1 μg/mL against P388 murine leukemia cells [26]. Not much else was known about these compounds as a synthesis based on organocatalytic intramolecular Diels–Alder (IMDA) reaction was developed (Scheme 8.12) [27]. The tetraene alcohol was initially oxidised with MnO_2 to the corresponding aldehyde, which underwent a highly diastereoselective IMDA when treated with the L-phenylalanine-derived chiral imidazolidinone. The synthesis was completed by coupling the aldehyde with an L-alanine-derived β-ketophosphonate followed by adjustments of oxidation levels, including diastereoselective reduction with modest diastereoselectivity (*ca.* 4:1) of an amino ketone to an amino alcohol (see also Section 5.4.1).

The examples were chosen to illustrate the power of both chiral pool-derived approach as well as the external asymmetric induction processes. The amaminol synthesis further emphasises the fact that although the chiral catalysis step is very efficient indeed (giving 98.1% *ee*), introduction of the remaining stereocenters at C2 and C3 still pose problems: the chiral information at C2 was derived from natural sources; and the stereocenter at C3 faces the same problem as the C15 center in prostaglandins (Scheme 7.5).

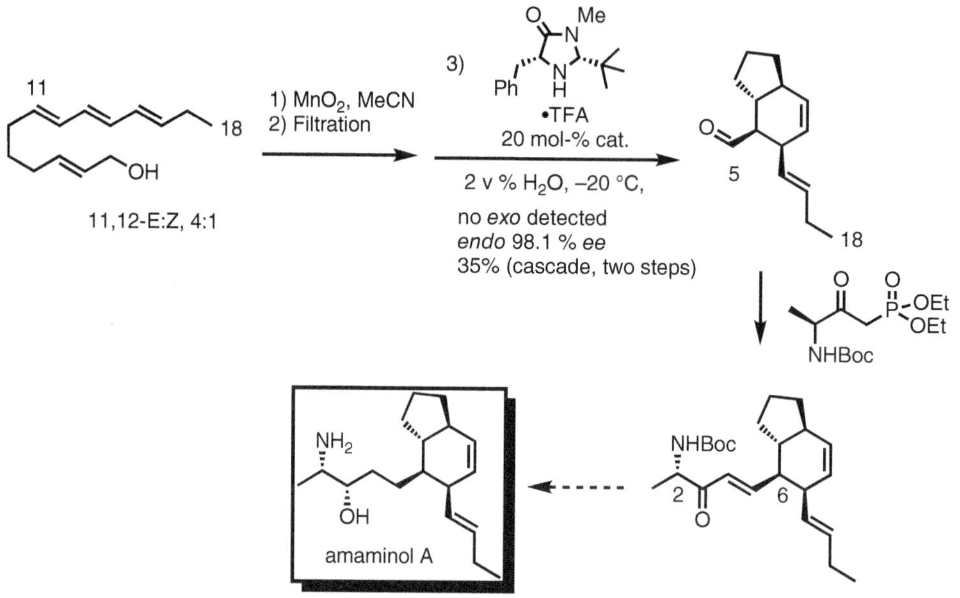

Scheme 8.12   *Synthesis of amaminol A*

Short synthesis of sphingosine derivatives in just seven steps developed by Davies utilised enantioselective conjugate addition of chiral amines to α,β-unsaturated esters. As an example, xestoaminol C was obtained from crotonic acid in 8 steps and 41% overall yield (Scheme 8.13) [28]. A highly diastereoselective *anti*-aminohydroxylation was achieved via conjugate addition of lithium (*S*)-*N*-benzyl-*N*-(α-methylbenzyl)amide and subsequent in situ enolate oxidation with (+)-(camphorsulfonyl)oxaziridine. The benzyl-type protecting groups were replaced in one step with *t*-butyl carbamate, followed by hemiaminal formation with dimethoxypropane (DMP). Standard reduction-oxidations then gave the aldehyde ready for a Wittig reaction to install the aliphatic side chain. Finally, reduction of the double bond and global removal of protections gave xestoaminol C.

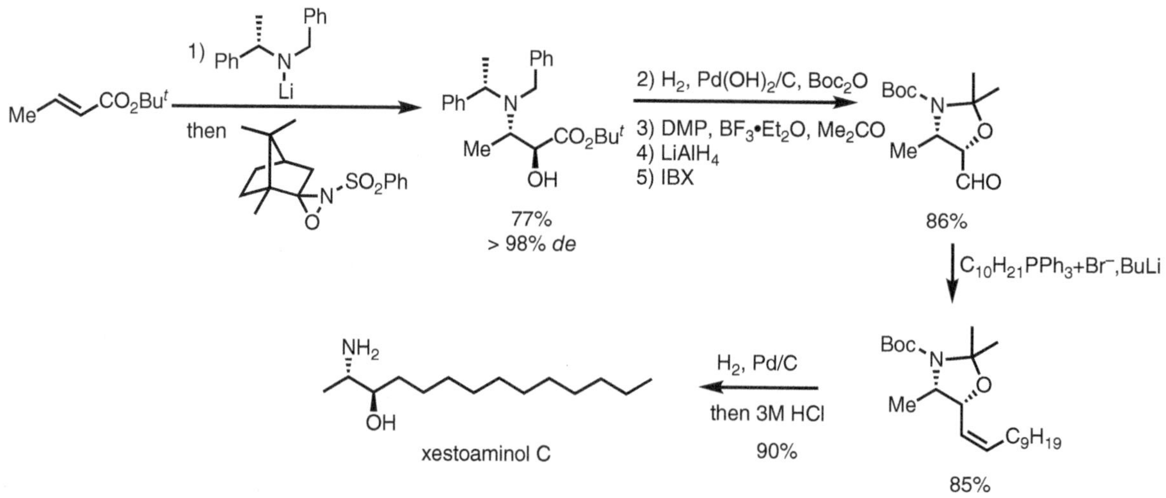

Scheme 8.13   *Xestoaminol synthesis using asymmetric conjugate addition*

8.3 Polypropionates

Polypropionates constitute the structurally widest subclass of polyketides. Several hundred naturally occurring polypropionates are known, isolated originally from bacterial sources, but recently marine organisms have provided an enormous source of new chemical structures [29]. For the purposes of discussion, we shall use the sub-classification of polypropionates into polyethers, macrolides, and spiroketals (Figure 8.2). Although these classes overlap in some cases with each other (e.g. spiroketal polyethers) and even with other natural product classes (e.g. steroidal spiroketals), this classification helps us in the following discussion.

8.3.1 Polyether Antibiotics

The first two polyether antibiotics, nigericin and lasalocid, were isolated from different Streptomyces species in 1951 [30]. To date, more than 50 microorganisms are known to produce polyether antibiotics, and more than 120 structures have been reported. Polyether antibiotics are rich in oxygen atoms, they contain one carboxylic acid group, tetrahydrofuran and tetrahydropyran rings, a ketone, and several hydroxyl groups. These structural features enable the molecules to bind metal cations and form lipid-soluble complexes that can cross cellular membranes. Since 1967, these compounds have also been called polyether ionophores. Seven carboxylic ionophores are marketed around the world as anti-coccidial drugs for poultry and growth promoters in ruminants. These include monensin, laidlomycin, lasalocid A, salinomycin, narasin, maduramycin, and semduramycin (Figure 8.12) [31].

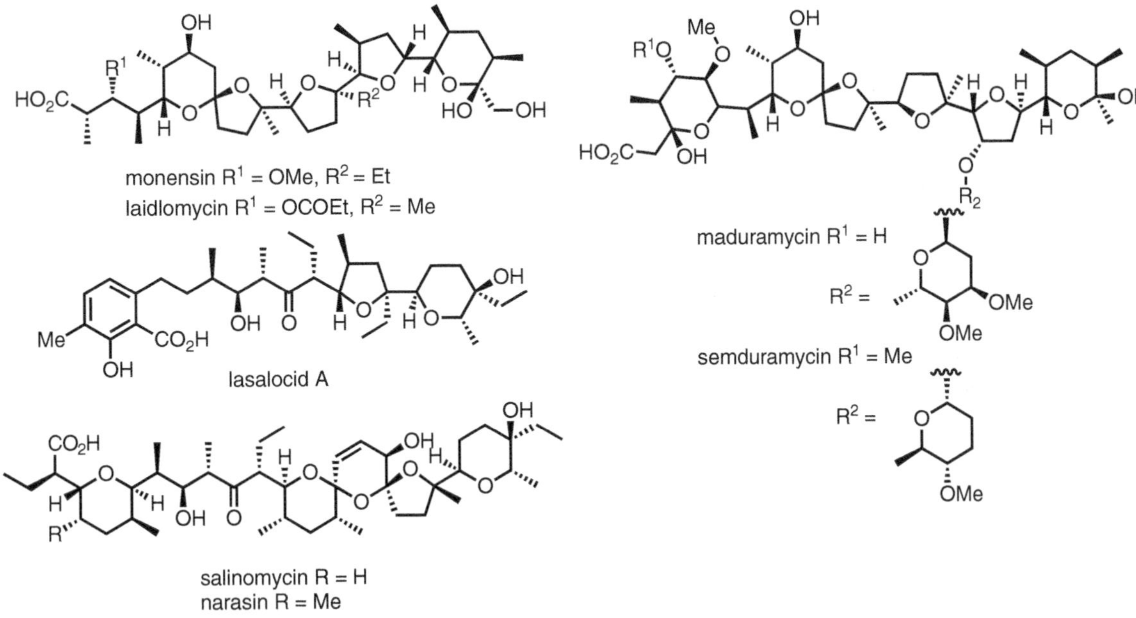

Figure 8.12 *Polyether antibiotics*

Marine sources have produced a number of toxins; suspect agents of many seafood-related poisonings. The structures of these molecules are among the most complex ones known at the moment (Figure 8.13). Ciguatera fish poisoning (CFP), or simply ciguatera, is an illness caused by eating reef fish whose flesh is contaminated with certain toxins including gambieric acid [32], brevetoxin A [33], ciguatoxin [34], and maitotoxin [35]. Due to their intriguing chemical structures, they have gained widespread synthetic interest, and truly monumental syntheses have been achieved [36].

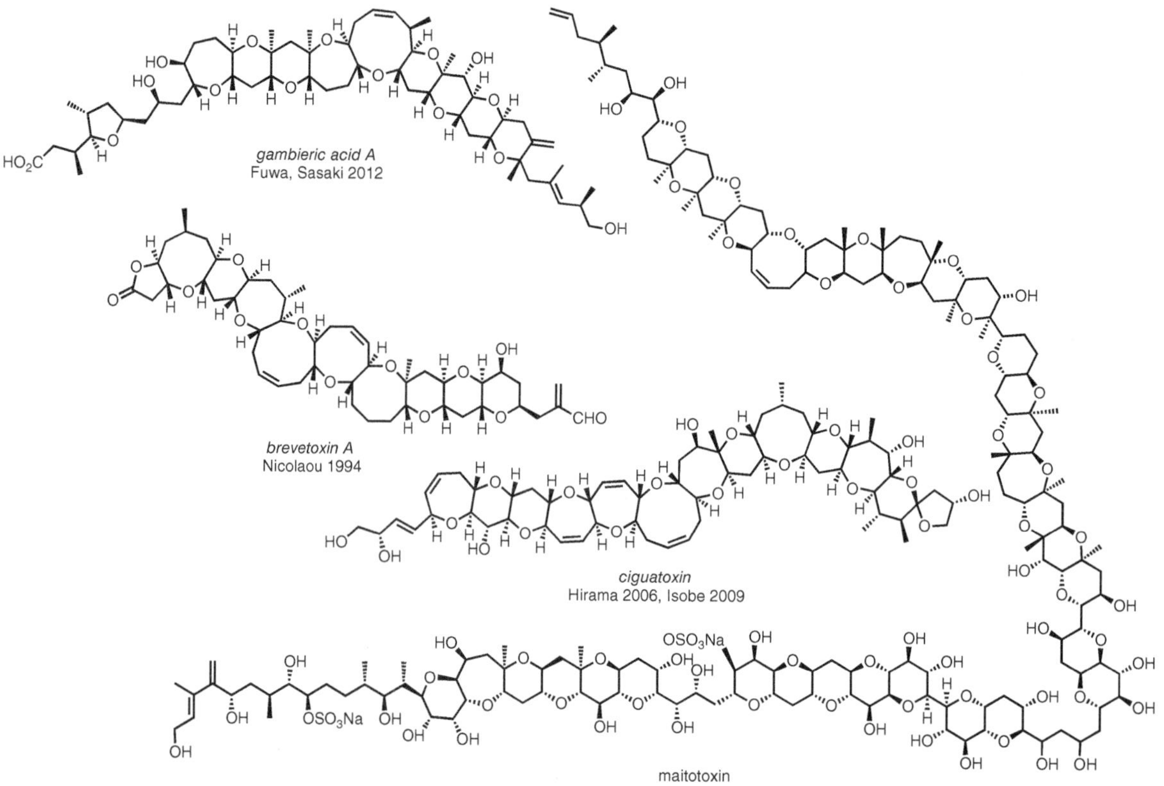

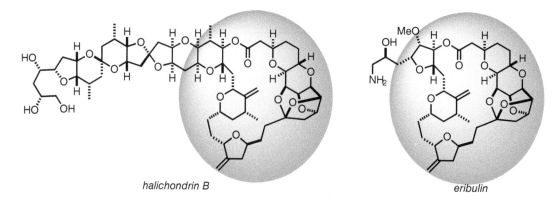

Figure 8.13 *Complex polyether marine toxins*

Halichondrin B[37] as a ciguateric toxin deserves a special mention here (Figure 8.14). It was originally isolated in the 1980s from the sea sponge *Lissodendoryx*, in quantities of 310 mg from 1 ton of the sponge. It was soon discovered that the compound is impressively active against several experimental cancer treatments. The compound contains 54 carbon atoms in its backbone together with 32 stereogenic centers, thus being one of the most daunting synthetic challenges. However, the total synthesis was achieved eventually, as well as a synthesis of a truncated compound called eribulin, which contains 19 stereogenic centers in a 36 carbon backbone, and was finally entered in the market for metastatic breast cancer in 2010 [38]. This can be considered a prime example of function-oriented synthesis.

halichondrin B   *eribulin*

Figure 8.14 *Halichondrin B and eribulin*

8.3.2 Macrolides

Macrolide antibiotics are a large group of polyketides, including well over one hundred metabolites, all characterised by the macrocyclic lactone moiety incorporated in their structures [39]. The first macrolide isolated was picromycin in 1950, soon overshadowed with erythromycin which entered the clinics in the mid 1950s. The term "macrolide" was coined in 1957 by R.B. Woodward as a portmanteau of *macro*lactone glyco*side*, the essential components of this class of antibiotics. Within the macrolides, one can distinguish four main types of structures; the polyoxo, polyene, ionophore, and ansamycin macrolides. Nearly all macrolides exhibit antibacterial activity, and research into their structures and chemistry has provided many compounds which are in medicinal use. These include the erythromycin derivatives shown in Figure 8.15. Erythromycin, obtained industrially by fermentation from *Saccharopolyspora erythraea*, is considered a first-generation macrolide. As a first-generation antibiotic, it suffers from a number of shortcomings, including degradation in the human stomach to a spiroketal motilin agonist, yielding painful stomach cramps. The second-generation macrolides, with the introduction of clarithromycin and azithromycin, enjoy improved pharmacokinetic and pharmacodynamic properties. Clarithromycin differs from erythromycin by a single methyl group installed on the C6 hydroxyl group, which improves the acid stability of the drug. Azithromycin, one of the most prescribed drugs, is an expanded 15-membered macrolide, synthesised from erythromycin through an oxime-mediated Beckmann rearrangement and subsequent methylation of the installed secondary amine. While the second generation mainly sought to improve the pharmacokinetic/pharmacodynamic properties, the third generation addressed acquired resistance. Macrolide resistance requires the presence of the L-cladinose sugar of erythromycin, and the third-generation macrolides known as "ketolides" have the L-cladinose cleaved and the resulting C3 hydroxyl group oxidised to a ketone.

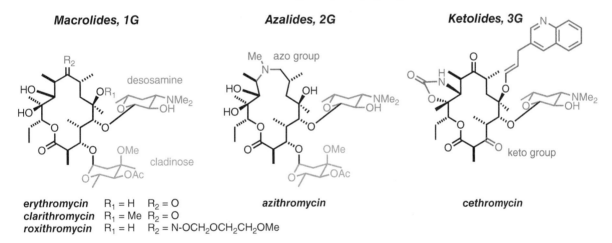

erythromycin	R_1 = H	R_2 = O
clarithromycin	R_1 = Me	R_2 = O
roxithromycin	R_1 = H	R_2 = N-OCH$_2$OCH$_2$CH$_2$OMe

Figure 8.15 *Erythromycin generations*

Figure 8.16 illustrates the wide variety of the structures of macrolides. Epothilones are myxobacterial tubulin-binding antibiotics, which soon received much synthesis interest because of their activity as anti-cancer drug leads [40]. Rapamycin, a soil microbe isolate, turned out to be a highly active immunosuppressant [41]. The third compounds shown are stambomycins, recently isolated macrolides with a 51-membered ring [42]. The compounds were identified as the metabolic products of a modular polyketide synthase (PKS) in *Streptomyces ambofaciens* through a genomics-driven approach involving rational genetic manipulation to induce transcription of the biosynthetic genes. The planar structures and stereochemistry of the stambomycins were predicted via sequence analysis of the modular PKS responsible for their biosynthesis. The stereochemical assignment of the fragment C12–C27 was recently substantiated by correlation of the NMR data of the synthetic fragment with those of the natural product [43].

Polyoxo macrolides (Figure 8.17) typically contain either a 12-, 14-, or 16-membered lactone ring, which is usually also oxygenated at several sites. Unsaturation is not uncommon, and the main carbon chain is typically derived from a mixture of acetate and propionate units.[30b] In the case of the 16-membered macrolides, one butyrate unit is also

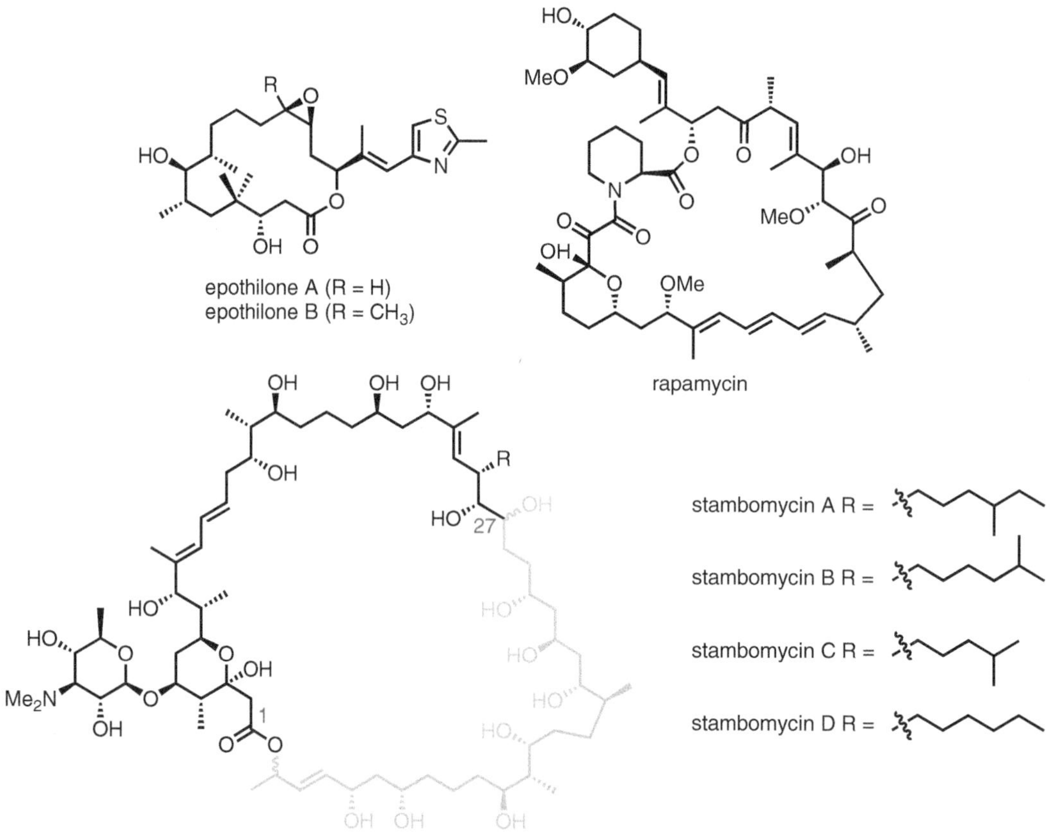

Figure 8.16   *Macrolides of varying ring sizes*

incorporated. The macrolide ring is also always connected to one or more carbohydrate units, often of the amino sugar type. The compound with the sugar residue detached, the aglycone, is called the corresponding -olide (e.g. tylonolide is the aglycone of tylosin).

The first macrolide discovered was pikromycin [44]. By the end of the 1950s, the structures of methymycin, erythromycin A and B, as well as carbomycin A (magnamycin) were elucidated through classical chemical degradation reactions. The structural and conformational variations in these natural products have provided several examples where X-ray crystallography, NMR, and mass spectrometry have been used ingeniously in the structure elucidation.

The biosynthesis of polyketides is very similar to that of fatty acids, and much of our current understanding of polyketide synthase (PKS) enzymes is thanks to the extensive studies on the fatty acid synthase (FAS) systems. They share a common evolutionary history. The biosynthesis of both polyketides and fatty acids begins with a starter acyl group, which is extended with units from activated acyl-CoA precursors and assembled head-to-tail into the growing chain through several enzymatic activities [45]. An example of the modular PKS organisation is seen in the multienzyme complex known as 6-deoxyerythronolide B synthase (DEBS), which synthesises the aglycone core of erythromycin (Scheme 8.14). DEBS is a modular PKS in which a loading module and six extension modules, divided among three large proteins (DEBS1, DEBS2, and DEBS3), work in an assembly-line fashion to produce 6-deoxyerythronolide B. Only the first complex DEBS1 is shown in the scheme.

Erythromycin A has gained wide synthetic interest, as it is currently used as an antiobiotic agent; especially useful for patients with penicillin allergies. However, as Mulzer stated, "The synthesis of […] erythromycin A […] is probably the most extensive single project in the history of synthetic organic chemistry. This phenomenon is not rational as [they] are accessible in large quantities from fermentation […]. It is the complexity of the molecule's structure, the plethora

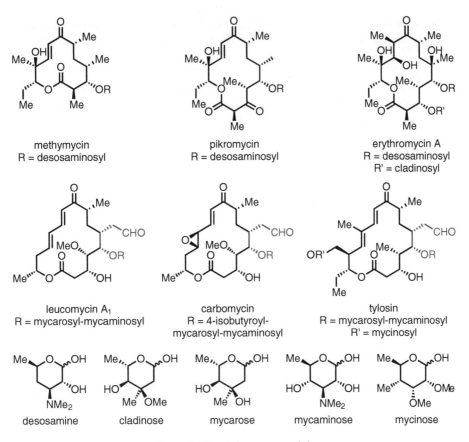

methymycin
R = desosaminosyl

pikromycin
R = desosaminosyl

erythromycin A
R = desosaminosyl
R' = cladinosyl

leucomycin A$_1$
R = mycarosyl-mycaminosyl

carbomycin
R = 4-isobutyroyl-
mycarosyl-mycaminosyl

tylosin
R = mycarosyl-mycaminosyl
R' = mycinosyl

desosamine

cladinose

mycarose

mycaminose

mycinose

Figure 8.17 *Polyoxo macrolides*

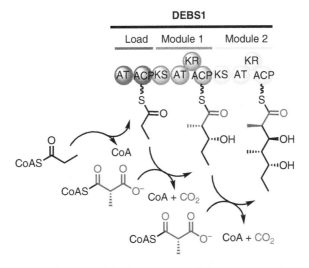

Scheme 8.14 *Initial stages of the biosynthesis of the aglycone of erythromycin A*

of stereocenters and functional groups and the magic of the medium ring that has fascinated about 15 large research groups worldwide for more than a decade."[46] As the number of syntheses is quite large, and many syntheses follow rather straightforward enantioselective strategies (chiral pool, aldol strategies, etc.), we will highlight only two in this section.

In 1956 Woodward wrote: "Erythromycin, with all our advantages, looks at present time quite hopelessly complex, particularly in view of its plethora of asymmetric centers."[47] In a monumental synthetic effort, 25 years later Woodward et al. reported the first (and also the only one to date) total synthesis of erythromycin A, in 1981 [48]. Thiopyranone ring strategy was the key for the successful total synthesis as well as for the synthesis of the *syn, syn, syn*-stereotetrad C2–C5 (Scheme 8.15) [49]. The racemic thiopyranone derivative was S-alkylated with the racemic mesylate to give a racemic ketoaldehyde. This intermediate was subjected to an intramolecular aldol reaction catalysed by D-proline (nowadays called organocatalytic), and a 1:1 mixture of enantiomerically enriched diastereomeric aldols were obtained (36% *ee* for both diastereomers). The desired diastereomer was separated, and the synthesis was continued by dehydration to the enantiomerically enriched enone. The desired enantiomer could be purified by crystallisation from the enantiomeric mixture and the synthesis was continued with optically pure material. The synthesis was 52 steps (longest linear sequence), and gave an overall yield of 0.0089%.

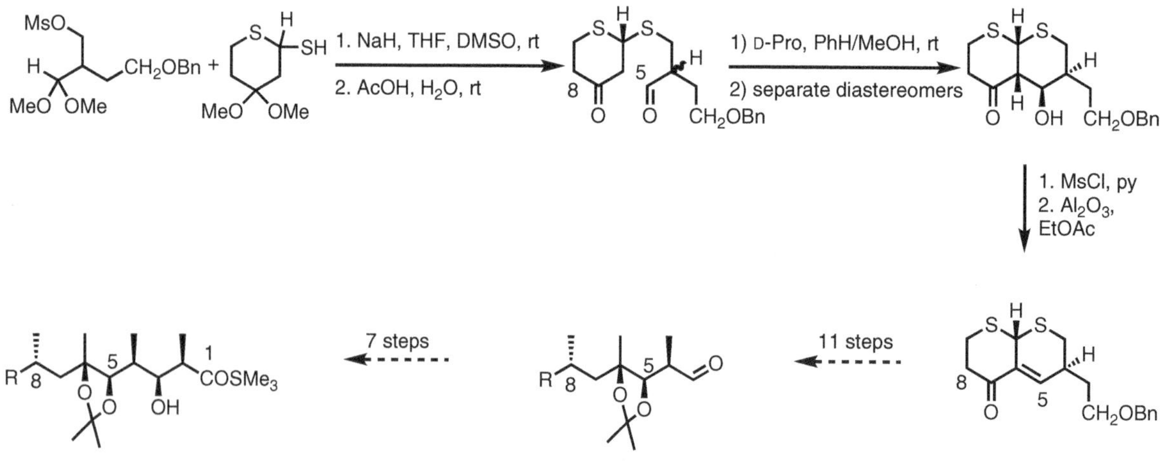

Scheme 8.15 Woodward synthesis of the erythromycin C1–C8 syn, syn, syn-stereotetrad

The second synthesis we illustrate here is that of the 14-membered macrolide 6-deoxy-erythronolide B by *Michael J. Krische* (1966–) [50]. Two different methods for the highly efficient enantioselective alcohol CH-crotylation via transfer hydrogenation are deployed for the first time in target-oriented synthesis. The synthesis of the C9–C15 fragment began with the hydro-hydroxyalkylation of butadiene employing *n*-propanol to form the product of *syn*-crotylation in 59% yield and 98% *ee* (Scheme 8.16). The double bond was osmylated to an aldehyde, and a Z-selective aldol reaction established the C10 and C11 stereocenters. A three-step sequence involving an Ohira–Bestmann alkyne formation gave the first coupling partner.

The synthesis of the second coupling partner, fragment C1–C8, began with the *anti*-diastereo- and enantioselective Ir-catalysed double crotylation of 2-methylpropane-1,3-diol to furnish the pseudo-C_2-symmetric diol (Scheme 8.17). The diol was obtained as a single enantiomer, as the minor enantiomer of the mono-adduct was converted to a pseudo-*meso*-diastereomer. The alkene termini were differentiated by iodoetherification (yellow highlight) and protection. The terminal double bond was oxidised to a carboxylic acid, and the iodo ether was reductively transformed (Zn dust) to the alkene. Inversion of the C3-stereocenter was achieved by conversion to the β-lactone followed by hydrolysis and protection of the resulting diol.

The two fragments were joined by Yamaguchi lactonisation (Scheme 8.18). Hoveyda–Grubbs enyne metathesis followed by oxidation of the C9-methylidene appendage gave an enone intermediate. Nickel-catalysed conjugate reduction

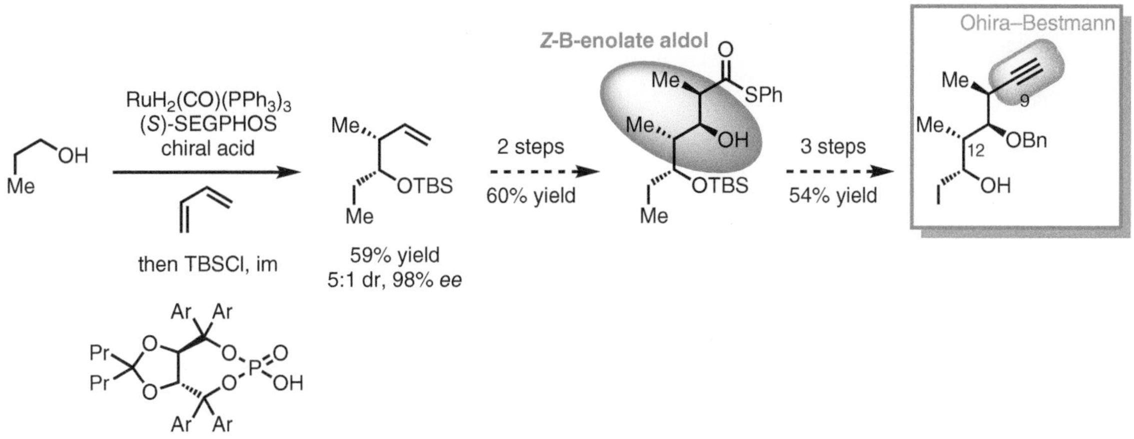

Scheme 8.16   Synthesis of the C9–C15 fragment

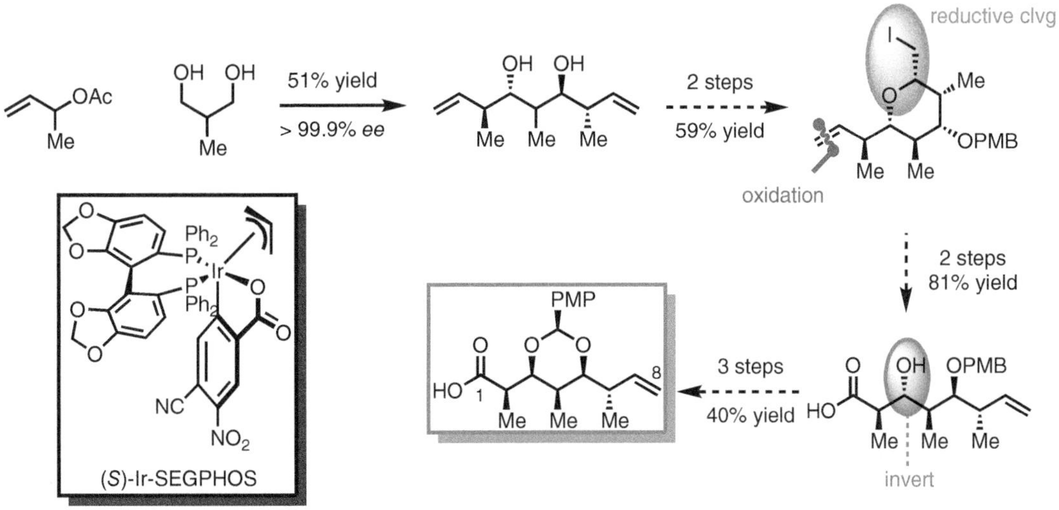

Scheme 8.17   Synthesis of the C1–C8 fragment

followed by enolate alkylation installed the C8-methyl group. The stereochemistry at C8 was inconsequential, as the slightly acidic conditions of Pd-catalysed homogeneous hydrogenation provided 6-deoxyerythronolide B in 93% isolated yield as a single diastereomer. 6-deoxyerythronlide B was prepared in 14 steps (longest linear sequence) and 20 total steps, in an overall yield of 2.5%.

Characteristic to *polyene macrolides* (Figure 8.18) is low degree of alkylation on the lactone ring and the presence of a conjugated polyene moiety. The ring size is correspondingly larger than in the polyoxomacrolides. Most of the polyene macrolides exhibit antifungal activity with reduced or no antibacterial activity. Natamycin, nystatin A_1, and amphothericin B (isolated from *Streptomyces* species) bind ergosterol and form ion-channels, which leads to leakage of K^+ ions and death of the fungal cell. Filipin is thought to be a simple membrane disrupter. Filipin III is highly fluorescent and has therefore found widespread use as a histochemical stain for cholesterol.

Synthesis of the C33–C37 dipropionate fragment of amphotericin B (a *syn, anti, anti*-stereotetrad) has been achieved several times [50, 51], and here we illustrate a synthesis relying on the use of the chiral auxiliary strategy (Scheme 8.19) [52]. Similar to the Woodward erythromycin synthesis (which contains a *syn, syn, syn* stereotetrad), this synthesis was based on the thiopyrane ring strategy. Commercially available tetrahydrothiopyran-4-one was first

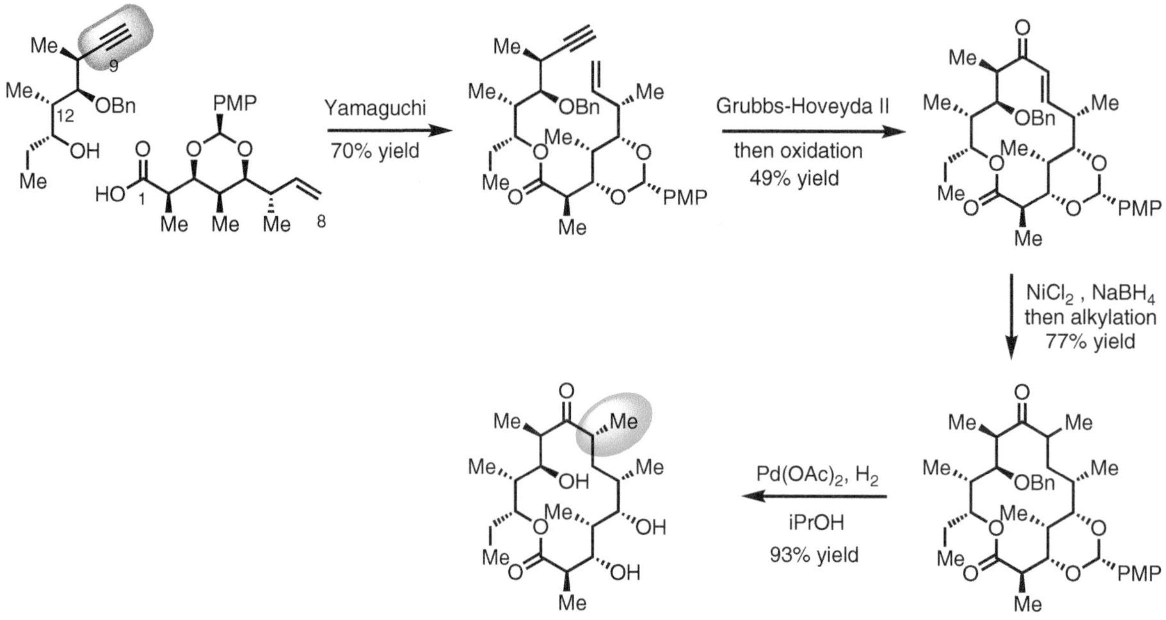

Scheme 8.18   *Fragment coupling and final steps*

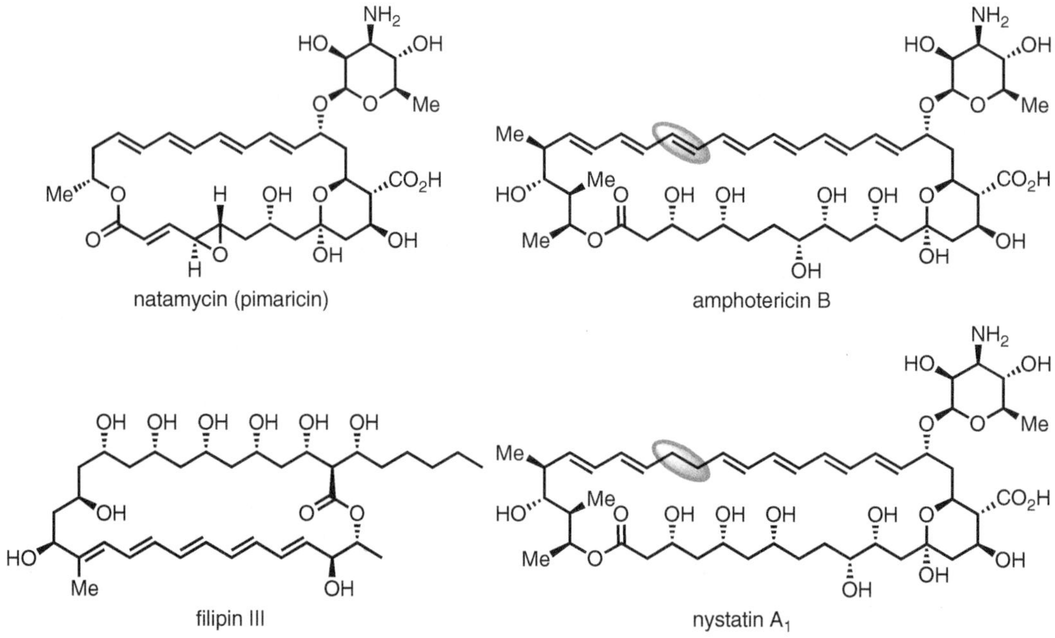

Figure 8.18   *Polyene antibiotics*

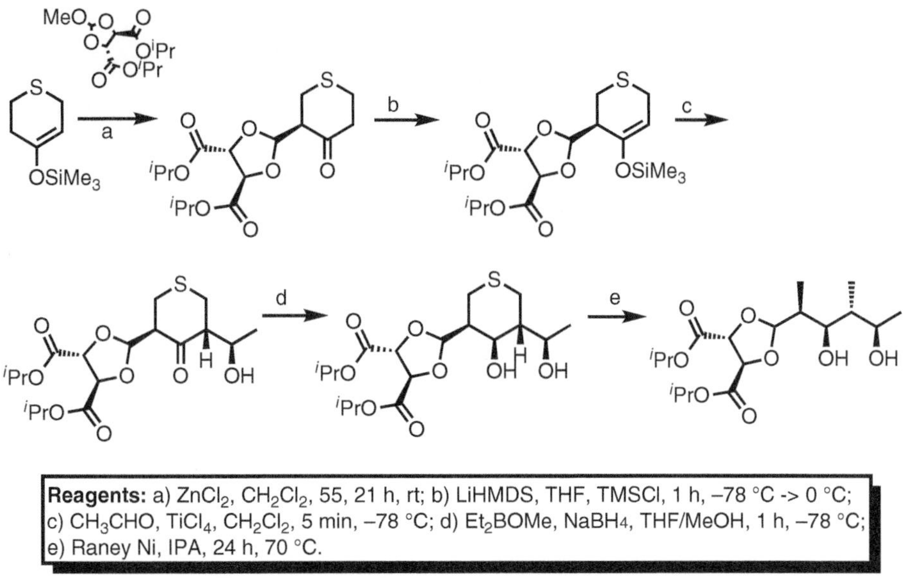

Reagents: a) ZnCl$_2$, CH$_2$Cl$_2$, 55, 21 h, rt; b) LiHMDS, THF, TMSCl, 1 h, −78 °C -> 0 °C; c) CH$_3$CHO, TiCl$_4$, CH$_2$Cl$_2$, 5 min, −78 °C; d) Et$_2$BOMe, NaBH4, THF/MeOH, 1 h, −78 °C; e) Raney Ni, IPA, 24 h, 70 °C.

Scheme 8.19 *Amphotericin B C33–C37 syn, anti, anti–stereotetrad*

converted to the corresponding silyl enol ether. The asymmetry was then introduced into the ring by alkylation with the tartrate-derived orthoester. The two diastereomers were obtained in a 3:1 ratio and the diastereomers were separated by crystallisation. A highly diastereoselective aldol reaction via the kinetic silyl enol ether gave the desired aldol product. Finally, 1,3-*syn* diol reduction and removal of the sulfur with Raney Nickel produced the enantiomer of the stereotetrad of amphotericin B. The sole source of asymmetry in the entire synthesis was the chiral tartrate-derived orthoester, which actually worked as a chiral auxiliary (masked aldehyde).

A sophisticated application of two-directional synthesis and highly enantioselective aldol addition with an innovative terminus differentiation is showcased in the synthesis of dermostatin, one of the more complex oxopolyene macrolides (Scheme 8.20) [53]. Thus, 3-hydroxyglutaraldehyde, readily derived from 3-hydroxyglutarate, was subjected to the sparteine-catalysed acetate aldol to give the meso triol in very high diastereoselectivity. This was converted to the bis-PMP acetal, which underwent clean terminus differentiation to the thermodynamically most stable bis-acetal, which was the only isomer observed. The C17–C25 segment was obtained in only seven straightforward steps.

A number of macrolides are composed of two or more ω-hydroxycarboxylic acids through the formation of an oligolactone. These are called macrodiolides, -triolides and -tetrolides, according to the number of acid units they contain. Thus, pamamycin is a macrodiolide, and nonactin is a macrotetrolide formed through cyclotetramerisation of nonactic acid. It is interesting to observe that the 32-membered macrocycle is composed of two alternating (+)- and (−)-nonactic acid units, thus making the natural product achiral [54]. These macrolides can strongly chelate metal cations (as shown in Figure 8.19), and thus transport the hydrophobic chelates through biological membranes quite efficiently.

Ansamycins (Figure 8.20) are atypical macrolides in the sense that they contain a large ring lactam moiety, a so-called ansa bridge (from Latin *ansa*: handle) [56]. Typical to them is also an aromatic nucleus: a benzene or benzoquinone ring as in maytansine and geldanamycin or a naphthoquinone moiety as in rifamycins and streptovaricins, bridged through non-adjacent carbon atoms by the ansa ring. Most of the ansamycins exhibit broad antibacterial and powerful antitumor activities. Following the clinical introduction of rifamycin SV (the reduced naphthalenic form of rifamycin S), extensive programs of semisynthesis eventually led to the commercial introduction of rifampicin. It shows more pronounced activity against Gram-positive bacteria, particularly mycobacteria; better activity against Gram-negative bacteria; and importantly, excellent oral bioavailability. It has become one of the mainstay agents in the treatment of tuberculosis, leprosy, and AIDS-associated mycobacterial infections.

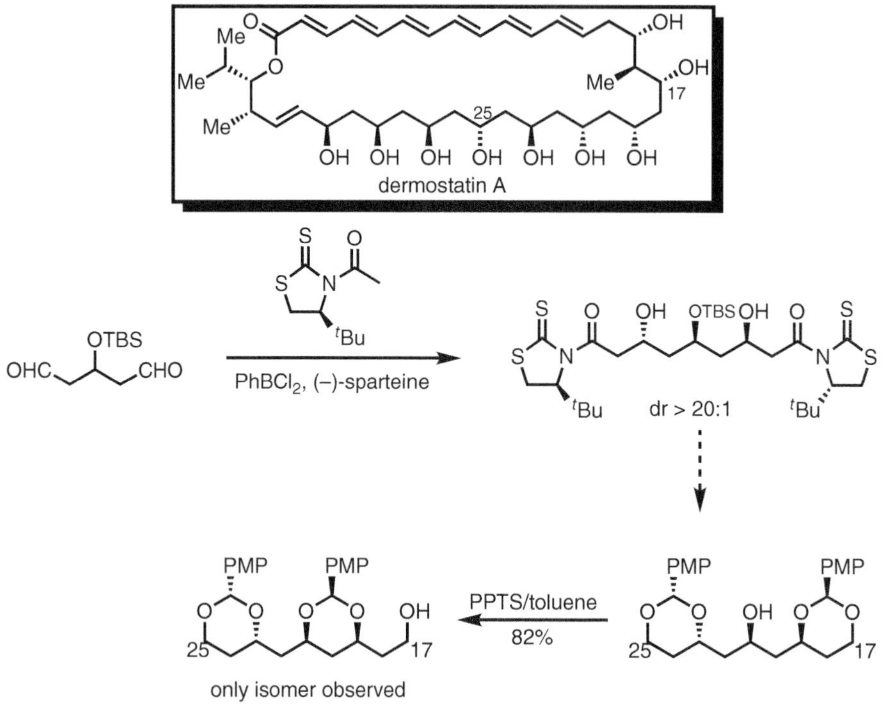

Scheme 8.20 *Dermostatin synthesis*

The synthesis of the C19–C28 segment of rifamycin S is a beautiful early example of the application of two directional synthesis (Scheme 8.21) [57]. Hydroboration reactions of allylic alcohols are highly diastereoselective with regard to the diastereofacial approach of the hydroborating reagent, and, of course, have the added bonus of being stereospecific with regards to the delivery of the hydrogen and hydroxyl groups. *W. Clark Still* (1946–, Columbia University) applied this method, further maximising the stereoselectivities by employing the allylic $A^{1,3}$-strain favorably to give the meso pentapropionate fragment.

Other compounds, such as zearalenone, pyrenophorin, and brefeldin A, also contain a macrolide structure (Figure 8.21). Zearalenone has growth-inducing properties and is widely used in animal breeding. Brefeldin specifically blocks the transport of proteins from the endoplasmic reticulum to the Golgi apparatus, which imparts brefeldin its antiviral activity.

The stereochemistry and conformation of the polyoxo macrolides are surprisingly constant. Although different ring sizes can be found, and the oxidation patterns vary, the relative stereochemistry of the chiral centers can often be predicted with the Celmer model (Figure 8.22) [58]. The chiral centers of methymycin (12-membered ring), erythromycin A (14), and tylosin (16), as well as the ansa chain of rifamycin, match with the Celmer model. This invariability has been used in some cases to correct the originally proposed (wrong) stereostructure of an isolated macrolide.

Chivosazole is a naturally occurring macrolide with a 31-membered ring (Figure 8.23). Application of the Celmer model to elucidate the stereochemistries at the distantly spaced stereocenters would have been impossible, and an ingenious combination of advanced NMR methods, molecular modeling, and genetic analysis together with total synthesis, was utilised to establish the final structure of this challenging compound [59]. Initially, the relative stereochemistries of the C32,C34-diol were assigned based on the application of NMR methods of the corresponding cyclic ketal [60]. This was followed by Monte Carlo molecular mechanics simulation of the macrocyclic ring and matching the predicted and observed NMR data to obtain the relative configurations, and finally, by using information from the gene analysis of the polyketide synthase producing chivosazole [61], the absolute stereochemistries at the secondary alcohol centers at C30, C32, and C34 could be ascertained, and thus the remaining centers could be assigned.

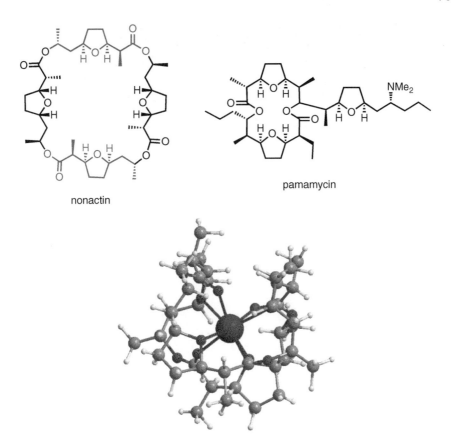

nonactin

pamamycin

Figure 8.19 *Cyclic ionophore antibiotics. Below is the X-ray structure of nonactin Ca²⁺ complex (CSDD code CAXHEO)[55]*

8.3.3 Spiroketals

The spiroketal unit enjoys a widespread occurrence in polyketide structures from fungi, insect, microbes, plants, and marine organisms. These natural products are also of wide pharmacological importance [62]. The spiroketal unit was originally identified in steroidal structures such as the saponins (glycosides) in which the aglycone (sapogenin) contains a steroid nucleus fused through its D-ring to a spiroketal moiety. Glycosylation usually occurred in the A-ring unit of the molecule. Tomatidine and hecogenin are typical structures of steroidal spiroketals (Figure 8.24). Both of these can be used as starting materials for steroid synthesis.

Several natural products from different structural groups have been identified to inhibit serine/threonine-specific protein phosphatases. The natural toxin inhibitors are also known as the *okadaic acid class inhibitors* (Figure 8.25). Okadaic acid, the causative agent of diarrhetic seafood poisoning, was the first of these inhibitors, discovered in 1981. It is a marine polyketide initially found from marine sponges *Halicondria okadai* and *Halicondria melanodocia*. Fostriecin, from *Streptomyces pulveraceus*, is the most selective protein phosphatase 2A inhibitor known to date. This may be advantageous for the development of protective agents for myocardial infarction [63]. Tautomycin, from *Streptomyces spiroverticillatus*, is more selective for protein phosphatase 1 relative to PP2A [64].

Calyculins, highly cytotoxic polyketides, originally isolated from the marine sponge *Discodermia calyx* by *Nobuhiro Fusetani* (1943–) and co-workers, belong to the lithistid sponges group. These molecules have become interesting targets for cell biologists and synthetic organic chemists. To date, eighteen calyculins and calyculin-related structures have been isolated in total. The fascinating structures of calyculins (Figure 8.26) have inspired various groups of synthetic organic chemists to develop total syntheses of the most abundant calyculins A and C. However, with fifteen

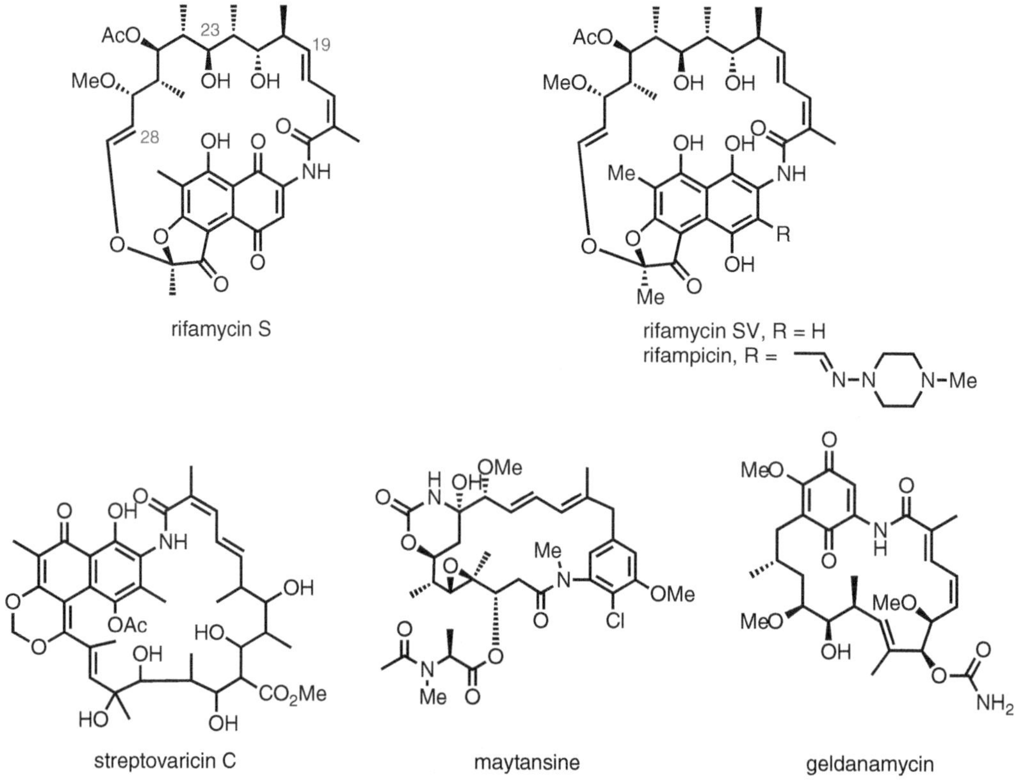

Figure 8.20   *Ansamycins*

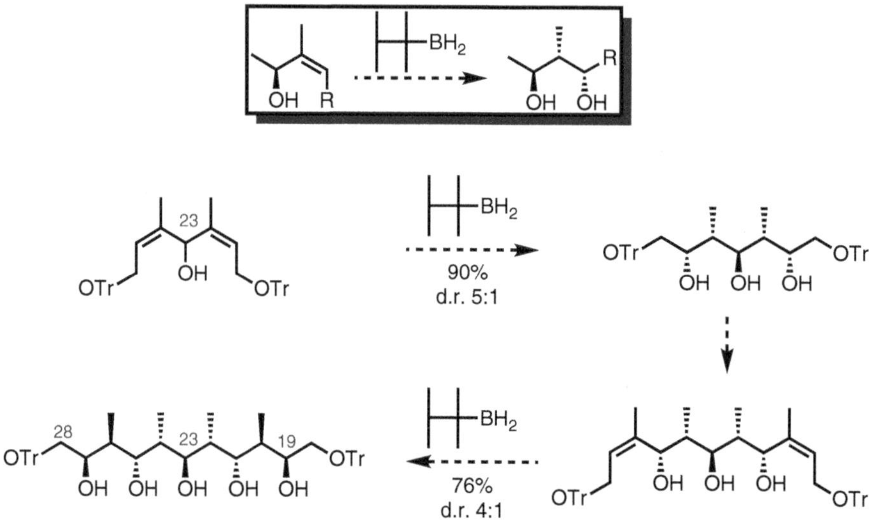

Scheme 8.21   *Two-directional synthesis of the C19–C28 segment of rifamycin S*

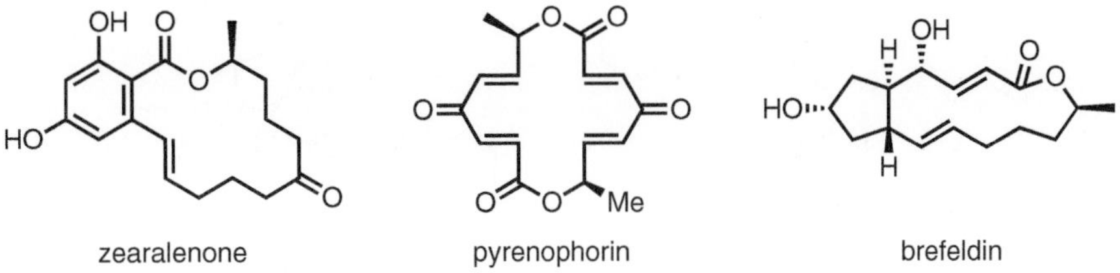

Figure 8.21   *Unsaturated macrolides*

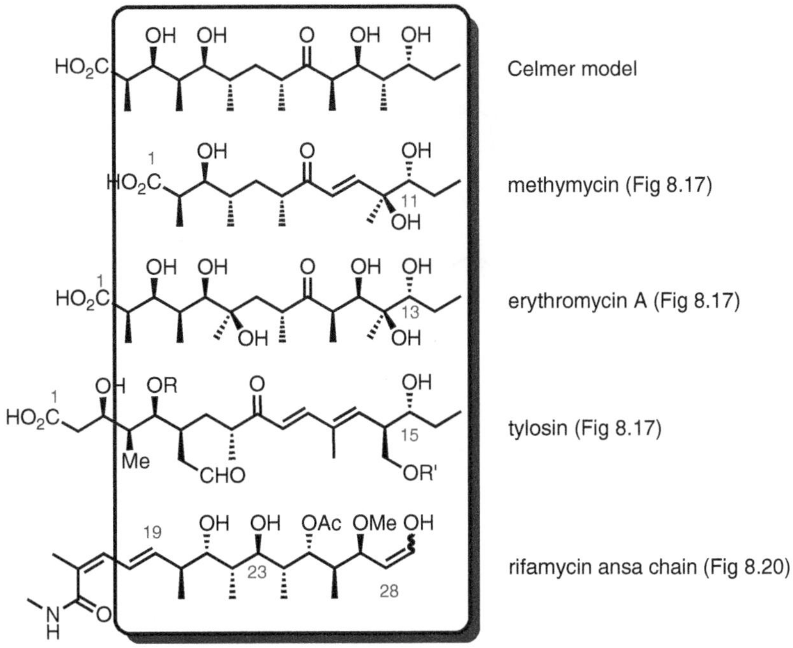

Figure 8.22   *The Celmer model for macrolide stereochemistry*

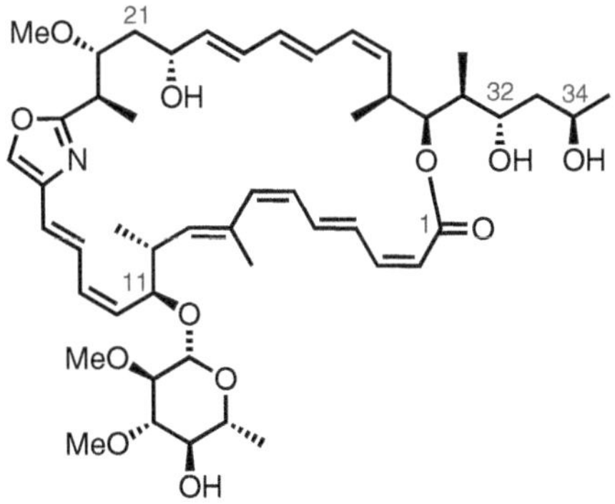

Figure 8.23   *Structure of chivosazole*

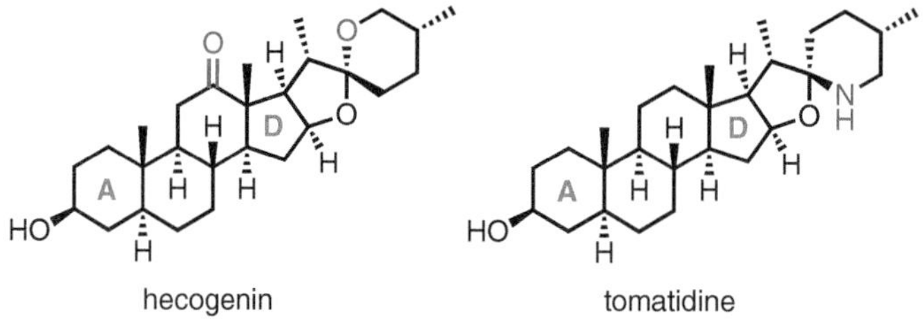

Figure 8.24   *Steroidal spiroketals*

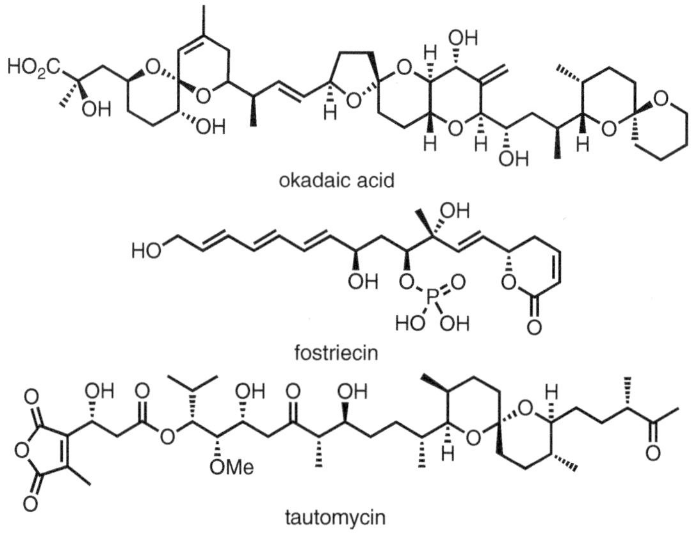

Figure 8.25   *Okadaic acid class inhibitors of protein phosphatases*<.FC>

chiral centers, a cyano-capped tetraene unit, a phosphate-bearing spiroketal, an *anti, anti, anti* dipropionate segment, an α-chiral oxazole, and a trihydroxylated γ-amino acid, calyculins reach versatility that only few natural products can surpass, and truly challenge modern chemists' asymmetric synthesis skills. Despite more than two decades of synthetic efforts, no synthesis capable of producing material for more detailed pharmacological profiling, let alone structure-activity studies, has emerged [65].

Milbemycins and avermectins are 16-membered macrolide antibiotics which also contain a spiroketal moiety (Figure 8.27). These compounds possess significant activity as anthelmintics (expelling parasitic worms, helminths), insecticides and acaricides (pesticides that kill pests of the *Acari* group, including ticks and mites). Females of *Onchocerca volvulus* microfilariae spread onchocerciasis, river blindness, a parasitic disease ultimately leading to blindness and affecting some 20–40 million people worldwide. The parasite is transmitted to humans by the bite of the black fly *Simulium yahense*. Ivermectin, 22,23-dihydroavermectin B_1 [66], kills the parasite by stimulating the release and binding of gamma-aminobutyric acid (GABA) at nerve endings, which blocks the neurotransmission. This results in increased influx of chloride ions into the cells, leading to hyperpolarisation and subsequent paralysis. GABA-ergic receptors are found at the neuromuscular junctions and the central ventral cords in nematodes, whereas in mammals they are found primarily in the brain. Ivermectin does not readily cross the blood–brain barrier in mammals at therapeutic doses and is therefore of low mammalian toxicity.

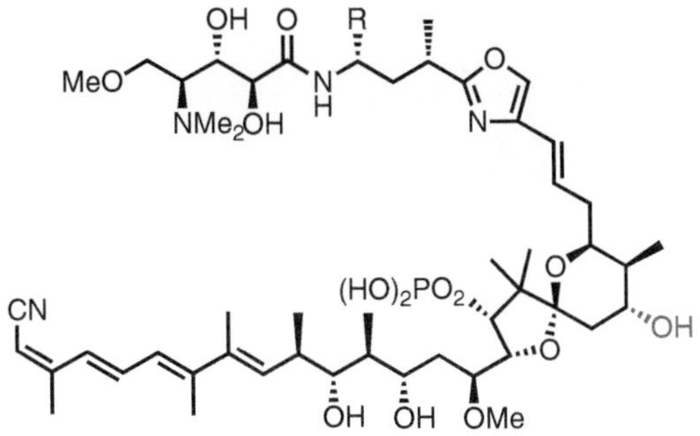

calyculin A, R = H
calyculin C, R = Me

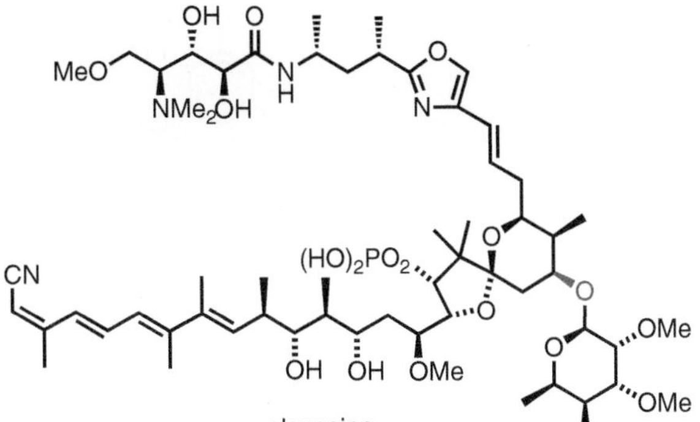

clavosine

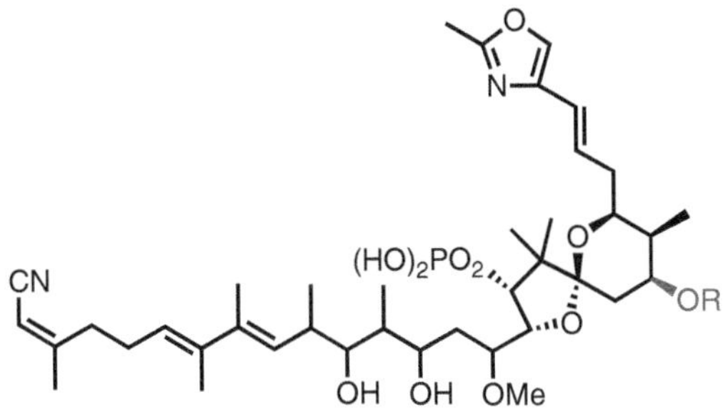

geometricin, R = H
swinhoeiamide A, R = Me

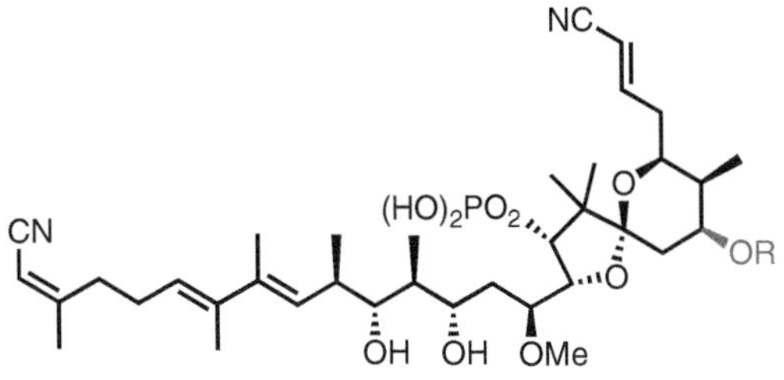

hemicalyculin A

Figure 8.26 *Structures of calyculins and related compounds*

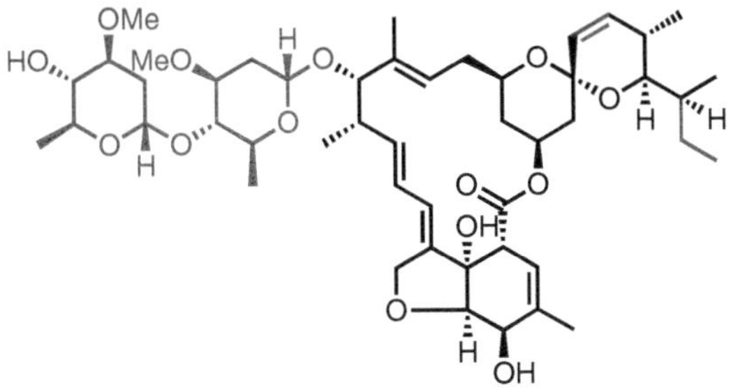

avermectin B₁ (abamectin)

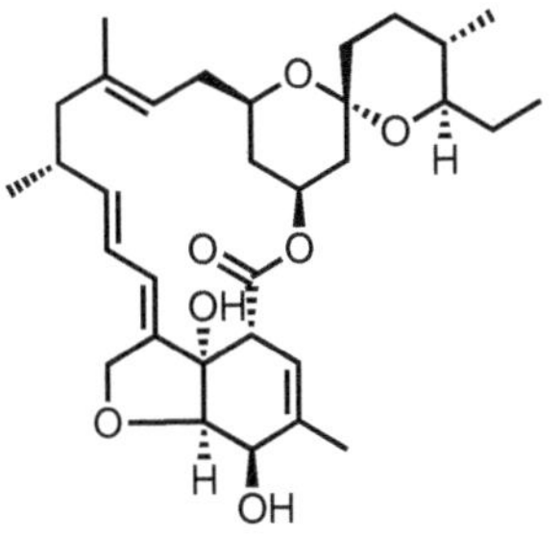

milbemectin

Figure 8.27 *Insecticidal mectins*

8.4 Aromatic Polyketides

The bright colors of several flowers and fruits are given by aromatic compounds, which are derived from polyketides (Figure 8.28). Flavones usually give rise to a yellow or orange color, and anthocyanidins produce a red or blue tinge. Aromatic polyketides arise by simple cyclocondensation reactions from acyclic precursors. After the cyclisation, many other chemical transformations may ensue, including halogenation, alkylation, and reduction.

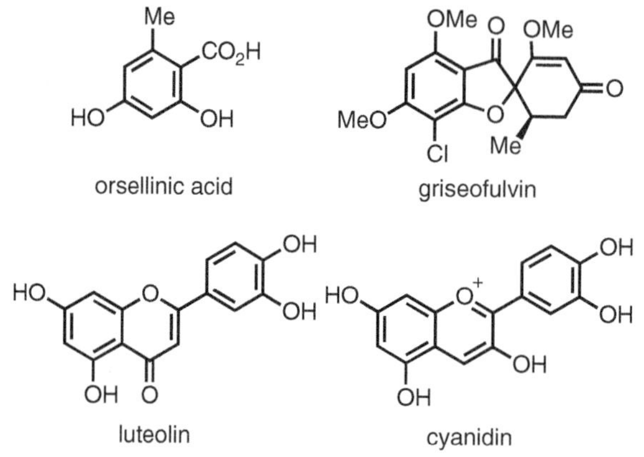

Figure 8.28   *Aromatic polyketides*

Tetracyclines, aromatic polyketides, are broad-spectrum, orally active antibiotics produced by cultures of *Streptomyces* species. Some of these are in medical use in the chemotherapy of cancer. Their antimicrobial activity arises from inhibition of protein synthesis by interfering with the binding of aminoacyl-tRNA to acceptor sites on the ribosome. Daunomycin is in use only for the treatment of acute leukemia, and its close congener adriamycin and other derivatives are being tested for the treatment of solid tumors (for structures, see Figure 4.27).

Aflatoxins are mycotoxins produced by *Aspergillus flavus* and *Aspergillus parasiticus* which can contaminate many commodities, including peanuts, Brazil nuts, pistachio nuts, and corn and grain sorghum during growth, harvesting, processing, storage, and shipment. They were first identified in 1962 as the causative agents of liver cancer in Christmas turkeys fattened on ground nuts which were infected with *A. flavus*. Aflatoxins are potent carcinogens, mutagens, and teratogens, thereby potentially causing major health and economic problems. Aflatoxins primarily affect the liver, causing fat deposition and necrosis, and also proliferation of the cells of the bile duct, which leads to irreversible loss of liver function. Consumption of food contaminated with aflatoxin B_1 may result in acute hepatitis at levels of only 0.1 ppm. Structurally, all these compounds contain an angularly-fused bisdihydrofuran system. Sterigmatocystin and versicolorin A also contain a similar structural unit, and these compounds are also active in mutagenicity tests. Their biosynthesis is shown in Scheme 8.22 [67]. Their mode of action is through oxidation (by liver cytochrome P_{450} oxidative enzymes) to an epoxide, which reacts with a guanine residue in DNA (with a mechanism similar to the one discussed for mitomycin, Scheme 6.1).

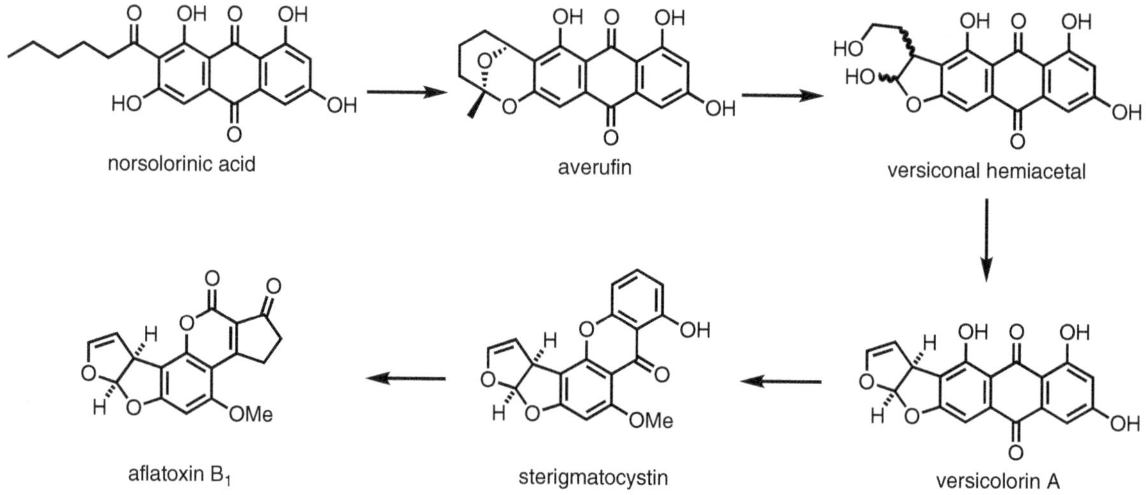

Scheme 8.22   *Biosynthesis of aflatoxins*

The recreational use of preparations from the Indian hemp, *Cannabis sativa*, are well known since the 1960s. The preparations are known with a variety of names. The most common term, hashish, is related to the word 'assassin': a group of hired Persian killers in the thirteenth century were paid in hashish. Material for drug use (ganja) is obtained from the flowering tops of the plant, whereas lower quality matter (bhang) is obtained from the leaves. Higher amounts are available in the resin (bhang), and cannabis oil can be up to 60% tetrahydrocannabinol (THC). Other common names are marijuana, kief, and dagga. Beside this illicit use, the medicinal use of *Cannabis* is known for millennia. According to legend, the Chinese emperor *Huang Ti* recommended the use of cannabis for rheumatic and menstrual pain as early as 2600 BC.

The biosynthesis of cannabinoids is initiated by hexanoate (Scheme 8.23). Three malonyl units are appended, and an intramolecular aldol reaction generates the aromatic core of olivetolic acid. C-Alkylation with geranyl pyrophosphate leads to cannabigerolic acid, which is oxidised to an allylic cation. Double cyclisation and decarboxylation gives tetrahydrocannabinol.

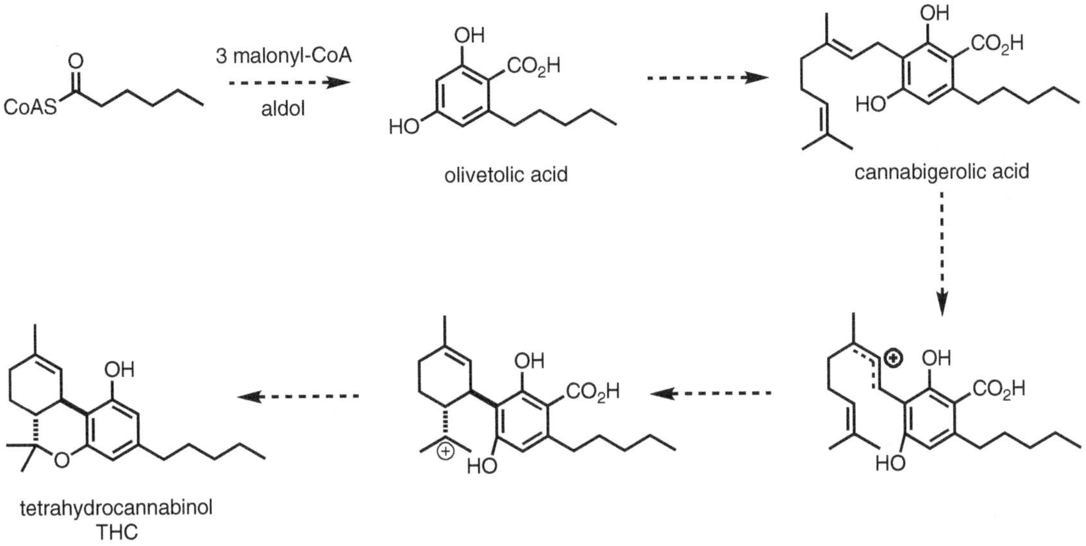

Scheme 8.23 *Biosynthesis of tetrahydrocannabinol*

Tetrahydrocannabinol was only isolated as a pure compound in 1964 [68], and despite its rather simple structure, the first catalytic asymmetric synthesis of tetrahydrocannabinol was achieved only in 1997 by David Evans (Scheme 8.24) [69]. The synthesis relies on an efficient Cu(bisoxazoline)-catalysed Diels–Alder reaction, which sets up the stereochemistry. Four simple transformations lead to the enantiomer of the natural tetrahydrocannabinol in 21% overall yield from the acrylate starting material. In principle, the natural product could also be synthesised using the same protocol, however, the enantiomer of the CuBOX catalyst is even more expensive than the one used in the published synthesis.

Ten years later, Trost demonstrated that all of the stereochemistry of (−)-Δ9-trans-tetrahydrocannabinol (Δ9-THC) could be derived from a single Mo-catalysed asymmetric allylic alkylation (AAA) reaction (Scheme 8.25) [70]. The allylic carbonate was readily prepared in five steps from olivetol in 79% yield. The carbonate reacted with sodium dimethyl malonate with [Mo(CO)$_3$C$_7$H$_8$] and the chiral ligand sluggishly, to give only the branched product in 95% yield and 94% enantiomeric excess. Alkylation or the malonate turned out to be trickier than expected, and a five-step sequence was required to arrive at the ring-closing metathesis precursor as a 2.4:1 mixture of diastereomers. The *syn* and *anti* diastereomers were separated and subjected to ring-closing metathesis with the second-generation Grubbs ruthenium carbene catalyst to provide the *syn* and *anti* cyclohexene compounds. The *syn* diastereomer could be recycled to the desired *anti* compound in 94% yield. Final elaboration to tetrahydrocannabinol proceeded in four steps in 54% yield. The steps involved dimethylation of the ester, cleavage of one of the methoxy methyl ethers, acid-catalysed ring closure, and final deprotection.

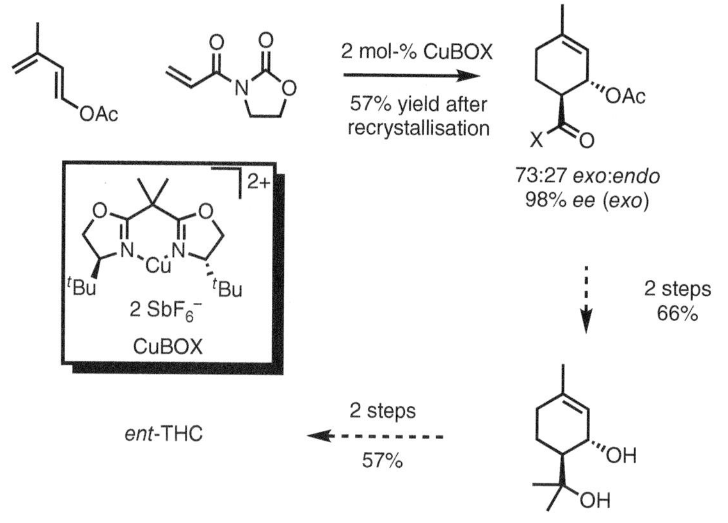

Scheme 8.24　*First catalytic asymmetric synthesis of* ent-*tetrahydrocannabinol*

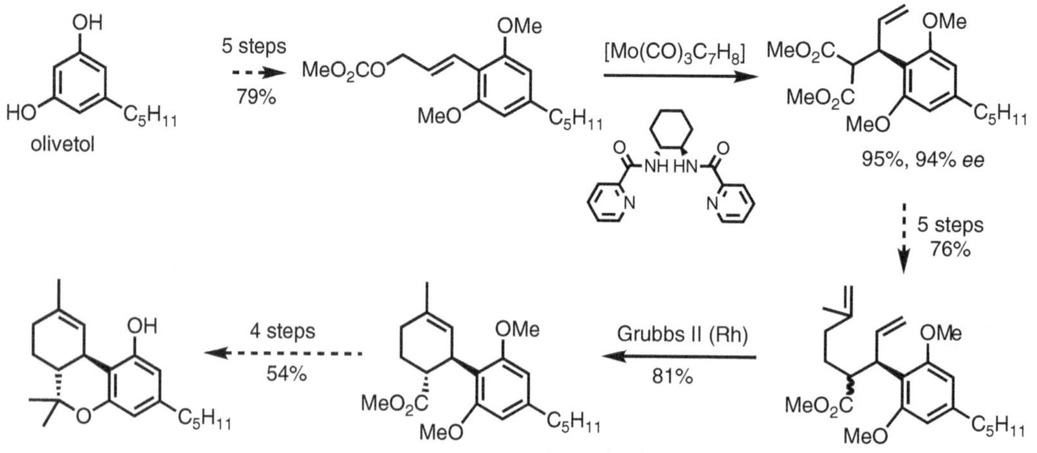

Scheme 8.25　*Trost synthesis of tetrahydrocannabinol*

Medicinally, cannabis is a mild analgesic and tranquilizer, which binds to specific receptors in the central nervous system. These cannabinoid receptors have recently gained attention as promising targets for pain relief [71]. Currently, Δ9-THC is administered as an antinauseant to patients undergoing chemotherapy. The natural ligands of cannabinoid receptors are also emerging, and at least anandamide (*ananda*, from Sanskrit meaning bliss) and 2-arachidonylglycerol (2-AG) (Figure 8.29) have been identified. They both belong to the arachidonic acid derivatives discussed in Section 8.2.1.

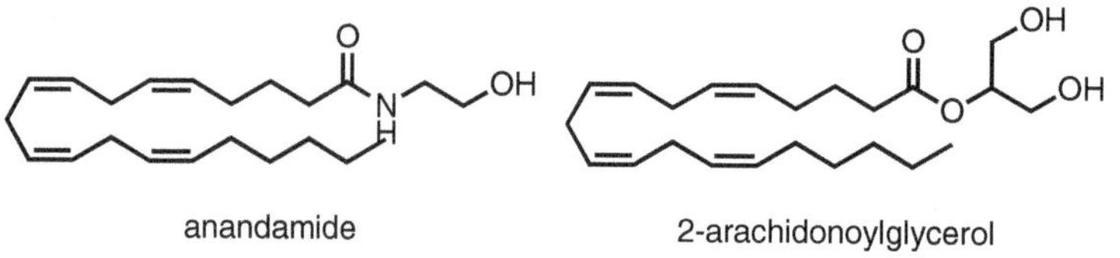

Figure 8.29　*Natural painkillers found in mammals*

References

1. a) Heathcock, C.H. *Science* **1981**, *214*, 395–400; b) Schetter, B., Mahrwald, R. *Angew. Chem. Int. Ed.* **2006**, *45*, 7506–7525; c) Carreira, E.M., Fettes, A., Marti, C. *Org. React.* **2006**, *67*, 1–216; d) Brodmann, T., Lorenz, M., Schäckel, R., Simsek, S., Kalesse, M. *Synlett* **2009**, 174–192.

2. a) Chang, S.I., Hammes, G.G. *Accts. Chem. Res.* **1990**, *23*, 363–369; b) Smith, S., Tsai, S.-C. *Nat. Prod. Rep.* **2007**, *24*, 1041–1072; c) Kwan, D.H., Schulz, F. *Molecules* **2011**, *16*, 6092–6115.

3. a) Behrouzian, B., Buist, P.H. *Curr. Opinion. Chem. Biol.* **2002**, *6*, 577–582; b) Behrouzian, B., Buist, P.H. *Prostaglandins, Leukotrienes and Ess. Fatty Acids* **2003**, *68*, 107–112.

4. a) von Euler, U.S. *Klin. Woch.* **1935**, *14*, 1182–1183; b) Goldblatt M.W. *J Physiol* **1935**, *84*, 208–18.

5. Bergstrom, S. *Science* **1967**, *157*, 382–391.

6. Corey, E.J. *Angew. Chem. Int. Ed. Engl.* **1991**, *30*, 455–465.

7. Das, S., Chandrasekhar, S., Yadav, J.S., Grée, R. *Chem. Rev.* **2007**, *107*, 3286–3337.

8. Corey, E.J. *Angew. Chem. Int. Ed.* **2002**, *41*, 1650–1667.

9. a) Suzuki, M., Yanagisawa, A., Noyori, R. *J. Am. Chem. Soc.* **1985**, *107*, 3348–3349; b) Suzuki, M., Yanagisawa, A., Noyori, R. *J. Am. Chem. Soc.* **1988**, *110*, 4718–4726; c) Suzuki, M., Koyano, H., Morita, Y., Noyori, R. *Synlett* **1989**, 22–23.

10. Kitamura, M., Kasahara, I., Manabe, K., Noyori, R., Takaya, H. *J. Org. Chem.* **1988**, *53*, 708–710.

11. a) Corey, E.J., Schaaf, T.K., Huber, W., Koelliker, U., Weinshenker, N.M. *J. Am. Chem. Soc.* **1970**, *92*, 397–398; b) Corey, E.J., Becker, K.B., Varma, R.K. *J. Am. Chem. Soc.* **1972**, *94*, 8616–8618.

12. Noyori, R., Tomino, I., Yamada, M., Nishizawa, M. *J. Am. Chem. Soc.* **1984**, *106*, 6717–6725.

13. Corey, E.J., Cheng, X.-M. *The Logic of Chemical Synthesis*. John Wiley & Sons, Inc., New York, **1989**, pp. 250–254.

14. a) Coulthard, G., Erb, W., Aggarwal, V.K. *Nature* **2012**, *489*, 278–281; b) Pelšs, A., Gandhamsetty, N., Smith, J.R., Mailhol, D., Silvi, M., Watson, A.J.A., et al. *Chem. Eur. J.* **2018**, *24*, 9542–9545.

15. Zhang, F., Zeng, J., Gao, M., Wang, L., Chen, G.-Q., Lu, Y., et al. *Nature Chem.* **2021**, *13*, 692–697.

16. Jing, C., Aggarwal, V.K. *Org. Lett.* **2020**, *22*, 6505–6509.

17. Thudichum, J.L.W. *A Treatise on the Chemical Constitution of the Brain* **1884**, Lontoo: Bailliere. New edition **1962**, Archon Books, Hamden, Connecticut, USA.

18. a) Hakomori, S. *J. Biol. Chem.* **1990**, *265*, 18713; b) Van Meer, G., Burger, K.N J. *Trends in Cell Biology* **1992**, *2*, 332.

19. Pruett, S.T., Bushnev, A., Hagedorn, K., Adiga, M., Haynes, C.A., Sullards, M.C., et al. *J. Lipid Res.* **2008**, *49*, 1621–1639.

20. a) Merrill, A.H. *Chem. Rev.* **2011**, *111*, 6397–6422; b) Futerman, A.H., Hannun, Y.A. *EMBO Reports* **2004**, *5*, 777–782.

21. Spiegel, S., Milstien, S. *J. Biol. Chem.* **2002**, *277*, 25851–25854.

22. Brunner, M., Koskinen, A.M.P. *Curr. Org. Chem.* **2004**, *8*, 1629–1645.

23. Strader, C.R., Pearce, C.J., Oberlies, N.H. *J. Nat. Prod.* **2011**, *74*, 900–907.

24. Koskinen, A.M.P. *Pure Appl. Chem.* **2011**, *83*, 435–443.

25. Passiniemi, M., Koskinen, A.M.P. *Org. Biomol. Chem.* **2011**, *9*, 1774–1783.

26. Sata, N.U., Fusetani, N. *Tetrahedron Lett.* **2000**, *41*, 489–492.

27. Kumpulainen, E.T.T., Koskinen, A.M.P., Rissanen, K. *Org. Lett.* **2007**, *9*, 5043–5045.

28. Abraham, E., Davies, S.G., Millican, N.L., Nicholson, R.L., Roberts, P.M., Smith, A.D. *Org. Biomol. Chem.*, **2008**, *6*, 1655–1664.

29. a) Morris, J.C., Phillips A.J. *Nat. Prod. Rep.* **2009**, *26*, 245–265; b) Morris, J.C., Phillips A.J. *Nat. Prod. Rep.* **2010**, *27*, 1186–1203; c) Morris, J.C., Phillips A.J. *Nat. Prod. Rep.* **2011**, *28*, 269–289; d) Morris, J.C. *Nat. Prod. Rep.* **2013**, *30*, 783–805; e) Yeung, K.-S., Paterson, I. *Angew. Chem. Int. Ed.* **2002**, *41*, 4632–4653; f) Nicolaou, K.C., Frederick, M.O., Aversa, R.J. *Angew. Chem. Int. Ed.* **2008**, *47*, 7182–7225; g) Blunt, J.W., Copp, B.R., Munro, M.G.H., Northcote, P.T., Prinsep, M.R. *Nat. Prod. Rep.* **2011**, *28*, 196–268; h) Blunt, J.W., Copp, B.R., Keyzers, R.A., Munro, M.G.H., Prinsep, M.R. *Nat. Prod. Rep.* **2012**, *29*, 144–222; i) Blunt, J.W., Copp, B.R., Keyzers, R.A., Munro, M.G.H., Prinsep, M.R. *Nat. Prod. Rep.* **2013**, *30*, 237–323; j) Blunt, J.W., Copp, B.R., Keyzers, R.A., Munro, M.G.H., Prinsep, M.R. *Nat. Prod. Rep.* **2014**, *31*, 160–258; k) Blunt, J.W., Copp, B.R., Keyzers, R.A., Munro, M.G.H., Prinsep, M.R. *Nat. Prod. Rep.* **2015**, *32*, 116–211; l) Blunt, J.W., Copp, B.R., Keyzers, R.A., Munro, M.G.H., Prinsep, M.R. *Nat. Prod. Rep.* **2016**, *33*, 382–435; m) Blunt, J.W., Copp, B.R., Keyzers, R.A., Munro, M.G.H., Prinsep, M. R. *Nat. Prod. Rep.* **2017**, *34*, 235–294; n) Blunt, J.W., Carroll, A.R., Copp, B.R., Davis, R.A., Keyzers, R.A., Prinsep, M.R. *Nat. Prod. Rep.* **2018**, *35*, 8–53; o) Carroll, A.R., Copp, B.R., Davis, R.A., Keyzers, R.A., Prinsep, M.R. *Nat. Prod. Rep.* **2019**, *36*, 122–173; p) Carroll, A.R., Copp, B.R., Davis, R.A., Keyzers, R.A., Prinsep, M.R. *Nat. Prod. Rep.* **2020**, *37*, 175–223.

30. a) *Polyether Antibiotics* (Westley, J.W., ed.) Marcel Dekker: New York, **1982**, vol. 1–2; b) O'Hagan, D. *Nat. Prod. Rep.* **1989**, 205–219; c) Dutton, C.J., Banks, B.J., Cooper, C.B. *Nat. Prod. Rep.* **1995**, 165–181.

31. a) Kevin II, D.A., Meujo, D.A.F., Hamann, M.T. *Expert Opin. Drug Discov.* **2009**, *4*, 109–146; b) Huczyński, A. *Bioorg. Med. Chem. Lett.* **2012**, *22*, 7002–7010.

32. a) Nagai, H., Torigoe, K., Satake, M., Murata, M., Yasumoto, T., Hirota, H. *J. Am. Chem. Soc.* **1992**, *114*, 1102–1103; b) Nagai, H., Murata, M., Torigoe, K., Satake, M., Yasumoto, T. *J. Org. Chem.* **1992**, *57*, 5448–5453.

33. Nicolaou, K.C. *Angew. Chem., Int. Ed. Engl.* **1996**, *35*, 589–607.

34. Murata, M., Legrand, A.M., Ishibashi, Y., Yasumoto, T. *J. Am. Chem. Soc.* **1989**, *111*, 8929–8931.

35. Nicolaou, K.C., Aversa, R.J. *Isr. J. Chem.* **2011**, *51*, 359–377.

36. a) Gambieric acid: Fuwa, H., Ishigai, K., Hashizume, K., Sasaki, M. *J. Am. Chem. Soc.* **2012**, *134*, 11984–11987; b) Ciguatoxin: Hamajima, A., Isobe, M. *Angew. Chem. Int. Ed.* **2009**, *48*, 2941–2945.

37. Jackson, K.L., Henderson, J.A., Phillips, A.J. *Chem. Rev.* **2009**, *109*, 3044–3079.

38. a) Kishi, Y. *Tetrahedron* **2002**, *58*, 6239–6258; b) Kishi, Y. *Annu. Rep. Med. Chem.* **2011**, *46*, 227–241; c) Chase, C.E., Fang, F.G., Lewis, B.M., Wilkie, G.D., Schnaderbeck, M.J., Zhu, X. *Synlett* **2013**, *24*, 323–326; d) Austad, B.C., Benayouda, F., Calkins, T.L., Campagna, S., Chase, C.E., Choi, H.-W., et al. *Synlett* **2013**, *24*, 327–332; e) Austad, B.C., Calkins, T.L., Chase, C.E., Fang, F.G., Horstman, T.E., Hu, Y., et al. *Synlett* **2013**, *24*, 333–337.

39. Masamune, S., Bates, G.S., Corcoran, J.W. *Angew. Chem.* **1977**, *89*, 602–624.

40. Feyen, F., Cachoux, F., Gertsch, J., Wartmann, M., Altmann, K.-H. *Accts Chem. Res.* **2008**, *41*, 21–31.

41. Nicolaou, K.C., Yue, E.W. *Pure & Appl. Chem.* **1997**, *69*, 413–418.

42. Laureti, L., Song, L., Huang, S., Corre, C., Leblond, P., Challis, G.L., et al *Proc. Nat. Acad. Sci. U.S.A.* **2011**, *108*, 6258–6263.

43. Lim, J., Chintalapudi, V., Gudmundsson, H.G., Tran, M., Bernasconi, A., Blanco, A., et al. *Org. Lett.* **2021**, *23*, 7439–7444.

44. a) Brockmann, H., Henkel, W. *Naturwissenschaften* **1950**, *37*, 138–139; b) Brockmann, H., Henkel, W. *Chem. Ber.* **1951**, *84*, 284–288.

45. Kwan, D.H., Schulz, F. *Molecules* **2011**, *16*, 6092–6115.

46. Mulzer, J. *Angew. Chem. Int. Ed.* **1991**, *30*, 1452–1454.

47. Woodward, R.B. 'Synthesis' in *Perspectives in Organic Synthesis*, Todd, A. (ed.) Interscience, New York, **1956**, 155–184.

48. Woodward, R.B., Logusch, E., Nambiar, K.P., Sakan, K., Ward, D.E., Au-Yeung, B.-W., et al. *J. Am. Chem. Soc.* **1981**, *103*, 3210–3213; 3213–3215; 3215–3217.

49. For a review on stereotetrad synthesis, see: Koskinen, A.M.P., Karisalmi, K. *Chem. Soc. Rev.* **2005**, *34*, 677–690.

50. Gao, X., Woo, S.K., Krische, M.J. *J. Am. Chem. Soc.* **2013**, *135*, 4223 4226.

51. Cereghetti, D.M., Carreira, E.M. *Synthesis* **2006**, 914–942.

52. Karisalmi, K., Rissanen, K., Koskinen, A.M.P. *Org. Biomol. Chem.* **2003**, *1*, 3193–3196.

53. Zhang, Y., Arpin, C.C., Cullen, A.J. ,Mitton-Fry, M.J., Sammakia, T. *J. Org. Chem.* **2011**, *76*, 7641–7653.

54. Fleck, W.F., Ritzau, M., Heinze, S., Gräfe, U. *J. Basic. Microbiol.* **1996**, *36*, 235–238.

55. Vishwanath, C.K., Shamala, N., Easwaran, K.R.K., Vijayan, M. *Acta Cryst. C: Cryst. Struct. Comm.* **1983**, *C39*, 1640–1643.

56. Floss, H.G., Tin-Wein Yu, T.-W. *Chem. Rev.* **2005**, *105*, 621–632.

57. Still, W.C., Barrish, J.C. *J. Am. Chem. Soc.* **1983**, 2487–2489.

58. Celmer, W.D. *Pure Appl. Chem.* **1971**, *28*, 413–453.

59. Janssen, D., Albert, D., Jansen, R., Müller, R., Kalesse, M. *Angew. Chem. Int. Ed.* **2007**, *46*, 4898–4901.

60. a) Evans, D.A., Rieger, D.L., Gage, J.R. *Tetrahedron Lett.* **1990**, *31*, 7099–7100; b) Rychnovsky, S.D., Rogers, B.N., Richardson, T.I. *Acc. Chem. Res.* **1998**, *31*, 9–17.

61. a) Reid, R., Piagentini, M., Rodriguez, E., Ashley, G., Viswanathan, N., Carney, J., et al. *Biochemistry* **2003**, *42*, 72–79; b) Caffrey, P. *ChemBioChem* **2003**, *4*, 654–657.

62. Perron, F., Albizati, K. *Chem. Rev.* **1989**, *89*, 1617–1661.

63. Boger, D.L., Ichikawa, S., Zhong, W. *J. Am. Chem. Soc.* **2001**, *123*, 4161–4167.

64. Oikawa, M., Ueno, T., Oikawa, H., Ichihara, A. *J. Org. Chem.* **1995**, *60*, 5048–5068.

65. Fagerholm, A. E., Habrant, D., Koskinen, A.M.P. *Marine Drugs* **2010**, *8*, 122–172.

66. Campbell, W.C. *Med. Res. Rev.* **1993**, *13*, 61–79.

67. Silva, J.C., Minto, R.E., Barry III, C.E., Holland, K.A., Townsend, C.A. *J. Biol. Chem.* **1996**, *271*, 13600–13608.

68. Gaoni, Y., Mechoulam, R. *J. Am. Chem. Soc.* **1964**, *86*, 1646–1647.

69. Evans, D.A., Shaughnessy, E.A., Barnes, D.M. *Tetrahedron Lett.* **1997**, *38*, 3193–3194.

70. Trost, B.M., Dogra, K. *Org. Lett.* **2007**, *9*, 861–863.

71. a) DiMarzo, V., Bifulco, M., De Petrocellis, L. *Nature Rev. Drug Disc.* **2004**, *3*, 771–784; b) DiMarzo, V. *Trends in Pharm. Sci.* **2006**, *27*, 134–140; c) DiMarzo, V., Bisogno, T., De Petrocellis, L. *Chem. Biol.* **2007**, *14*, 741–756.

9

Isoprenoids or Terpenes

Although there are many useful guidelines that can be followed during the planning stage, there are simply no general rules to apply when synthesizing a terpene.

Phil Baran, 2007

In the early history of natural product chemistry, many strongly odorant plant compounds were observed to be formed from C_5 units called isopentenyl or isoprene units, thus they are often called isoprenoids. These compounds were termed terpenes, the term derived from the terebinth tree, *Pistacia terebinthus*, which exuded a viscous pleasantly smelling balsam called balsamum terebinthinae. The pleasant odors led to the utilisation of the extracts or steam distillates of conifers, balm trees, eucalyptus, lavender, lemon grass, caraway, peppermint, roses, rosemary, sage, thyme, and many other plants in fragrances. These essential or ethereal oils became important products of merchandise in many Mediterranean countries. According to a legend, in 1380, Queen Joan d'Anjou visited the Certosa di San Giacomo in Capri, and the prior of the monastery decided to honor the event by gathering the most beautiful and fragrant wild flowers on the island, and started the production of a fragrance famous to date.

Terpenes occupied a central role in the chemical research in the beginning of the twentieth century. The first compounds to be studied were the small volatile ones, and in the 1920s professor *Leopold Ruzicka* (1887–1976) and his co-workers at the University of Zurich developed the methods needed for the study of larger terpenes. Ruzicka also systematised the then-known concepts in the biosynthesis of terpenes in a theory, which is known as the isoprene rule [1]. It states that terpenes are formed from 2-methylbutane, or *isoprene*, (C_5) units linked together from head (isopropyl) to tail (ethyl). Thus, for instance, limonene can be constructed by a formal Diels–Alder reaction by joining the head of one isoprene unit with the tail of another one (Scheme 9.1). Although this generalisation fails to be true in some cases, it has proven to be very useful in the majority of cases; especially the longer terpenes (tri- and tetraterpenes) contain one tail-to-tail linkage in the middle of the structure. The isolation of mevalonic acid in 1956 helped to understand the various rearrangement, methylation, and dealkylation reactions involved in the biogenesis of more complex terpenes [2].

Terpenes are classified according to the number of pairs of these units present in the molecule (Figure 9.1): monoterpenes, C_{10}; sesquiterpenes, C_{15}; diterpenes, C_{20}; sesterterpenes, C_{25}; triterpenes, C_{30}, tetraterpenes C_{40}, and polyterpenes $(C_5)_n$, n > 8. Often one or more carbon atoms are excised from the molecule, and these *terpenoids* are sometimes called nor-derivatives (e.g. a norditerpene contains 19 carbon atoms in its skeleton).

Camphor was known as a flavoring agent in the pre-Islamic Arabia, and in the nineth century, the Arab chemist *Abu Yūsuf Yaʻqūb ibn 'Isḥāq aṣ-Ṣabbāḥ al-Kindī* (801–873) described the first recipe for the production of this valued perfume. Camphor has been used for medicinal purposes, for the treatment of heart symptoms and fatigue. Camphor

Asymmetric Synthesis of Natural Products, Third Edition. Ari M.P. Koskinen.
© 2023 John Wiley & Sons Ltd. Published 2023 by John Wiley & Sons Ltd.

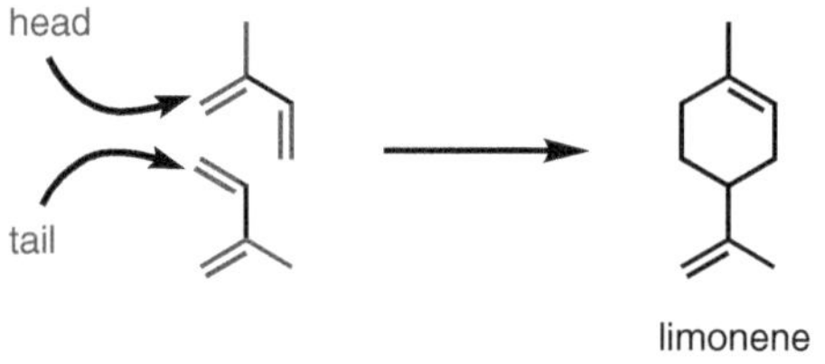

Scheme 9.1 *Isoprene rule*

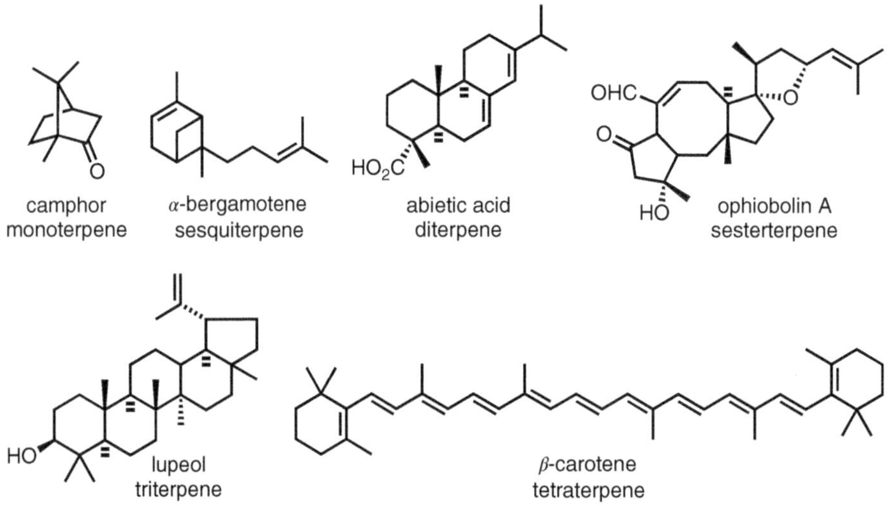

Figure 9.1 *Typical terpenes*

is also included in some products to give relief for respiratory tract congestion, although its action probably is mainly through acting as a weak local anesthetic. It still finds use in plastics, especially in the form of celluloid, considered to be the first thermoplastic. Celluloid is a product of nitrocellulose and camphor, and it became available in the 1870s. The flexible celluloid film soon found use in photography, manufacture of table tennis balls, and guitar picks. Although the structure of camphor is relatively simple by modern standards, altogether nearly thirty structural proposals were presented before *Julius Bredt* (1855–1937) showed the correct structure for camphor in 1895 [3]. The Finnish chemist *Gustav Komppa* (1867–1949) first prepared fully synthetic camphor in 1903, and this is regarded as the first total synthesis of a complex natural product [4]. The British chemist W.H. Perkin presented the synthesis of terpineol in the following year, 1904 [5].

In the beginning of the nineteenth century, hydrochloric acid had been shown to convert turpentine to a crystalline salt (!), $C_{10}H_{16} \bullet HCl$ ("artificial camphor," Kindt, 1802). *Marcellin Berthelot* (1827–1907) heated this at 260 °C with sodium benzoate or sodium stearate and obtained pure camphene (the base hydrocarbon). When he further heated camphene under air in the presence of platinum black, he obtained camphor (Scheme 9.2) [6]. Berthelot's synthesis of camphor thus constitutes an early example of a total synthesis of a natural product. In current terms, we would call this a formal total synthesis, since it begins with an isolated natural product (turpentine is approx. 95% chemically pure α-pinene). Berthelot also showed another interesting episode in his camphor synthesis: when artificial camphor was prepared from terebenthine (l-pinene), one obtained l-terecamphene (l-camphene). When artificial camphene was prepared from australene (d-pinene), the product was d-austracamphene (d-camphene).

The Komppa synthesis was a true total synthesis, although strictly speaking a relay one, because it ended with camphoric acid, which had been converted to camphor by Bredt. The Komppa synthesis was commercialised in 1908, with annual production reaching 200 kg. However, the turpentine oil route soon allowed a German company to reach annual volumes of up to 800 tons.

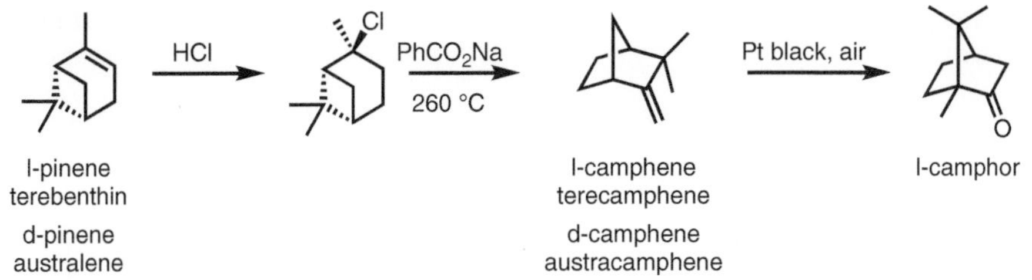

Scheme 9.2 *Berthelot synthesis of camphor*

The sesquiterpene bergamotene occurs in several essential oils (e.g. water pepper, kumquat, and basil). Bergamot oil is widely used in the perfumery industry. Pine resin contains large quantities of the diterpene abietic acid, which is therefore one of the least expensive sources of optically active compounds. Abietic acid is a resin acid occurring in rosin, which has been widely used for caulking ships. The triterpene lupeol (from e.g. mango and acacia) has anti-inflammatory properties. Ophiobolin is a sesterterpene with a rare eight-membered ring. It is found in *Helmithospora* species and exhibits a broad spectrum of antibacterial, antifungal, and anthelmintic activity. The tetraterpene carotene (from e.g. carrots, sweet potato, mangoes, apricots, and cantaloupe melons) is converted to retinol (vitamin A) which is a precursor of retinal, essential for vision. The carotene on the market is produced by one of two total synthesis routes employing either a Wittig reaction or a Grignard reaction.

9.1 Terpenes

Over 55 000 terpenes (or terpenoids) have been isolated and characterised, mainly from plant origins [7]. However, terpenes occur in all forms of life, and even archaebacteria produce most fascinating tetraterpene structures.

Isoprene itself is not involved in the biosynthesis of terpenes as a reactive species, but is utilised in the form of isopentenyl and dimethylallyl pyrophosphates (IPP or DMAP, respectively, Scheme 9.3). The most common biosynthetic route in plants and some insects is depicted on the left, and in this route, mevalonic acid (MVA) is first formed through the polyacetate pathway by acylation of the acetoacetyl chain followed by reduction. Phosphorylation followed by ATP-assisted loss of water and carbon dioxide gives isopentenyl pyrophosphate. Isomerisation of the double bond through the action of a thiol-containing isomerase gives dimethylallyl pyrophosphate [8]. In the late 1990s and early 2000s, a second pathway to the monomeric precursors was uncovered in bacteria and in the chloroplasts of green algae and higher plants by *Duilio Arigoni* (1928–2020, ETH) and Rohmer [9]. In this deoxyxylylose phosphate (DOX) or methylerythitol phosphate (MEP) pathway, pyruvic acid and glyceraldehyde 3 phosphate are condensed with concomitant loss of carbon dioxide. Subsequent isomerisation and reduction results in MEP. Further transformations lead to IPP [10].

9.1.1 Monoterpenes

Many familiar fragrances are terpenes with relatively small size and high volatility (Figure 9.2). The odor typical to lemons (*Citrus lemon*) is mainly caused by (*R*)-limonene, whereas its *S*-enantiomer is present in turpentine. The main component in caraway seeds (*Carum carvi*) and dill seed oil (*Anethum graveolens*) is (*S*)-carvone (up to 40%), whereas the *R*-isomer is present in spearmint oil (*Mentha spicata*).

Turpentine, which is distilled oil from resin produced by some trees, typically pines, has been used for centuries as a solvent. The name derives from the Greek word *terebinthine* for the Mediterranean terebinth tree (*Pistacia terebinthus*). The main component of turpentine oil is α-pinene (up to 95%, the rest mainly being β-pinene). Interestingly, the α-pinene from European turpentine oils is levorotatory, whereas in North American oils the dextrorotatory form is found. Curiously, wild celery (*Angelica archangelica*) and *Eucalyptus* oils contain racemic α-pinene. 3-Carene is also found in the *Pinus sylvestris* (Scots pine) turpentine (up to 42%). *Antoine Laurent Lavoisier* (1743–1794) studied the behavior and chemistry of turpentine, and the first studies of optical activity by Jean-Baptiste Biot were conducted

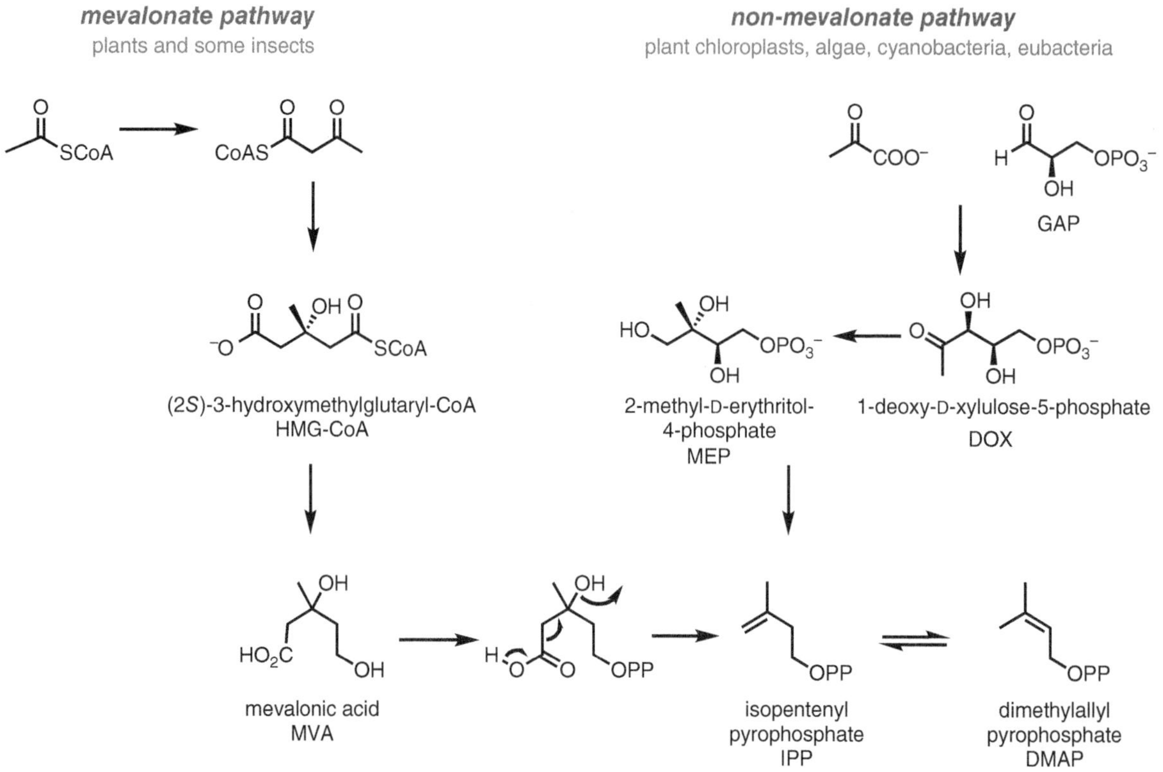

mevalonate pathway
plants and some insects

non-mevalonate pathway
plant chloroplasts, algae, cyanobacteria, eubacteria

(2S)-3-hydroxymethylglutaryl-CoA
HMG-CoA

2-methyl-D-erythritol-
4-phosphate
MEP

1-deoxy-D-xylulose-5-phosphate
DOX

GAP

mevalonic acid
MVA

isopentenyl
pyrophosphate
IPP

dimethylallyl
pyrophosphate
DMAP

Scheme 9.3   Biosynthesis of terpene building blocks IPP and DMAP

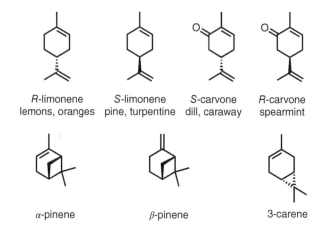

R-limonene
lemons, oranges

S-limonene
pine, turpentine

S-carvone
dill, caraway

R-carvone
spearmint

α-pinene

β-pinene

3-carene

Figure 9.2   Monoterpenes

on turpentine in the early nineteenth century. Furthermore, a large number of nitrogen-containing compounds derived from terpenes are classified as alkaloids, because their physical properties more closely resemble those of alkaloids (see Chapter 10).

The biosynthesis of monoterpenes involves the dimerisation of two isoprene units, in a head-to-tail fashion, to form geranyl pyrophosphate (Scheme 9.4). Isomerisation to the *cis* olefin, neryl pyrophosphate, sets the stage for cationic cyclisation to give the menthane skeleton. Loss of a proton and hydration of the endocyclic olefin gives menthol.

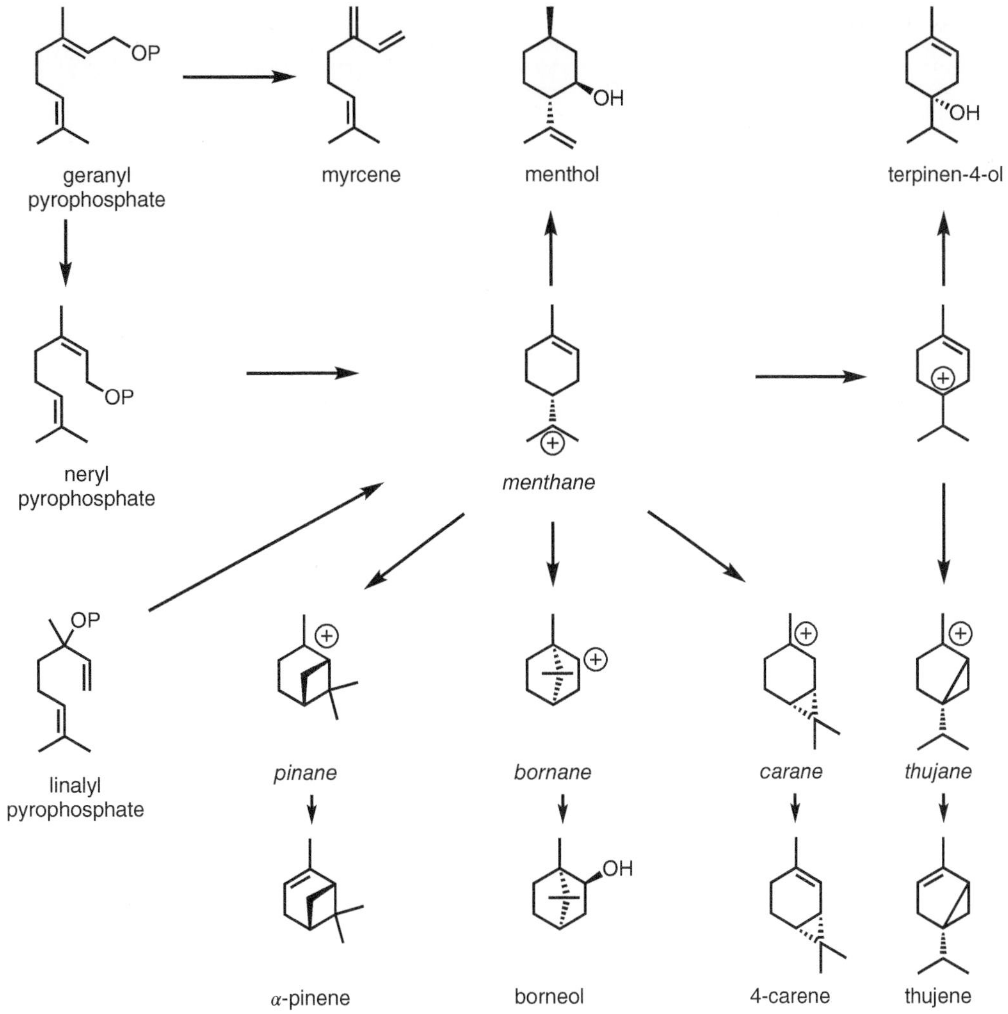

Scheme 9.4 *Biosynthesis of monoterpenes*

Further cationic cyclisations of the menthane cation lead to the pinane, bornane, and carane skeleta. A Wagner–Meerwein 1,2-shift of the menthane cation gives a rearranged cation which can be trapped by hydroxide to give terpinen-4-ol. An alternative cationic cyclisation with the participation of the ring double bond leads to the thujane skeleton.

Iridoids are a large family of bicyclic monoterpenes which contain a cylopentanopyran. They possess a range of potential pharmacological activities, including anticancer, anti-inflammatory, antifungal, and antibacterial activities. Certain iridoids are sex pheromones of aphids, indicating new potential for pest-management strategies. Whereas the biosynthesis of canonical terpenoids begins by activation of the linear precursors by either loss of pyrophosphate or protonation, the biosynthesis of iridoids in plants begins by hydrolysis and oxidation of geranyl pyrophosphate GPP to 8-oxogeranial. A two-step activation-cyclisation process then leads to the iridoid skeleton (Scheme 9.5) [11].

Nepetalactones are volatile iridoids produced by plants of the genus *Nepeta*, notably catnip (*N. cataria*) and catmint (*N. mussinii*). Besides being aphid sex pheromones and insect and mosquito repellets, nepetalactones also cause euphoria in cats. A recent enantiospecific synthesis of nepetalactones has been achieved utilising one-step oxidative NHC (*N*-heterocyclic carbene) catalysis (Scheme 9.6) [12]. Commercially available (*S*)-citronellol can be converted in

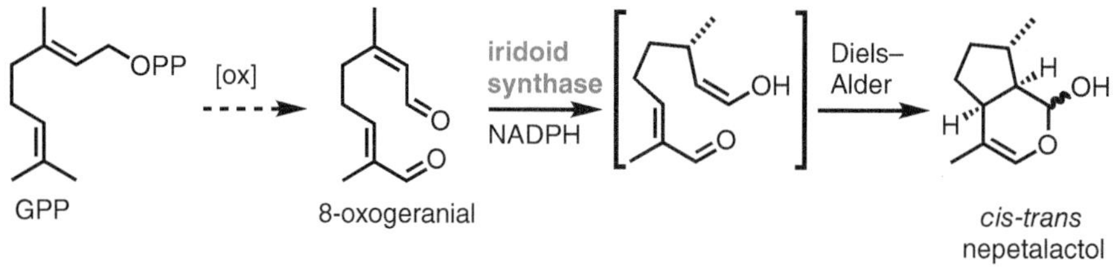

Scheme 9.5   *Biosynthesis of iridoids*

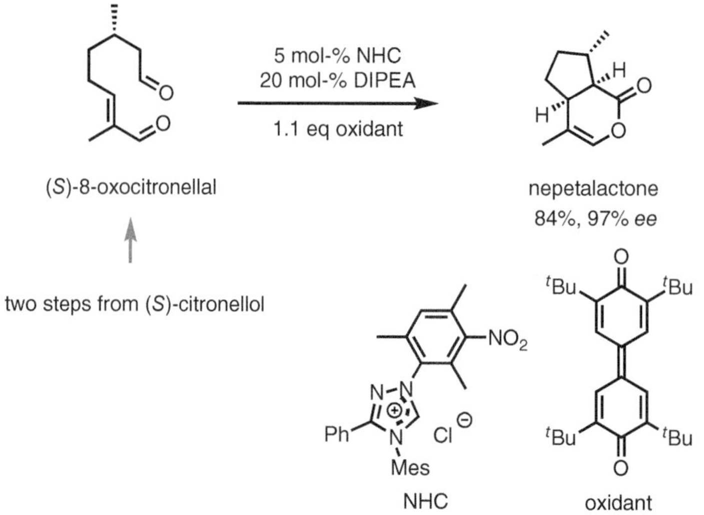

Scheme 9.6   *One-step synthesis of nepetalactone*

two steps to (*S*)-8-oxocitronellal. When this is treated with the NHC, the saturated aldehyde is converted to a Breslow intermediate, which is then oxidised in situ to a ketone. Enolisation and conjugate addition leads to the formation of the five-membered ring as a single diastereomer. Formation of the lactone ring liberates the carbene. As both stereoisomers of citronellol are commercially available, the synthesis allows the preparation of both enantiomers of nepetalactone.

Secologanin plays an important role in the biosynthesis of monoterpenoid indole alkaloids, which will be discussed in Section 10.1.3.1. The biosynthesis of the monoterpene loganin and its further oxidation product secologanin is shown in Scheme 9.7. A number of oxidative transformations of geraniol are followed by glucosylation and cyclisation, and further oxidation leading to loganin. This is further oxidatively cleaved to give secologanin.

9.1.2   Sesquiterpenes

Sesquiterpenes contain 15 carbon atoms, and are formed from three isoprene units by way of *cis-trans*-farnesyl pyrophosphate through cationic cyclisation similar to the formation of the menthane cation. Approximately 10 000 sesquiterpenes have been isolated from nature, and it is no surprise that the structures vary quite broadly. The simplest ones, farnesanes (Figure 9.3), are acyclic compounds, as exemplified by *α*- and *β*-farnesenes, flavors and natural coatings of apples, pears, and other fruits, and components of several ethereal oils, including those of camomile, citrus, and hops. The derived aldehydes, such as *α*-sinensal, contribute to the flavor of the oil of orange from the fresh peel of ripe fruits of *Citrus sinensis* and mandarin peel oil from *Citrus reticulata* and *C. aurantium*. The oil of neroli from orange flowers contains (*S*)-(+)-nerolidol, similar to farnesol from *Acacia farnensiana* and the oils of bergamot, hibiscus, jasmine, and rose, and blossoms of lily of the valley. (*S*)-2,3-Dihydrofarnesol, known as terrestrol, is the marking pheromone of the male bumble bee *Bombus terrestris*.

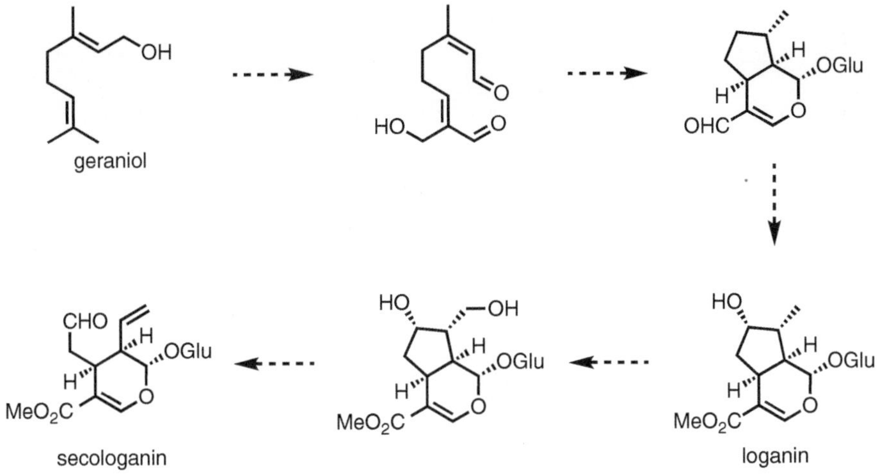

Scheme 9.7 *Biosynthesis of secologanin*

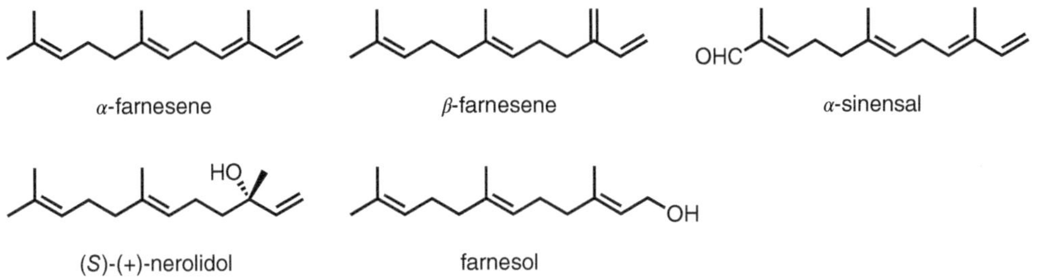

Figure 9.3 *Farnesanes*

Farnesyl pyrophosphate is the biosynthetic precursor of a large number of cyclic derivatives. The bisabolyl cation can undergo similar cyclisation reactions as the menthane cation to give different ring systems (Scheme 9.8). More than 100 bisabolane derivatives are known, including fragrant bisabolol itself, and the related zingiberene, a major constituent of oil of ginger. Sirenin, formed by ring closure of the cyclopropane (route a), is a sperm attractant in the marine mold *Allomyces*. Campherenol, from nucleophilic attack of the double bond to the carbocation (route b), occurs in lemon peelings, and is found in limoncello, an Italian lemon liqueur. An alternative mode of nucleophilic attack of the double bond (route c) gives bergamotene, isolated from the rind of the bergamot orange (*Citrus bergamia*). Since the 1820s, this oil of bergamot has been used in England to flavor black teas to imitate the aromas of the most expensive Chinese teas. This flavored tea is commonly known as the Earl Grey blend, and several varieties with different additives have been developed.

An alternative cyclisation of nerolidyl pyrophosphate gives rise to the germacradienyl cation (Scheme 9.9). This can simply hydrate to form the corresponding *E,Z*-germacradienol, a major component of the sesquiterpene fraction of the resin of several pine varieties. Larvae of the European pine sawfly (*Neodiprion sertifer*) sequester pine resin and secrete 1,6-germacradien-5-ol as a defense secretion against its predators (ants, spiders, wasps, and some birds). Further processing of the germacradienyl cation can lead to aristolactone with a ten-membered ring or the bicyclic eudesmane skeleton, e.g. the nerve growth-promoting dictyophorine B, the first eudesmanes isolated from fungi, *Phallus indusiatus*.

Humulyl cation is formed from the macrocyclisation of farnesyl pyrophosphate and gives rise to humulene (*α*-caryophyllene). Humulene derives its name from its major source, *Humulus lupulus* (hops). It also gives its aroma to Vietnamese coriander and contributes to the characteristic aroma of *Cannabis sativa*.

Yet another mode of cyclisation of the *trans,trans*-farnesyl pyrophosphate leads to caryophyllene (Figure 9.4). *β*-Caryophyllene itself is a major constituent of clove oil, and the derived aldehyde is prevalent in sage oil. Further

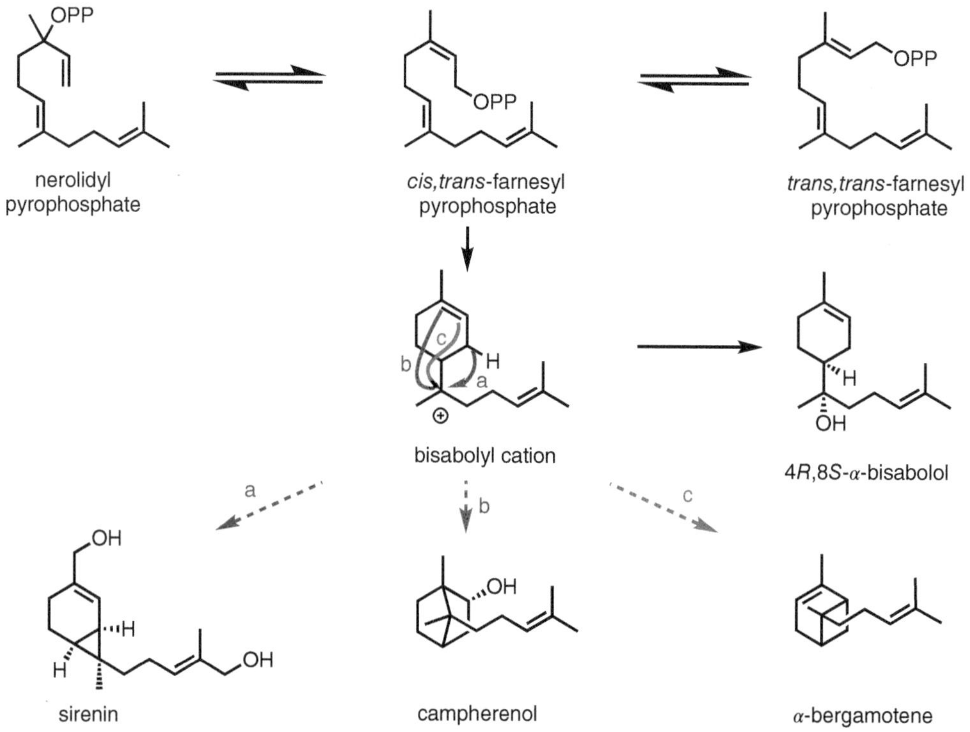

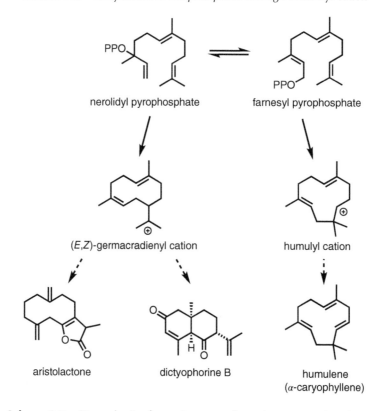

Scheme 9.8 *Biosynthesis of sesquiterpenes through bisabolyl cation*

Scheme 9.9 *Biosynthesis of sesquiterpenes through germacradienyl cation*

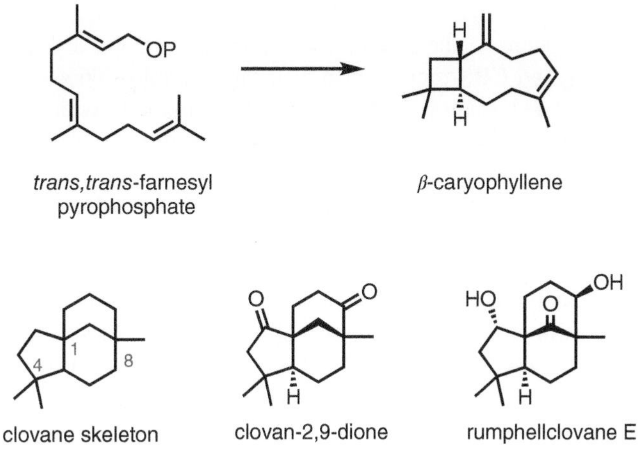

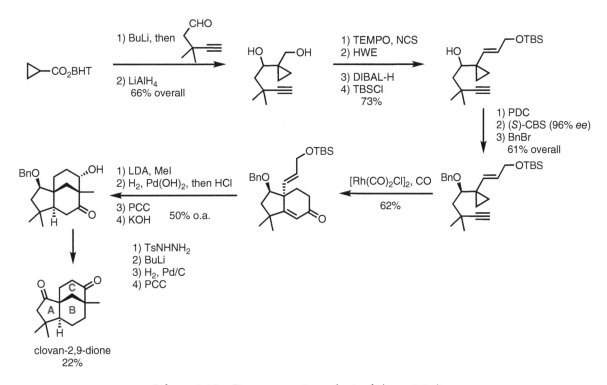

Figure 9.4 *Clovane biosynthesised through caryophyllene*

ring closure with rearrangement of the four-membered ring leads to a class of sesquiterpenes known as clovanes, isolated from the gorgonioan coral *Rumphella antipathies*. Clovane-type natural products contain a tricycle [6.3.1.0 [1, 5]]dodecane skeleton with three quaternary carbon centers at C1, C4, and C8. Biological evaluations have indicated that some clovane derivatives inhibit production of superoxide anion and elastase released by human neutrophils, and inhibit growth of embryonic hippocampal and cortical neurons.

The first asymmetric total synthesis of clovane-2,9-dione was reported in 2017, employing a unique Rh-catalysed [3 + 2 + 1]-cycloaddition reaction to simultaneously construct rings A and B efficiently (Scheme 9.10) [13]. The

Scheme 9.10 *First asymmetric synthesis of clovan-2,9-dione*

synthesis began with the aldol coupling of cyclopropane carboxylic acid and dimethyl pentynal. The resulting racemic alcohol was converted through five steps to the racemic extended alcohol, and asymmetry was then introduced by oxidation of the secondary alcohol and CBS reduction to give the alcohol in 96% *ee* ready for the key cyclisation reaction. The key Rh(I)-catalysed [3 + 2 + 1] cycloaddition occurred smoothly with 5 mol-% rhodium dimer catalyst under CO atmosphere. The reaction afforded the desired *trans*-product in 62% isolated yield after chromatographic separation of the minor diastereomer. The remaining steps involved methylation of the ketone, and reduction of the enone, followed by intramolecular aldol reaction to construct the C ring. Final adjustments of the oxidation levels lead to the natural product.

Recently, an eight-step synthesis of rumphelloclovane E has been reported starting with commercially available carvone (Scheme 9.11) [14]. Rh-catalysed cyclopropanation with diethyl diazomalonate followed by ring closure of the A ring and alkylation with acrolein proceeded in 56% overall yield to give the key precursor for the closure of the C ring. This was achieved by iron-catalysed reductive aldol reaction and protection of the alcohol. The cyclopropane ring was reductively opened to give the *gem*-dimethyl group, and the remainder consisted of reductive operations and protecting-group removal.

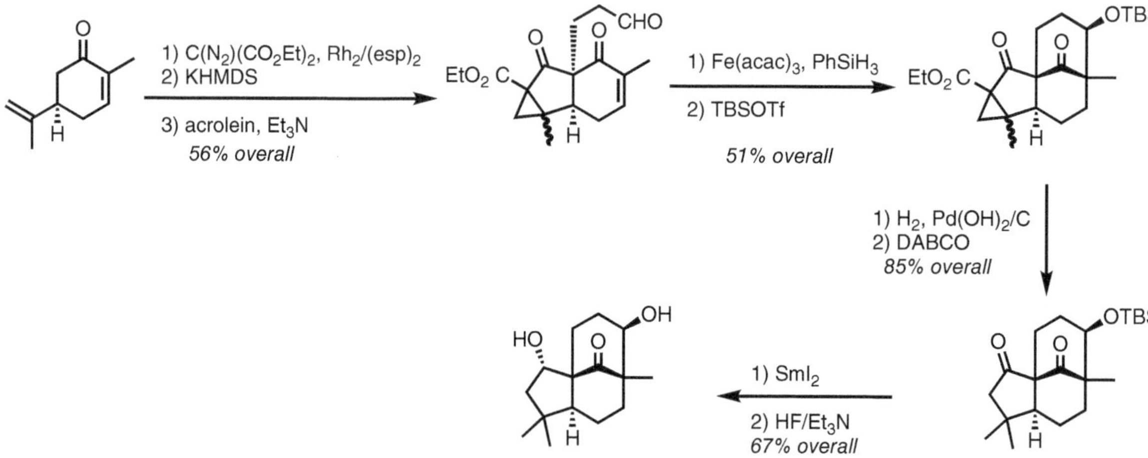

Scheme 9.11   *Chiral-pool synthesis of rumphelloclovane E*

The C10 double bond can also participate in the cyclisation reactions, leading to new ring systems. Cationic rearrangements are common in terpene chemistry [15], and the biosynthesis of widdrol exhibits a particularly intriguing example (Figure 9.5). A quick look at the structure of this sesquiterpene would suggest the incorporation of the isoprene units in accordance with the isoprene rule, head-to-tail, as shown in arrangement A. However, labeling studies showed that this is not the case. Arrangement B represents the incorporation of the three isoprene units. Thus, the central unit must undergo a rearrangement at some stage of the biosynthesis.

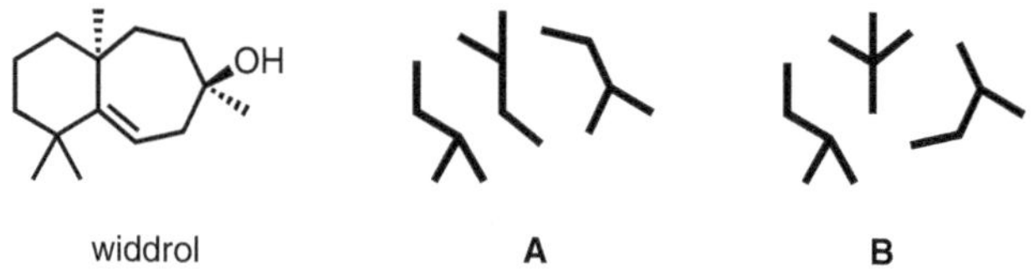

widdrol   A   B

Figure 9.5   *Possible isoprene assemblies of widdrol*

The postulated biosynthesis of widdrol is shown in Scheme 9.12. Cyclisation of the bisabolyl cation gives a new cation, which can form the conjugated diene through loss of a proton (cuparenene). 1,2-Shift of the *cis* alkyl group leads to the spirocyclic chamigrane skeleton, which undergoes another alkyl shift to give a 6,7-fused ring system. The isomerisation of the double bond finally occurs through a ring contraction–ring-enlargement pathway involving two cyclopropanoid intermediates.

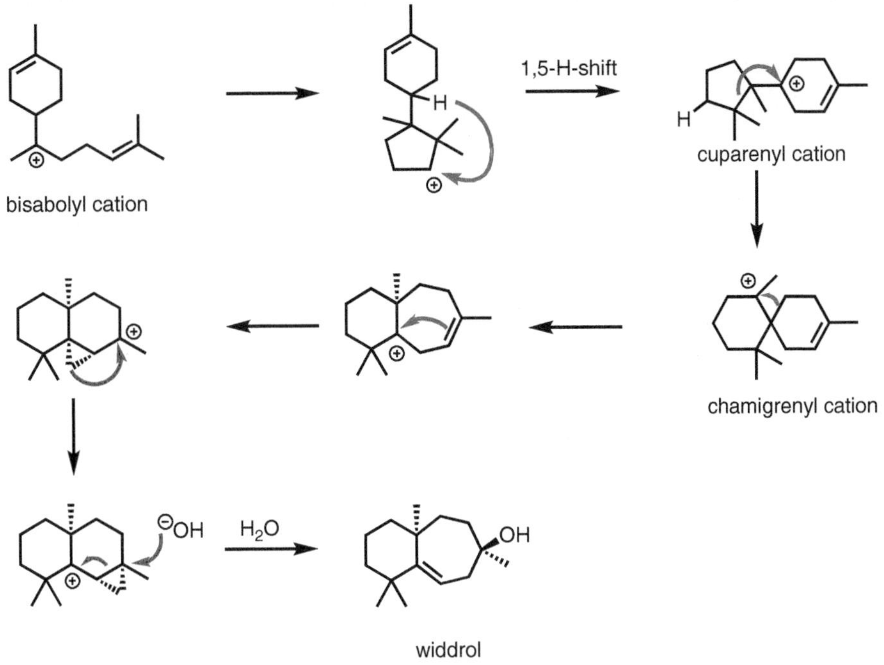

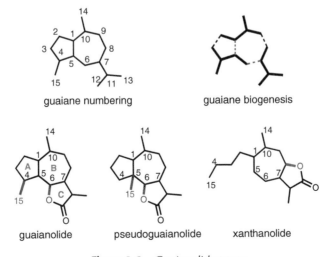

Scheme 9.12   *Biosynthesis of widdrol*

The guaiane skeleton contains a 5,7-fused decahydroazulene ring system. Approximately 500 guaianes are known in nature (Figure 9.6). The basic guaiane structure is formed by head-to-tail coupling of three isoprene units, in other words cyclisation of the farnesene structure (red in Figure 9.6). The guaianolides contain a third ring; an α-methyl-γ-lactone fused to the seven-membered ring. In pseudoguaianolides, C15 has moved to C5. Ring opening of the A ring of guaianolides and simultaneous isomerisation of the lactone from C6 to C8 gives xanthanolides.

Figure 9.6   *Guaianolide types*

The actual biosynthesis starts from germacrdiene cation, which is oxidised to unsaturated carboxylic acid and lactonised to costunolide (Scheme 9.13). Epoxidation to parthenolide followed by ring closure of the A ring leads to guaianolides [16].

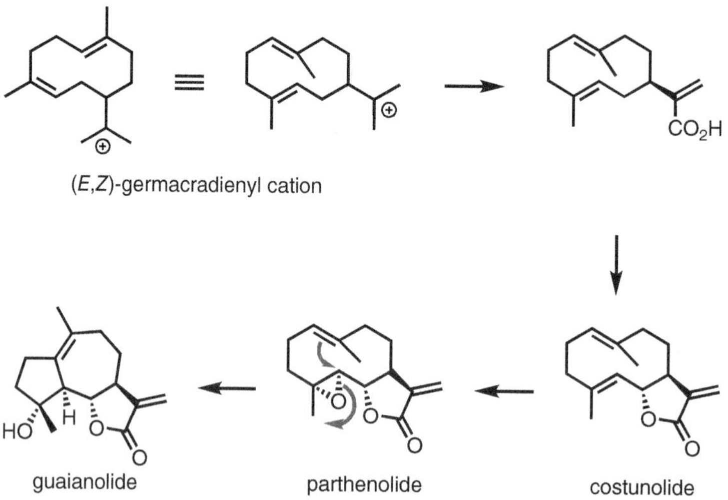

(*E,Z*)-germacradienyl cation

guaianolide   parthenolide   costunolide

Scheme 9.13   *Biosynthesis of guaianolides*

Several guaianes (Figure 9.7), including damsin and confertin, possess interesting pharmacological activities, including promising anti-inflammatory, antibacterial, and antineoplastic effects, as well as insect anti-feedant properties, and they occupied a pronounced position in the development of synthetic methods during the 1970s and early 1980s. Helenalin is a bitter-tasting compound which occurs in several *Helenium* species. It irritates the mucous membrane, causes indigestion, and paralyses the heart muscle. It has also aroused interest because of its insecticidal and anthelmintic activity. Thapsigargin (Tg) is a highly oxygenated guaianolide abundant in the common Mediterranean weed *Thapsia garganica*, commonly known as "deadly carrot" due to its pronounced toxicity to sheep and cattle. Recently, Tg has been shown to be a promising antineoplastic agent.

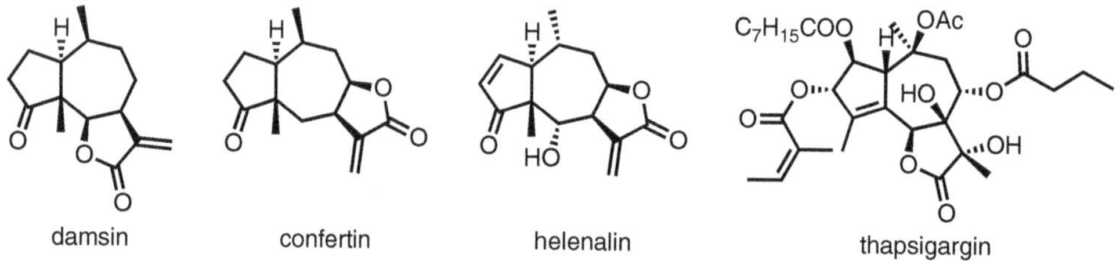

damsin   confertin   helenalin   thapsigargin

Figure 9.7   *Guaianes with biological interest*

The enantioselective syntheses of the guaianes usually rely on the use of chiral starting materials from the terpene pool. In these strategies, a suitably substituted cyclopentane derivative is first obtained, and this intermediate is then elaborated to the hydrazulenoid intermediate using different cyclisation protocols. One of the earliest synthetic protocols was developed based on camphor as the source of the five-membered ring (Scheme 9.14) [17]. D-Camphor was sequentially and selectively brominated at C9 and C10 [18]. Treatment with aqueous base triggered an efficient cleavage of the bicyclic ring to give a cyclopentane derivative with an *exo*-methylene group. Further elaboration through standard operations then gave the cyclopentane intermediate which had the correct substitution and stereochemistry for the synthesis of, e.g. confertin, damsin, and helenalin.

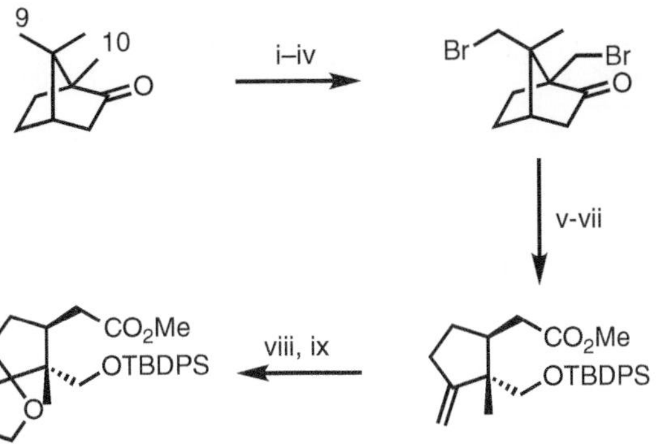

REAGENTS: i) Br_2, HOAc, 80 °C, 80%; ii) Br_2, $ClSO_3H$, 5 h, 75%; iii) Br_2, $ClSO_3H$, 5 d; iv) Zn, HOAc – Et_2O, 60%; v) KOH, DMSO – H_2O, 90 °C, 85%; vi) K_2CO_3, DMF, MeI, 94%; vii) TBDPSCl, im, DMF, 95%; viii) O_3, CH_2Cl_2 – MeOH, –78 °C; Me_2S, –78 °C to rt, 92%; ix) $(CH_2OH)_2$, *p*-TsOH, PhH, rfx, 92%.

Scheme 9.14 *Synthetic degradation of camphor*

(*S*)-Carvone can be converted in just four operations to the tetrasubstituted intermediate shown in Scheme 9.15 [19]. Thus, Weitz–Scheffer epoxidation of carvone gave the epoxide which was opened with lithium chloride under acidic conditions to give a *trans*-chlorohydrin. Protection of the alcohol was followed by Favorskii rearrangement to deliver the functionalised cyclopentane as a single diastereomer. The enantiomer of the intermediate was used by Ley to access thapsigargins [20].

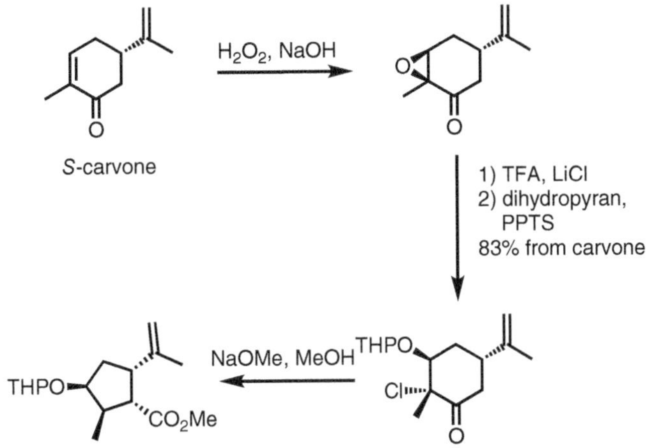

Scheme 9.15 *Synthetic degradation of carvone*

Lee used the intermediate in an intriguing application of the Ueno–Stork radical cyclisation [21] to the synthesis of cladantholide (Scheme 9.16) [22]. The aldehyde derived by DIBAL–H reduction reacted with vinylmagnesium bromide in Felkin sense to give the allylic alcohol as the sole diastereomer. Conversion to the Ueno–Stork precursor followed by treatment with tributylstannane and 2,2-azobisisobutyronitrile (AIBN), in benzene, afforded the hydroazulene in 99% isolated yield. The initial 5-*exo* cyclisation follows the established Beckwith rules of radical ring closure [23], and after the 7-*endo* cyclisation, hydrogen abstraction by the tertiary radical occurred from the sterically less encumbered R face to give the correct stereochemistry at the methyl-bearing C10 center. The exclusive formation of the seven-membered ring over the six-membered one, along with the high diastereoselectivity are noteworthy features of this transformation.

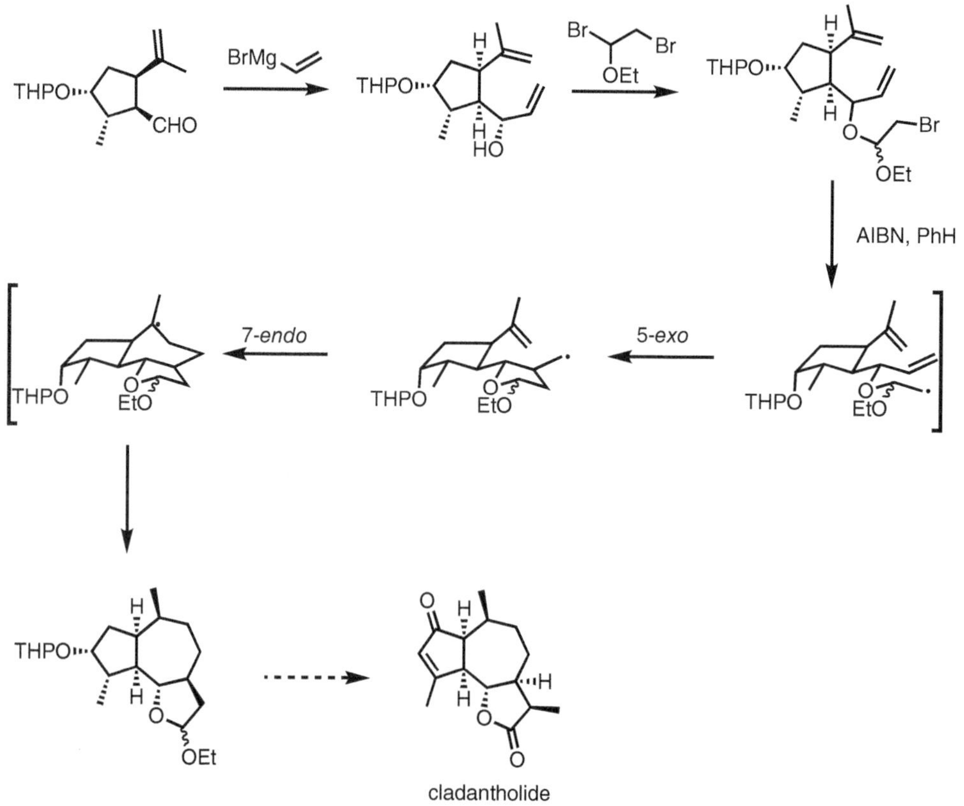

Scheme 9.16 *Ueno–Stork radical cyclisation route to cladantholide*

(*R*)-Carvone served as the starting material for the synthesis of (+)-chinensiolide B (Scheme 9.17). The TBS-protected aldehyde was subjected to Lewis acid-catalysed tandem allylboration/lactonisation, which gave the desired steoisomer in excellent diastereoselectivity. The *trans*-diastereoselectivity in the allylboration step was explained with the usual six-membered chairlike transition state of the reaction. The remarkable diastereofacial selectivity on aldehyde was rationalised according to the Felkin model with the large vinyl-bearing carbon placed opposite to the approach of the allylboron reagent. The primary alcohol was deprotected and converted to terminal alkene via Grieco elimination (selenation/oxidation) in three steps and 60% overall yield. Ring-closing metathesis with the second-generation Grubbs catalyst gave the seven-membered ring in excellent yield. Final stages of the synthesis involved diasteoselective epoxidation of the endocyclic double bond, regioselective opening of the epoxide to give the Markovnikov product, and final conversion of the TBS-protected alcohol to a ketone. The first enantioselective total synthesis of (+)-chinensiolide B was achieved in 15 steps for the longest linear sequence with an overall yield of 6.7% starting from inexpensive and readily available (*R*)-carvone [24].

Spurred by the promising biological activity and the fact that viable fermentation routes are not yet available, the synthetically challenging structure of thapsigargin has intrigued several organic chemists, and a number of syntheses have been developed [25]. In many of the syntheses, carvone or dihydrocarvone have been used as the starting materials, not much unlike in the previous synthesis of, e.g. confertin and cladantholide.

Polyquinane terpenes form a structurally intriguing class of natural products (Figure 9.8) [26]. The linear and angular fusion of cyclopentane rings lead to a wide variation in their structures. Formal connection of the bonds C3–C7, C2–C9, and C1–C11 in farnesane, and subsequent shifts of the methyl groups C14 and C15 lead to the triquinane skeleton of hirsutanes. Hirsutene, isolated from cultures of the fungus *Coriolus consors*, is biogenetically derived from humulene by cyclisation. The linear hirsutene is the biogenetic precursor for hirsutic acid and the antibiotic and antineoplastic

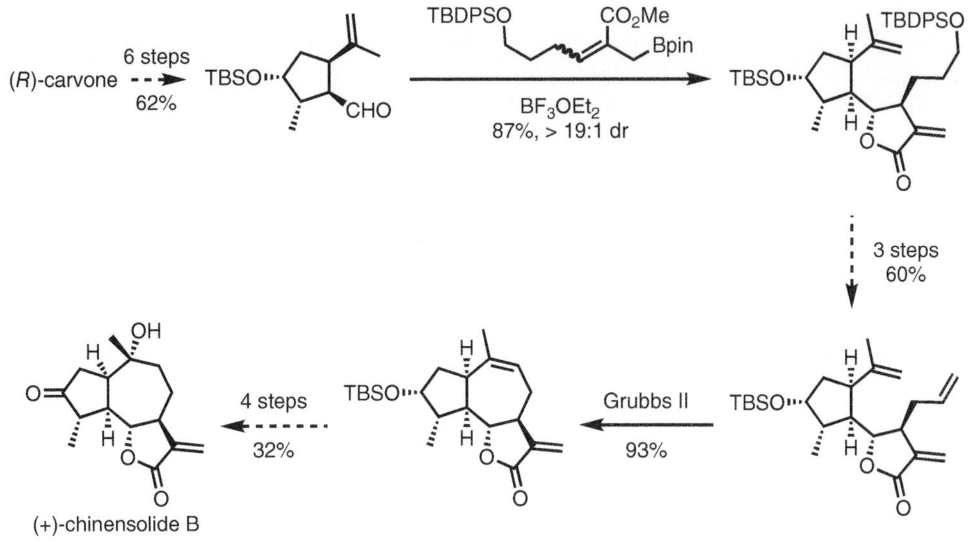

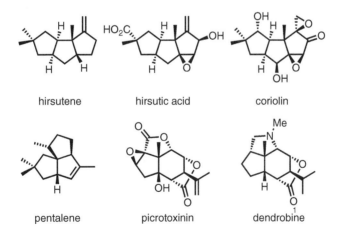

Scheme 9.17 *Asymmetric synthesis of (+)-chinnensolide B from (R)-carvone*

compound coriolin from the same fungus. Pentalene, isolated from *Streptomyces griseochromogenes*, is the parent hydrocarbon of the pentalenolactone family of antibiotic fungal metabolites. Picrotoxinin is one of the most toxic compounds of plant origin ($LD_{50} = 3.0\,\mathrm{mg/kg}$). It is a specific antagonist of the neurotransmitter γ-aminobutyric acid. Note the structural similarity to dendrobine, an alkaloid from an Orchidaceae plant *Dendrobium nobile*, a constituent of the Chinese medicinal herbal remedy Chin-Shih-Hu, used as a tonic and antipyretic.

Figure 9.8 *Polyquinane sesquiterpenes*

Hirsutene has attracted considerable synthetic interest, more than 25 total and formal syntheses have been reported, most of them giving the racemic product. Here we highlight an organocatalytic asymmetric total synthesis of (+)-hirsutene (Scheme 9.18) [27]. The synthesis commenced with two-directional elaboration of commercially available 3,3-dimethylpentane-1,5-diacid. Borane reduction and Swern oxidation of the diacid to the dialdehyde followed by double Wittig reaction gave the dienoate, which underwent a reductive cyclisation upon treatment with magnesium metal in methanol to give the desired cyclopentane in 88% yield as a 1.1:1 mixture of *cis:trans* isomers. The cyclic diester was then converted in three steps to the corresponding bis-α-diazoketone, which cyclised to the fused cyclooctenedione upon treatment with CpRuCl(PPh$_3$)$_2$ in refluxing CH$_2$Cl$_2$. At this stage, the diastereomers were isolated, and the *cis*-isomer was reduced to the pivotal cyclooctanedione. When the diketone was treated with *trans*-4-fluoro proline (10 mol-%) in DMSO at room temperature, the aldol reaction was complete in 15 h and furnished *cis/anti/cis* β-hydroxy ketone 24 in 84% yield and with 98:2 *er*. The relative and absolute stereochemistry was predicted based on the transition state model and confirmed by conversion to the natural enantiomer of hirsutene in three steps (base-catalysed elimination, conjugate reduction of the enone followed by in situ alkylation of the enolate, and final Wittig methylenation.

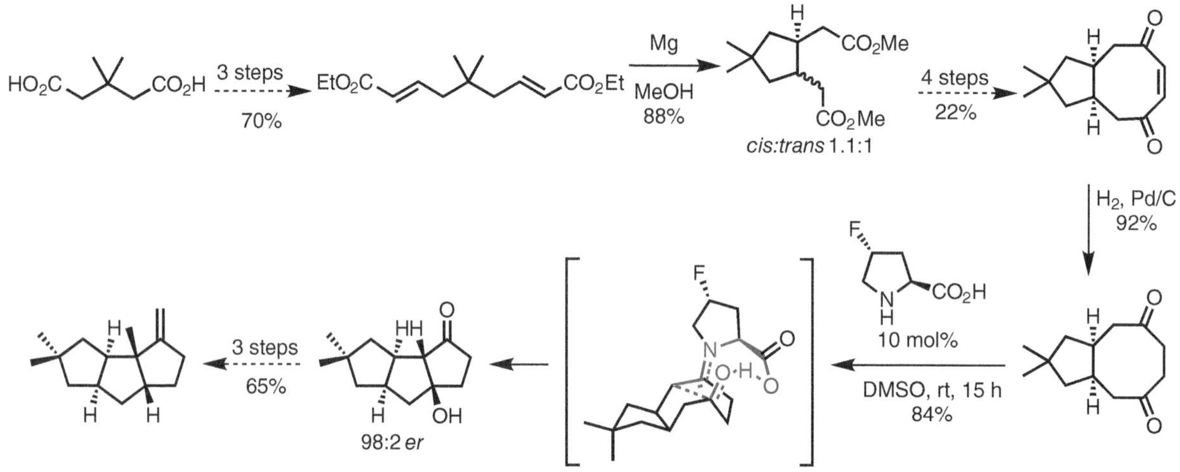

Scheme 9.18   Organocatalytic asymmetric synthesis of hirsutene

9.1.3 Diterpenes

Diterpenes contain a C$_{20}$ skeleton, which is formed from four isoprene units. The common precursor is the linear *trans-trans*-geranylgeranyl pyrophosphate whose cyclisation can be effected in many ways (Scheme 9.19). The cyclisation normally proceeds directly to a bicyclic *trans*-decalin system, which then undergoes a variety of different transformations. Straightforward loss of a proton gives labdadienyl pyrophosphate followed by another cationic cyclisation to give 8-pimarenyl cation, which functions as the intermediate to several structural types of diterpenes. Simple loss of a proton gives pimaradiene, which is transformed to abietic acid. Alternatively, the 8-pimarenyl cation can undergo a different set of oxidative transformations to give rosenonolactone.

Gibberellins are important plant growth hormones which control developmental processes such as seed germination and stem elongation. These compounds were first isolated from the fungus *Gibberella fujikuroi*, which is a parasite of rice causing the straw cells to grow too long and thin, thereby making the straw less stiff. The infection is a very serious threat for the rice crops in rice-producing countries. Gibberellins have subsequently been found in several plants in small quantities, ostensibly to act as plant growth hormones [28].

The biosynthesis of gibberellins involves the cyclisation of labdadienyl pyrophosphate to *ent*-8-pimarenyl cation through loss of pyrophosphate (Scheme 9.20). Further loss of the proton from the angular methyl group is accompanied by contraction of the C-ring through alkyl shift, which terminates in the closure of the bridge of *ent*-kaurene. Oxidation gives the hydroxy carboxylic acid, which upon further oxidation at the 6-position, undergoes a ring-contractive rearrangement of the B-ring to furnish the gibberellin skeleton (gibberellin A$_{12}$ aldehyde). Final oxidative transformations then lead to the various gibberellins.

Taxane diterpenes, isolated from various yew (*Taxus*) species, have gained widespread interest mainly due to the singular anti-tumor activity of paclitaxel (taxol). Taxol is produced by the Pacific yew tree, *Taxus brevifolia*, and the Mediterranean cousin, *Taxus baccata*, produces baccatin III, structurally closely related to taxol. The history of this compound is dreadful. Pliny the elder in ancient Rome described the Mediterranean cousin of the plant producing

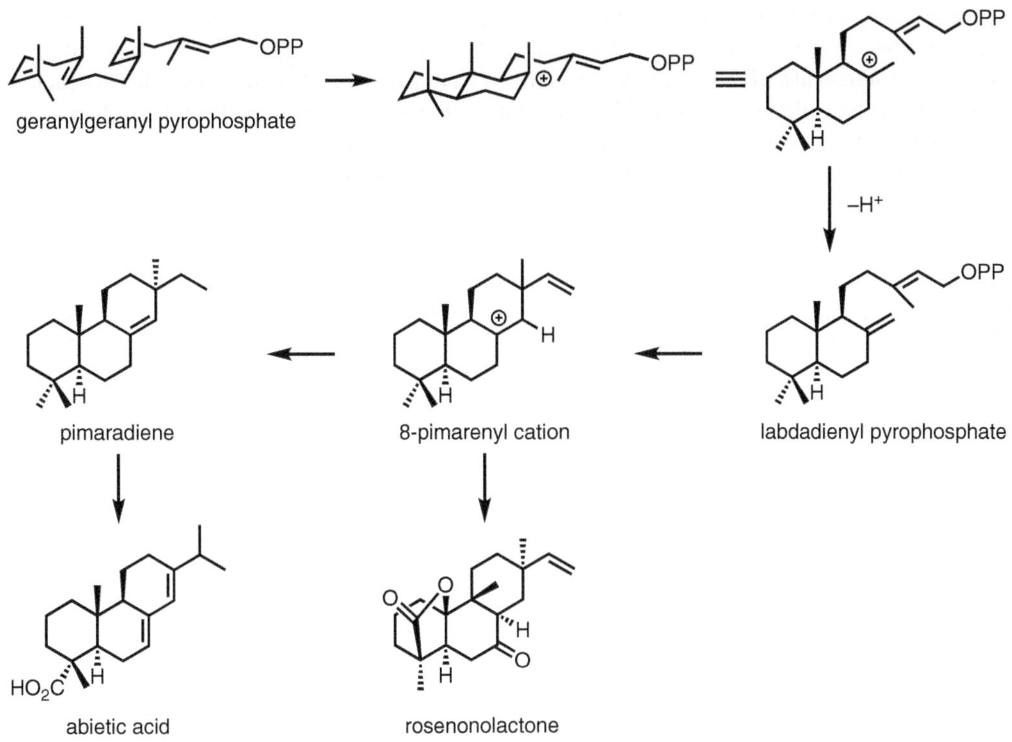
OPP
geranylgeranyl pyrophosphate
OPP
OPP
H
–H+
OPP
H
H
H
pimaradiene
8-pimarenyl cation
labdadienyl pyrophosphate
O
O
H
O
HO2C
H
H
abietic acid
rosenonolactone

Scheme 9.19 *Diterpene biosynthesis*

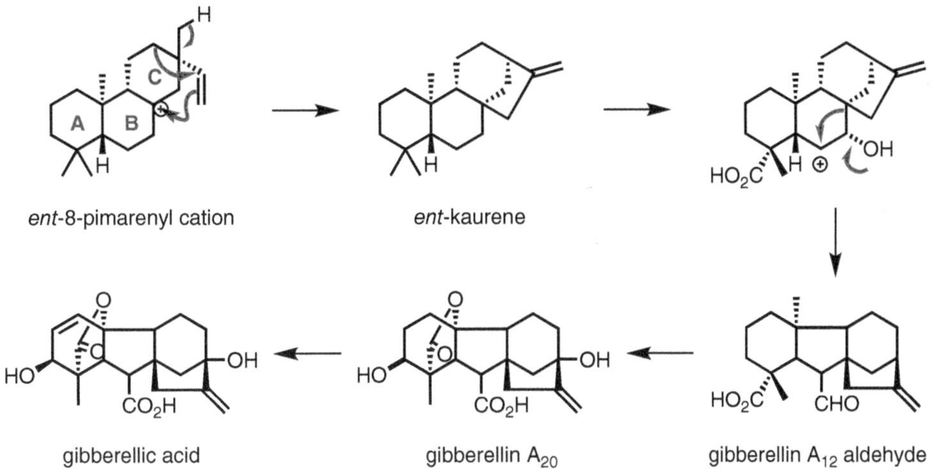
H
C
A
B
H
H
OH
HO2C
H
ent-8-pimarenyl cation
ent-kaurene
O
O
O
O
HO
OH
HO
OH
CO2H
CO2H
HO2C
CHO
gibberellic acid
gibberellin A20
gibberellin A12 aldehyde

Scheme 9.20 *Biosynthesis of gibberellins*

taxol to be so toxic that 'already its shadow kills.' Taxol interacts with tubulin during the mitotic phase of the cell cycle and thus prevents the disassembly of the microtubules thereby interrupting the cell division. This phenomenon is being widely applied in the development of new anti-cancer agents based on the taxol structure, and promising results have already been obtained for the treatment of ovarian, breast and lung cancers. The plant source, however, was insufficient to satisfy even the need for clinical trials, and massive efforts to achieve total synthesis of this challenging compound played a dominant role in the synthetic efforts during the 1990s. Tens of research groups participated in these efforts, and semisynthetic paclitaxel and its chemical relative taxotere are used in the clinic for the treatment of various cancers [29].

Taxol is a diterpene which has been proposed to be derived from geranylgeranyl pyrophosphate (Scheme 9.21) [30]. Cyclisation leads to the bicyclic intermediate of verticillene skeleton, and further cyclisation of the C-ring, through a formal 1,5-hydrogen shift, concludes the assembly of the tricyclic core of the taxanes. Taxol itself is suggested to be formed through a number of oxidative transformations from the verticillene structure.

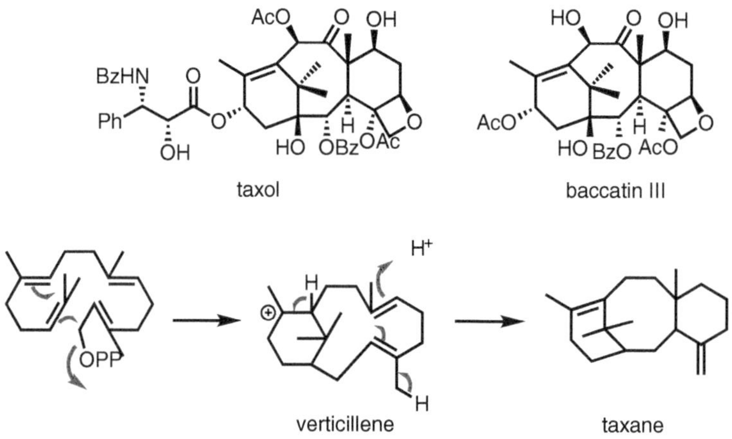

taxol

baccatin III

verticillene

taxane

Scheme 9.21   *Biosynthesis of the taxane skeleton*

9.1.4   Higher Terpenes.

Before turning our attention to two individual classes of higher terpenes, the carotenoids and steroids, let us examine the construction of a tail-to-tail joined C_{30} hydrocarbon, squalene (Figure 9.9). This functions as the intermediate to the steroids, and the formation of the tail-to-tail linkage is instructive of the mechanism for the formation of phytoene, the C_{40} biogenetic precursor of carotenoids.

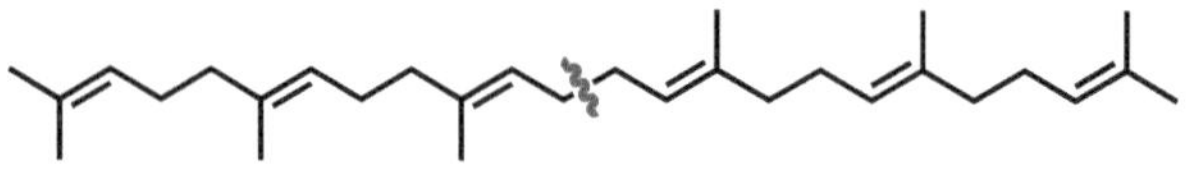

Figure 9.9   *Structure of squalene*

Squalene was originally isolated from shark liver (*Squalus* species), but was later found to be widely distributed being produced by practically all plants and animals. Olive oil is also rich in squalene, and partly a reason why the Mediterranean diet has therefore been suggested to be chemoprotective against cancer. Squalene is formed by joining two farnesyl groups tail-to-tail (Scheme 9.22). The mechanism of this transformation remained a challenging problem until the isolation of a cyclopropane-containing intermediate, presqualene pyrophosphate [31]. The two farnesyl units are joined together, and the carbocation undergoes cyclisation to the cyclopropane through a stereospecific loss of the pro-*S* proton as indicated. The pyrophosphate in the cyclopropylmethanol functions as a powerful leaving group, giving rise to a rearranged carbocation with a cyclobutane skeleton. Being a high-energy species, this undergoes rapid ring opening to the much more stable allylic cation, which is finally trapped by nicotinamide adenine dinucleotide phosphate (NADPH), the biological hydride reductant, to give squalene.

The flexible squalene molecule can be folded in a number of ways. The enzymes that catalyse the cyclisation of squalene exhibit remarkable specificity in folding the chain into proper orientation before cyclisation, and the following scheme show the form incorporating mainly chair forms. This is the usual pathway observed in photosynthetic plants, which produce a number of triterpenes after a wealth of rearrangement steps. For an essentially cationic polycyclisation, squalene needs activation, and this is achieved by epoxidation of the terminal double bond (Scheme 9.23) [32].

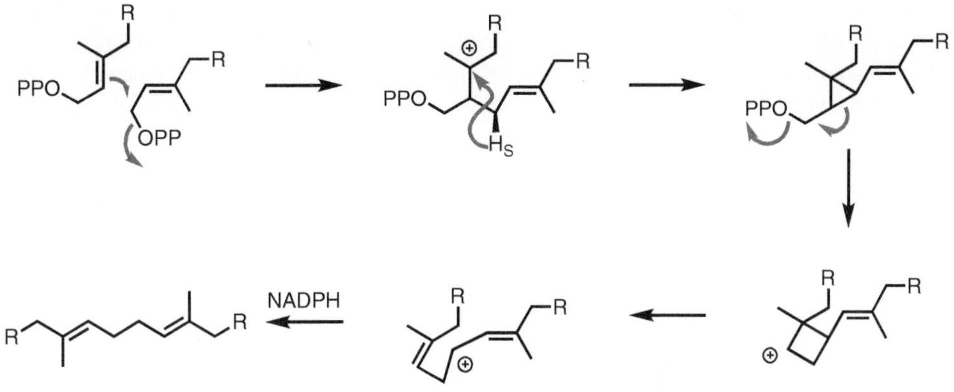

Scheme 9.22 *Two farnesyl groups are joined tail-to-tail to form squalene*

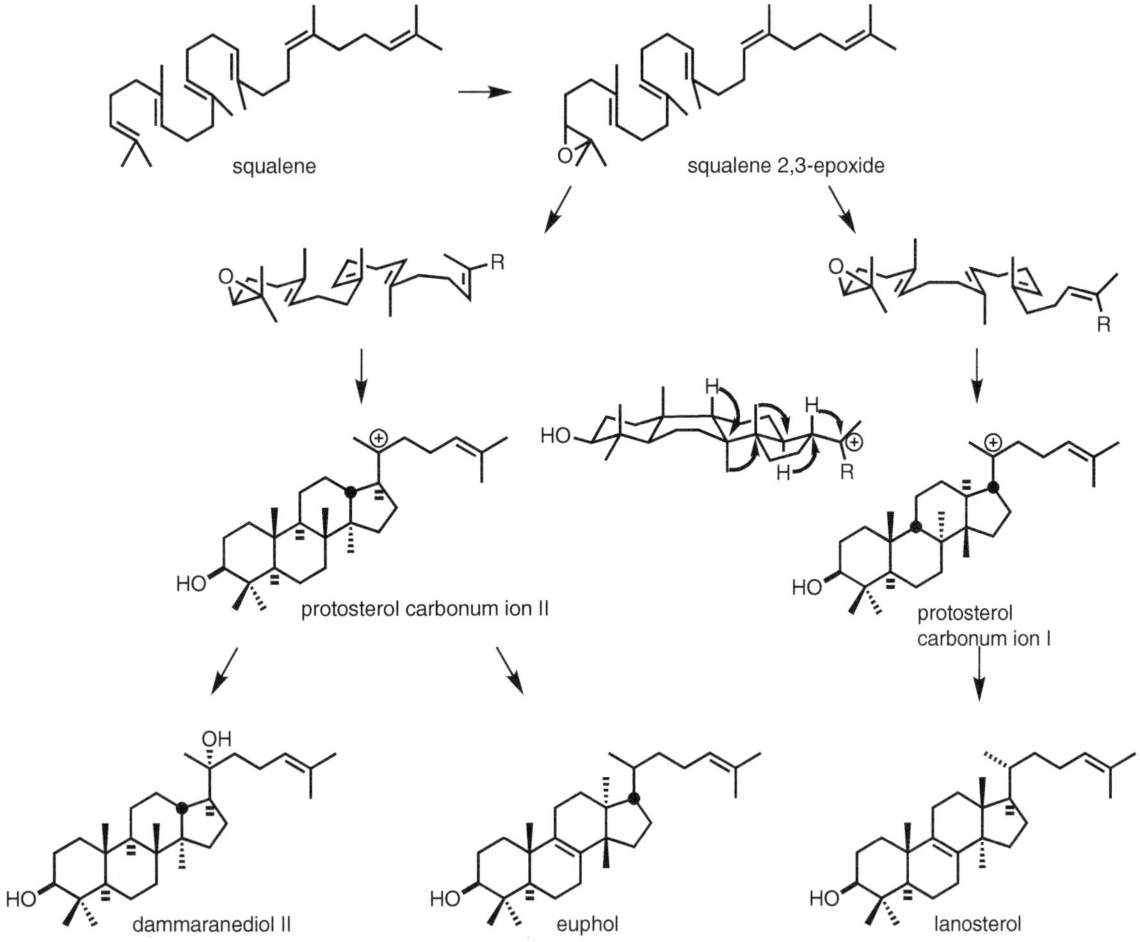

Scheme 9.23 *Squalene cyclisation*

Acid-catalysed opening of the epoxide gives the protosterol carbonium ion I or II, which are converted into a number of plant sterols through cationic rearrangement cascades. Protosterol carbonium ion I is the intermediate towards lanosterol, the precursor of steroids. In animals, squalene epoxide adopts a chair-boat-chair arrangement for the forming ABC rings, and the oxidative cyclisation via the epoxide had already been proposed for the mechanism of transformation of squalene to lanosterol in the 1950s by Albert Eschenmoser and Gilbert Stork [33]. This so called Stork–Eschenmoser hypothesis stated that the ring-junction stereochemistry can be predicted from the starting olefin geometry; *E*-alkenes give *trans*-ring junctions and *Z*-alkenes give *cis*-ring junctions. The epoxidase has been isolated and is known to utilise molecular oxygen and NADPH [34]. The direct enzymatic formation of protosterol with the side chain at C17 β-oriented has been proved [35].

Based on the biosynthetic considerations, *William Summer Johnson* (Stanford University, 1913–1995) proposed and demonstrated the biomimetic cyclisations to be a powerful tool for the synthesis of tetracyclic terpenes, terpenoids, and steroids [36]. Corey's synthesis of dammarenediol II beautifully demonstrates the utility of these biomimetic cyclisations for the enantioselective synthesis of this natural product [37]. The chiral information is derived from a cinchona alkaloid-mediated dihydroxylation of farnesyl acetate to a diol, which then is converted to the epoxy allyl bromide shown in Scheme 9.24. Chain elongation in a few steps gives the crucial epoxy triene ready for Lewis acid-catalysed cyclisation to a tricyclic intermediate. The natural product, with five quaternary carbon centers generated during the synthesis, is constructed in only 12 steps from the epoxy bromide, with an overall yield of 6%.

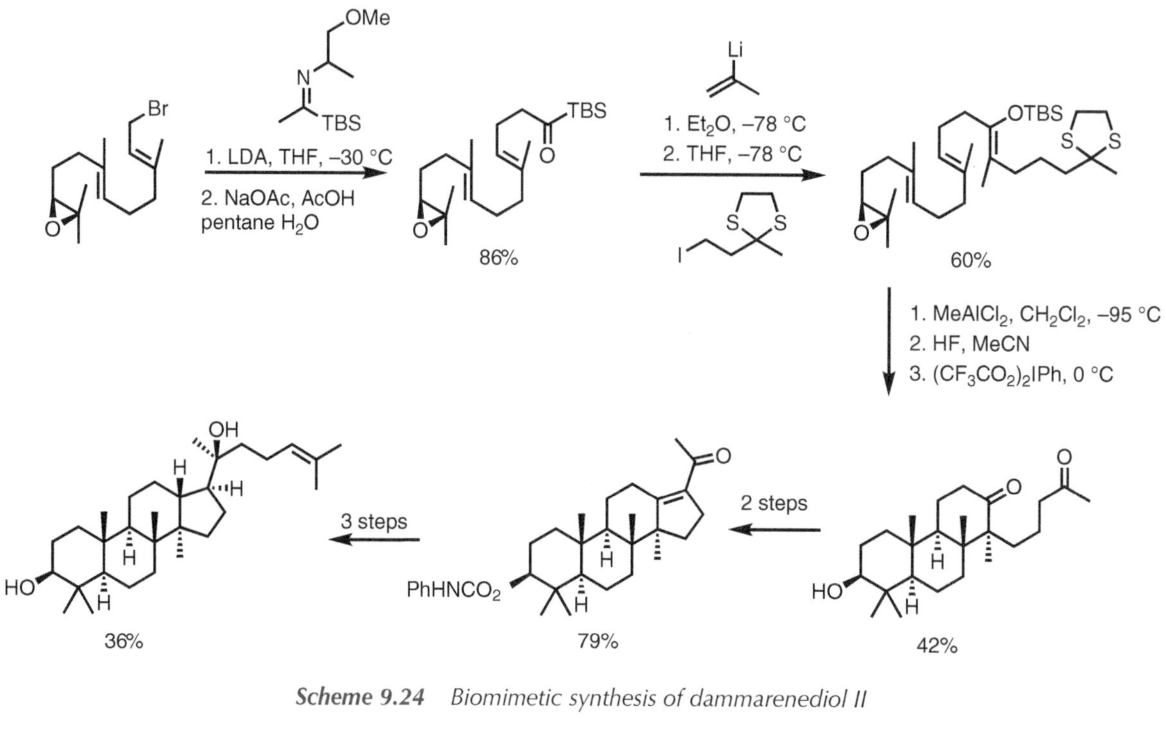

Scheme 9.24   *Biomimetic synthesis of dammarenediol II*

Protosterol carbonium ion I also serves as the precursor to cucurbitacins, which are the toxic principle of cucurbitaceous plants (e.g. pumpkins, gourds, squash, and melons, Scheme 9.25). Cucurbitacin E (elaterin) was used in medieval medicine as a strong purgative.

Protosterol carbonium ion II can undergo ring enlargement and further cyclisation to pentacyclic triterpenes lupeol and β-amyrin, as shown in Scheme 9.26.

Fernene, a fern triterpene lacking the C3 hydroxyl group and the protozoan metabolite tetrahymanol (hydroxylated at C21, Figure 9.10) are produced from squalene by direct protonation rather than through squalene epoxide. Squalene

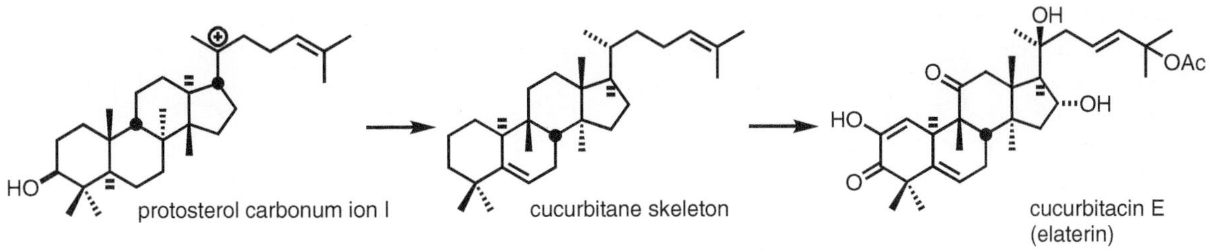

Scheme 9.25 *Biosynthesis of cucurbitacin*

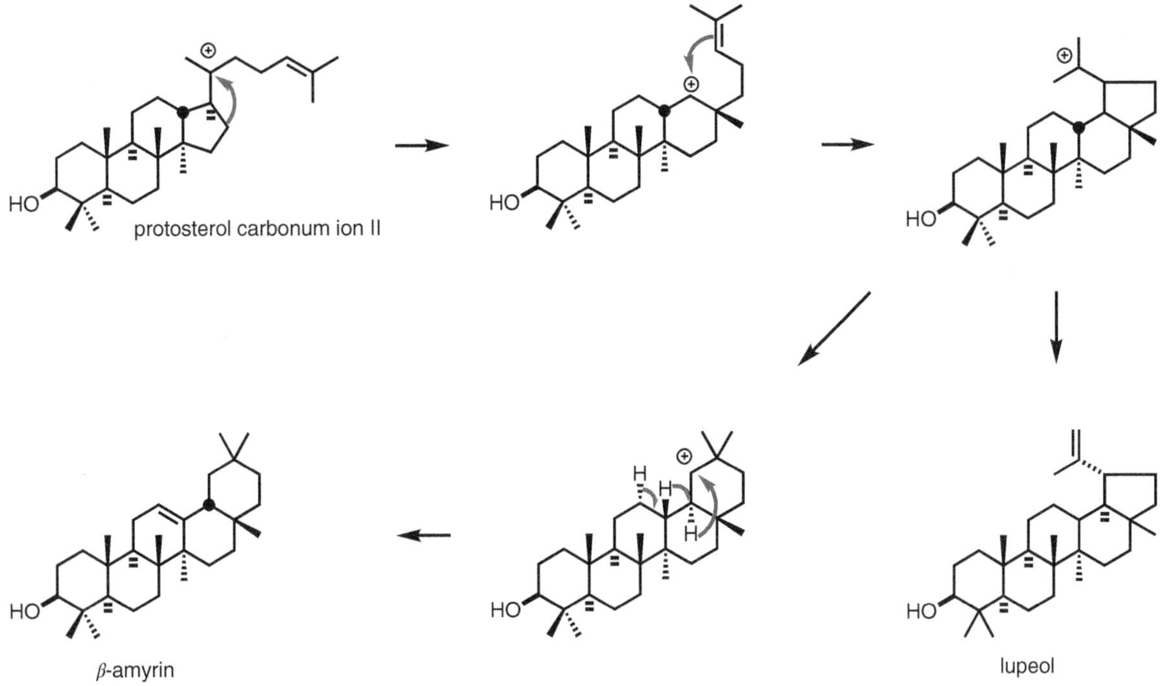

Scheme 9.26 *Biosynthesis of pentacyclic triterpenes*

adopts an all-chair conformation which, after protonation, undergoes a number of 1,2-Wagner–Meerwein shifts to give the products.

Glycyrrhetinic acid (also known as enoxolone) was first obtained by Tschirsch and Cederberg [38] as the hydrolysis product of glycyrrhizic acid (glycyrrhizin), the main glycoside of liquorice roots (*Glycyrrhiza glabra*). The structure of glycyrrhetinic acid was eventually ascertained [39], stereochemistry elucidated [40], and structure confirmed by synthesis of *β*-amyrin from it [41].

Glycyrrhizin contains two molecules of glucuronic acid and is the main sweet-tasting compound in licorice (some 50 times sweeter than sucrose). Since glycyrrhizin inhibits cortisol metabolism in the kidneys, excess consumption can lead to hypertension and edema. EU recommendations suggest a maximum daily intake of 100 mg of glycyrrhizin per day (equivalent to *ca.* 50 g licorice sweets). The aglycone glycyrrhetinic acid inhibits PGE-2- and PGF-2*α*-metabolising enzymes, thus leading to increased levels of PGs. PGs inhibit gastric acid secretion, thus glycyrrhetinic acid has potential as an agent to treat peptic ulcers.

Osladine is a saponin isolated from the rhizomes of *Polypodium vulgare* (common polypody). It has a distinctively bittersweet taste and is some 500 times sweeter than sucrose. Due to its flavoring chracteristics, it is used in confectionary

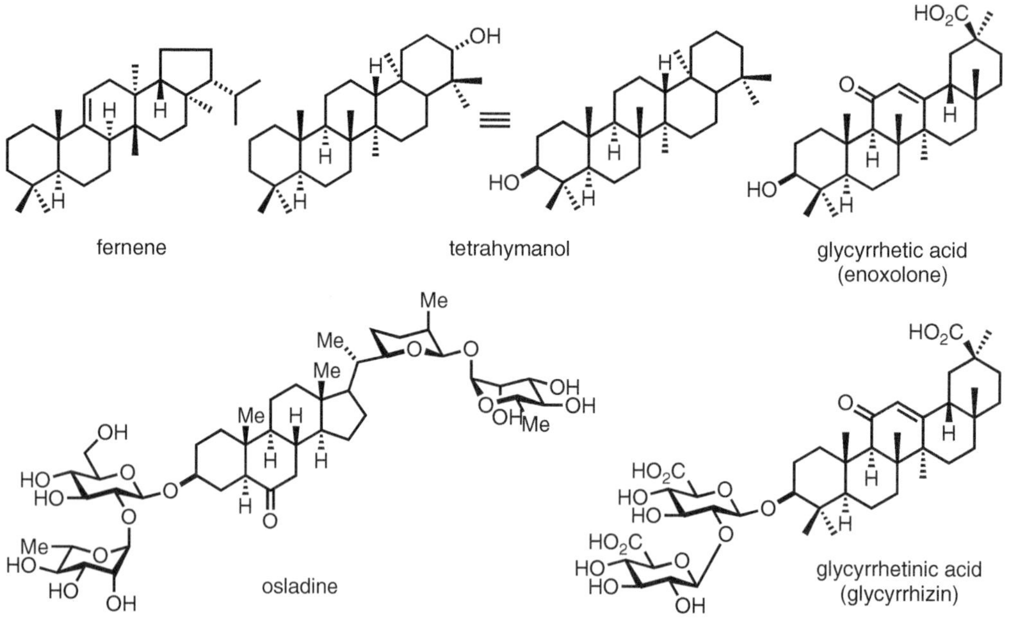

Figure 9.10 *Triterpenes*

such as making nougat. *Polypodium vulgare* is related to *Polypodium virginianum* (rock polypody). Licorice fern is also known as *Polypodium glycyrrhiza*.

The triterpene nucleus can undergo several modifications after its synthesis. Oxidation of the existing hydroxyl groups, oxygenations to introduce further oxygen functionalities, cationic rearrangements involving Wagner–Meerwein shifts, and *S*-adenosyl methionine-mediated alkylations provide the possible routes to the large number of structurally varied triterpenes called limonoids (Figure 9.11). Limonoids derive their name from limonin, and were first obtained from the bitter principles of citrus fruits in 1841, but their structure was only established only in 1960 [42]. Approximately 1500 limonoids are known presently. Limonoids mainly occur in *Meliaceae* and *Rutaceae*, less commonly in the *Cneoraceae* family. Quassinoids [43], the bitter principles of *Simaroubaceae*, have undergone extensive degradation and recyclisation reactions, as have the bitter principles of citrus species, as well as cneorins, the bitter principles of the Mediterranean *Cneoria* species [44]. Azadirachtin is a highly oxygenated tetranortriterpene isolated from the neem tree *Azadirachta indica*. It is extremely potent as an insect antifeedant and growth regulator and has prompted extensive synthetic effort toward simpler analogues still retaining these desired properties [45]. Synthesis of the limonoids has been reviewed [46].

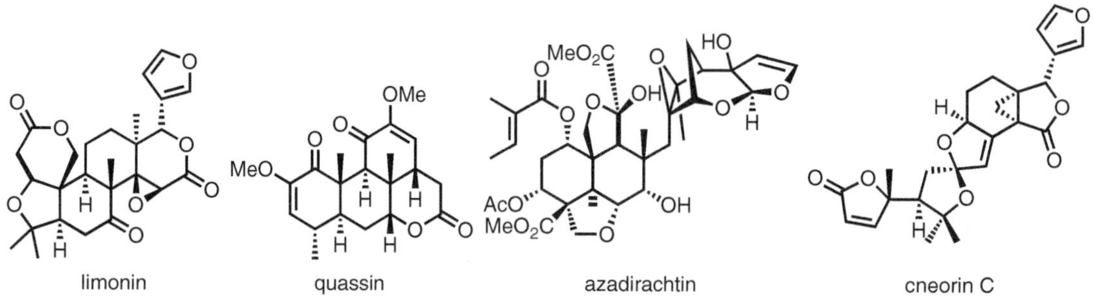

Figure 9.11 *More elaborate triterpenes, limonoids*

The biosynthesis of limonoids is complex, going through the dammarenyl cation (Scheme 9.27). Muliple oxidative steps lead to melianol, a protolimonoid. The basic limonoid skeleton of 26 carbon atoms is reached at this stage with the loss of four carbons. The exact biosynthesis is still largely unveiled, but recently, some enzymes and genes related to the biogenetic path have been identified [47]. As some limonoids are toxic to insects yet harmless to mammals, they have potential in crop protection and other applications. However, their current availability depends on extraction from limonoid-producing plants. Therefore, the elucidation of the enzymes and genes involved can pave the way for metabolic engineering to generate crop plants with enhanced insect resistance.

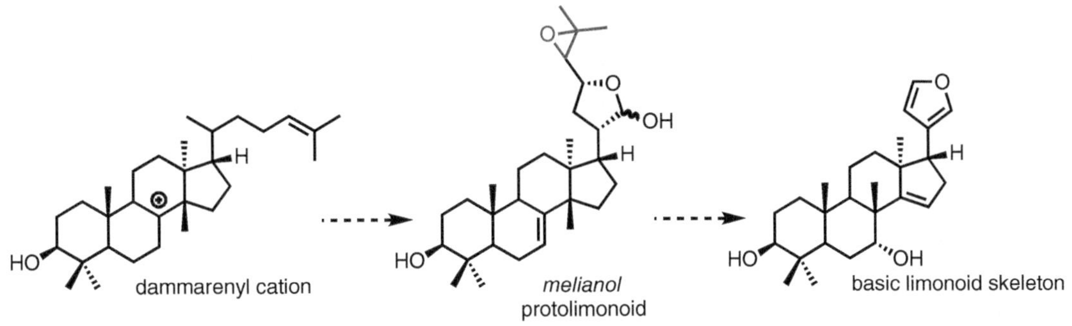

Scheme 9.27   *Biosynthesis of limonoids*

9.2   Carotenoids

The intense colors of carrots, egg yolk, tomatoes, yellow autumn leaves, and algae are caused by carotenoids; tetraterpenes with long stretch of conjugated polyenes, which cause the colored properties of these compounds. In photosynthetic organisms, the carotenoids function as supplemental light-absorbing molecules passing the excitation energy to chlorophyll.

The most important tetraterpene derivative is retinol (vitamin A_1, Figure 9.12) which is necessary for night vision. Retinol is formed in the liver from β-carotene through oxidative degradation. This is further oxidised to the corresponding aldehyde (retinal), which reacts with the opsin protein present in the rod cells of the retina to form a covalent bond. This new protein complex, rhodopsin, contains the *cis*-double bond form of retinal, which absorbs light at 500 nm. Light absorption causes the isomerisation of the double bond to *trans* configuration, and the *trans*-retinal is released from the opsin complex. Simultaneously, a nerve signal is transmitted to the center of vision in the brain. Subsequently, *trans*-retinal is enzymatically isomerised back to the *cis* form, and the cycle can begin again. Rhodopsin absorbs light with a very high sensitivity, and the absorption of a few light quanta leads to the isomerisation. The sensitivity of the rod cells towards blue–green light is also explained by the aforementioned events, light at 500 nm is blue–green light. Because we humans cannot produce β-carotene itself, we must obtain it from food. The best sources are carrots, spinach, and lettuce. The lycopene of tomatoes is an intermediate in the biosynthesis of β-carotene.

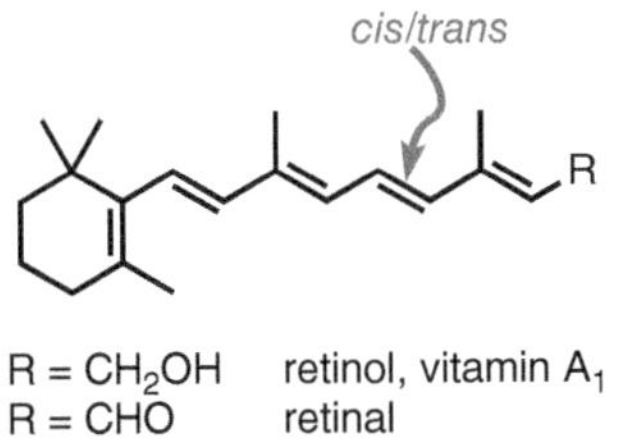

R = CH$_2$OH   retinol, vitamin A$_1$
R = CHO   retinal

Figure 9.12   *Vitamin A and retinal*

Some bacteria can produce even longer carotenoids and terpenes, and continuation of the prenylation process in plants produces a polymeric latex. Rubber is obtained from the rubber tree, *Hevea brasiliensis*, and is a high-molecular-weight polymer of isoprene (*ca.* 2000 units). Nearly all the double bonds in rubber are *cis*, whereas the harder Gutta-percha (from *Palaquium* species) contains *trans*-double bonds.

In the biosynthesis of β-carotene (Scheme 9.28), two geranylgeranyl pyrophosphates are joined tail-to-tail to give phytoene in a fashion not unlike the one we saw for the synthesis of squalene (i.e. involving a cyclopropane intermediate). Oxidation to the highly conjugated (11 conjugated double bonds) lycopene is followed by cationic cyclisation at each end of the chain to form the carotene molecule.

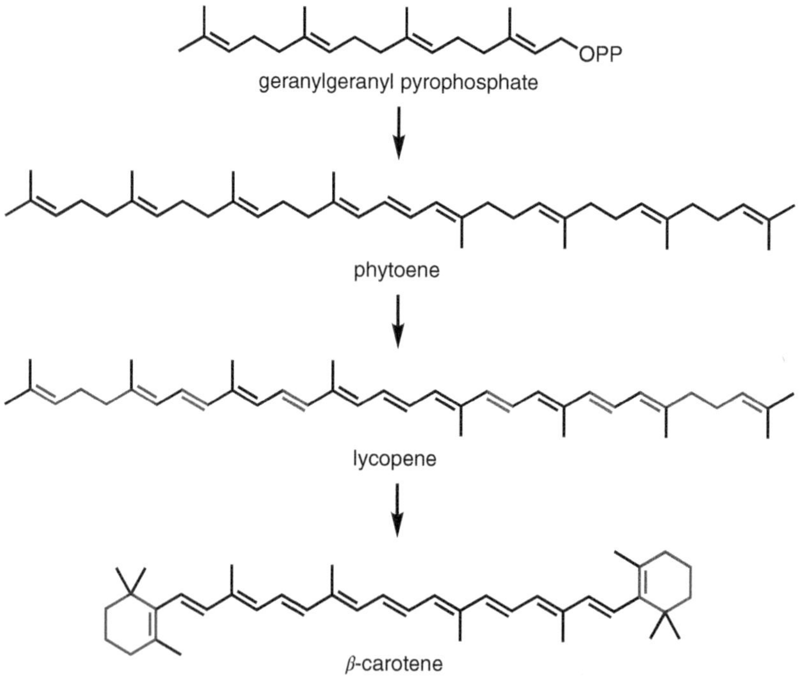

geranylgeranyl pyrophosphate

phytoene

lycopene

β-carotene

Scheme 9.28 *Biosynthesis of β-carotene*

9.3 Steroids

When plant or animal cells are extracted with diethyl ether, chloroform, or benzene, one can isolate a mixture of several compounds containing lipids. This extract can be divided into two fractions based on the behavior of the compounds towards basic hydrolysis. The hydrolysable lipids produce water-soluble compounds, and the remaining fraction contains, among other compounds, steroids. Steroids are a class of compounds that resemble the solid alcohols, sterols (from Gr. *stereos*, solid, and -ol for alcohol).

Steroids occur widely as building blocks for all cell membranes. Cholesterol was first isolated from gall stones nearly two centuries ago by Chevreul. For instance, the dry material of the human brain is 17% free cholesterol. The structure of cholesterol is very rigid, and its incorporation into the cell membrane has a rigidifying effect. Cholesterol and lipid bilayers were discussed in Section 8.2 (Figure 8.6).

Whereas cholesterol is the structural rigidifying component in eukaryotic cell membranes, in prokaryotes a higher terpene, bacteriohopanetetrol (Figure 9.13), with a C_{35} carbon skeleton, occupies a similar role [48]. The dimensions of bacteriohopanetetrol (7.7 Å by 18.45 Å to the tetraol portion) are very similar to those of cholesterol and can thus easily accommodate a similar positioning in the lipid bilayer. Prokaryotes also contain a high proportion of longer (C_{40}) terpenes which crosslink the two lipophilic surfaces of the bilayer.

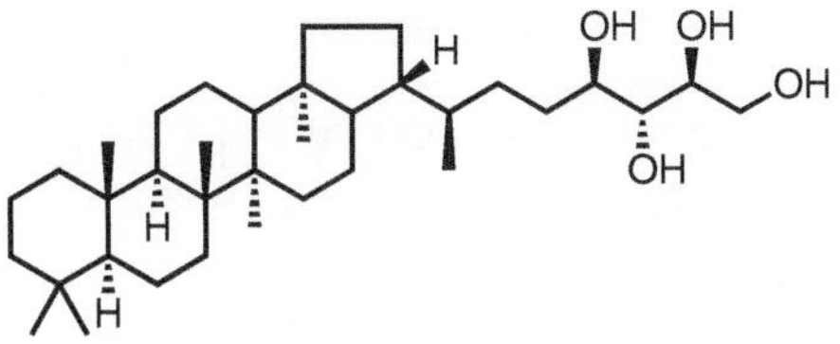

Figure 9.13   *Bacteriohopanetetrol*

The elucidation of the structures of the steroids took several decades, and in 1928 *Heinrich Otto Wieland* (1877–1957) of Munich and *Adolf Otto Reinhold Windaus* (1876–1957) of Göttingen obtained the Nobel Prize in chemistry for their studies on the structures of these important compounds. However, the structure they proposed for cholesterol (Figure 9.14) [49] was based on oxidative degradation studies and the application of Balc's rule, which states that pyrolysis of substituted glutaric acids will yield an anhydride, whereas adipic acids will be transformed to a cyclopentanone derivative. These rules, and thus also the structure of cholesterol, turned out to be wrong. The structure was corrected through the aid of the first applications of X-ray diffraction studies in natural product structure elucidation a few years later by *John Desmond Bernal* (1901–1971) [50]. However, this structure was also wrong, and the final solution of the structure had to await until 1940 [51].

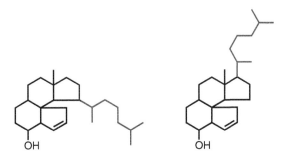

Wieland–Windaus structure 1928   Bernal structure 1932

Figure 9.14   *Original structures proposed for cholesterol*

The isolation and structural studies of the first steroid hormones give a good indication of the difficulties encountered by natural product chemists in the early part of the twentieth century. The first male sex hormone to be isolated and characterised was the excreted hormone androsterone, which is less active than the primary hormone testosterone. For the structure elucidation of the first, *Adolf Friedrich Johann Butenandt* (1903–1995, Nobel Prize in Chemistry in 1939 with Leopold Ruzicka "for his work on sex hormones") extracted 17000 liters of male urine to obtain only 50 mg of pure androsterone [52].

Steroids are classified according to their structures (Figure 9.15). Cholesterol is an example of simple sterols, and bile acids are related structures. The hormone preparations are the widest group of these compounds and have found wide use as drugs. The steroid hormones include the female sex hormones estrogens and gestagens, the male sex hormones androgens, and (adreno)corticoids secreted by the cortex of the adrenal gland. The cardiac glycosides (cardenolides) are also steroid derivatives. These include the highly toxic strophantidine from lily of the valley (*Convallaria majalis*). Sapogenins (e.g. digitoxigenin) and some steroid alkaloids (e.g. cyclopamine, or 11-deoxojervine) are more elaborate structures belonging to these natural products.

A common structural feature for steroids is the cyclopentaphenanthrene (androstane) skeleton where the rings are joined together highly regularly (Figure 9.16). Only the fusion between the A and B rings varies, and even here regularities can be observed. In the bile acids and cardenolides, the fusion is *cis*, whereas the standard steroids have either a *trans*-fusion or an aromatic A-ring. A third structural variation with *cis*-A/B, *trans*-B/C and *cis*-C/D ring junctions occurs in cardiac glycosides.

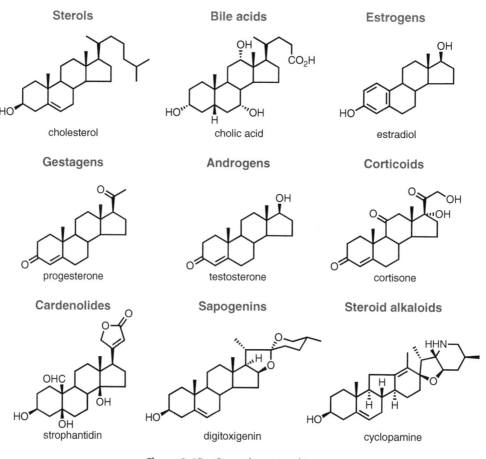

Figure 9.15   *Steroid structural types*

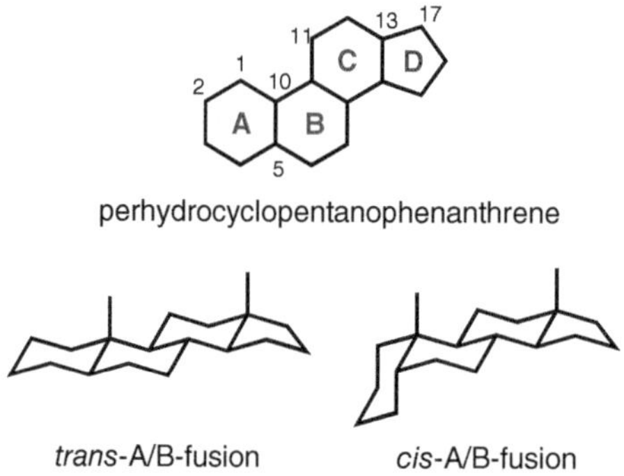

Figure 9.16   *Ring fusions and numbering in steroids*

The methyl groups at C10 and C13 in natural steroids are always β (above the plane of the paper), except of course in those cases where the A-ring is aromatic and the angular methyl group is thus missing. Natural steroids also often have a side chain at C17 which is β-oriented, as shown in Figure 9.15 for cholesterol. The numbering scheme for the atoms in steroids is also shown.

The steroid hormones are biosynthesised from cholesterol. After liberation from its storage site in the cell walls, cholesterol is transformed into pregnenolone, which functions as the starting material for the remaining steroid hormones. The ovary gland produces estrogens which regulate the release of the egg (ovulation) and the development of the external sexual properties of females. Through the response to adenohypophysis (anterior pituitary gland), estrogen inhibits the secretion of follicle-stimulating hormone (FSH), which results in the lack of ovulation when the estrogen effect is strong. Estrone is the oldest known female sex hormone, and estradiol has the strongest effects (Figure 9.17). Several molecular modifications of these compounds are in medicinal use, such as ethinylestradiol (contraceptive) and various stilbene derivatives, which also bind to the estrogen receptor.

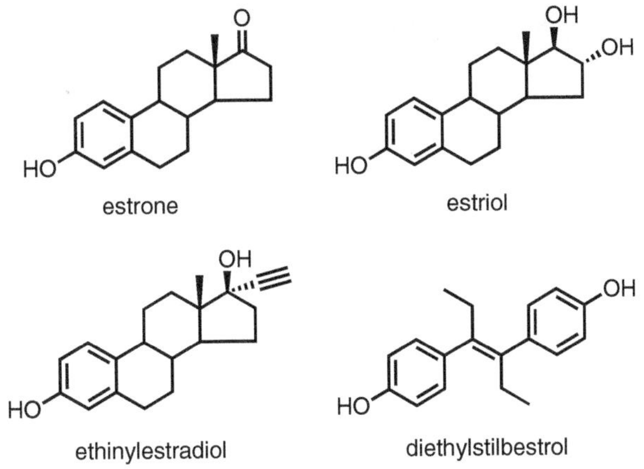

Figure 9.17   *Estrogens*

When the release of estrogens has lowered the levels of FSH, the ovary gland (follicle) is turned into an endocrinous gland (*corpus luteum*) which, during the first third of pregnancy, secretes progesterone. After this period, the placenta takes care of the excretion of both progesterone and estrogen. In later phases of pregnancy, progesterone prohibits ovulation, and this action has been used in the development of oral contraceptives since the early 1950s. Progesterone has an inhibitory effect on the secretion of luteinising hormone, and it also prepares the membranes of the womb to support fertilised pregnancy. Progesterone itself is rapidly metabolised in the liver, and it has little practical medicinal value. Progestins are synthetic hormones which function as progesterone does, but are slowly metabolised in the liver and can be used in oral contraceptive formulations (Figure 9.18).

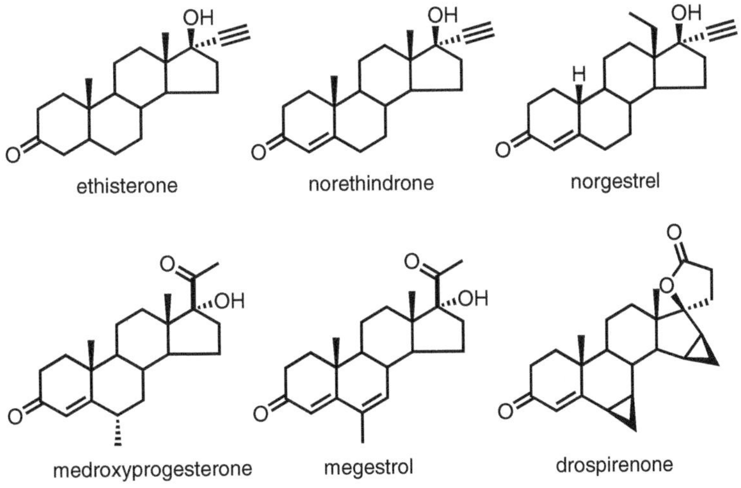

Figure 9.18   *Progestins*

The male sex hormones, androgens (for structure of testosterone, see Figure 9.15), affect the development of the external sexual properties of males, the production of sperm (spermatogenesis), and the growth of muscle tissues (anabolic effect). Androgens are used for the treatment of male sterility and impotency, as well as female breast and genital cancers.

In the anabolic effect, the excretion of nitrogenous compounds is decreased, and the biosynthesis of structural proteins is increased. This is externally manifested as the growth of muscle mass. With molecular modifications, the androgenic component can be eliminated, and the anabolic steroids are used to speed up the tissue growth after operations, to treat growth abnormalities in muscles and bones, and also in the notorious doping cases in sports. Androsterone is an example of androgens, and metenolone is an anabolic steroid (Figure 9.19).

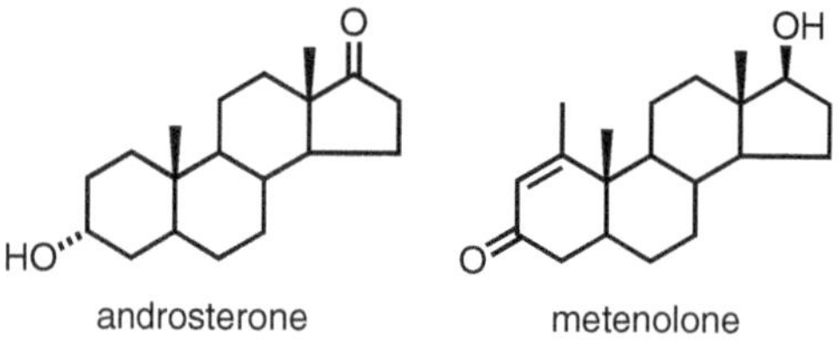

androsterone   metenolone

Figure 9.19 *Androgens*

The external regions of the kidneys (adrenal cortex) secrete corticosteroids, which are of two main classes (Figure 9.20). The glucocorticoids act on the carbohydrate, lipid, and protein metabolism, and the mineralocorticoids regulate the secretion of Na^+ ions in the kidneys. The main medicinal effect of the glucocorticoids is on the irritation and prevention of rheumatic effects. The use of steroidal anti-inflammatory drugs is declining, because new more effective drugs with less severe side effects are being developed. Of the hormones, only cortisone and hydrocortisone are used medicinally. Synthetic molecular modifications include prednisone (five times as effective as hydrocortisone) and dexamethasone (30 times). The mineralocorticoids regulate the Na^+/K^+ equilibrium. The only hormone with this ability to promote water retention is aldosterone, the last of the steroid hormones to have its structure determined, in 1954. Its antidiuretic action has found only minor use in medical practice in the treatment of hypertension. Desoxycortone and fludrocortisone are synthetic mineralocorticoids used in the treatment of Addison's disease, a rare disorder where the adrenal glands do not produce enough glucocorticoids.

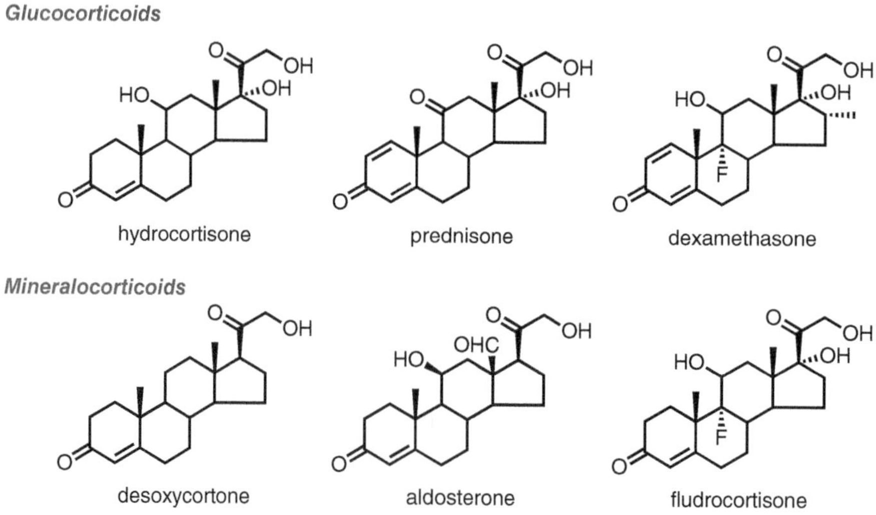

Glucocorticoids

hydrocortisone   prednisone   dexamethasone

Mineralocorticoids

desoxycortone   aldosterone   fludrocortisone

Figure 9.20 *Corticoids*

Steroids occur also in plants and invertebrates. In the early 1930s the British entomologist *Sir Vincent Brian Wigglesworth* (1899–1994) made significant contributions to our understanding of the metamorphosis of insects [53]. He showed that metamorphosis and molting (ecdysis) are under hormonal control. The neurosecretory cells in the brain release a brain hormone (prothoracitropic hormone, PTTH), which stimulates the release of molting hormone (ecdysteroids) in the prothorax. Whether the insect molts into another larval stage, into a pupa, or into an adult is determined by the presence of a third hormone, juvenile hormone. The structure of the brain hormone has remained elusive, although in the early 1960s, M. Kobayashi isolated from the brains of 220 000 *Bombyx mori* silkworms 4 mg of a crystalline compound which demonstrated brain hormone activity. However, on further examination, this substance turned out to be cholesterol.

Juvenile hormones are commonly acyclic unsaturated epoxyesters related to methyl farnesate, structurally far removed from steroids (Figure 9.21). They ensure the growth of the larvae while preventing metamorphosis. They were discovered in 1965, and their structures were elucidated in 1967.

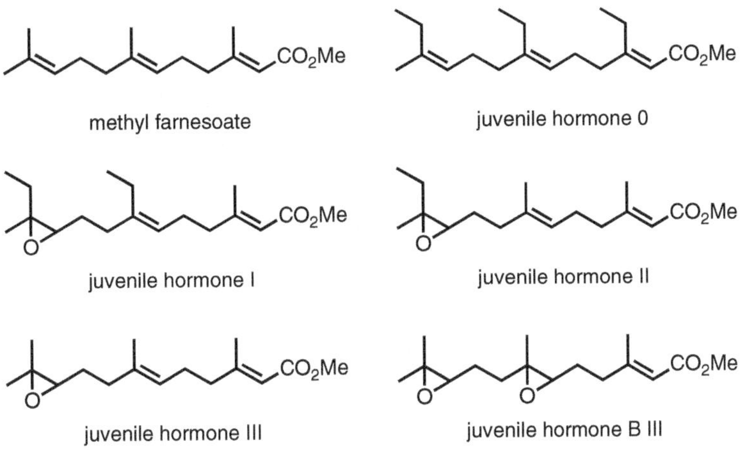

Figure 9.21   *Juvenile hormones*

In 1954, Butenandt extracted 500 kg of *B. mori*, and obtained 25 mg of molting hormone [54]. The structure of ecdysone was elucidated through X-ray crystallography in 1965 (Figure 9.22) [55]. Ecdysone is actually a prohormone, which is oxidised in the prothoracic gland, ring gland, gut, and fat bodies of larvae to 20-hydroxyecdysone (20E, ecdysterone or crustecdysone). 20E is also found in seawater crayfish Cape rock lobster *Jasus lalandii*, in which it is one of the most abundant molting hormones.

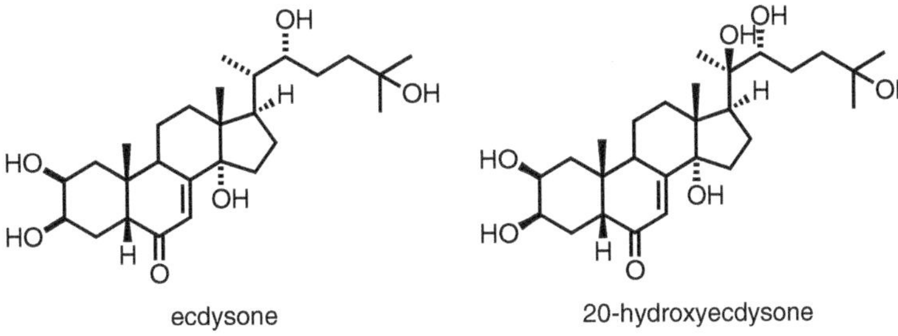

Figure 9.22   *Ecdysteroids*

9.3.1   Biosynthesis of steroids

The steroids are biosynthesised from squalene by way of lanosterol, which still has to undergo a number of degradative steps to reach cholesterol, the common precursor to the rest of the steroid hormones. The side-chain double bond needs to be reduced, the endocyclic double bond has to be isomerised, and altogether three methyl groups must be removed. The hydrogenation of the side-chain double bond is the most straightforward transformation; a NADPH mediated *cis*-hydrogenation gives the fully saturated precursor.

The next event is the oxidative removal of the angular C14 methyl group (Scheme 9.29). This occurs through sequential oxidation of the methyl group. The methyl carbon is lost as formic acid. It is also known that the 15α proton is selectively lost and replaced during this oxidation. The emerging double bond is then hydrogenated (NADPH) to give the demethylated compound.

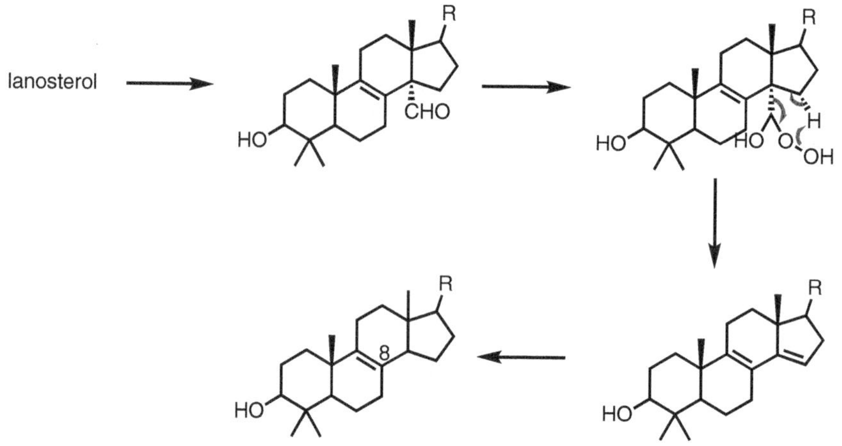

Scheme 9.29 *Biosynthetic removal of C14*

The demethylation of the geminal dimethyl grouping at C4 is one of the last events to occur in the biosynthesis of cholesterol (Scheme 9.30). This occurs via oxidation of the methyl on the lower face of the ring system (α) to a carboxylic acid and oxidation of the C3 hydroxyl to a ketone. The β-keto ester undergoes decarboxylation and equilibration of the remaining methyl group to the α face. Repetition of the oxidation-decarboxylation cycle is followed by NADPH mediated reduction to give the demethylated C3 alcohol.

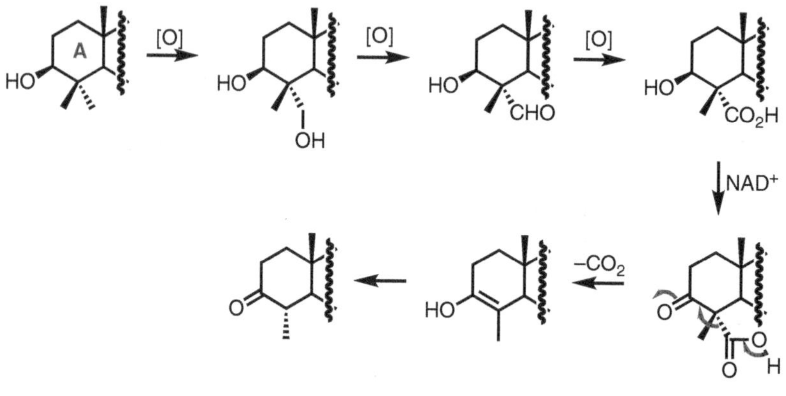

Scheme 9.30 *Removal of gem-dimethyl group at C4*

The mechanism (and indeed the timing) of the isomerisation of the C8 double bond is not as clear. Whether it occurs relatively early (even before the oxidative removal of the C14 methyl group) or late in the biosynthesis is still under debate. A plausible suggestion was introduced [56] based on the occurrence of certain natural products, that before elimination of the C14 methyl group, the original double bond is isomerised to give the C7 olefin.

Calcitriol (Scheme 9.31) is a hormone that circulates in the blood, and its function is to regulate the concentration of calcium in the blood, thus affecting healthy growth and remodeling of bones. Calcitriol is formed in the kidneys by oxidation of cholecalciferol (vitamin D_3). Vitamin D_3 is produced in the body from cholesterol; a 5,7-diene (7-dehydrocholesterol) is converted to precalciferol by a light-induced retrocyclisation of the B-ring, and further thermal isomerisation leads to calciferol (vitamin D_3). Industrially, vitamin D_3 is produced by UVB-irradiation of 7-dehydrocholesterol, which is obtained from the wool grease (lanolin) of sheep.

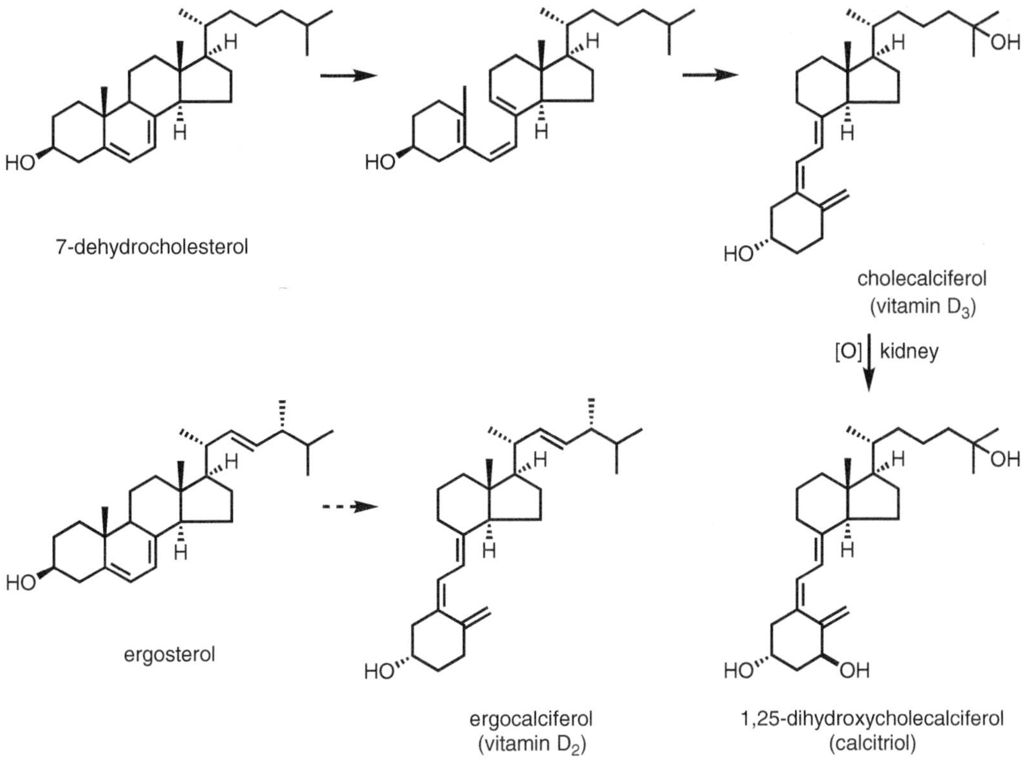

Scheme 9.31 *Biosynthesis of vitamins D*

Deficiency of vitamin D can lead to rickets or osteoporosis (weakening of bone structure), which can be prevented by supplementing the diet with vitamin D-rich nutrients. Fish liver oil is a traditional source of vitamin D_3. Ergosterol was first isolated from ergot, but is also readily available in yeast. It is also processed in the body in a similar way as dehydrocholesterol to produce ergocalciferol (vitamin D_2), which can substitute vitamin D_3 in its functions.

9.3.2 Asymmetric Synthesis of Steroids

Because of their importance in medicinal use, steroids have enjoyed a tremendous amount of synthetic activity, and much of this is already incorporated into the basic organic chemistry texts. Some recent developments in the asymmetric synthesis methodology have facilitated the development of new strategies for the elaboration of the tetracyclic nucleus. The common routes depend on either the formation of a hydrindanone system (CD rings) or the construction of the BC rings through a stereoselective cyclisation. The asymmetric syntheses have been thoroughly reviewed [57], and here we shall only point out the so-called Hajos–Parrish–Eder–Sauer–Wiechert process (Scheme 9.32), which paved the way for the renaissance of amino acid-catalysed enantioselective reactions, and thus the conception of organocatalysis (see also Chapter 3).

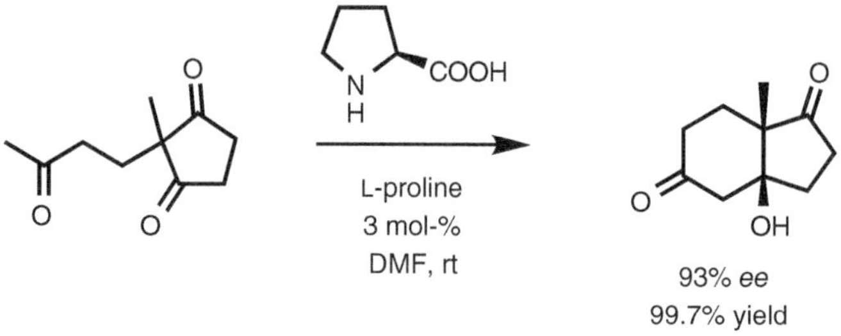

Scheme 9.32 *Hajos–Parrish–Eder–Sauer–Wiechert process*

In the early 1970s, chemists at Roche and Schering nearly simultaneously developed a remarkably facile catalytic aldolisation process to construct the hydrindanone in practically enantiopure form [58]. Starting from the C_2 symmetric cyclopentanedione (itself easily obtained by a Michael addition of 2-methyl-1,3-cyclopentanedione to methyl vinyl ketone), catalysis with natural L-proline gives the CD ring fragment with the natural stereochemistry.

References

1. Ruzicka, L. *Experientia*, **1953**, *9*, 357–367.
2. Wolf, D.E., Hoffman, C.H., Aldrich, P.E., Skeggs, H.R., Wright, L.D., Folkers, K. *J. Am. Chem. Soc.* **1956**, *78*, 4498–4499.
3. Bredt, J., von Rosenberg, M. *J. Liebigs Ann. Chem.* **1895**, *289*, 1–14.
4. Komppa, G. *Ber. Deutsch. Chem. Ges.* **1903**, *36*, 4332–4335.
5. Fisher, K., Perkin, W.H., Jr. *J. Chem. Soc., Trans.* **1908**, *93*, 1871–1876.
6. Berthelot, M. *Comptes rendus*, **1852**, *35*, 136–138; (b) *Comptes rendus*, **1858**, *47*, 266–268.
7. a) Breitmeier, E. *Terpenes*, Wiley-VCh: Berlin, **2006**; b) Maimone, T.J., Baran, P.S. *Nature Chem. Biol.* **2007**, *3*, 396–407.
8. Lynen, F. *Pure Appl. Chem.* **1967**, *14*, 137–167.
9. a) Rohmer, M., Knani, M., Simonin, P., Sutter, B., Sahm, H. *Biochem. J.* **1993**, *295*, 517–524; b) Arigoni, D., Sagner, S., Latzel, C., Eisenreich, W., Backer, A., Zenk, M.H. *Proc. Natl. Acad. Sci. USA* **1997**, *94*, 10600–10605.
10. Dickschat, J.S. *Nat. Prod. Rep.* **2011**, *28*, 1917–1936.
11. a) Geu-Flores, F., Sherden, N., Courdavault, V., Burlat, V., Glenn, W.S., Wu, C., et al. *Nature* **2012**, *492*, 138–142; b) Lichman, B.R., Kamileen, M.O., Titchiner, G.R., Saalbach, G., Stevenson, C.E.M., Lawson, D.M. et al. *Nat. Chem. Biol.* **2019**, *15*, 71–79.
12. Harnying, W., Neudörfl, J.-M., Berkessel, A. *Org. Lett.* **2020**, *22*, 386–390.
13. Yang, J., Xu, W., Cui, Q., Fan, X., Wang, L.-N., Yu, Z.-X. *Org. Lett.* **2017**, *19*, 6040–6043.
14. Liu, G., Zhang, Z., Fu, S., Liu, B. *Org. Lett.* **2021**, *23*, 290–295.
15. Tantillo, D.J. *Nat. Prod. Rev.* **2011**, *28*, 1035–1053.
16. Drew, D.D., Krichau, N., Reichwald, K., Simonsen, H.T. *Phytochem. Rev.* **2009**, *8*, 581–599.
17. Money, T., Wong, M.K.C. *Tetrahedron*, **1996**, *52*, 6307–6324.
18. Dadson, W.M., Lam, M., Money, T., Piper, S.E. *Can. J. Chem.* **1983**, *61*, 343–346.
19. Lee, E., Yoon, C.H. *J. Chem. Soc., Chem. Commun.* **1994**, 479–481.
20. Andrews, S.P., Ball, M., Wierschem, M., Cleator, E., Oliver, S., Högenauer, K., et al. *Chem. Eur. J.* **2007**, *13*, 5688–5712.
21. a) Ueno, Y., Chino, K., Watanabe, M., Moriya, O., Okawara, M. *J. Am. Chem. Soc.* **1982**, *104*, 5564–5566; b) Stork, G., Mook, R. Jr., Biller, S.A., Rychnovsky, S.D. *J. Am. Chem. Soc.* **1983**, *105*, 3741–3742; c) Salom-Roig, X.J., Dénès, F., Renaud, P. *Synthesis 2004*, **1903**–1928.
22. Lee, E., Lim, J.W., Yoon, C.H., Sung, Y.-S., Kim, Y.K., Yun, M., et al. *J. Am. Chem. Soc.* **1997**, *119*, 8391–8392.
23. Beckwith, A.L.J., Schiesser, C.H. *Tetrahedron* **1985**, *41*, 3925–3941.
24. Elford, T.G., Hall, D.G. *J. Am. Chem. Soc.* **2010**, *132*, 1488–1489.
25. Jaskulska, A., Janecka, A.E., Gach-Janczak, K. *Int. J. Mol. Sci.* **2021**, *22*, 4.
26. Paquette, L.A., Doherty, A.M. *Polyquinane Chemistry: Syntheses and Reactions*, Springer Verlag: Berlin, **1987**.
27. Chandler, C.L., List, B. *J. Am. Chem. Soc.* **2008**, *130*, 6737–6739.
28. Mander, L.N. *Chem. Rev.* **1992**, *92*, 573–612.
29. Nicolaou, K.C., Chen, J.S., Dalby, S.M. *Bioorg. Med. Chem.* **2009**, *17*, 2290–2303.

30. Harrison, J.W., Scrowston, R.M., Lythgoe, B. *J. Chem. Soc. C 1966*, **1933**–1945.

31. a) Epstein, W.W., Rilling, H.C. *J. Biol. Chem.* **1970**, *245*, 4597–4605; b) Popjak, G., Edmond, J., Wong, S.-M. *J. Am. Chem. Soc.* **1973**, *95*, 2713–2714.

32. a) Corey, E.J., Russey, W.E., de Montellano, P.P.O. *J. Am. Chem. Soc.* **1966**, *88*, 4750–4751; b) van Tamelen, E.E., Willett, J.D., Clayton, R.B., Lord, K.E. *J. Am. Chem. Soc.* **1966**, *88*, 4752–4754.

33. a) Stork, G., Burgstahler, A.W. *J. Am. Chem. Soc.* **1955**, *77*, 5068–5077; b) Eschenmoser, A., Ruzika, L., Jeger, O., Arigoni, D. *Helv. Chim. Acta* **1955**, *38*, 1890–1904; c) Eschenmoser, A., Arigoni, D. *Helv. Chim. Acta* **2005**, *88*, 3011–3050.

34. a) Yamamoto, S., Bloch, K. *J.Biol. Chem.* **1970**, *245*, 1668–1672; b) Ebersole, R.C., Godtfredsen, W.O., Vangedal, S., Caspi, E. *J. Am. Chem. Soc.* **1973**, *95*, 8133–8140.

35. Corey, E.J., Virgil, S.C. *J. Am. Chem. Soc.* **1991**, *113*, 4025–4026.

36. a) Johnson, W.S. *Angew. Chem. Int. Ed. Engl.* **1976**, *15*, 9–17; b) Johnson, W.S., Lindell, S.D., Steele, J. *J. Am. Chem. Soc.* **1987**, *109*, 5852–5853.

37. Corey, E.J., Lin, S. *J. Am. Chem. Soc.* **1996**, *118*, 8765–8766.

38. a) Tschirch, A., Cederberg, H. *Arch. Pharm.* **1907**, *145*, 97; (b) Voss, W., Klein, P., Sauer, H. *Ber.* **1937**, *70*, 122–132.

39. Ruzicka, L., Jeger, O., Ingold, W. *Helv. Chim. Acta*, **1943**, *26*, 2278–2282.

40. a) Barton, D.H.R., Holness, N.J. *J. Chem. Soc.*, **1952**, 78–92; b) Beaton, J.M., Spring, F.S. *J. Chem. Soc.* **1955**, 3126–3129.

41. Corey, E.J., Cantrall, E.W. *J. Am. Chem. Soc.* **1959**, *81*, 1745–1751.

42. Roy, A., Saraf, S. *Biol. Pharm. Bull.* **2006**, *29*, 191–201.

43. Spino, C. *Synlett* **2006**, 23–32.

44. Mondon, A., Epe, B. *Progr. Chem. Org. Nat. Prod.* **1983**, *44*, 101–187.

45. a) Ley, S.V. *Pure Appl. Chem.* **1994**, *66*, 2099–2102; b) Ley, S.V. *Pure Appl. Chem.* **2005**, *77*, 1115–1130; c) Veitch, G.E., Boyer, A., Ley, S.V. *Angew. Chem. Int. Ed.* **2008**, *47*, 9402–9429.

46. Heasley, B. *Eur. J. Org. Chem.* **2011**, 19–46.

47. a) Hodgson, H., De La Peña, R., Stephenson, M.J., Thimmappa, R., Vincent, J.L., Sattel, E.S., et al. *PNAS* **2019**, *116*, 17096–17104; b) Pandreka, A., Chaya, P.S., Kumar, A., Aarthy, T., Mulani, F.A., Bhagyashree, D.D., et al. *Phytochemistry* **2021**, *184*, 112669.

48. Ourisson, G., Rohmer, M., Poralla, K. *Annu. Rev. Microbiol.* **1987**, *41*, 301–333.

49. Windaus, A., Rosenbach, A., Riemann, T. *Z. Physiol. Chem.* **1923**, *130*, 113–125; see also Nobel lectures by Wieland and Windaus in 1928.

50. Bernal, J.D. *Nature* **1932**, *129*, 277–278.

51. Bernal, J.D., Crowfoot, D., Fankuchen, I. *Trans. Roy. Soc. London* **1940**, *A239*, 135–182.

52. Butenandt, A., Tscherning, K. *Z. Physiol. Chem.* **1934**, *229*, 167–184.

53. Wigglesworth, V.B. *Quart. J. Microsc. Sci.* **1934**, *77*, 191–223.

54. Butenandt, A., Karlson, P. *Z. Naturforsch.* **1954**, *9b*, 389–391.

55. Huber, R., Hoppe, W. *Chem. Ber.* **1965**, *98*, 2403–2424.

56. Schroepfer, G.J. Jr., Lutsky, B.N., Martin, J.A., Huntoon, S., Fourcans, B., Lee, W.-H., et al. *Proc. Roy. Soc.* **1972**, *B180*, 125–146.

57. a) Chapelon, A.-S., Moraléda, D., Rodruguez, R., Ollivier, C., Santelli, M. *Tetrahedron*, **2007**, *63*, 11511–11616; b) Khatri, H.R., Carney, N., Rutkoski, R., Bhattarai, B., Nagorny, P. *Eur. J. Org. Chem.* **2020**, 755–776.

58. a) Hajos, Z.G., Parrish, D.R. *J. Org. Chem.* **1974**, *39*, 1615–1621; b) Eder, U., Sauer, G., Wiechert, R. *Angew. Chem., Int. Ed. Engl.* **1971**, *10*, 496–497.

10

Alkaloids

One wonders how people in primitive societies, with no knowledge of chemistry or physiology, ever hit upon a solution to the activation of an alkaloid by a monoamine oxidase inhibitor.

Richard Evans Schultes

Plant-derived extracts and concoctions have been used since ancient times for both healing purposes and recreational use. Mesopotamians are known to have used medicinal plants as early as 2000 BC. According to Plato's *Phaedon*, Socrates died of poisoning with a drink containing coniine, an alkaloid in poison hemlock (*Conium maculatum*) (Figure 10.1). Morphine has played a central role in the early days of natural product chemistry; according to Odyssey by Homer, Helen of Troy received a gift from the Egyptian queen, a drug bringing oblivion. This probably was opium, the extract of opium poppies (*Papaver somniferum*). In the early sixteenth century the famous alchemist from Basel, *Paracelsus* (*Theophrastus Bombastus von Hohenheim*, 1493–1541) claimed to be in possession of a secret remedy to all maladies, and he called it laudanum, an alcoholic extract of poppies, containing up to 10% opium, or 1% morphine.

Use of natural products and their derivatives as drugs was quite different even at the beginning of the twentieth century from today's targeted drug therapy. Strychnine was celebrated to be the cure for almost anything, and it was even used as a doping agent to increase strength and stamina for marathon runners in the 1904 St. Louis Olympic Games [1]. Cocaine was used for 'recreational purposes,' and at one point it was so popular that it was added to wine (the Mariani red wine was one of the most popular reds in the 1890s – one bottle contained some 100 mg of cocaine). Besides, cocaine was also good for fighting dandruff [2]! Morphine was advertised as a cough suppressant and prescribed to teething children as a soothing syrup! In fact, at one point, diacetylmorphine (heroin) was considered a painkiller superior to aspirin! Luckily, the company decided not to follow that line of research [3].

Actual chemical research began with the isolation of narcotine (now called noscapine) by the French pharmacist *Charles Derosne* (1780–1846) from opium [4]. A pharmacist's apprentice in Paderborn (later a pharmacist himself in Hameln from 1822), *Friedrich Wilhelm Adam Sertürner* (1783–1841) isolated the soporific principle of opium, which he called morphium in honor of the Greek god of dreams, Morpheus [5]. Other compounds soon followed; *Pierre Joseph Pelletier* (1788–1842) and *Joseph Bienaimé Caventou* (1795–1877) isolated strychnine in 1818 [6] and quinine in 1820 [7]. Although the compounds were isolated, their chemical structures were not known. The structure of coniine was not established until 1870 [8], and synthesis of racemic coniine was achieved only in 1886 by *Albert Ladenburg* (1842–1911) [9]. The first asymmetric synthesis for coniine had to wait another century, when it was reported in 1983 by Henri-Philippe Husson [10]. The correct structure for strychnine was proposed in 1946 by *Sir Robert Robinson* (1886–1975) [11], but it took more than a decade before it was elucidated by X-ray crystallography in 1956 [12].

Asymmetric Synthesis of Natural Products, Third Edition. Ari M.P. Koskinen.
© 2023 John Wiley & Sons Ltd. Published 2023 by John Wiley & Sons Ltd.

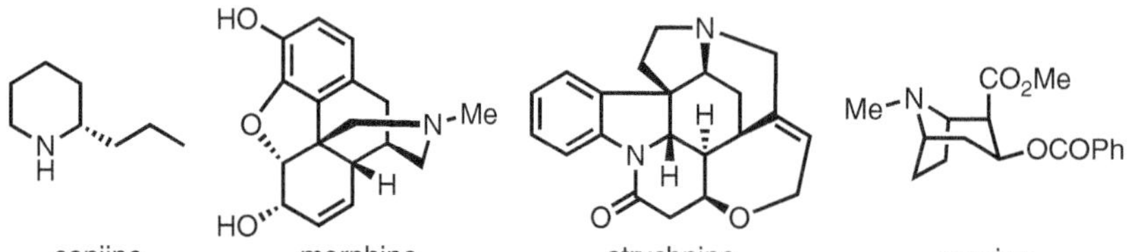

Figure 10.1   *Alkaloid structures*

This was two years after it was synthesised by R.B. Woodward in racemic form in 1954 [13]. The first asymmetric synthesis of strychnine by Larry Overman remained in the shadows yet another 40 years [14]!

The term alkaloid was originally coined by the German chemist *Carl Friedrich Wilhelm Meissner* (1792–1853), in 1819, to cover natural compounds similar in behavior to alkalis, basic compounds [15]. The word alkaloid derives from Arabic: *al qalay* = to roast, and *eidoç-oidos* from Greek meaning 'appearance.' However, not all alkaloids are notably basic in their character. It is often very difficult to distinguish naturally occurring compounds from non-natural ones. The simple definition of alkaloids has been questioned for a long time. Manfred Hesse presented a modified definition for alkaloids: nitrogen-containing compounds derived from plants or animals [16]. Even this definition has its shortcomings as, for instance DNA, RNA, and peptides are not considered alkaloids. On the other hand, we shall see that there exists a definitive group of peptide alkaloids, a group of compounds containing the ergot alkaloids and cyclopeptide alkaloids, both of which are medicinally important.

At present, some 20 000 natural alkaloids are known and can be sub-divided according to their chemical structures into the following groups: i) heterocyclic alkaloids; ii) alkaloids with an exocyclic nitrogen atom; iii) polyamines; iv) peptide alkaloids; and v) terpene alkaloids. Typical examples of each group will be discussed in the next sections in terms of their structures, biosynthesis, and synthesis.

10.1   Heterocyclic Alkaloids

In its most common usage, the term alkaloid is used to refer to heterocyclic alkaloids. Many of these have had and still have an important role to play in the healing practices of most cultures. Typical structural groups are (Figure 10.2) indole alkaloids (e.g. reserpine which possesses blood pressure reducing properties), pyrrolidine alkaloids (e.g. mesembrine), tropane alkaloids (cocaine and its relatives), quinoline alkaloids (the malaria drug quinine and morphine for pain relief), and the izidine alkaloids (securinine is an indolizidine). Pilocarpine, a histidine derivative used for the treatment of glaucoma, also belongs to the broad class of alkaloids.

Medicinally, the alkaloids have played a key role for millennia. Natural products and compounds inspired by natural products represent 25–50% of commercial drugs [17], and since drugs typically contain nitrogen atoms, alkaloids or their structural modifications or analogues are well represented. In drug design, the search for new chemical entities is still heavily dependent on natural compounds, and as long as new structures can be found from the plant or animal kingdom, this process can continue. A major concern is the destruction of rain forests and other sources of chemical diversity presented by nature. It has been estimated that there are still tens, and perhaps hundreds of thousands of unclassified and unexplored species, which are under serious threat due to human intervention.

Next, we shall discuss the alkaloids based on their biogenetic precursors, the original amino acids that have given rise to the biogenic amines from which the alkaloids emanate.

10.1.1   Pyrrolidine and Tropane Alkaloids

Structures of typical tropane alkaloids are shown in Figure 10.3. Both atropine and scopolamine are anticholinergic agents. Since the tropane skeleton itself is not chiral, atropine is, in fact, a mixture of L- and D-hyoscyamines, the physiological activities mainly deriving from L-hyoscyamine.

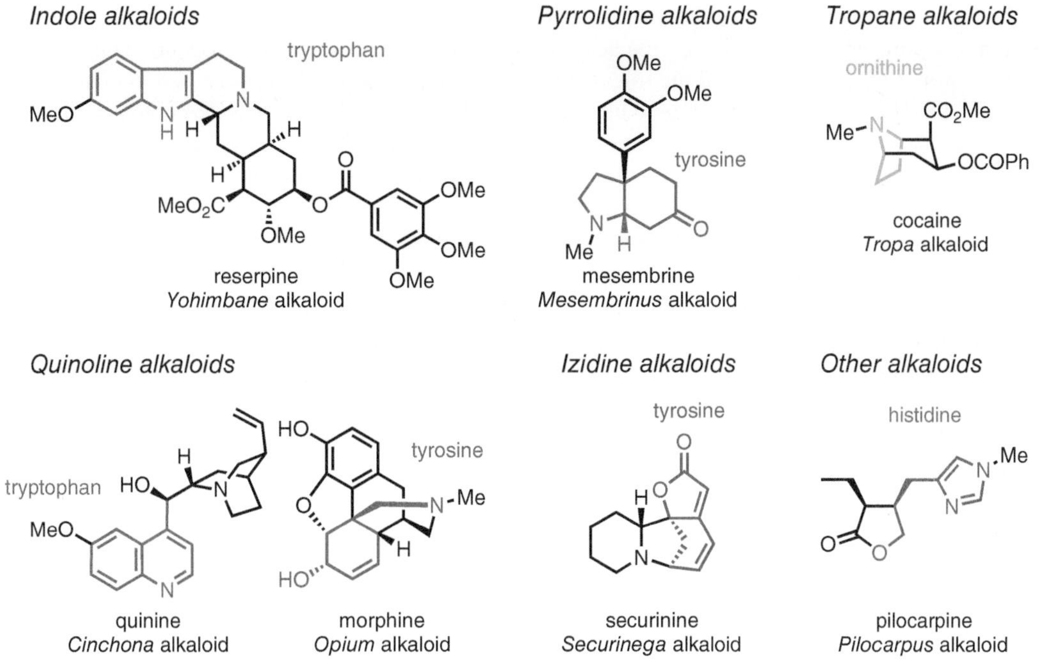

Indole alkaloids

tryptophan

reserpine
Yohimbane alkaloid

Pyrrolidine alkaloids

tyrosine

mesembrine
Mesembrinus alkaloid

Tropane alkaloids

ornithine

cocaine
Tropa alkaloid

Quinoline alkaloids

tryptophan

tyrosine

quinine
Cinchona alkaloid

morphine
Opium alkaloid

Izidine alkaloids

tyrosine

securinine
Securinega alkaloid

Other alkaloids

histidine

pilocarpine
Pilocarpus alkaloid

Figure 10.2   *Structural classes of alkaloids*

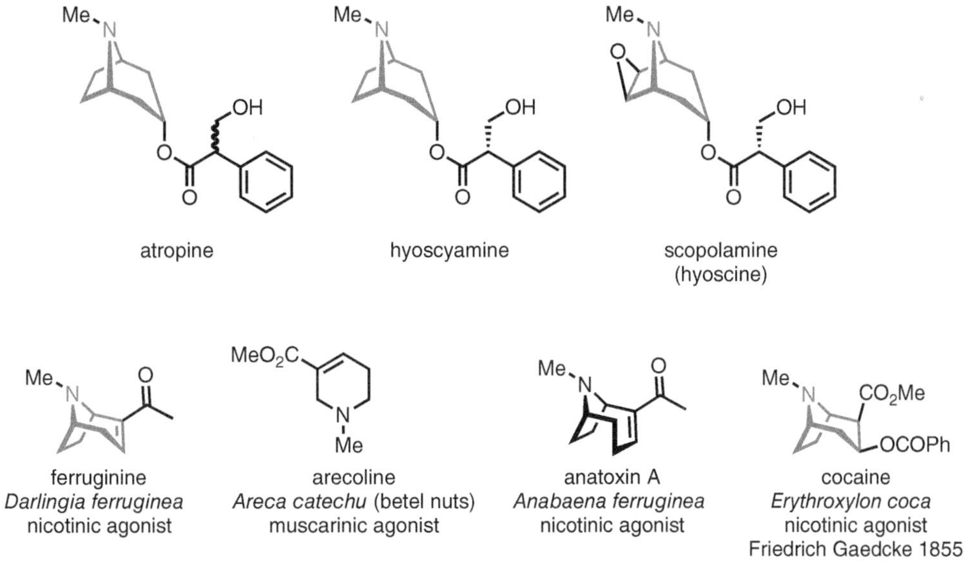

atropine

hyoscyamine

scopolamine
(hyoscine)

ferruginine
Darlingia ferruginea
nicotinic agonist

arecoline
Areca catechu (betel nuts)
muscarinic agonist

anatoxin A
Anabaena ferruginea
nicotinic agonist

cocaine
Erythroxylon coca
nicotinic agonist
Friedrich Gaedcke 1855

Figure 10.3   *Tropane alkaloids*

Hyoscyamine and scopolamine have been isolated from species of the family *Solanaceae*, nightshades. The association of fertility to mandrake (*Mandragora officinarum*) is alluded to in the Book of Genesis (Chapter 30). Use of mandrake as an anesthetic was described by Dioscorides in 60 AD, and in 1431, Jeanne d'Arc was accused of having a mandrake about her person, and was condemned to the stake. Thanks to the powerful madragora – a tincture associated with magic, witchcraft, and the supernatural – mandrake roots became valuable items of trade and the target of myths to mystify it. One of the myths tells that this man-shaped root is inhabited by a demon, and that attempts to uproot the plant will lead to a horrible deadly shriek by the demon [18]. Mandrake grows in the Mediterranean regions, whereas henbane (*Hyoscyamus niger* L.) occupied the roles of mandrake in Britain. Henbane was used in many magic potions, and it is possible that the cursed hebenon poured into the ear of Hamlet's father was an extract of henbane. Henbane use by the ancient Greeks was documented by Discorides, and *Pliny the Elder* (AD 23–79) likened its action to be 'of the nature of wine and therefore offensive to the understanding.' Henbane was known in Greece as "Herba Appolinaris" and taken by the Priestesses of Apollon for producing their oracles. The oracle was named "Pytho" and the Priestess "Pythia." The Priestesses of the Delphic oracle were said to have inhaled smoke from smoldering henbane. The deadly nightshade (*Atropa belladonna* L.) has allegedly been used to dilate the pupils of the eye by ancient Greeks and Egyptians – hence the name belladonna, beautiful woman. Thorn apple or Jimson weed (*Datura stramonium* L.) has been used as an herbal medicine to relieve the symptoms of asthma, as well as an analgesic in surgeries.

Pyrrolidine alkaloids are biogenetically formed from a diamino acid, ornithine (Scheme 10.1). Pyridoxal phosphate functions as an activator in these reactions, first leading to decarboxylation and simultaneous formation of the electrophilic imine (or iminium ion after protonation). This reacts with an activated acyl coenzyme A derivative, releasing pyridoxamine. Decarboxylation finally gives hygrine. However, hygrine is not directly involved in the biosynthesis of tropane alkaloids. Instead, putrescine is formed, methylated, and then converted oxidatively to the *N*-methylpyrrolidinium ion.

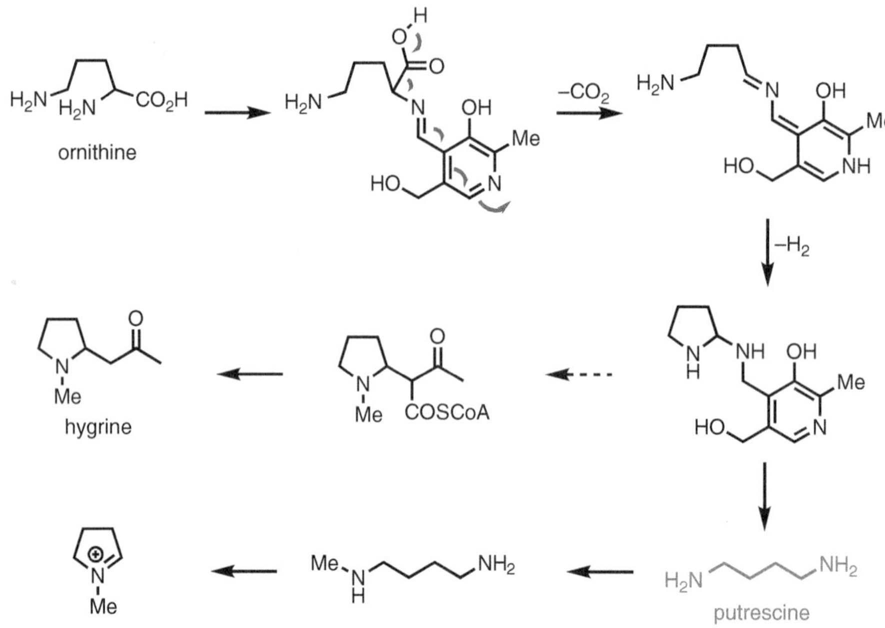

Scheme 10.1   *Biosynthesis of hygrine and iminium ion towards tropane alkaloids*

Participation of the doubly activated acyl coenzyme A leads to a β-keto acid derivative, which further cyclises to give the tropane skeleton (Scheme 10.2) [19]. The cyclisation is presumably preceded by an oxidative transformation of the amino group into an iminium ion. Tropinone is reduced to tropine, which is the precursor for many more tropane alkaloids, cocaine amongst others.

The tropane alkaloids occur mainly in two plant families, *Erythroxylonaceae* and *Solanaceae*. The dried leaves of the South American bush *Erythroxylon coca* Lam. have been chewed by the natives of Peru to improve endurance and promote a sense of well-being. In 1850, the Austrian *Carl von Scherzer* (1821–1903) imported 40 lb of leaves of the South American bush *Erythroxylon coca*, enough to permit the isolation of cocaine by *Friedrich Gaedcke* (1828–1890) [20]. In 1884, prompted by Sigmund Freud's experiments with cocaine, the Austrian *Carl Koller* (1857–1944) tested it

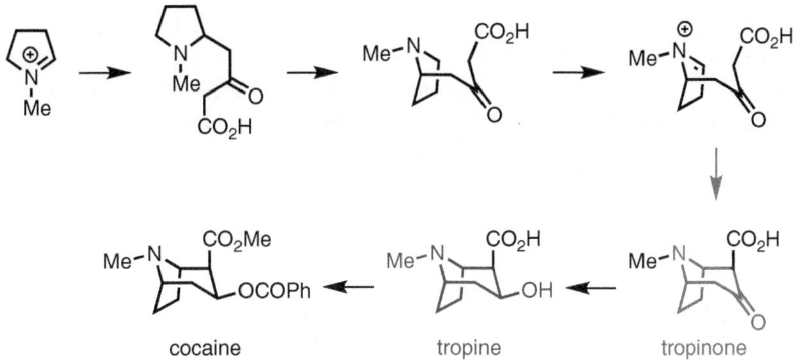

Scheme 10.2 *Biosynthesis of tropane alkaloids*

with the eyes of a frog, rabbit, dog, and eventually his own eye. He had earlier tested chloral, bromide, and morphine for local anesthesia of the eye – no painless operations. Cocaine's toxicity was soon evidenced by many deaths among both patients and (addicted) medical staff. Synthesis came to rescue here: total synthesis of cocaine by *Richard Willstätter* (1872–1942) in 1901 [21] was soon followed by the synthesis of several 'caines,' e.g. amylocaine (1903), procaine (1905), lidocaine (1943), and bupivacaine (1957). (*S*)-Ropivacaine, which showed less side effects, was the first pure enantiomer of this class to enter the market in 1996 (Figure 10.4).

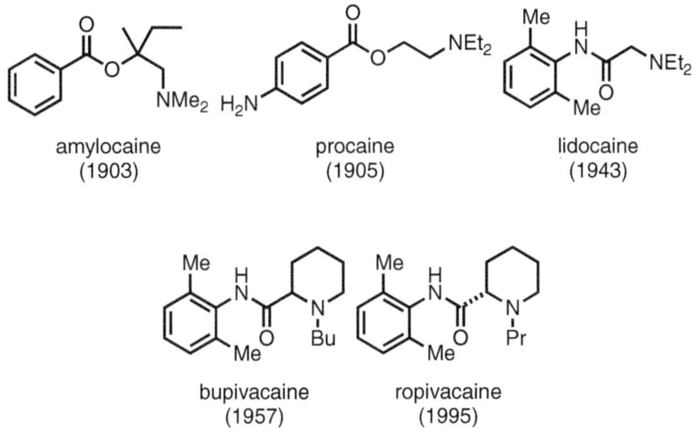

amylocaine
(1903)

procaine
(1905)

lidocaine
(1943)

bupivacaine
(1957)

ropivacaine
(1995)

Figure 10.4 *Local anesthetics*

   Sir Robert Robinson (1886–1975, Nobel prize in 1947) succeeded in the synthesis of tropinone (Scheme 10.3) in 1917. The synthesis was formally a remarkably simple Mannich-type condensation of succinaldehyde, methylamine, and acetone. In the event, succinaldehyde was liberated from dimethoxytetrahydrofuran in an aqueous solution of acetonedicarboxylic acid and methylamine, in an efficient synthesis, much ahead of its time [22]. In modern terms, this reaction was organocatalytic and represented very green chemistry. Different strategies for the enantioselective synthesis of tropane alkaloids have been reviewed [23].

10.1.2 Izidine Alkaloids

Izidine alkaloids include some 1000 compounds containing a pyrrolizidine, indolizidine, or quinolizidine skeleton (Figure 10.5). Several of them have interesting physiological and pharmacological activities. Some of these compounds also bear close resemblance to the amino sugars discussed in Section 4.1.3.

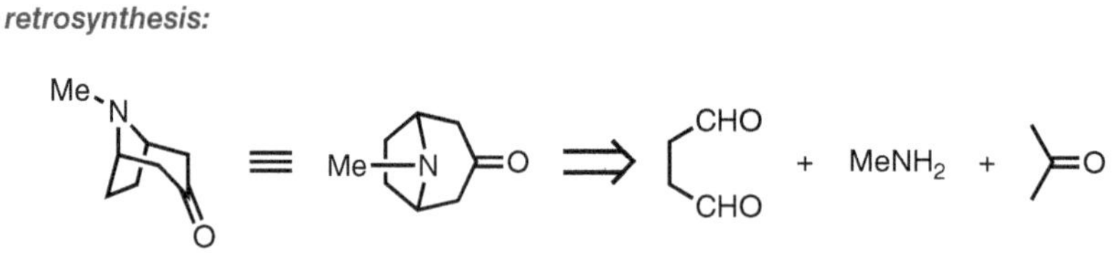

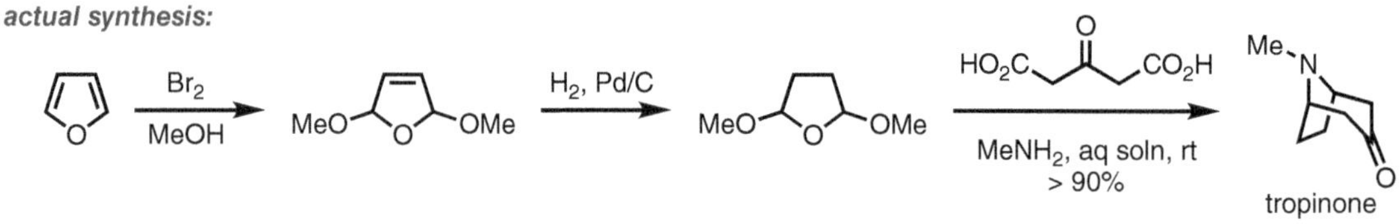

retrosynthesis:

actual synthesis:

Scheme 10.3   *Robinson tropinone synthesis*

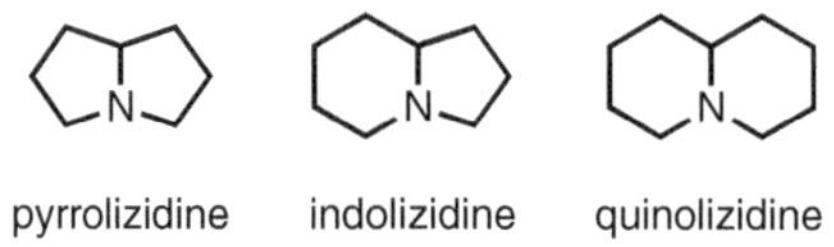

pyrrolizidine   indolizidine   quinolizidine

Figure 10.5   *Izidine alkaloids*

Some 700 pyrrolizidine alkaloids have been isolated and characterised from nearly 6000 plant species, which corresponds to *ca.* 3% of known flowering plants. Several invasive noxious weeds, mainly plants of *Boraginaceae* (borages, forget-me-not, and heliotropes), *Asteraceae* (asters, or daisy, or sunflower family, especially tribes *Senecioneae* and *Eupatorieae*), and *Fabaceae* or *Legumoinosae* (genus *Crotalaria*) are especially rich in these toxic alkaloids, which makes them the most common poisonous plants affecting livestock, wildlife, and humans. The pyrrolizidines are activated in the liver to reactive pyrrole derivatives, which can either be eliminated through conjugation with glutathione or react with liver tissue causing fatal damage (Scheme 10.4).

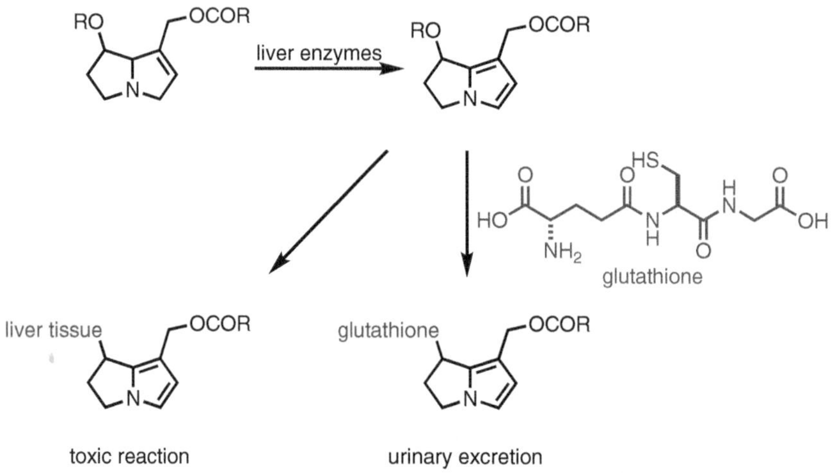

liver tissue

toxic reaction

glutathione

urinary excretion

Scheme 10.4   *Liver toxicity of pyrrolizidines*

The pyrrolizidines typically contain a necine base (Figure 10.6) esterified with a carboxylic acid. Heliotridine, retronecine, and platynecine are capable of forming cyclic diesters with a dicarboxylic acid. *N*-Oxides of each of these

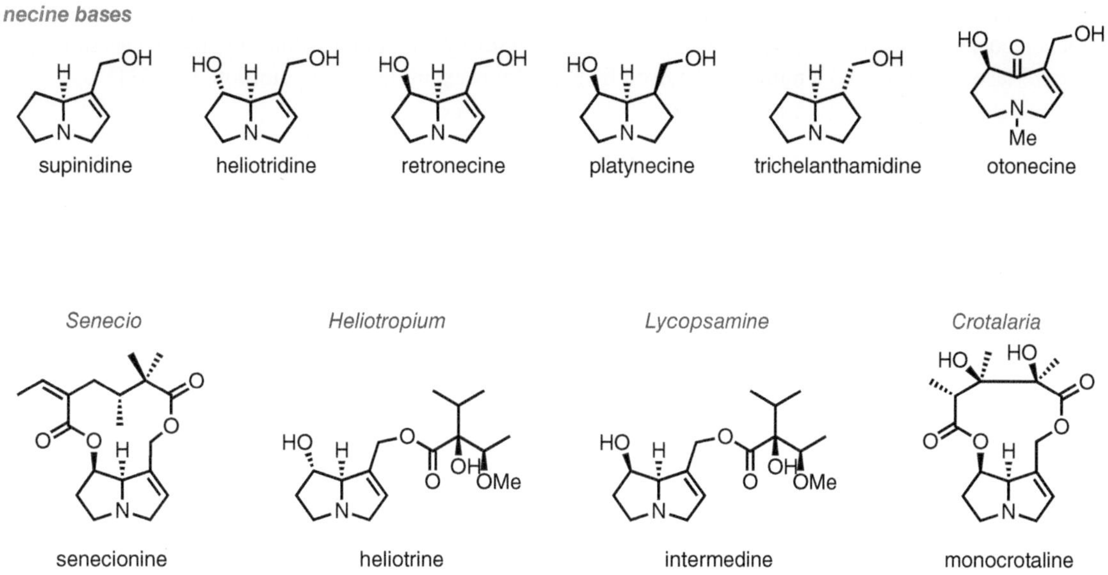

necine bases

supinidine heliotridine retronecine platynecine trichelanthamidine otonecine

Senecio *Heliotropium* *Lycopsamine* *Crotalaria*

senecionine heliotrine intermedine monocrotaline

Figure 10.6   *Pyrrolizidine alkaloids*

types are also known. Typical pyrrolizidine alkaloids include the *Senecio*, *Heliotropium*, *Lycopsamine*, and *Crotalaria* alkaloids.

Pyrrolizidine alkaloids are biosynthesised from putrescine derived from L-ornithine (or L-arginine) (Scheme 10.5). Dimerisation to homospermidine is followed by oxidative ring closure and intramolecular Mannich-type cyclisation. Oxidation state adjustments lead to retronecine. The senecic acid in senecionine is derived from L-isoleucine.

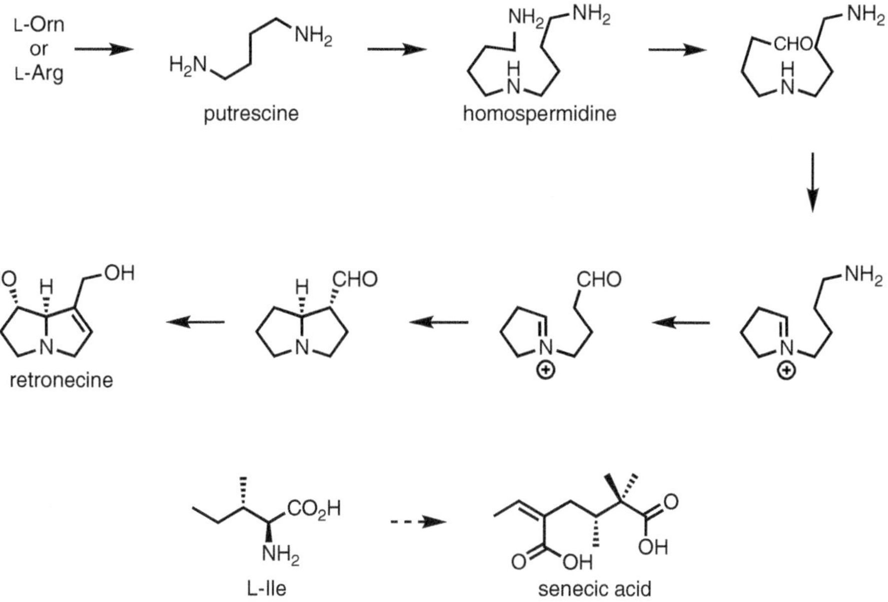

L-Orn or L-Arg putrescine homospermidine

retronecine

L-Ile senecic acid

Scheme 10.5   *Pyrrolizidine alkaloid biosynthesis*

The indolizidine alkaloids (Figure 10.7) are a broad and varied class of compounds, including such alkaloids as slaframine, elaeocanine, securinine, tylophorine, and the polyhydroxylated indolizidines related to castanospermine, a potent glycosidase inhibitor, and a number of poison frog alkaloids exemplified by pumiliotoxin 251D.

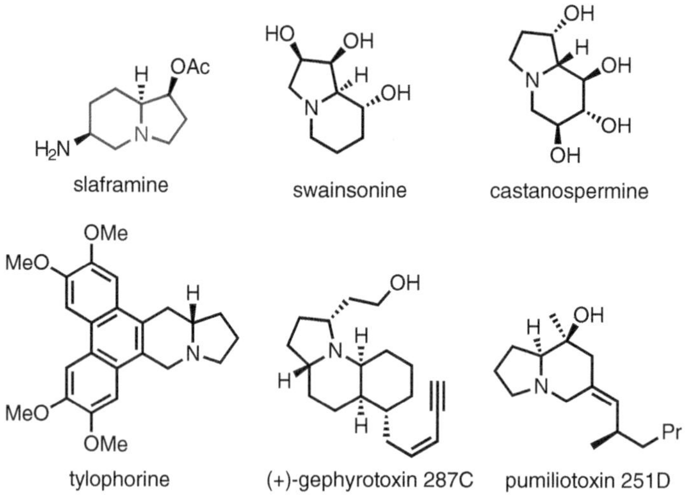

Figure 10.7   *Indolizidine alkaloids*

The Overman methodology on electrophilic Mannich cyclisations of formaldiminium ions is exemplified by the synthesis of pumiliotoxin 251D (Scheme 10.6) [24]. The epoxycarbamate derived from L-proline was reacted with the vinylalanate derived from the silylalkyne to give the cyclic carbamate. This was in turn hydrolysed and treated with formalin to yield a cyclopentaoxazoline. Heating this intermediate in ethanol with camphorsulfonic acid (CSA) cleaved the aminal, and the resulting iminium ion cyclised to pumiliotoxin 251D in high yield. The vinylsilane underwent electrophilic cyclisation with retention of configuration [25], thus securing the double-bond geometry in the product.

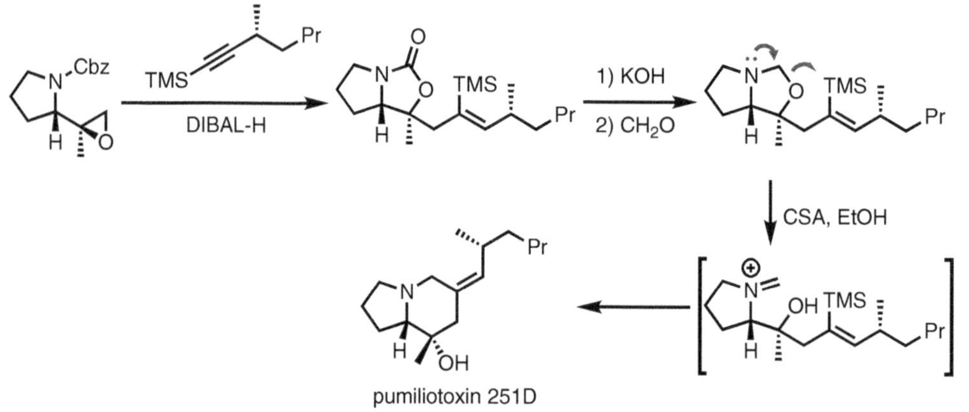

Scheme 10.6   *Synthesis of pumiliotoxin 251D*

Unlike other amphibian alkaloids such as pumiliotoxin C and histrionicotoxins, gephyrotoxin 287C is relatively nontoxic, exhibits weak activity as a muscarinic antagonist, and acts as a moderate blocker of nicotinic acetylcholine receptor channels. Amat synthesis of (+)-gephyrotoxin 287C utilised phenylglycinol as the source of chiral information (Scheme 10.7) [26]. Thus, the known bromo enone was converted in two steps to the required keto ester which, on

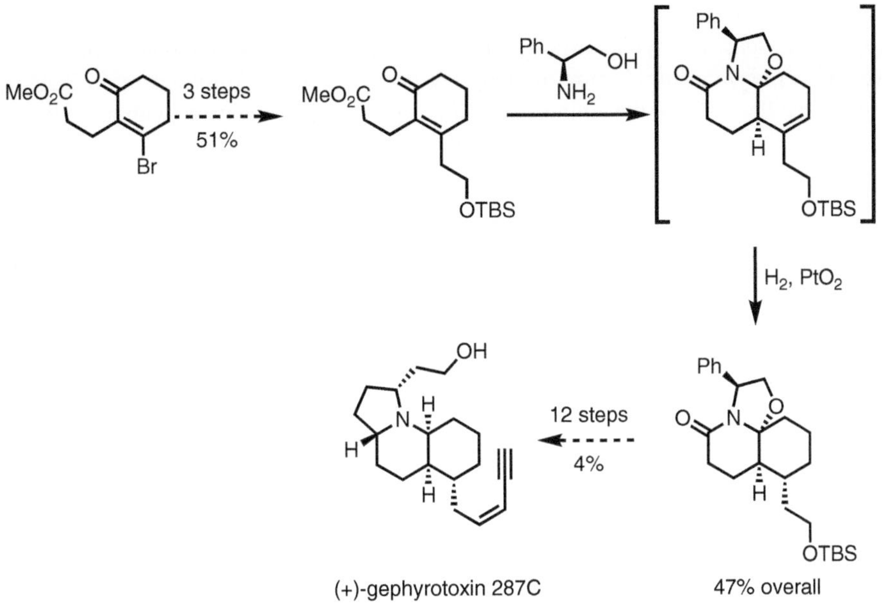

Scheme 10.7 *Relayed asymmetric induction strategy to (+)-gephyrotoxin 287C*

treatment with (*S*)-phenylglycinol, formed the tricyclic lactam with *cis* fusion of the hexahydroquinolone. Catalytic hydrogenation of the double bond set the stereochemistry of the remaining center. In the remaining steps, the rest of the tricycle and side chains were assembled using standard chemistry.

The above cases have highlighted internal asymmetric induction and relayed asymmetric induction to introduce the asymmetric information to the product. Methods employing kinetic resolution have also been used in the formation of the pyrrolidine ring, as exemplified by the synthesis of L-hydroxyindolizidines, which are the biosynthetic precursors to slaframine and swainsonine (Scheme 10.8) [27]. Sharpless asymmetric epoxidation of racemic *N*-benzyloxycarbonyl-3-hydroxy-4-pentenylamine led to kinetic resolution with high efficiency. Intramolecular amidomercuration and radical Michael addition as key steps led to the chain-elongated precursor for the final cyclisation.

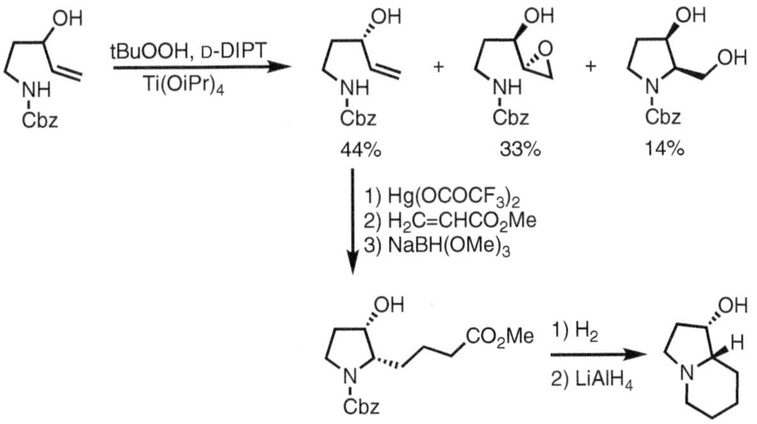

Scheme 10.8 *Kinetic resolution in the synthesis of hydroxyindolizidine*

Epilupinine is structurally one of the simplest quinolizidine alkaloids, but still possesses potent and diverse biological activities. In 1960 *Eugene van Tamelen* (Stanford University, 1925–2005) reported a biomimetic synthesis of racemic epilupinin starting with an acyloin ring-closure of diethyl *N*-benzyl-*N*,*N'*-bis-(*ω*-*n*-valerate) (Scheme 10.9) [28]. The acyloin was converted in a few steps to the amino diol, which upon a glycol cleavage with periodic acid spontaneously generated the cyclised product. Lithium aluminium hydride reduction then gave racemic epilupinine.

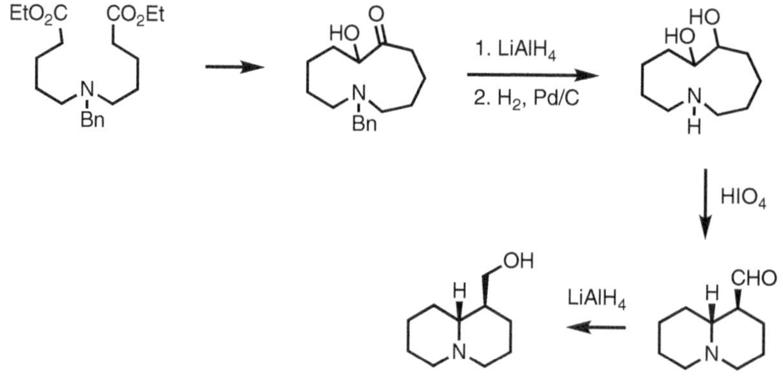

Scheme 10.9 *van Tamelen biomimetic synthesis of* rac-*epilupinine*

Some 30 racemic syntheses and 15 enantioselective syntheses of epilupinine have been developed. Recently Namba applied an organocatalytic Mannich cyclisation to the key reaction in the van Tamelen synthesis (Scheme 10.10) [29]. The dialdehyde was synthesised by dialkylating nosylamide with hexenyl bromide followed by ozonolysis. In a one-pot operation, the nosyl group was removed with thiophenol, the organocatalytic Mannich reaction was induced with L-proline, and finally, the aldehyde was reduced with sodium borohydride. Although the Mannich reaction only preceded in 83% *ee*, the final product could be recrystallised in its triphenylacetic salt form to give enantiopure (+)-epilupinine.

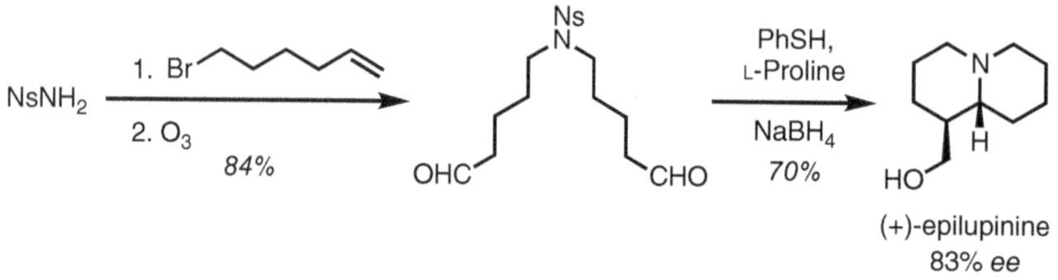

Scheme 10.10 *Organocatalytic synthesis of (+)-epilupinine*

The synthesis of croomine, an alkaloid isolated from Chinese herbal tea used to treat tuberculosis, bronchitis, pertussis, and other symptoms, provides another example of the utilisation of electrophilic cyclisation (Scheme 10.11) [30]. The substrate was constructed from two units of methyl 3-hydroxy-2-methylpropionate (one of each enantiomer) and Sharpless asymmetric epoxidation to generate an intermediate epoxide for the formation of the initial azepine ring. Chain elongation gave the alkene, which upon treatment with iodine gave rise to an intermediate iodoamine. The *anti* stereochemistry of the iodine and the amino group in this reaction are secured by the involvement of an iodonium intermediate. Neighboring-group participation of the amino group through the formation of an aziridine intermediate was also the reason for a subsequent transformation to give the *anti* array between the amino and acyloxy functions.

More than 6000 quinolizidine alkaloids known as lycopodium alkaloids are recognised (Figure 10.8). Lycodine was first isolated from club moss *Lycopodium complanatum* [31]. These alkaloids exhibit diverse physiological activities including diuretic, analgesic, hemostyptic, antispasmodic, cardiovascular, and neuromuscular effects. Huperzine

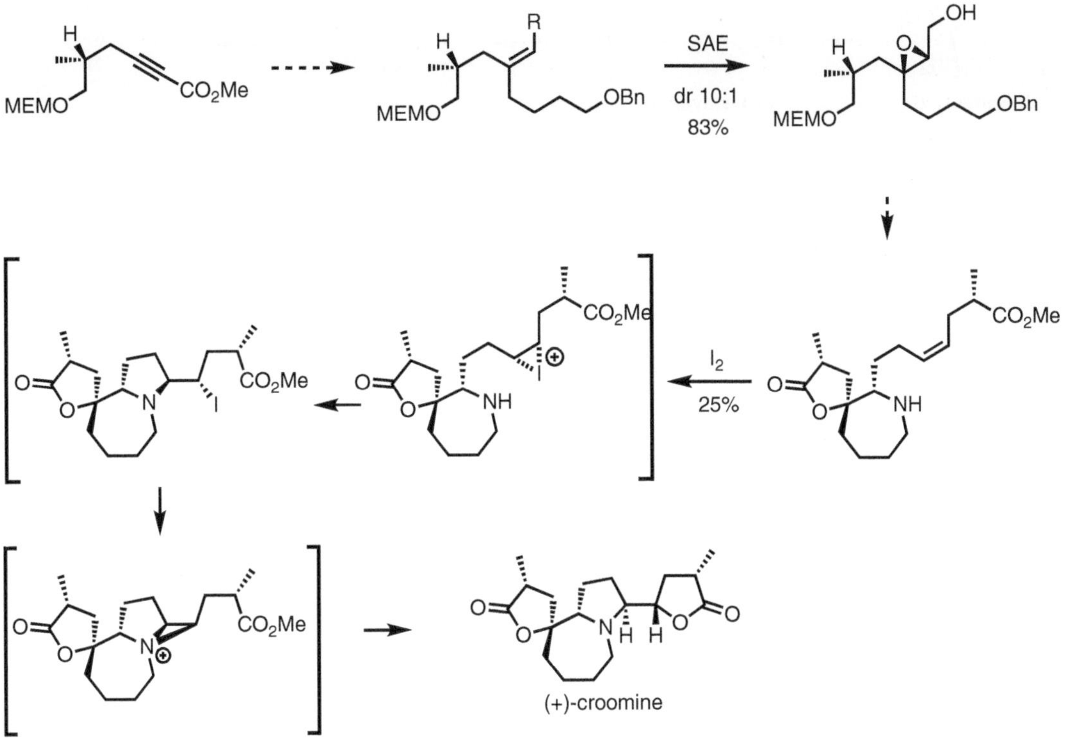

Scheme 10.11   Synthesis of croomine

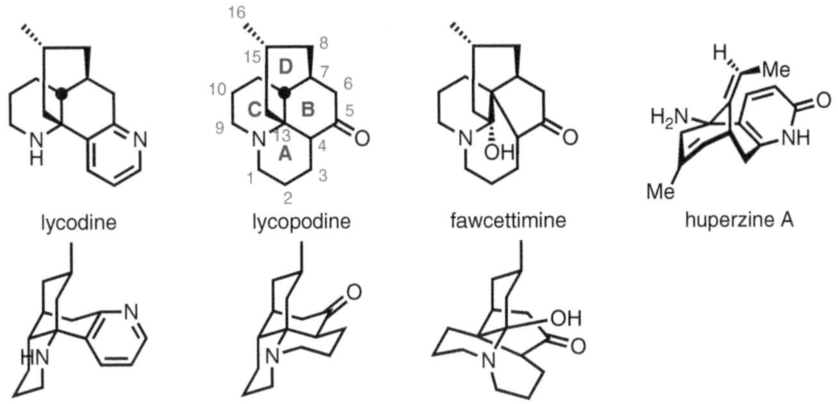

Figure 10.8   Lycopodium alkaloids

A (HupA) is reported to increase efficiency for learning and memory in animals, and it shows promise in the treatment of Alzheimer's disease (AD).

The biosynthesis of lycopodine (Scheme 10.12) is very similar to the biosynthesis of other alkaloids derived from aliphatic amino acids. L-Lysine is decarboxylated to cadaverine, which is then oxidised and cyclised to an iminium ion. Chain extension twice with malonyl-CoA followed by decarboxylation gives pelletierine, which in turn is dimerised. Decarboxylation and elimination of water leads to a tetracyclic intermediate, which after hydrolysis, ring-closure, and reduction steps leads to lycopodine.

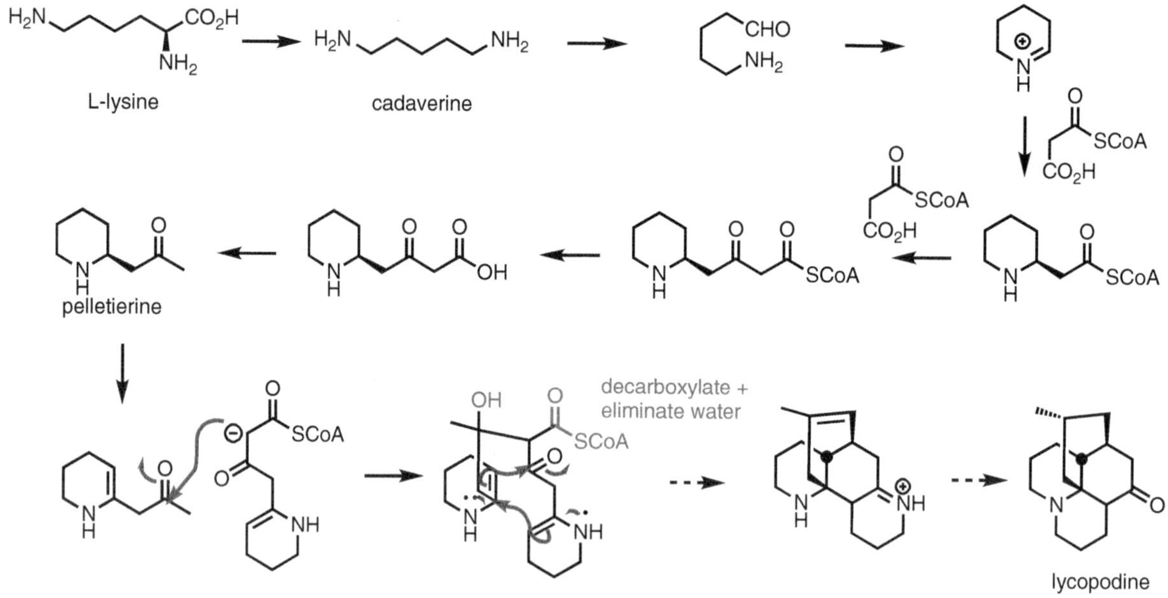

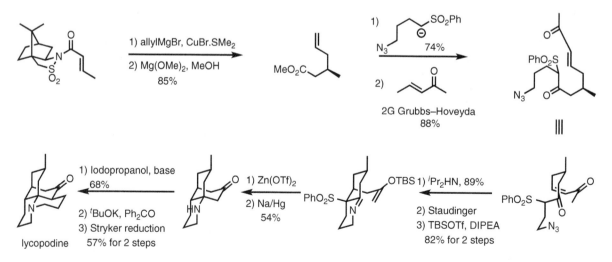

Scheme 10.12   *Biosynthesis of lycopodine*

An example of the application of the Oppolzer diastereoselective conjugate-addition reaction is shown in the first asymmetric synthesis of lycopodine (Scheme 10.13) [32]. The commercially available acyl sultam was converted in two steps to optically active methyl (*R*)-3-methylhex-5-enoate. [(4-azidobutyl)sulfonyl]benzene was deprotonated with LiTMP and treated with the chiral ester to yield the ketosulfone, which was then subjected to a cross-metathesis reaction to give the key cyclisation precursor. Treatment of the keto sulfone with i-Pr$_2$NH led to clean conjugate addition to a single product that crystallised out of the reaction in 89% yield. After reduction of the azide and conversion of the methyl ketone to the TBS-enol ether, Lewis acid-mediated Mannich ring closure followed by reductive cleavage of the sulfone gave the penultimate tricycle. The fourth ring was constructed by first alkylating the nitrogen with

Scheme 10.13   *First asymmetric synthesis of lycopodine*

iodopropanol. Oxidation of the alcohol gave the corresponding aldehyde, which under the basic conditions, underwent an intramolecular aldol condensation. Finally, Stryker reduction of the enone gave lycopodine.

The total synthesis of Lycopodium alkaloid (−)-huperzine A was accomplished from commercially available (*R*)-pulegone (Scheme 10.14) [33]. Buchwald–Hartwig coupling reaction of the pulegone-derived vinyl triflate gave vinyl carbamate, which was alkylated with the bromomethyl pyridine and reduced to give the ring-closure precursor as an inconsequential mixture of diastereomers. Both isomers underwent the Heck reaction with equal efficiency, and under optimal conditions, the crude alcohol gave the depicted Heck product after Ley–Griffith oxidation in an overall yield of 63%. Grignard addition of ethyl magnesiumbromide followed by elimination and final deprotection with concomitant isomerisation of the double bond gave (−)-huperzine A in 10 steps with 17% overall yield.

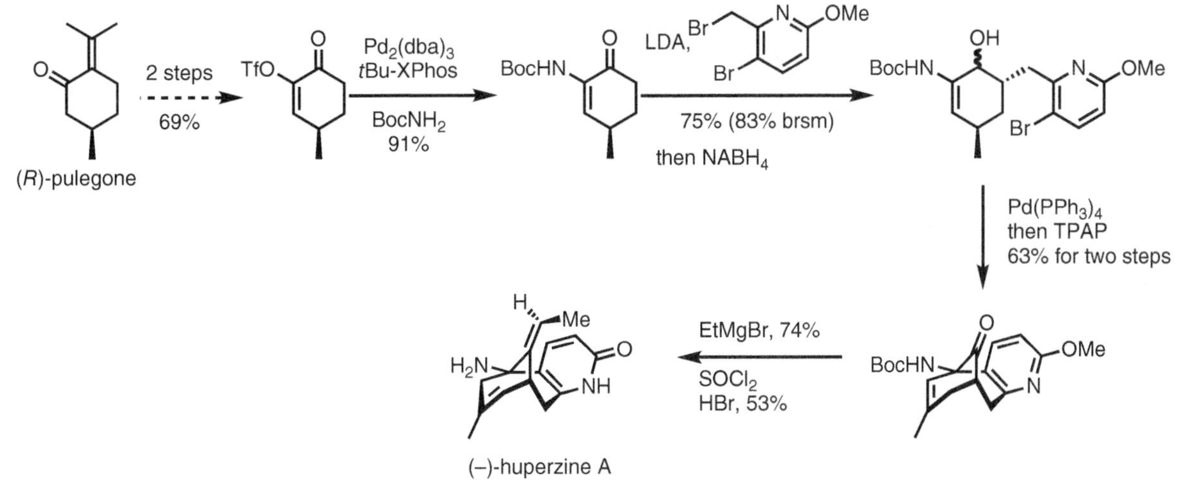

Scheme 10.14   *Asymmetric synthesis of (−)-huperzine A from (R)-pulegone*

10.1.3   Monoterpene Indole Alkaloids

Monoterpene indole alkaloids, or shortly indole alkaloids, comprise a group of plant-derived compounds which represents some 20 000 natural products. Many indole alkaloids have been traditionally used as medicinal agents (Figure 10.9). Vincamine promotes blood circulation in the brain, and is used for the treatment of stroke in many European countries. The commercial production of vincamine is based on both fermentation methods as well as total synthesis. Another indole alkaloid with remarkable medicinal impact is the bisindole alkaloid vinblastine. This was the first compound observed to heal Hodgkin's disease, a form of lymphoid cancer. Vincristine, a close structural relative of vinblastine, is claimed to cure up to 70% of acute lymphocytic leukemia cases in children. Vinblastine is produced by the Madagascan periwinkle, *Catharanthus roseus*. The plant is related to the *Vinca* plants, which are widely used as ornamental plants, and these alkaloids are sometimes confusingly referred to as the *Vinca* alkaloids. Vinblastine and vincristine inhibit cell propagation by inhibiting the formation of microtubules during mitosis. Ajmaline, a member of the sarpagine–ajmaline alkaloids [34], is used for the treatment of cardiac arrhythmia. Indole alkaloids are biosynthesised from tryptophan (indicated with blue color in the scheme).

The chemistry of indole alkaloids is strongly influenced by the unique reactions occurring to the indole moiety. Two of these reaction types, Pictet–Spengler (Scheme 10.15) and Bischler–Napieralski (Scheme 10.17) cyclisations, are employed in the biosynthesis in the formation of the polycyclic alkaloid skeleta, namely the condensation reactions of indole with aldehydes and amides. These two reactions form the basis of many biogenetic reactions and these reactions have also been very efficiently utilised in the so-called biomimetic syntheses of several indole alkaloids.

Condensation of indolylethylamine (tryptamine) with an aldehyde provides an example of the Pictet–Spengler cyclisation (Scheme 10.15) [35]. The reaction is initiated by the formation of the Schiff base, which after protonation (or

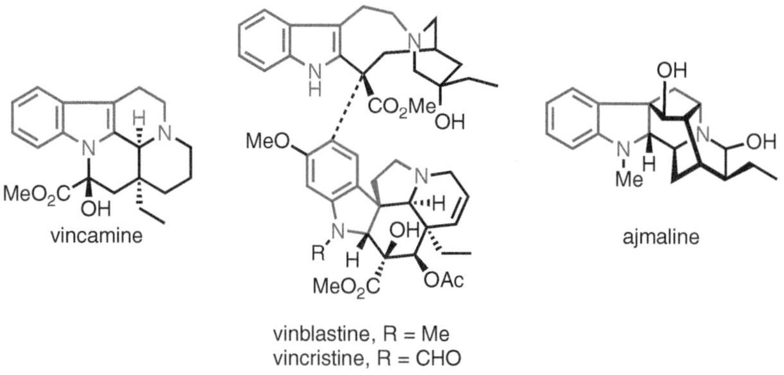

Figure 10.9 *Representative indole alkaloids*

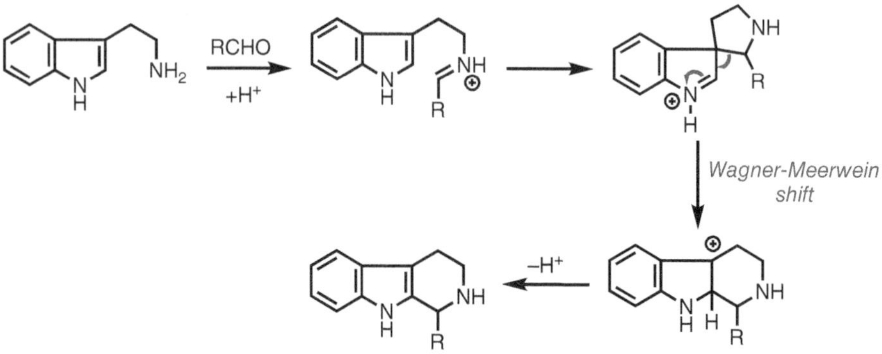

Scheme 10.15 *Pictet–Spengler cyclisation*

the action of a Lewis acid) gives an electrophilic iminium ion prone to cyclisation with the indole unit. Much work was expended on finding out whether the 3- or 2-position of the indole would perform as the nucleophile. The end result is 2-alkylation, although similar reactions with simple pyrroles tend to be directed to the 3-position. It was finally Jackson's isotope-labeling studies which confirmed that the reaction occurs by initial alkylation at C3, followed by a Wagner–Meerwein-type 1,2-shift [36].

If the newly forming C-ring contains a substituent at the carbon atom bearing the amino group (e.g. tryptophan), both *cis-* or *trans*-substituted products can be formed (Scheme 10.16). The stereochemical aspects of the cyclisation have been studied [37]. Under kinetic conditions, the attack of the indole nucleophile to form the spiroindolenine intermediate would favor the formation of the *cis*-substituted product (both substituents equatorial in the final six-membered ring). Substitution of the indole nitrogen, or the N_α nitrogen (R''' and R'', respectively) will increase allylic $A^{1,2}$ strain in the transition state leading to the *cis* product, and therefore the *trans* cyclisation is observed.

Reaction of the amide derived from tryptamine and a carboxylic acid can be induced to cyclise by using dehydrating conditions. This reaction is known as the Bischler–Napieralski reaction (Scheme 10.17). One usually employs phosphorus oxychloride, tosyl chloride, or similar (acidic) dehydrating agents and the mechanism has been rationalised as involving a chloroiminium intermediate. The initial product is the corresponding unsaturated tricycle, which can be hydrogenated to the saturated compound.

10.1.3.1 Biosynthesis of indole alkaloids

Biogenetically, indole alkaloids are derived from tryptophan and secologanin. The earliest proposals for the origins of indole alkaloids were put forth by William Henry Perkin and Robert Robinson in 1919 [38], when they postulated the role of tryptophan as the source of the indole moiety in these alkaloids. The monoterpene unit secologanin was the subject of much debate until late 1970s. The Barger–Hahn hypothesis suggested that the carbon

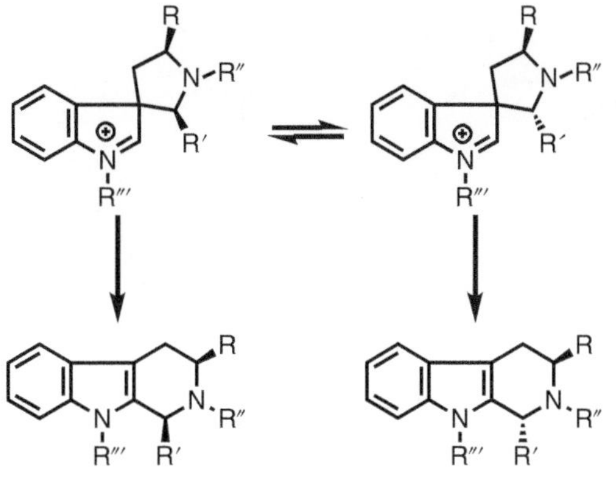

Scheme 10.16 *Spiroindolenine intermediate in Pictet–Spengler cyclisation*

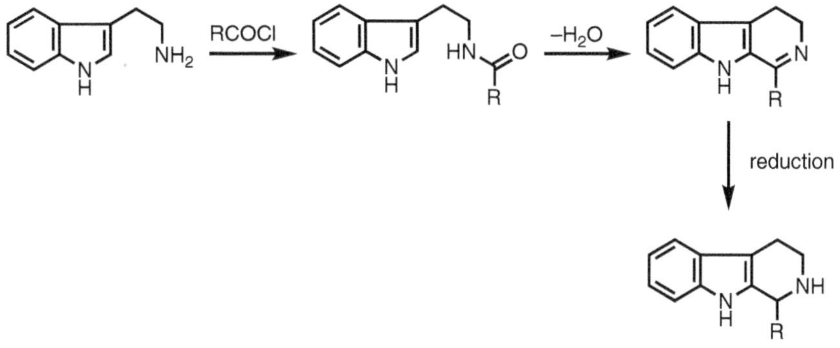

Scheme 10.17 *Bischler–Napieralski cyclisation*

skeleta of many of the indole alkaloids is derived from tryptophan, phenylalanine, and formaldehyde [39]. Woodward refined this model and substituted 3,4-dihydroxyphenylalanine for the aromatic precursor [40]. According to this theory, the *ortho*-hydroquinone moiety would facilitate the cleavage and further processing of the carbon structure to produce the *Strychnos* and other indole alkaloids. The similarity of the non-tryptophan moiety of the indole alkaloids with several non-alkaloidal glucosides led to the suggestion that the C_{10} unit arises from two mevalonate units [41]. This Thomas–Wenkert hypothesis was much disputed when first presented, but it was later shown by radioactive labeling studies to be a correct representation of the events. In recent years, plant cell cultures and isolated enzyme preparations have enabled the identification of several new intermediates in the biogenetic pathways, and thus our understanding of the biosynthesis of alkaloids has increased considerably in detail [42]. According to their biogenetic origins, the indole alkaloids can be divided into the following classes (Scheme 10.18):

1. secologanin unrearranged (*Corynanthé-Strychnos*, e.g. ajmalicine);
2. secologanin rearranged (route a, *Aspidosperma-Hunteria*, e.g. vincadifformine);
3. secologanin rearranged (route b, *Iboga*, e.g. catharanthine);
4. indole alkaloids not derived from secologanin;
5. bis-indole alkaloids.

In the biosynthesis of the *Corynanthé-Strychnos* alkaloids (Scheme 10.19), coupling of tryptophan (tryptamine) and secologanin through Schiff base formation and Pictet–Spengler cyclisation, with concomitant loss of the carboxyl function, gives strictosidine, the common intermediate for all indole alkaloids. After cleavage of the hemiacetal, a second Schiff base formation leads to an intermediate, which, after dehydration and reduction, gives geissoschizine. Alternatively, iminium ion–enamine tautomerization followed by cyclisation gives cathenamine.

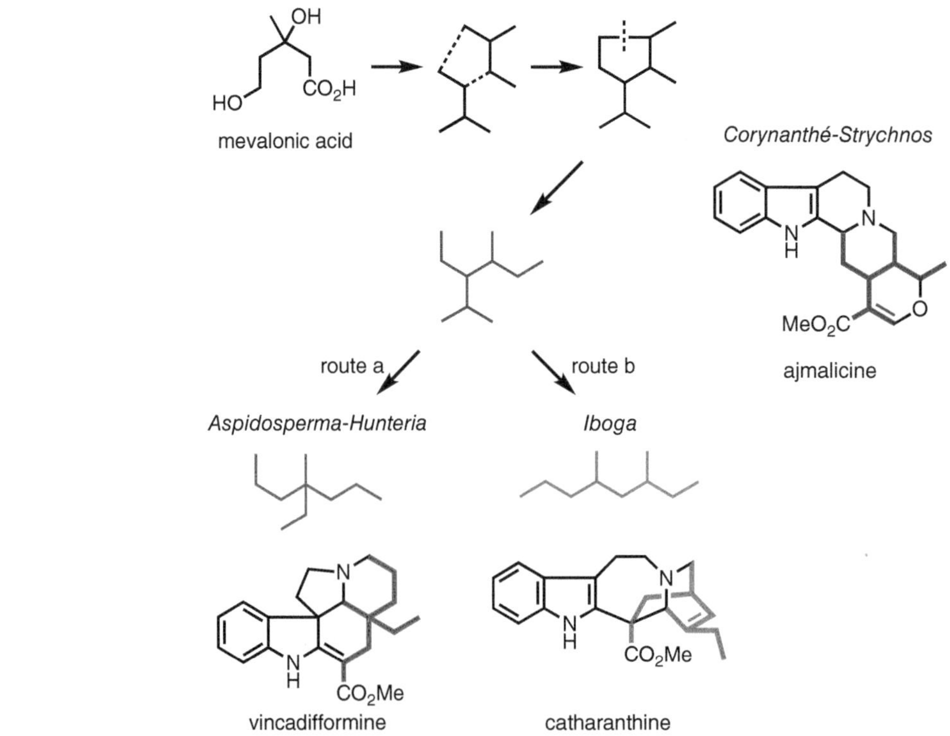

Scheme 10.18 *Thomas–Wenkert hypothesis for the biosynthesis of indole alkaloids*

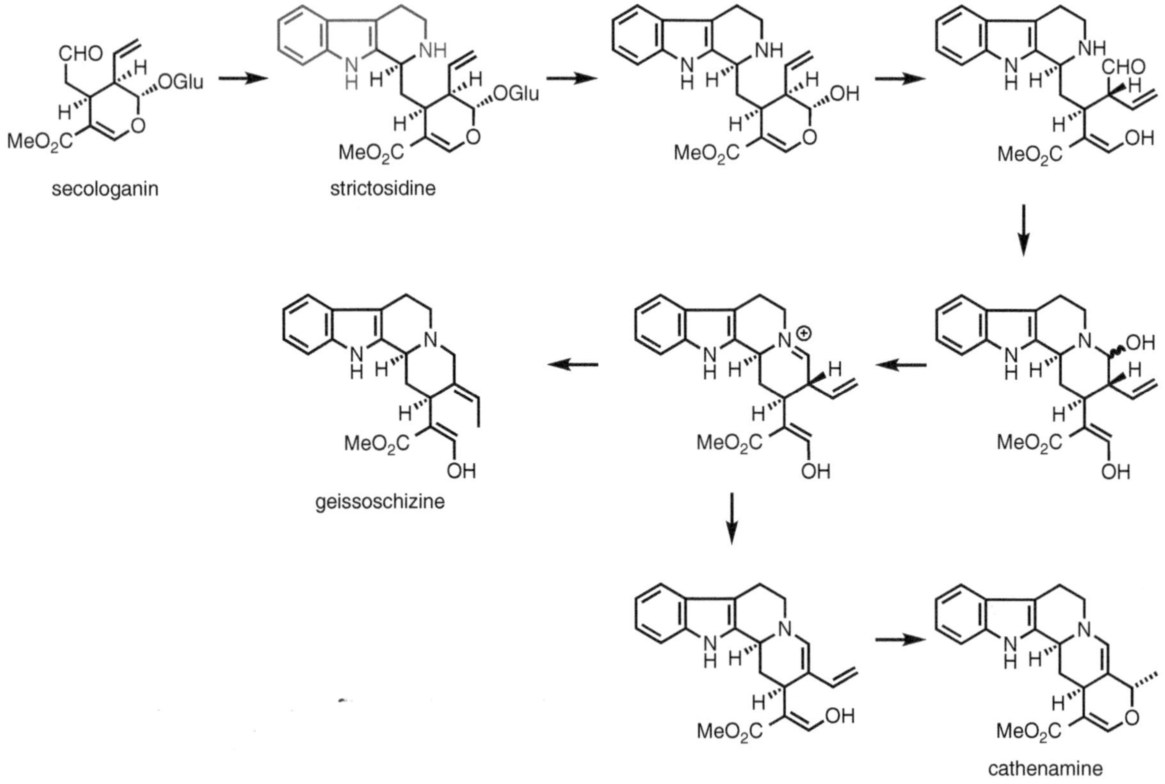

Scheme 10.19 *Biosynthesis of* Corynanthé-Strychnos *alkaloids*

The pentacyclic *Corynanthé-Strychnos* alkaloids can be grouped into structurally related classes: yohimbanes contain a carbocyclic E-ring; whereas in the heteroyohimbanes the E-ring is heterocyclic. The E-ring-cleaved alkaloids are secoyohimbanes (corynanes). The basic ring structures are shown in Figure 10.10, as well as the biogenetic ring numbering which is commonly used in alkaloid chemistry [43].

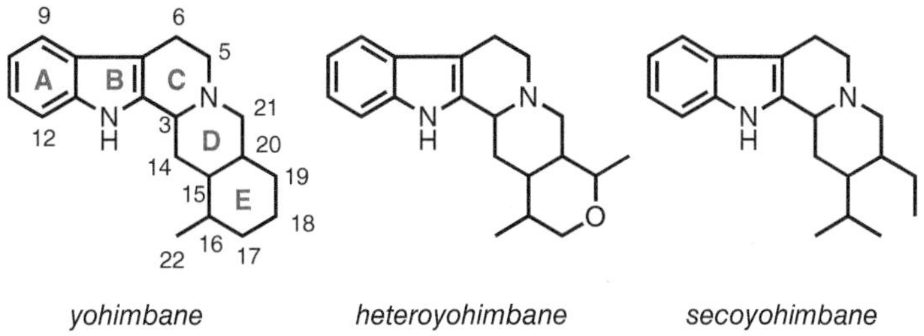

yohimbane   *heteroyohimbane*   *secoyohimbane*

Figure 10.10   *Yohimbane skeleta*

The three chiral centers in the ring carbon atoms of yohimbanes give rise to the possibility of eight stereoisomers. The hydrogen atom at C15 is always down (α), reducing the number of naturally occurring isomers to four. All four structures (Figure 10.11), yohimbane ($3\alpha,20\beta$), pseudoyohimbane ($3\beta,20\beta$), alloyohimbane ($3\alpha,20\alpha$), and epialloyohimbane ($3\beta,20\alpha$) have been found in naturally occurring alkaloids.

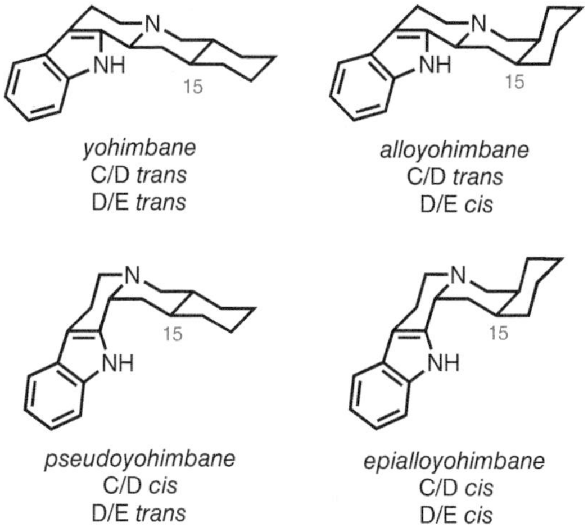

yohimbane
C/D *trans*
D/E *trans*

alloyohimbane
C/D *trans*
D/E *cis*

pseudoyohimbane
C/D *cis*
D/E *trans*

epialloyohimbane
C/D *cis*
D/E *cis*

Figure 10.11   *Yohimbane isomers*

The biogenesis of both the *Aspidosperma-Hunteria* and the *Iboga* alkaloids requires more deep-seated rearrangements. There is overwhelming evidence that *Strychnos* alkaloids are formed from strictosidine through the *Corynanthé* alkaloids (Scheme 10.20). These are the precursors for both *Aspidosperma* and *Iboga* alkaloids. The biogenesis proceeds via geissochizine, which, through oxidation of the indole ring, is cleaved and recyclised to preakuammicine. This, in turn, functions as the intermediate for both akuammicine and stemmadenine, the former leading to the *Strychnos* alkaloid skeleton, and the latter to the *Aspidosperma-Hunteria* alkaloids.

Formation of the *Aspidosperma-Hunteria* alkaloids from stemmadenine is depicted in Scheme 10.21. Isomerisation of the ethylidene double bond is followed by fragmentation with participation of the nitrogen lone pair (1,6-Grob–type fragmentation) to give secodine, which is the common intermediate for both types of alkaloids. A formal Diels–Alder reaction leads to the *Iboga* alkaloids. An alternative cyclisation mode of the enamine portion onto the acrylate unit gives rise to an iminium ion, whose further cyclisation gives the *Aspidosperma* skeleton.

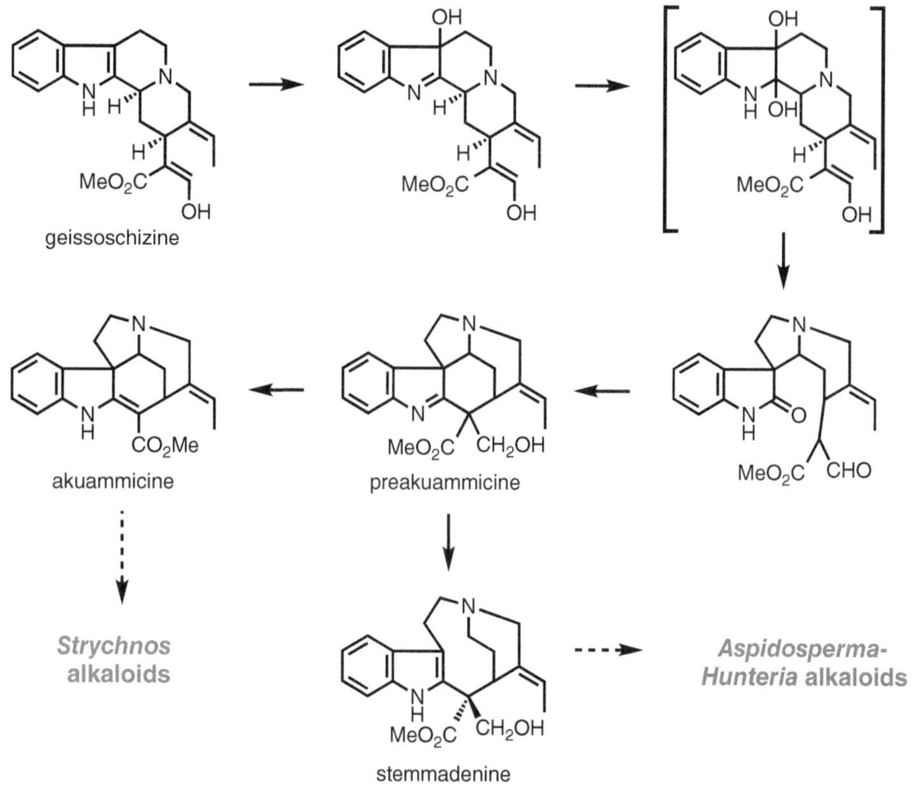

Scheme 10.20   *Biosynthesis of* Aspidosperma-Hunteria *and the* Strychnos *alkaloids*

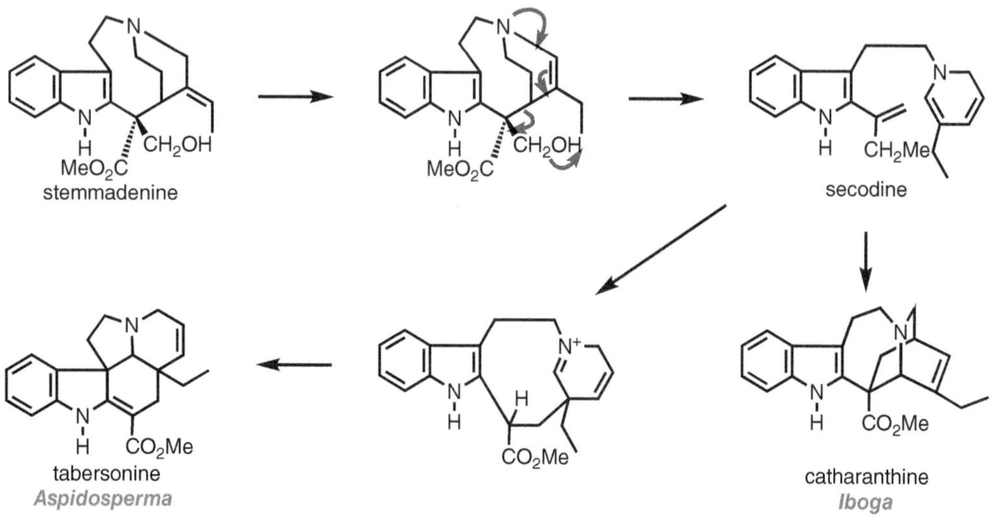

Scheme 10.21   *Biosynthesis of* Aspidosperma-Hunteria *and* Iboga *alkaloids*

Bisindole alkaloids are biosynthesised from their monomeric constituents (Scheme 10.22). Thus, catharanthine is oxidised by a cytochrome P450 enzyme to the *N*-oxide, which undergoes addition reaction with vindoline as the nucleophile with simultaneous Grob-type fragmentation of the isoquinuclidine ring. Hydration of the double bond gives vinblastine.

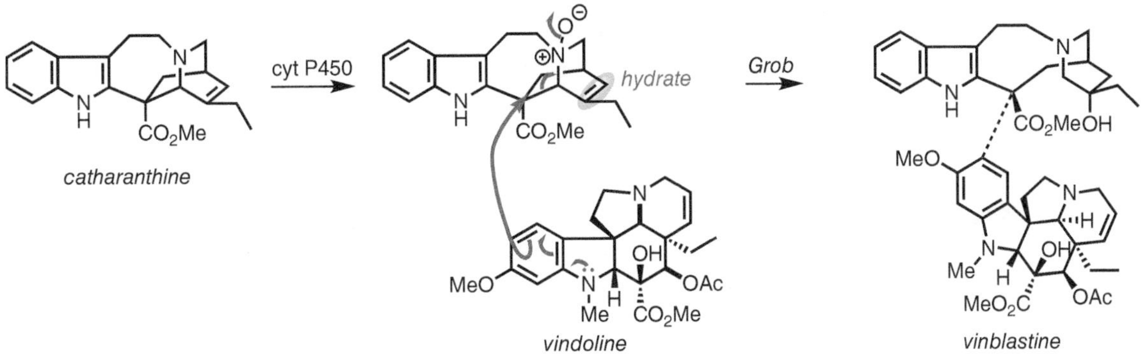

Scheme 10.22   *Biosynthesis of vinblastine*

Camptothecin is a quinoline alkaloid found in the barks of the Chinese *Camptotheca acuminata* tree ("Happy tree"). It and its close chemical relatives are the only known naturally-occurring DNA topoisomerase I inhibitors. Camptothecin is one of the newest chemotherapy drugs; it and some of its chemical relatives are in clinical trials to treat breast and colon cancers, malignant melanoma, small-cell lung cancer, and leukemia. Two camptothecin derivatives, irinotecan and topotecan (Figure 10.12), are currently on the market for the treatment of colon, ovarian, and lung cancers.

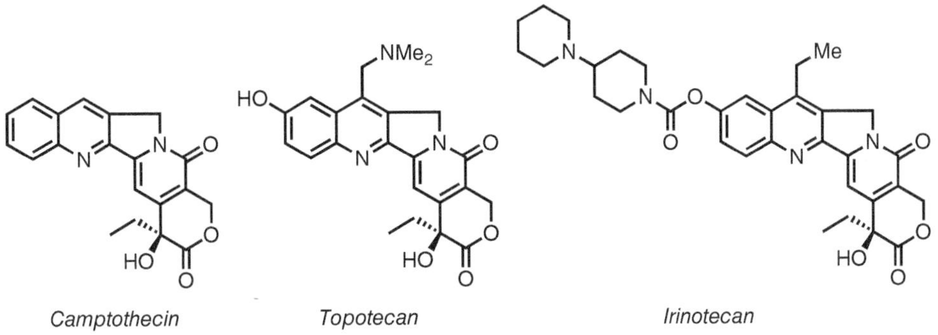

Figure 10.12   *Clinically useful tecans*

Camptothecin is biosynthesised (Scheme 10.23) through strictosidine followed by lactam formation [44]. The indole of strictosamide is oxidatively cleaved and the resulting ketoamide undergoes an intramolecular aldol condensation to pumiloside, a quinolone. The exact nature of the rest of the biosynthesis from pumiloside to camptothecin is not known, but it obviously involves subsequent reduction to deoxypumiloside, deglycosylation, and other metabolic conversions to lead to the end product, camptothecin.

10.1.3.2   *Asymmetric synthesis of indole alkaloids*

Since the Pictet–Spengler-type cyclisation reactions are highly stereoselective, most of the generally used strategies for asymmetric synthesis of indole alkaloids rely on this powerful process. The asymmetric information can be introduced either through the use of tryptophan, or via an aldehyde containing the desired chirality.

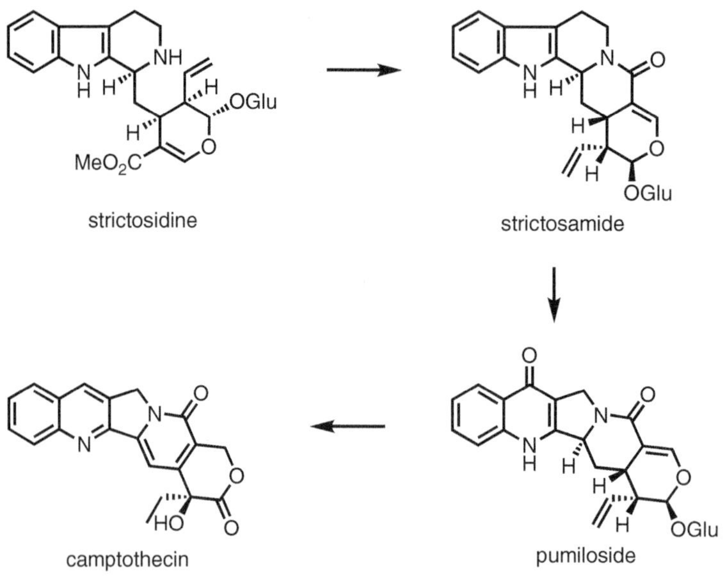

Scheme 10.23   Biosynthesis of camptothecin

Our first example shows the use of enantioselective deprotonation as the source of asymmetric induction (Scheme 10.24). Deprotonation of the bicyclo[3.3.0]octane-3,7-dione monoketal with chiral lithium amide base followed by alkylation gave the enantiomerically enhanced methylated product [45]. Baeyer–Villiger oxidation of the cyclopentanone ring led to the E-ring of the heteroyohimbanes. DIBAL-reduction of the lactone to the lactol followed by acid-catalysed dehydration sets the unsaturation in the E-ring precursor and also liberates the ketone. Cleavage of the remaining five-membered ring was finally achieved through conversion of the ketone to the enol ether followed by oxidative scission of the enol double bond.

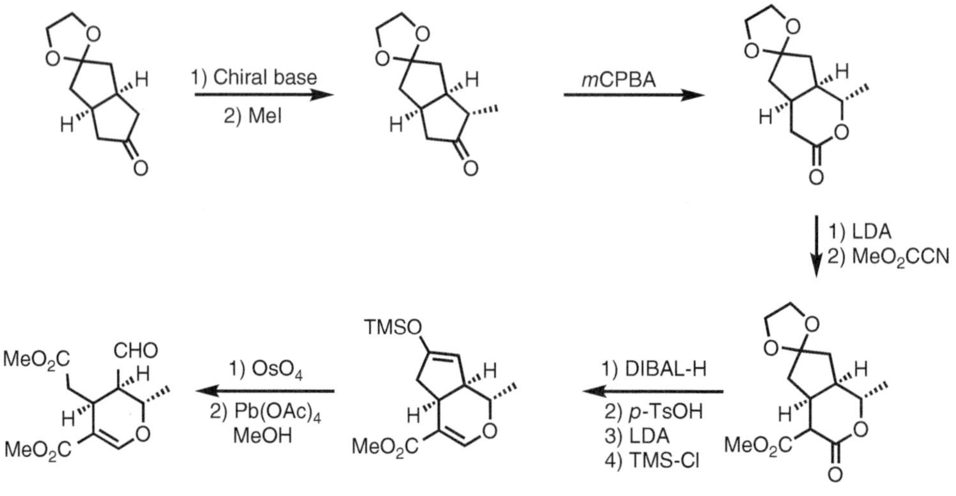

Scheme 10.24   Enantioselective deprotonation in indole alkaloid synthesis

The synthesis of ajmaline–sarpagine alkaloids has prompted many synthetic studies on using tryptophan as the source of chirality for the indole alkaloid skeleton. Many research groups have been working on this intriguing problem, and that of *James M. Cook* (1945–) has provided a clear-cut optimised solution to a wide range of

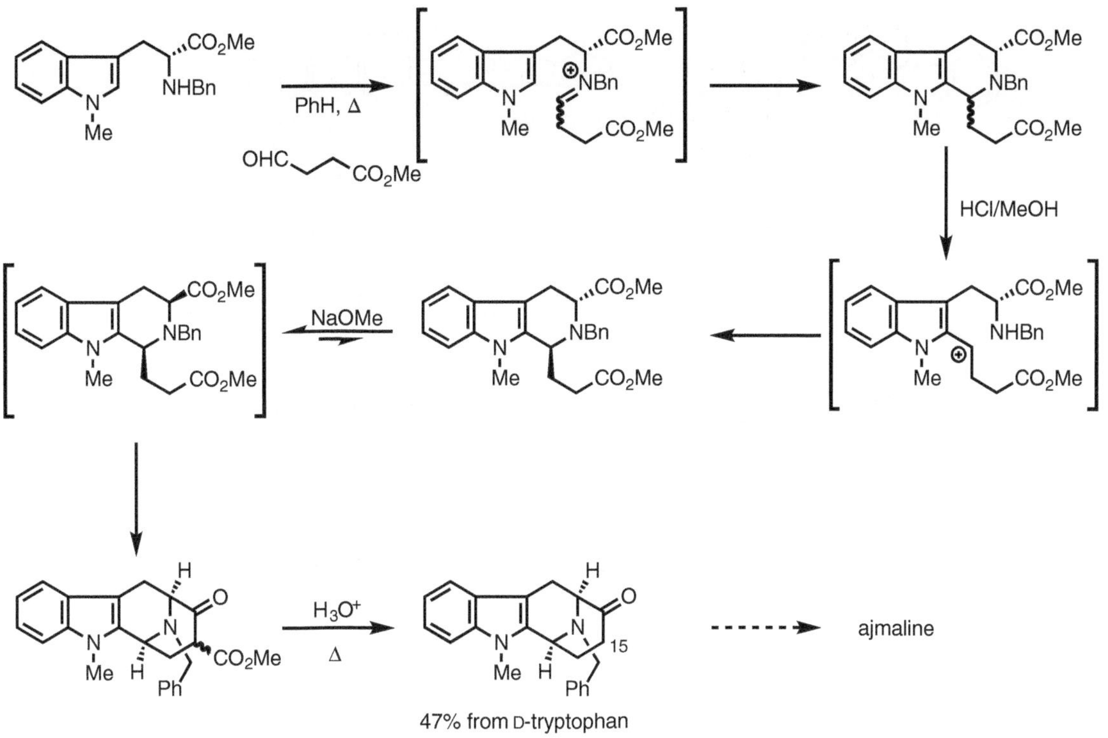

Scheme 10.25 *Pictet–Spengler reaction in ajmaline synthesis.*

targets (Scheme 10.25) [46]. D-Tryptophan was first condensed with succinic monoaldehyde in a Pictet–Spengler reaction to give a mixture of diastereomers at C3. The stereochemistry at C3 could be corrected by acid-catalysed ring-opening/ring-closure of the C-ring to give the *trans* diastereomer as the sole product. Base-induced intramolecular Dieckmann condensation first epimerised the original D-tryptophan stereocenters before ring closure to the key tetracycle. Acid hydrolysis and decarboxylation gave the versatile tetracyclic ketone. Cook has used this intermediate (or its structural relatives) for the synthesis of not only ajmaline, but a large number of related alkaloids, as well.

A very similar strategy was used by *Philip D. Magnus* (1943–) in a total synthesis of koumine (Scheme 10.26), the principal medicinal constituent of the Chinese plant *Gelsemium elegans* [47]. Pictet–Spengler cyclisation of the protected tryptophan with ketoglutaric acid gave the *trans* product as the major isomer, as expected. Dieckmann cyclisation was preceded by epimerisation at the carboxylate-bearing carbon, to give the tetracyclic intermediate after appropriate adjustment of functionalities. Installation of the propargylic acid side chain completed the assembly of the skeleton for final ring closures. Ring closure of the fifth ring was achieved by a Michael-type process triggered by pyrrolidine and trifluoroacetic acid (TFA) in refluxing benzene. Homologation of the ketone using the Tebbe reagent and adjustment of the oxidation states gave (*E*)-(+)-(16*S*)-12-benzyl-10-desoxy-18-hydroxysarpagine. Removal of the benzyl protecting group was followed by carbomethoxylation of the quinuclidine nitrogen. This intermediate underwent spontaneous cyclisation closely resembling the proposed biosynthetic route to assemble the koumine skeleton [48].

A highly versatile synthetic strategy has been developed for the synthesis of sarpagine alkaloids (Scheme 10.27) [49]. The 3-hydroxypyridinium salt was treated with base to generate an ylide, which underwent a [5 + 2] oxidopyridinium cycloaddition with Aggarwal's chiral ketene equivalent. The *S*-oxides were removed and the enone reduced to a ketone in 58% overall yield over three steps. Intramolecular Heck reaction then established the missing third ring. In only five straightforward steps this intermediate was transformed to the key tricyclic ketone, which underwent facile Fischer indole synthesis to give (+)-normacusine, (+)-lochnerine, (+)-affinisine, as well as ten non-natural synthetic sarpagine derivatives in good to excellent yields.

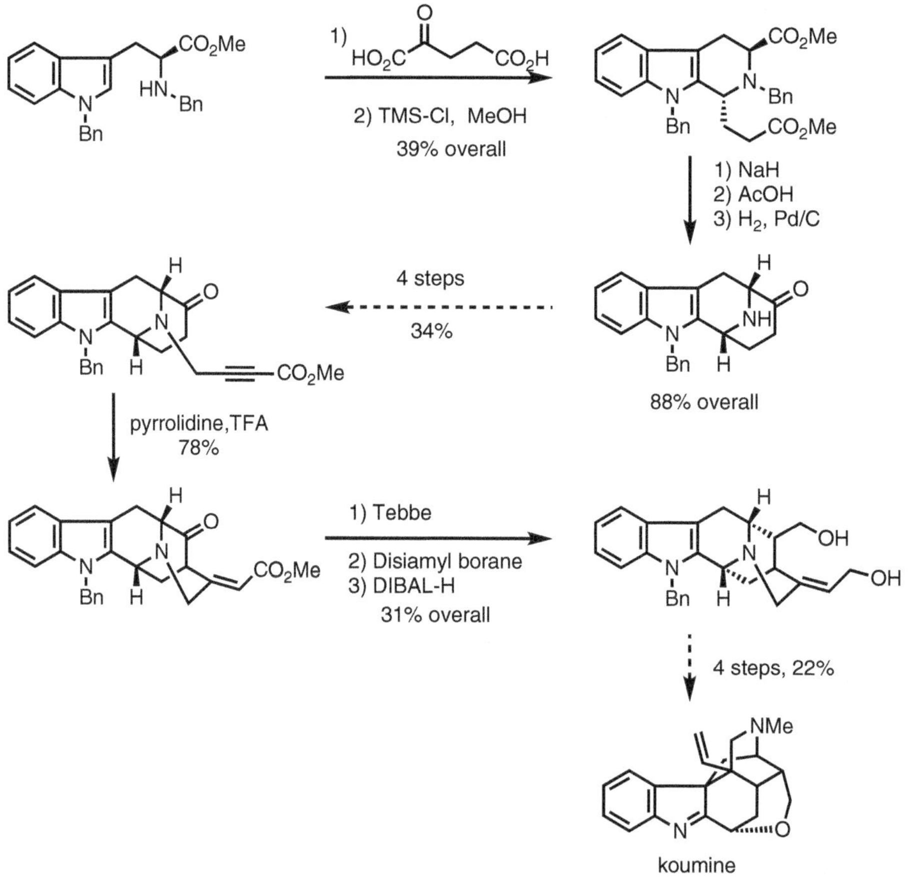

Scheme 10.26 *Koumine synthesis through Pictet–Spengler cyclisation*

Two organocatalytic syntheses of yohimbine are shown in Scheme 10.28. They are both based on an initial asymmetric Pictet–Spengler reaction to introduce chirality in the molecule, and later intramolecular Diels–Alder reaction to set up the three stereocenters in the DE rings. Historically, the first asymmetric synthesis of yohimbine was achieved by *Eric Jacobsen* (1960–, Harvard) [50], using a chiral thiourea catalyst. The Pictet–Spengler reaction gave a good 94% *ee*. The second example by *Henk Hiemstra* (1952–) utilised chiral binolphosphoric acid as the Pictet–Spengler catalyst [51]. In the event, the crucial reaction gave the product in 84% *ee*. The intermediate was converted to the IMDA precursor in a few straightforward steps.

Chiral auxiliary-based strategy was employed by *Albert I. Meyers* (Colorado State University, 1932–2007) in the synthesis of corynantheidol (Scheme 10.29) [52]. The Meyers group has developed the chemistry of chiral amidines, especially in asymmetric alkylation reactions [53]. Metalation of the β-carboline equipped with the leucine-derived formamidine auxiliary, followed by trapping of the anion with chloroacetonitrile, gives rise to the C3 chiral center.

"STRYCHNINE! The fearsome poisonous properties of this notorious substance attracted the attention of sixteenth century Europe to the *Strychnos* species which grow in the rain forests or the Southeast Asian Archipelagos and the Coromandel Coast of India, and gained for the seeds and bark of those plants a widespread use for the extermination of rodents, and other undesirables, as well as a certain vogue in medical practice – now known to be largely unjustified by any utility [13]".

Strychnine has certainly concerned the minds of synthetic chemists for the past century. Woodward's classical racemic synthesis, which was reported in a preliminary communication two years prior to the confirmation of strychnine

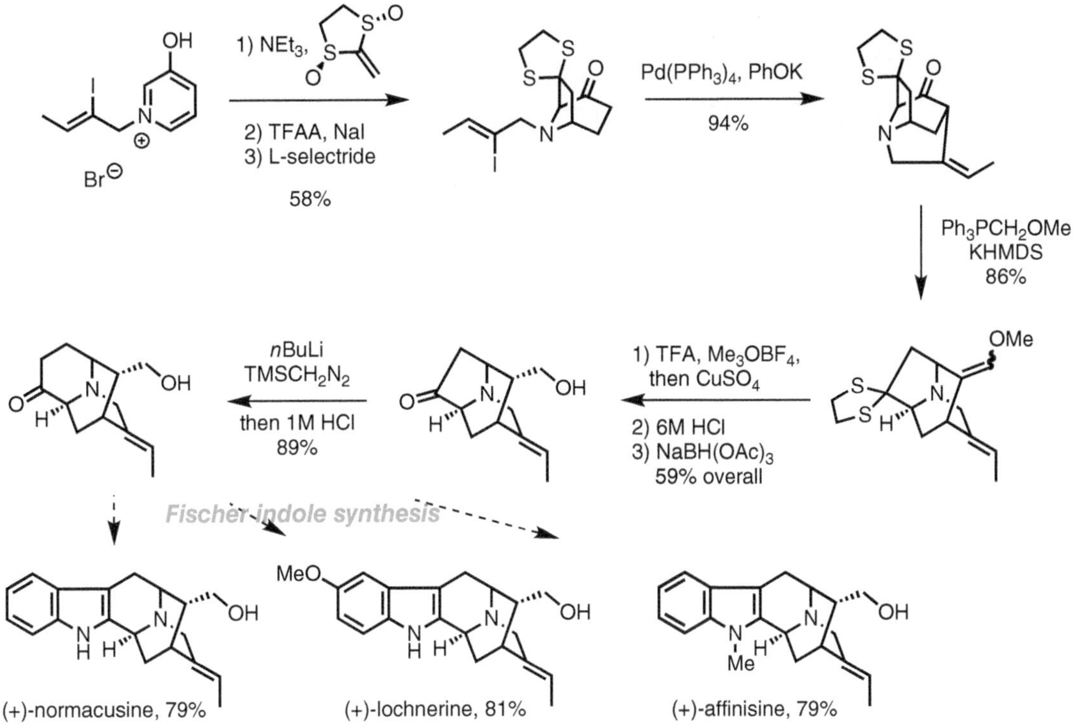

Scheme 10.27 Divergent asymmetric synthesis of sarpagine alkaloids

structure by X-ray crystallography, has been followed by a number of other syntheses, including the first asymmetric synthesis by Overman in 1993 [54].

A strategically different synthesis was recently reported by *David W. MacMillan* (1968–) (Scheme 10.30) [55]. The synthesis is based on an efficient organocatalytic cascade construction of the tetracyclic core employing an imidazolidinone-mediated formal Diels–Alder/conjugate-addition cascade. This sequence delivered the key chiral intermediate in 82% chemical yield and 97% *ee*. Only another eight steps were needed to construct strychnine in 6.4% overall yield.

The synthesis of camptothecin has been intensively studied because of the commercial value of these compounds. Gilbert Stork showed in 1971 that the Friedländer condensation is a viable route to camptothecin [56]. Since this pioneering synthesis, the enantioselective synthesis of the pyridine lactone has become a major challenge. Although the commercial syntheses often proceed through enzymatic resolution steps, here we illustrate a short synthesis through asymmetric dihydroxylation (Scheme 10.31) [57]. The starting lactone was converted to a cyclic enol ether, which on Sharpless asymmetric dihydroxylation with DHDQ-PYR ligand, gave the enantiomerically enriched (84% *ee*) lactol in acceptable yield. Simple oxidation to lactone, cleavage of the ketal protection and standard Friedländer condensation gave camptothecin.

10.1.4 Ergot Alkaloids

Ergot alkaloids are indole alkaloids, which are biosynthesised from L-trypthophan and a single unit of dimethylallyl pyrophosphate [58]. The best known producers are fungi of the phylum Ascomycota, e.g. *Claviceps*, *Epichloë*, *Penicillium*, and *Aspergillus* species.

Ergine (Figure 10.13) is the dominant alkaloid in the psychedelic seeds of *Rivea corymbosa* (ololiuhqui), *Argyreia nervosa* (Hawaiian baby woodrose), and *Ipomoea tricolor* (morning glories, tlitliltzin). In the Nahuatl language of

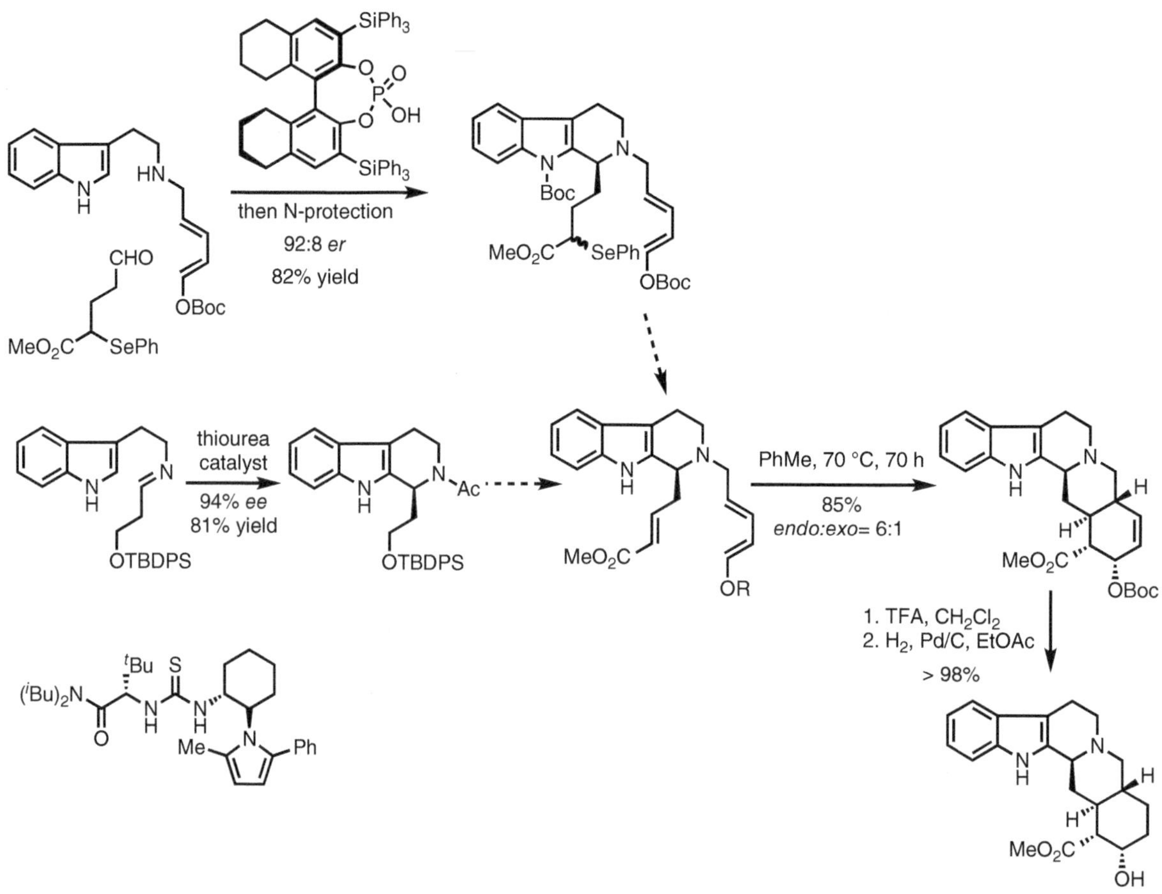

Scheme 10.28   *Organocatalytic asymmetric Pictet–Spengler cyclisations to yohimbine*

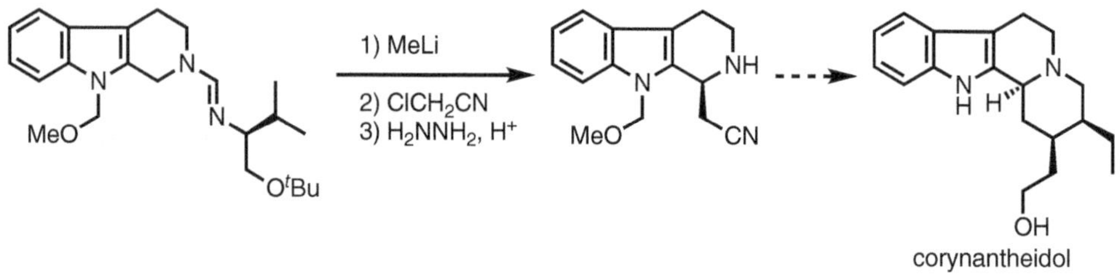

Scheme 10.29   *Chiral auxiliary route to* Corynanthé *alkaloids*

Aztecs, the word *ololiuqui* means "round thing," and refers to the small, brown, oval seeds of the morning glory. The plant itself is called *coaxihuitl*, "snake-plant." The seeds are called *semilla de la Virgen* (seeds of the Virgin Mary).

The ergot alkaloids gain their name from the first known producer, the fungus *Claviceps purpurea* that grows upon rye and other grains. Ergotamine is a typical example of a peptide alkaloid, although most of the known analogues of ergot alkaloids are their non-peptide derivatives. The ergot alkaloids are highly toxic, and as early as 600 BC, an Assyrian tablet described a 'noxious pustule in the ear of rye.' In one of the sacred books of Parsees, ergot is also alluded to: 'Among the evil things created by Angro Maynes are noxious grasses that cause pregnant women to drop the womb and die in childbed.' The Greeks and Romans rejected rye, and it was only in the Middle Ages when rye

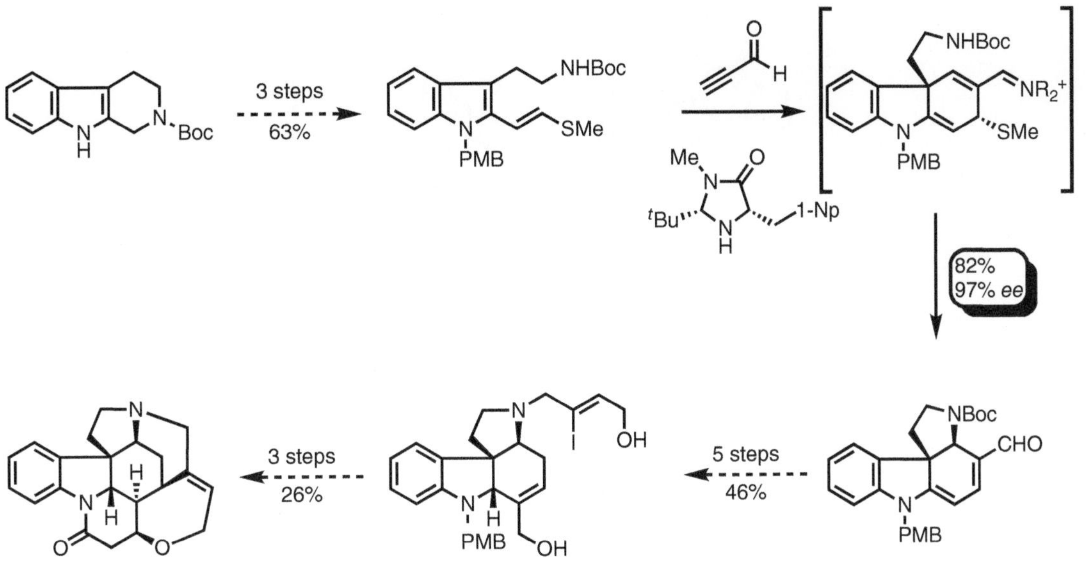

Scheme 10.30 *Short organocatalytic cascade synthesis of strychnine*

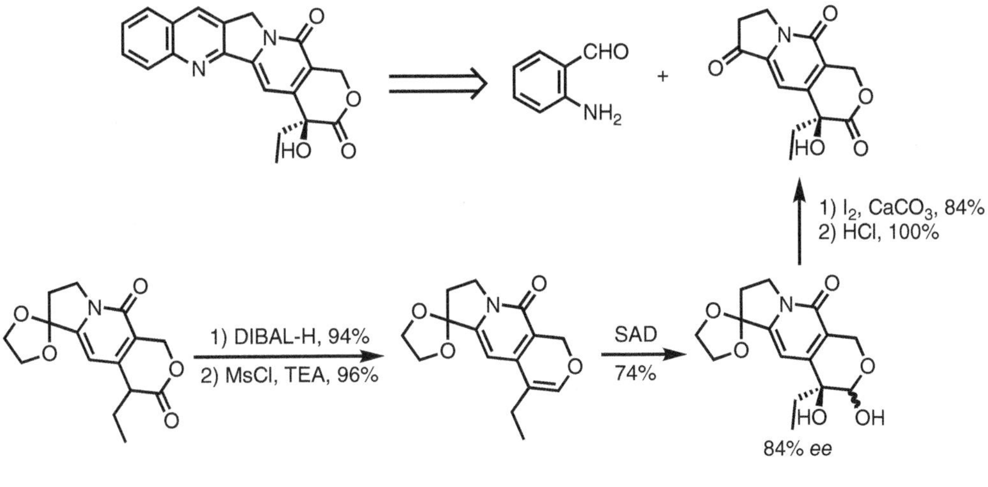

Scheme 10.31 *Asymmetric dihydroxylation in the synthesis of camptothecin*

was introduced into Southwest Europe. Strange epidemics were described, such as gangrene of the limbs. In severe cases, the tissue became dry and black, and the mummified limbs separated off without bleeding. The limbs were said to be consumed by the Holy Fire, and the disease was commonly called St. Anthony's Fire. Ergot was also used as an obstetrical herb, producing pains in the womb and thereby speeding childbirth. The interest towards Ergot alkaloids started to widen in the 1920s, and this led to the development of lysergic acid diethylamide, LSD, for the treatment of schizophrenia. LSD and many other agents acting similarly are hallucinogens, which in small doses cause psychedelic effects. Although LSD has been defended because it does not cause physiological changes, it is evident that its effects on the psyche can be irreparable. It is known that a person who has used LSD only once, has taken a 'short trip' decades later, without having taken the compound ever again!

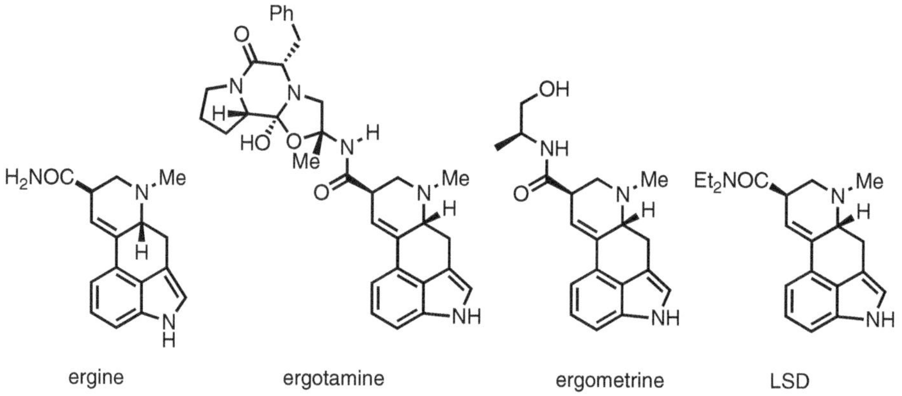

Figure 10.13 *Ergot alkaloids*

10.1.5 Quinine

Approximately 1000 members of this broad class of quinoline alkaloids are currently known. Their structures vary quite widely, but biosynthetically they are mostly derived from phenylalanine or tryptophan. The proposed biogenetic pathway leading to quinine exemplifies these sometimes quite remarkable transformations (Scheme 10.32) [59]. Oxidation of the pregeissoschizine derivative yields an iminium ion, which is cleaved and recyclised to the quinuclidine. Further oxidative cleavage of the pyrrole ring of the indole unit gives an amino aldehyde which ring closes to quinine.

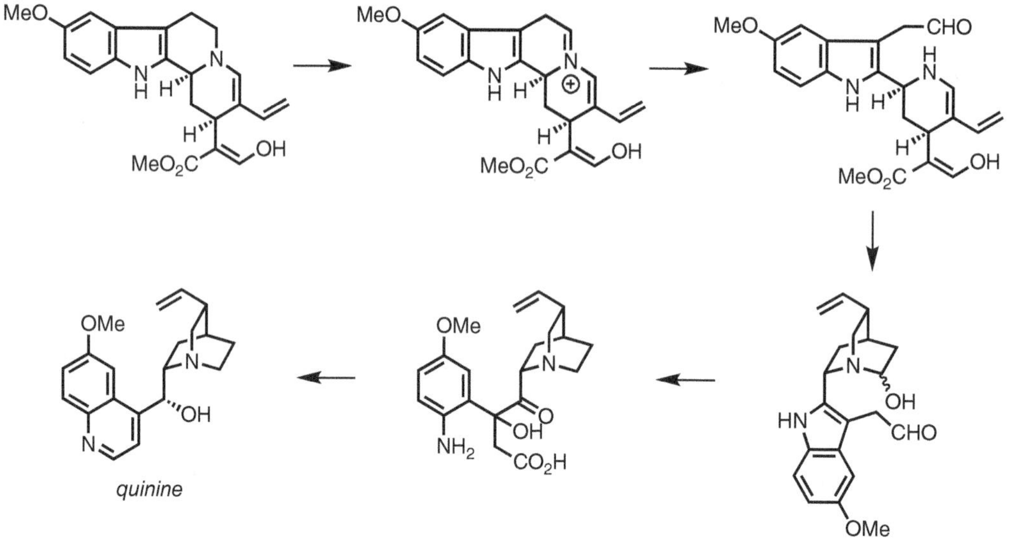

Scheme 10.32 *Biosynthesis of quinine*

Quinine was originally introduced to European and Western medicine by Spanish sea-farers and Jesuit monks. A popular story tells that the bark of the quina tree (*Cinchona officinalis*) was used to treat fevers and tertians (malaria fevers occurring every three days), and that the bark was used in 1638 to treat Countess *Ana de Osorio*, the wife of the 4th Count of Chinchón who was the viceroy to Peru. The bark extract was the sole source of cure for malaria for nearly two centuries, until *Pierre Joseph Pelletier* (1788–1842) and *Joseph Bienaimé Caventou* (1795–1877) were able to isolate the pure compounds quinine and cinchonine from cinchona in 1820. The structure of quinine was proposed by the German chemist *Paul Rabe* (1869–1952) at the beginning of the twentieth century [60].

Quinine has had an enormous impact on human well-being, and therefore it is surprising how little we know about its structure–activity relationships. Quinine and its derivatives have received much attention lately in enantioselective organocatalysis as well as phase-transfer catalysis, and again the lack of an efficient synthetic route for quinine and its analogues has prevented detailed molecular optimisations of these catalysts. The molecule has held a strong resistance to synthetic access [61]. As early as 1856, the then 18-year-old William Henry Perkin decided to synthesise quinine simply by oxidatively dimerizing allyl toluidine ($2\,C_{10}H_{13}N + 3\,O = $ quinine $C_{20}H_{24}N_2O_2 + H_2O$). Of course, the end product was not quinine, but the quest of an enthusiastic young mind was eventually rewarded: he discovered mauve and other aniline dyes, which made a permanent change in our daily lives [62]. The compound was actually synthesised for the first time in 1944 by Woodward and Doering (Scheme 10.33) [63], but the first asymmetric synthesis remained elusive until Gilbert Stork's achievement in 2001 [64].

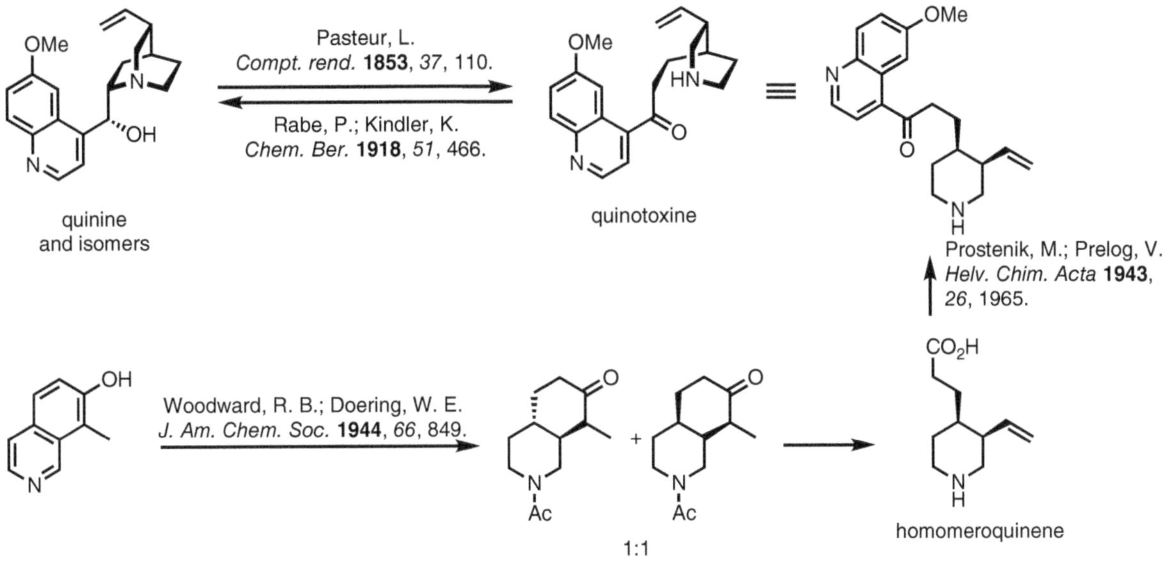

Scheme 10.33 *Evolution of the synthesis of quinine*

Jacobsen has presented a synthetic route where the initial chiral information is introduced by a catalytic asymmetric addition of a malononitrile onto an α,β-unsaturated imide (Scheme 10.34) [65]. The conjugate addition proceeded in 91% chemical yield and 92% *ee*. The intermediate was then converted to a boronic acid derivative, which was coupled with a bromoquinoline. Final elaboration of the alkene precursor interestingly relied on dihydroquinidine-based Sharpless dihydroxylation.

Isothiocineole can be prepared by the solvent-free reaction between elemental sulfur and limonene in the presence of γ-terpinene at 110 °C. Simple distillation gives essentially pure isothiocineole in 36% yield and with high *er* (99:1). Both (+)- or (−)-isothiocineole is readily available on a large scale. Aggarwal applied this chiral sulfide in an asymmetric sulfur ylide-mediated epoxidation in the synthesis of quinine (Scheme 10.35) [66]. The known quinolinecarboxylic acid was first reduced to a primary alcohol, then converted to the pyridinium carbamate and the iminium ion again reduced to give the 1,2-dihydropyridine primary allylic alcohol in 54% overall yield. Activation of the alcohol as a triflate followed by reaction with (+)-isothiocineole gave the sulfonium salt. After treatment with base, the ylide reacted with the chiral meroquinene aldehyde to give the epoxides as an 89:11 mixture of diastereomers. Subsequent treatment with CsF under microwave conditions triggered the reaction cascade involving deprotections of the silyl carbamates and ring closure. Final oxidation was achieved simply by stirring the crude reaction mixture under oxygen.

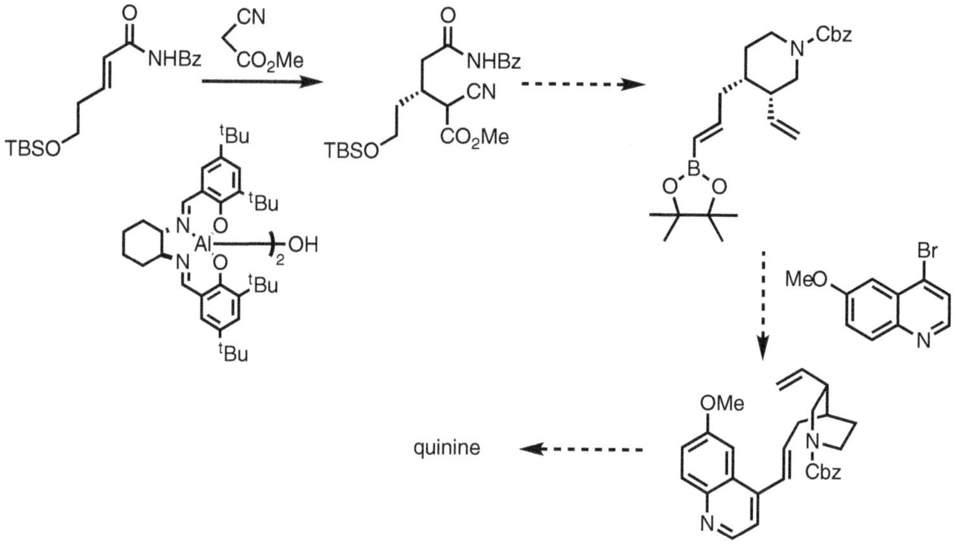

Scheme 10.34   *Catalytic asymmetric conjugate addition in the synthesis of quinine*

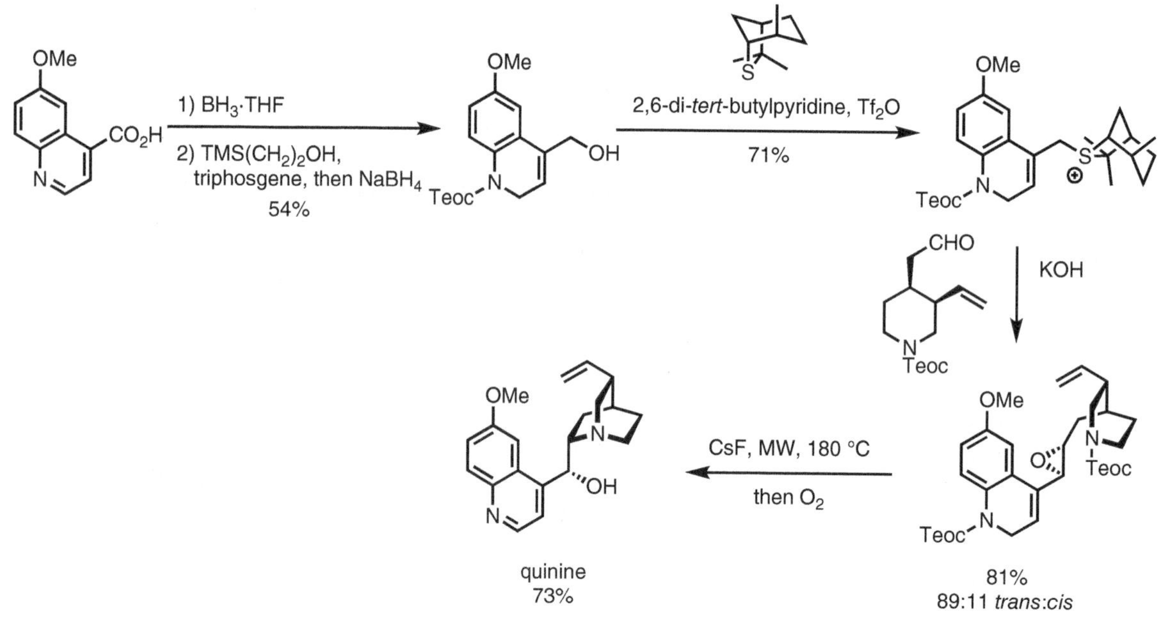

Scheme 10.35   *Asymmetric sulfur ylide-mediated epoxidation in the synthesis of quinine*

10.1.6   Morphine

Morphine alkaloids and structurally rather similar hasubanan and homomorphine alkaloids (e.g. androcymbine) also belong to the isoquinoline alkaloids (Figure 10.14). Morphine alkaloids are known as opium alkaloids due to their natural origin: the seeds of the opium poppy, *Papaver somniferum*. The Sumerians (3000 BC) called the plant *hul gil* = plant of happiness, and the latex of opium poppy was used in ancient Greece (twelfth century BC). The dried latex and the seed capsules contain some two dozen alkaloids, of which morphine covers nearly 10%. Although pure morphine has been available since 1803 (isolated by *Friedrich Sertürner*), its chemical structure remained elusive until 1925

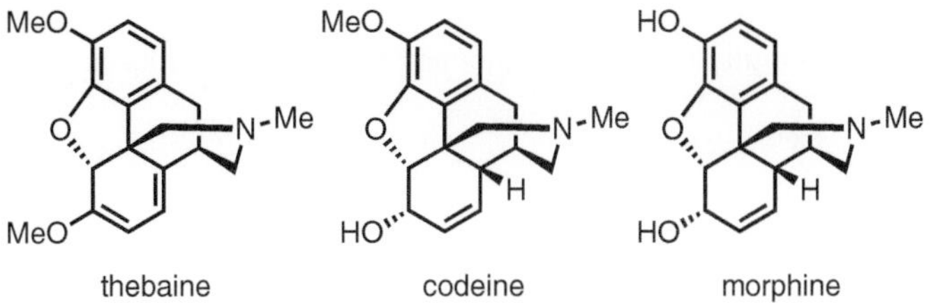

Figure 10.14 *Morphine alkaloids*

when Sir Robert Robinson and *Heinrich Otto Wieland* (1877–1957) independently proposed the structure for codeine (morphine methyl ether) and thereby implicitly for morphine [67], which was eventually confirmed by total synthesis by Gates in 1956 [68].

Morphine itself is a powerful painkiller, and its action is based on its ability to bind to specific opiate receptors both within and outside the central nervous system. Distinct opioid receptor classes elicit different actions: μ_1-receptors mediate euphoria, confusion, dizziness, and nausea, whereas activation of the μ_2-receptors is followed by respiratory depression, miosis, and urinary retention. Activation of the δ-receptor leads to spinal analgesia, cardiovascular depression, and decreased brain and heart oxygen demand. κ-Receptors also mediate spinal analgesia and additional dysphoria and psychotomimetic effects.

It has been suggested [69] that the tyramine ring of morphine and the tyrosine ring of Leu-enkephalin coincide, giving the alkaloid good affinity for the receptor. The principal action is through inhibition of adenylyl cyclase, which produces cyclic AMP (cAMP), a common second messenger. Reduced levels of cAMP will be compensated by the cell by increased production of acetylcholine. This will still cause the production of sufficient levels of cAMP to maintain normal functioning of the cell. However, if the opiate treatment has been used for a prolonged period and is then suddenly interrupted, the acetylcholine-induced cAMP production cannot accommodate rapidly enough, and the suddenly increased cAMP levels can trigger a multitude of withdrawal symptoms, which may prove fatal. The dangerous addictiveness of morphine in continued use is the reason why a large number of analogues have been synthesised and tested for their analgesic ability with the hope of lesser side effects. Levorphanol, a morphinan with the furan ring cleaved, is a typical example of such narcotic analgesics, and etorphine, an oripavine type compound, is among the most potent opiate agonists (Figure 10.15). Etorphine has a 10 000 times higher potency than morphine, and therefore it is used primarily for the immobilisation of large animals like elephants. Oxymorphone and buprenorphine are also strong opioid

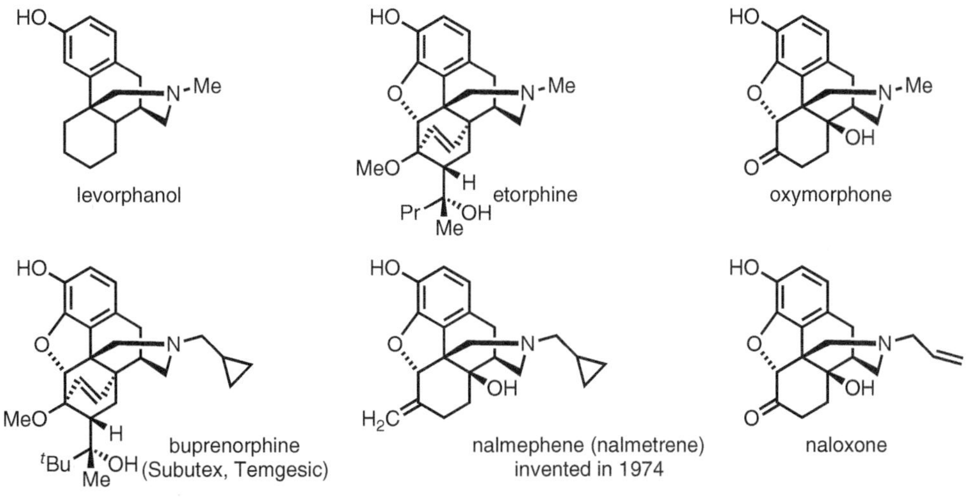

Figure 10.15 *Strong opiate analgesics*

analgesics. Buprenorphine and nalmephene are used to treat withdrawal syndromes and addictions. Naloxone is used to block the effects of opioids. It is commonly used to counter decreased breathing in opioid overdose.

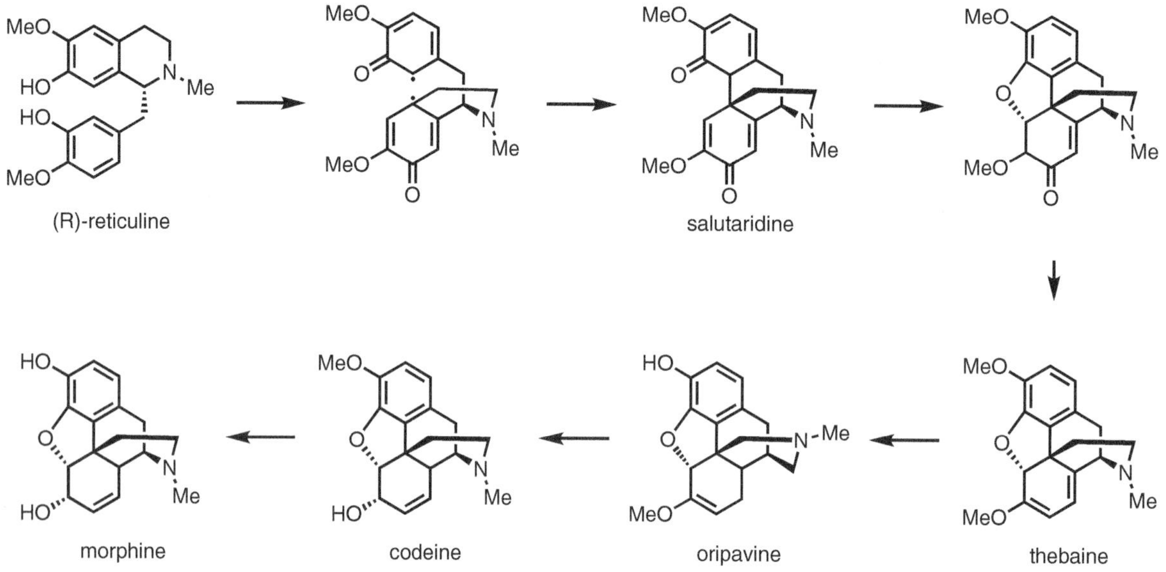

(R)-reticuline   salutaridine

morphine   codeine   oripavine   thebaine

Scheme 10.36   *Biosynthesis of morphine*

Morphine is biosynthesised from reticuline (Scheme 10.36) which is oxidised to a phenolic diradical species. Different modes of cyclisation will lead to a number of other alkaloids, but the *para*, *ortho*-coupling leads to a bridged ring system. The furanoid ring is formed by conjugate addition of the enol to the dienone. Adjustment of the oxidation levels and methylations then lead to thebaine, codeine, and eventually morphine.

Overman's synthesis of morphine relies on an efficient intramolecular cyclisation of an allylsilane iminium ion to furnish the central isoquinoline ring (Scheme 10.37) [70]. The asymmetric information was introduced into the allyl-silane through its synthesis from 2-allylcyclohexenone, itself readily obtained from *ortho*-anisic acid. CBS-reduction with an *R*-oxazaborolidine catalyst gave the allyl alcohol with high enantioselectivity. The alcohol was converted to a carbamate, and the alkene oxidised to a diol and protected. In preparation for the crucial ring closure, the allyl carbamate was transformed to the silane, and the nitrogen was introduced as a dibenzosuberylamine (DBS) group. Iminium ion formation with an arylacetaldehyde followed by ring closure provided the octahydroisoquinoline in 91% *ee*. Further elaboration with standard operations then led to the first asymmetric synthesis of morphine.

Trost's asymmetric synthesis of codeine and morphine serves as an example of a synthesis based on the introduction of chirality by asymmetric allylic alkylation of a phenol (Scheme 10.38) [71]. Thus, the racemic allylic carbonate was treated with palladium in the presence of chiral bis-phosphine ligand to give the chiral phenol ether in 82% *ee*. The next key step was the Heck cyclisation of the derived nitrile to the benzofuran intermediate, which again was Pd catalysed. Heck vinylation of the derived vinyl bromide effected the closure of the fourth ring. Note how the double bond of the original racemic allylic carbonate 'walks' around the cyclohexane ring to facilitate the metal-mediated ring-closure steps. The final heterocyclic ring was constructed through intramolecular hydroamination under basic conditions and irradiation with a tungsten lamp. Codeine was obtained in 15 steps and with an overall yield of 7%. Morphine was obtained by *O*-demethylation. Synthetic approaches to morphine have been reviewed [72].

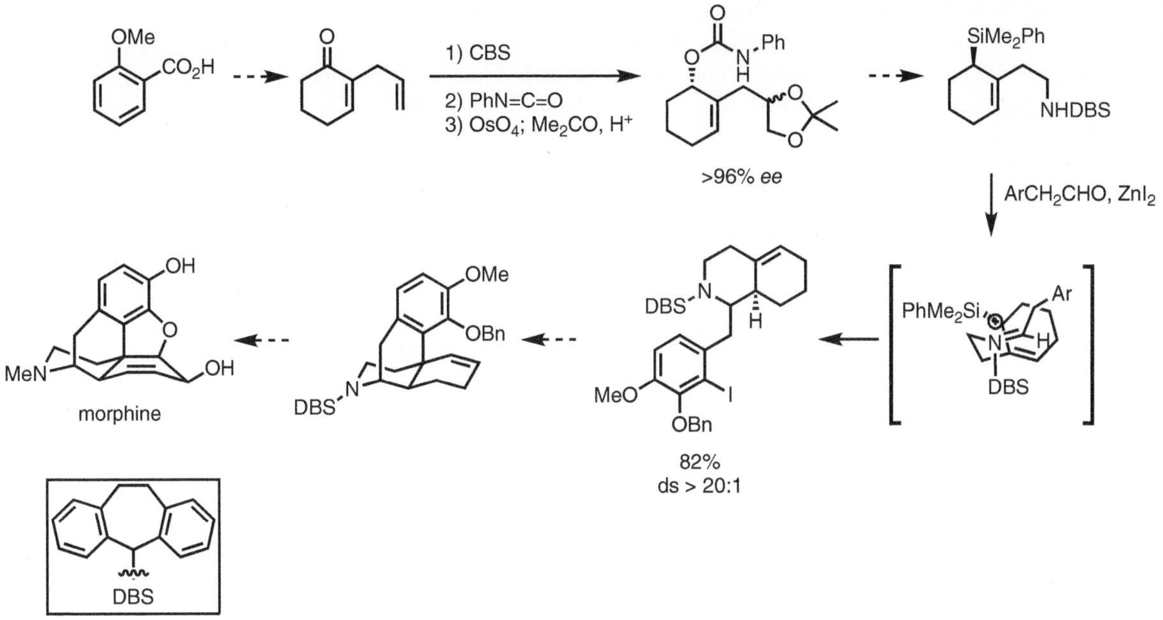

Scheme 10.37 Morphine synthesis utilising CBS-reduction

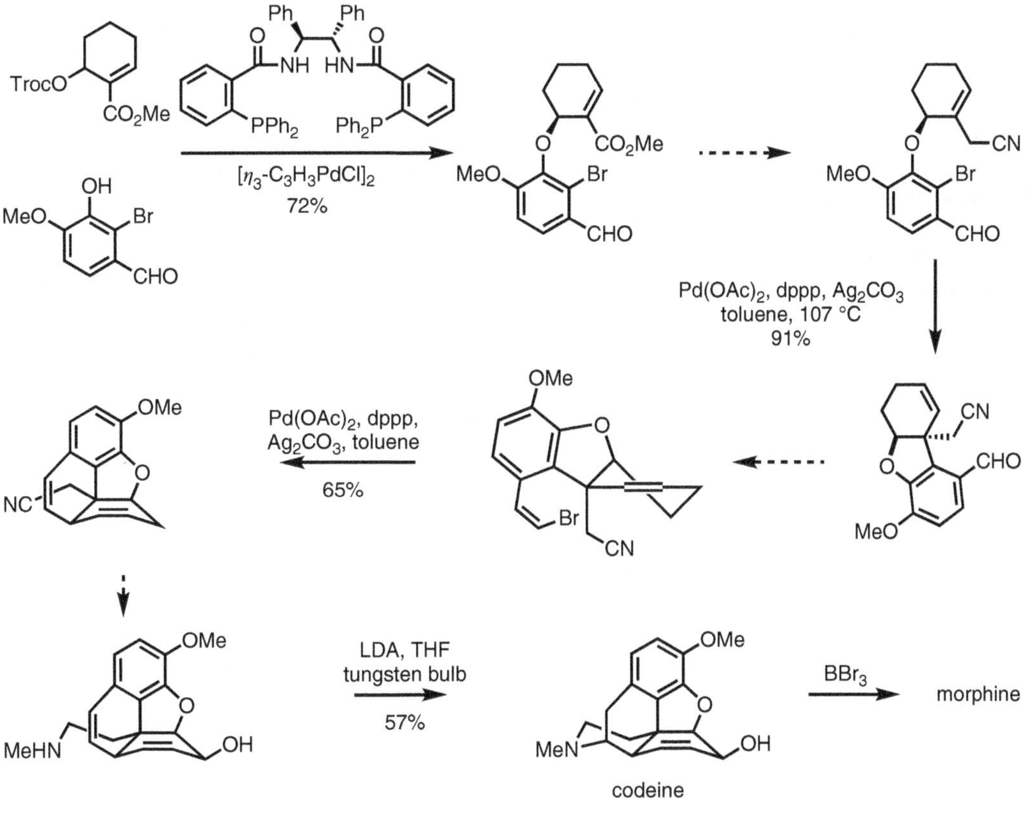

Scheme 10.38 Trost synthesis of morphine employing asymmetric allylic alkylation

10.2 Alkaloids with an Exocyclic Nitrogen

This rather diverse group of natural products contains compounds which are biogenetically derived using several of the routes (mainly polyketide and terpene pathways) discussed in previous chapters. The diterpene derivative taxol is also sometimes considered to belong to this class of alkaloids.

The phenylethylamine derivatives also belong to this class of alkaloids (Figure 10.16). These include the hallucinogenic mescaline (from the peyote cactus *Lophophora williamsii*), psilocybin (from *Psilocybe* mushrooms), and the highly toxic colchicine (from the autumn crocus, *Colchicum autumnale*), which is used for the treatment of gout.

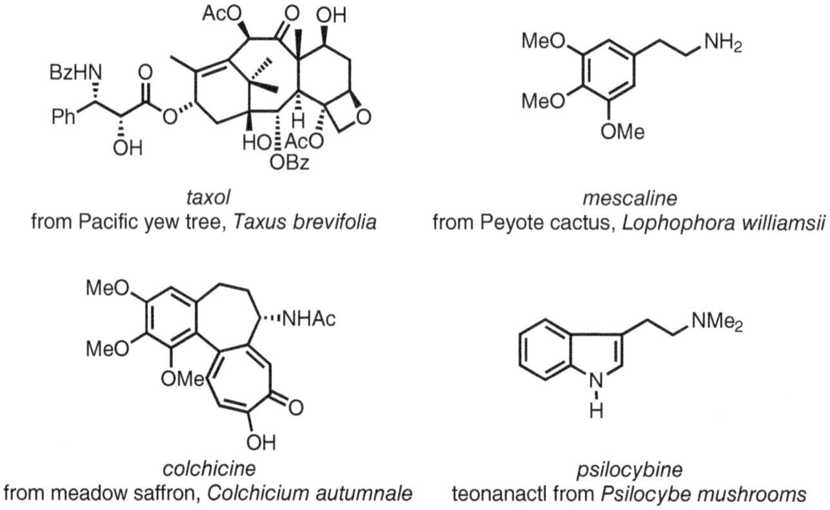

taxol
from Pacific yew tree, *Taxus brevifolia*

mescaline
from Peyote cactus, *Lophophora williamsii*

colchicine
from meadow saffron, *Colchicium autumnale*

psilocybine
teonanactl from *Psilocybe mushrooms*

Figure 10.16 *Miscellaneous alkaloids with exocyclic nitrogen*

10.3 Polyamine Alkaloids

Putrescine, spermidine, and spermine are diamines and members of the so-called biogenic amines (Figure 10.17). They occur as such and are also incorporated into more complex structures, such as chaenorrhine.

10.4 Diterpene Alkaloids

Monoterpene indole alkaloids are representative for alkaloids where the biosynthetic paths cross; a part of the molecule derives from amino acids, a part from mevalonic pathway. Diterpene alkaloids are another such class. They occur mainly in *Delphinium*, *Aconitum*, and *Garrya* species (Ranunculaceae), and they are biosynthesised by condensing L-serine with either *ent*-atisane or *ent*-kaurane diterpenes to form four types of diterpene alkaloids; those of the veatchine-, atisine-, lycoctonine-, and heteratisine-type alkaloids (Figure 10.18). Altogether *ca.* 100 C_{18}-, 700 C_{19}-, and 300 C_{20}-alkaloids are known [73].

In steroidal alkaloids, the steroid skeleton is further transformed into an alkaloid structure by adding a unit containing the nitrogen atom (Figure 10.19). Holarrhimine is the most obvious example of these alkaloids. Conessine (from *Holarrhena* species) and buxamine a (from *Buxus* species) are examples where more profound changes in the steroid skeleton are needed.

Solanum species, such as potato and tomato, contain poisonous solanum alkaloids, such as solasodine and solanidine. Both compounds are widely used for the production of synthetic steroids, especially in the Eastern European countries.

$$H_2N(CH_2)_3NH(CH_2)_4NH(CH_2)_3NH_2 \quad \text{spermine}$$

$$H_2N(CH_2)_4NH(CH_2)_3NH_2 \quad \text{spermidine}$$

$$H_2N(CH_2)_4NH_2 \quad \text{putrescine}$$

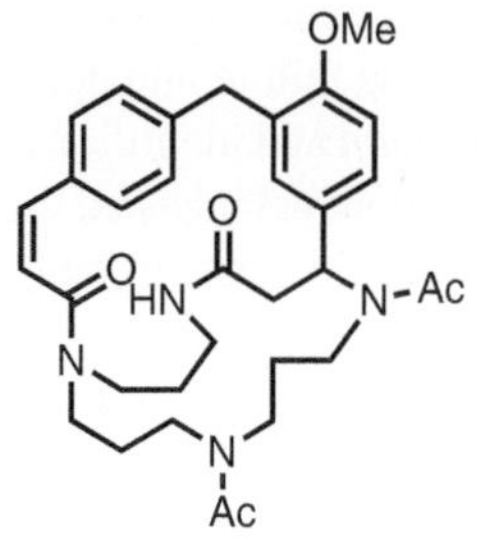

chaenorrhine

Figure 10.17 *Polyamine alkaloids*

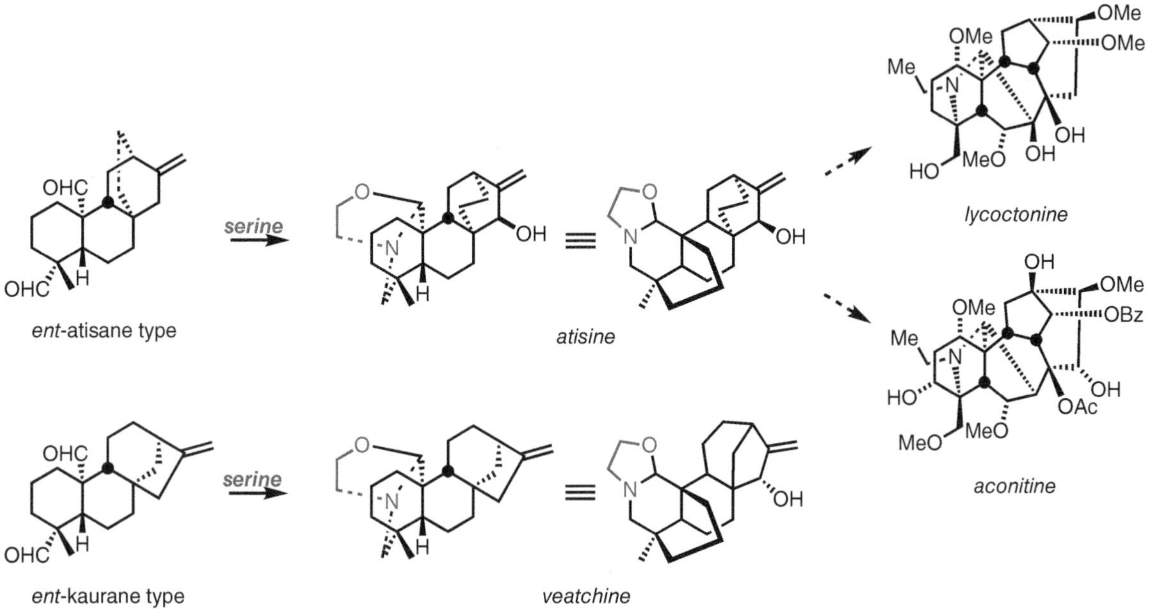

Figure 10.18 *Diterpene alkaloids*

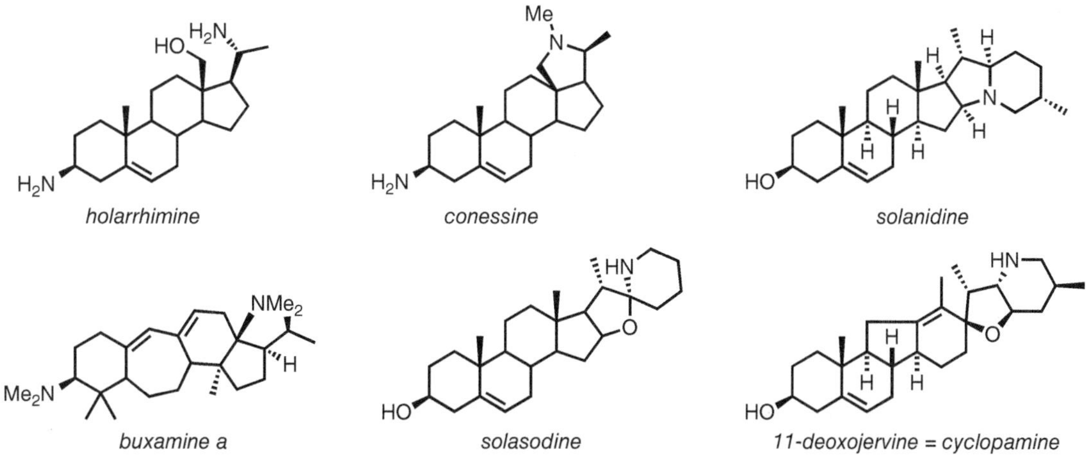

Figure 10.19 *Steroidal alkaloids*

Veratrum species (false hellebores or corn lilies) contain highly toxic alkaloids that activate sodium channels, and thereby cause rapid cardiac failure and death. Cyclopamine (11-deoxyjervine) [74] was isolated in the 1960s from the plant *Veratrum californicum* and identified as the teratogen responsible for craniofacial birth defects, including cyclops in the offspring of sheep that had ingested the plant during gestation. Cyclopamine inhibits the hedgehog (Hh) signaling pathway, which plays a critical role in embryonic development. Recently, aberrant Hh signaling has been implicated in several types of cancer, and inhibitors of the Hh signaling pathway, including cyclopamine derivatives, have been targeted as potential treatments for certain cancers.

References

1. Buckingham, J. *Bitter Nemesis, The Intimate History of Strychnine.* CRC Press, Boca Raton, FL, **2008**.
2. Karch, S.B. *A Brief History of Cocaine.* 2nd ed. CRC Press, Boca Raton, FL, **2006**, 188 pp.
3. Jeffreys, D. *Aspirin: The Story of a Wonder Drug.* Bloomsbury, London, **2004**, pp. 71–118.
4. Derosne, C. *Ann. Chim. (Paris)* **1803**, *45*, 257.
5. Sertürner, F.W. *Ann. Chim. Phys.* **1817**, *2*, 21.
6. Pelletier, P.J., Caventou, J.B. *Ann. Chim. Phys.* **1818**, *8*, 323–324.
7. Pelletier, P.J., Caventou, J.B. *Ann. Chim. Phys.* **1820**, *15*, 337–365.
8. Geiger, J. *Berzelius' Jahresber.* **1870**, *12*, 220.
9. Ladenburg, A. *Berzelius' Jahresber.* **1886**, *19*, 439.
10. Guerrier, L., Royer, J., Grierson, D.S., Husson, H.-P. *J. Am. Chem. Soc.* **1983**, *105*, 7754–7755.
11. a) Briggs, L.H., Openshaw, H T., Robinson, R. *J. Chem. Soc.* **1946**, 903–908; b) Robinson, R. *Experientia* **1946**, *2*, 28–29; c) Openshaw, H.T., Robinson, R. *Nature* **1946**, *157*, 438.
12. Peerdeman, A.F. *Acta Crystallogr.* **1956**, *9*, 824.
13. a) Woodward, R.B., Cava, M.P.; Ollis, W.D.; Hunger, A., Daeniker, H.U., Schenker, K. *J. Am. Chem. Soc.* **1954**, *76*, 4749–4751; b) Woodward, R.B., Cava, M.P., Ollis, W.D., Hunger, A., Daeniker, H.U., Schenker, K. *Tetrahedron* **1963**, *19*, 247–288.
14. Knight, S.D., Overman, L.E., Pairaudeau, G. *J. Am. Chem. Soc.* **1993**, *115*, 9293–9294.
15. Meissner, W. *J. Chem. Phys.* **1819**, *25*, 377–381.
16. Hesse, M. *Alkaloid Chemistry*, John Wiley & Sons: New York, **1981**.
17. Newman, D.J., Cragg, G.M. *J. Nat. Prod.* **2007**, *70*, 461–477.
18. Carter, A.J. *J. R. Soc. Med.* **2003**, *96*, 144–147.
19. a) Humphrey, A.J., O'Hagan, D. *Nat. Prod. Rep.* **2001**, *18*, 494–502; b) Huang, J.-P., Wang, Y.-J., Tian, T., Wang, L., Yana, Y., Huang, S.-X. *Nat. Prod. Rep.* **2021**, *38*, 1634–1658.
20. Gaedcke, F. *Arch. Pharm.* **1855**, *132*, 141–150.
21. Willstätter, R., Bode, A. *Chem. Ber.* **1901**, *34*, 1457–1461.
22. Robinson, R. *J. Chem. Soc., Trans.* **1917**, *111*, 762–768.
23. Rodriguez, S., Uria, U., Reyes, E., Prieto, L., Rodríguez-Rodríguez, M., Carrillo, L., et al. *Org. Biomol. Chem.* **2021**, *19*, 3763–3775.
24. Overman, L.E., Bell, K.L., Ito, F. *J. Am. Chem. Soc.* **1984**, *106*, 4192–4201.
25. Chan, T.H., Fleming, I. *Synthesis*, **1979**, 761–786.
26. Piccichè, M., Pinto, A., Griera, R., Bosch, J., Amat, M. *Org. Lett.* **2017**, *19*, 6654–6657.
27. Takahata, H., Banba, Y., Momose, T. *Tetrahedron: Asymmetry* **1990**, *1*, 763–764.
28. van Tamelen, E.E., Foltz, R.L. *J. Am. Chem. Soc.* **1960**, *82*, 502–503.
29. Tsutsumi, T., Karanjit, S., Nakayama, A., Namba, N. *Org. Lett.* **2019**, *21*, 2620–2624.
30. Williams D.R., Brown, D.L., Benbow, J.W. *J. Am. Chem. Soc.* **1989**, *111*, 1923–1925.
31. Bödeker, K. *Justus Liebig's Ann. Chem.* **1881**, *208*, 363–367.
32. Yang, H., Carter, R.G., Zakharov, L.N. *J. Am. Chem. Soc.* **2008**, *130*, 9238–9239.
33. Ding, R., Sun, B.-F., Lin, G.-Q. *Org. Lett.* **2012**, *14*, 4446–4449.
34. Lewis, S.E. *Tetrahedron* **2006**, *62*, 8655–8681.
35. a) Pictet, A., Spengler, T. *Ber. Dtsch. Chem. Ges.* **1911**, *44*, 2030–2036; reviews; b) Cox, E.D., Cook, J.M. *Chem. Rev.* **1995**, *95*, 1797–1842; c) Youn, S.W. *Org. Prep. Proc. Int.* **2006**, *38*, 505–591; d) Lorenz, M., van Linn, M.L., Cook, J.M. *Curr. Org. Synth.* **2010**, *7*, 189–223; e) Stöckigt, J., Antonchick, A.P., Wu, F., Waldmann, H. *Angew. Chem., Int. Ed.* **2011**, *50*, 8538–8564.
36. Jackson, A.H., Smith, P. *J. Chem. Soc., Chem. Commun.* **1967**, 264–266.

37. a) Ungemach, F., Cook, J.M. *Heterocycles* **1978**, *9*, 1089–1119; b) Bailey, P.D. *Tetrahedron Lett.* **1987**, *28*, 5181–5184; c) Bailey, P.D., Hollinshead, S.P. *J. Chem. Soc. Perkin Trans. I* **1988**, 739–745.

38. Perkin, W.H., Jr.; Robinson, R. *J. Chem. Soc.* **1919**, *115,* 933–967.

39. a) Barger, G., Scholz, C. *Helv. Chim. Acta*, **1933**, *16*, 1343–1354; b) Hahn, G., Werner, H. *Justus Liebigs Ann. Chem.* **1933**, *520*, 123–133.

40. a) Woodward, R.B. *Nature* **1948**, *162*, 155–156; b) Woodward, R.B. *Angew. Chem.* **1956**, *68*, 13–20.

41. a) Thomas, R. *Tetrahedron Lett.* **1961**, 544–553; b) Wenkert, E. *J. Am. Chem. Soc.* **1962**, *84*, 98–102.

42. Stöckigt, J. In *Indole and Biogenetically Related Alkaloids* (Phillipson, J.D., Zenk, M.H., eds.) Academic Press: New York, **1980**, Chapter 6.

43. LeMen, J., Taylor, W.I. *Experientia* **1965**, *21*, 508–510.

44. Sadre, R., Magallanes-Lundback, M., Pradhan, S., Salim, V., Mesberg, A., Daniel Jones, A., et al. *Plant Cell* **2016**, *28*, 1926–1944.

45. a) Leonard, J., Ouali, D., Rahman, S.K. *Tetrahedron Lett.* **1990**, *31*, 739–742; b) Izawa, H., Shirai, R., Kawasaki, H., Kim, H., Koga, K. *Tetrahedron Lett.* **1989**, *30*, 7221–7223.

46. a) Li, J., Wang, T., Yu, P., Peterson, A., Weber, R., Soerens, D., et al. *J. Am. Chem. Soc.* **1999**, *121*, 6998–7010; b) Yin, W., Kabir, M.S., Wang, Z., Rallapalli, S.K., Ma, J., Cook, J.M. *J. Org. Chem.* **2010**, *75*, 3339–3349.

47. a) Magnus, P., Mugrage, B., DeLuca, M., Cain, G.A. *J. Am. Chem. Soc.* **1989**, *111*, 786–789; b) Magnus, P., Mugrage, B., DeLuca, M., Cain, G.A. *J. Am. Chem. Soc.* **1990**, *112*, 5220–5230.

48. Lounasmaa, M., Koskinen, A. *Planta Medica* **1982**, *44*, 120–125.

49. Rebmann, H., Gerlinger, C.K.G., Gaich, T. *Chem. Eur. J.* **2019**, *25*, 2704–2707.

50. Mergott, D.J., Zuend, S.J., Jacobsen, E.N. *Org. Lett.* **2008**, *10*, 745–748.

51. Herlé, B., Wanner, M.J., van Maarseveen, J.H., Hiemstra, H. *J. Org. Chem.* **2011**, *76*, 8907–8912.

52. Beard, R.L., Meyers, A.I. *J. Org. Chem.* **1991**, *56*, 2091–2096.

53. a) Meyers, A.I. *Aldrichimica Acta* **1985**, *18*, 59–68; b) Meyers, A.I., Guiles, J. *Heterocycles* **1989**, *28*, 295–301.

54. Bonjoch, J., Solé, D. *Chem. Rev.* **2000**, *100*, 3455–3482.

55. Jones, S.B., Simmons, B., Mastracchio, A., MacMillan, D.W.C. *Nature* **2011**, *475*, 183–188.

56. Stork, G., Schultz, A.G. *J. Am. Chem. Soc.* **1971**, *93*, 4074–4075.

57. Jew, S.-S., Ok, K.-D., Kim, H.-J., Kim, M. G., Kim, J. M., Hah, J. M., et al. *Tetrahedron: Asymmetry* **1995**, *6*, 1245–1248.

58. Gerhards, N., Neubauer, L., Tudzynski, P., Li, S.-M. *Toxins* **2014**, *6*, 3281–3295.

59. Leete, E. *Accts. Chem. Res.* **1969**, *2*, 59–64.

60. Rabe, P. *Ber.* **1907**, *40*, 3655–3658.

61. a) Kaufman, T.S., Rúveda, E.A. *Angew. Chem. Int. Ed.* **2005**, *44*, 854–885; b) Seeman, J.I. *Angew. Chem. Int. Ed.* **2007**, *46*, 1378–1413.

62. Garfield, S. *Mauve: How One Man Invented a Color That Changed the World.* Norton & Company, New York, **2001**.

63. Woodward, R.B., Doering, W.E. *J. Am. Chem. Soc.* **1944**, *66*, 849.

64. Stork, G., Niu, D., Fujimoto, A., Koft, E.R., Balkovec, J.M., Tata, J.R., et al. *J. Am. Chem. Soc.* **2001**, *123*, 3239–3242.

65. Raheem, I.T., Goodman, S.N., Jacobsen, E.J. *J. Am. Chem. Soc.* **2004**, *126*, 706–707.

66. Illa, O., Arshad, M., Ros, A., McGarrigle, E.M., Aggarwal, V.K. *J. Am. Chem. Soc.* **2010**, *132*, 1828–1830.

67. a) Initial suggestion with a cyclobutane ring: Guilland, J.M., Robinson, R. *J. Chem. Soc., Trans.* **1923**, *123*, 980–998; b) Modification: *Mem. Manchester Phil. Soc.* **1925**, *69*, No. 10; c) van Duin, C.F., Robinson, R., Smith, J.C. *J. Chem. Soc.* **1926**, *129*, 903–908; d) Wieland, H., Kotake, M. *Ann.* **1925**, *444*, 69–93.

68. Gates, M., Tschudi, G. *J. Am. Chem. Soc.* **1956**, *78*, 1380–1393.

69. Aubry, A., Birlirakis, N., Sakarellos-Daitsiotis, M., Sakarellos, C., Marraud, M. *J. Chem. Soc., Chem. Commun.* **1988**, 963–964.

70. Hong, C.Y., Kado, N., Overman, L.E. *J. Am. Chem. Soc.* **1993**, *115*, 11028–11029.

71. Trost, B.M., Tang, W. *J. Am. Chem. Soc.* **2002**, *124*, 14542–14543.

72. a) Zezula, J., Hudlicky, T. *Synlett* **2005**, 388–405; b) Reed, J.W., Tomas Hudlicky, T. *Accts Chem. Res.* **2015**, *48*, 674–687.

73. Wang, F.P. *The Alkaloids* **2010**, *69*, 1–609.

74. a) Lee, S.T., Welch, K.D., Panter, K.E., Gardner, D.R., Garrossian, M., Chang, C.-W.T. *J. Agric. Food Chem.* **2014**, *62*, 7355–7362; b) Mousavizadeh, F., Meyer, D., Giannis, A. *Synthesis* **2018**, *50*, 1587–1600.

Index

Asymmetric Synthesis of Natural Products, Third Edition. Ari M.P. Koskinen.
© 2023 John Wiley & Sons Ltd. Published 2023 by John Wiley & Sons Ltd.